COLOUR REPRODUCTION IN ELECTRONIC IMAGING SYSTEMS

COLOUR REPRODUCTION IN ELECTRONIC IMAGING SYSTEMS

PHOTOGRAPHY, TELEVISION, CINEMATOGRAPHY

Michael S Tooms
Principal Consultant, MTA, UK

Library of Congress Cataloging-in-Publication Data

Tooms, Michael S.
  Colour reproduction in electronic imaging systems : photography, television, cinematography / Michael S Tooms.
    pages cm
  Includes bibliographical references and index.
  ISBN 978-1-119-02176-6 (cloth)
  1. Color display systems.   2. Color television.   3. Color cinematography–Equipment and supplies.   4. Color photography–Digital techniques.   5. Color sensitometry (Photography)   I. Title.   II. Title: Color reproduction in electronic imaging systems.
  TK6670.T66 2016
  621.382–dc23

                                                                                    2015019419

A catalogue record for this book is available from the British Library.

ISBN: 9781119021766

Set in 10/12pt Times by Aptara Inc., New Delhi, India

1   2016

Contents

Preface

Many excellent books are available which deal in one way or another with image reproduction, either generically for example across photography or television, or in a more specialist manner for a particular technology such as digital television. Invariably colour is discussed as an integral part of the reproduction process, often as an adjunct to the technology of the media being covered. However, few books leave the reader who is specifically interested in the *reproduction of colour* entirely happy that the colour reproduction process has been fully understood at a fundamental level.

Often the more obscure elements of colorimetry are glossed over, which is fine for those who require only a superficial understanding in this area but is frustrating for those who need to understand fully the derivation and interpretation of the various chromaticity diagrams in use. In contrast, the aim of this book is to cover comprehensively the complete process of colour reproduction from capturing the scene to rendering the final image, whether it be in the form of a display or a print. Emphasis is placed on the analysis and processing of the colour components, rather than the technology associated with generating and conveying the components representing the original image from the camera to the point where the image is rendered.

Nevertheless, aspects of colorimetry, such as the derivation of the CIE[1] x,y chromaticity diagram, for example, are fundamentally quite mathematical. So in order not to detract from the flow of the text, the development of the subject is described in a heuristic manner and the supporting mathematics are constrained to the appendices for those who wish to grasp the complete picture at a deeper fundamental level.

The material for this book evolved initially from notes used to support the 'Television Image Formation, Analysis and Reproduction' lecture given by the author in a series of annual engineering lectures for the Royal Television Society (RTS) and subsequently developed into a chapter for the planned RTS book on television engineering. He was also able to build upon the material used more recently in a presentation to the Image Science Section of The Royal Photographic Society on colour management as a means of achieving accurate colour reproduction in photography. As a member of the Society of Motion Picture and Television Engineers (SMPTE), he has watched with interest his colleagues driving the evolution of international standards for colour reproduction in digital cinematography and has drawn upon this material for the part dealing with this topic; this step completed the transformation of the three major picture media systems: television, photography and cinematography to electronic-based systems.

---

[1] Commission Internationale de l'Eclairage or International Commission on Illumination, the international body responsible for specifying the measurement of colour.

In recent years, electronic image reproduction has made tremendous strides not only in television, to embrace digital, high definition and 3D television, but also by expanding into photography and latterly into 'digital cinema', where digital cameras have virtually replaced film at the shooting stage and digital large screen projectors have become the norm in most cinemas. Despite these advances, the fidelity of colour reproduction continues to have limitations and there remain enticing approaches to extending the range of colours which may be reproduced as described in the book.

The book sets out to provide an in-depth analysis of colour, its measurement and its reproduction at a fundamental level before going on to provide a comprehensive coverage of its application in uniquely different ways to television, photography and cinematography, respectively.

The application of colour reproduction theory to practical systems is addressed from a historical perspective, since the application of electronics to each media system has always been built on the technology developed for digitising the previous media. Much of the groundwork of applying the then relatively new understanding and standardisation of colour analysis and measurement of the 1930s, through the work of the CIE, was brought to bear by the members of the National Television System Committee (NTSC) of the United States in the early 1950s, when the first practical colour television system was introduced. The European television systems (Phase Alternating Line (PAL) and Sequential Colour with Memory (SECAM)) which followed in the 1960s were essentially based upon the same colour fundamentals. However, they differed from the earlier system in that they evolved methods of encoding the colour signals for transmission that were less prone to the effects of the distortions apparent in the electronic systems of the day, which in its early days had given the NTSC system a poor reputation.

In electronic terms, these pioneer colour television systems reigned supreme for some 40 years before developments in technology in the 1980/90s gave rise to the possibility of adopting a new world-wide standard for television which included a tightening up of the standards associated with the specifications for colour reproduction. Some would claim the opportunity was missed at that time to introduce standards which would embrace developments in colour reproduction which had already been foreseen, and are now waiting on the side-lines for the opportunity to come to the fore.

During broadly the same period of the early 1990s, these developments also saw the evolution of the standardisation of digital video signals, the JPEG digital compression system and cost-effective solid-state image sensors, making practical digital stills cameras available at increasingly affordable prices for general consumer use. Finally in the early 2000s, with the adaptation of specialised television cameras of high resolution for recording sequences for the cinema, and the availability of suitably bright and high-resolution projectors, the way was open in the late 2000s for experts within the cinematographic standards bodies[2] to set the colour specifications for the digital cinema. Part 5 of this book describes that journey in terms of the colour techniques and specifications adopted by each of the three media: television, photography and cinematography.

---

[2] The Academy of Motion Picture Arts and Sciences, the American Society of Cinematographers, the Digital Cinema Initiatives and the Society of Motion Picture and Television Engineers.

Acknowledgements

Without the encouragement I have been fortunate enough to receive from so many people, I would not have been in a position to write this book. I am resolved therefore to acknowledge not only those who have assisted me in preparing the book but also those who in one way or another have encouraged my interest and enthusiasm for colour from the beginning, albeit that some are no longer with us.

It was evident to me as a young child that my interest in the exciting range of colours around me went well beyond the norm of those in my circle of family and friends, with the possible exception of my mother who did her best to assist me with the difficult topic of colour naming. I remember well one particular, rather unusual colour that when questioning was told it was 'cerise'.

My formal introduction to colour came whilst serving as a radar technician in the RAF when I selected as a birthday present, John W. Wentworth's book 'Colour Television Engineering'. This was read avidly during the plentiful non-shift time available and became my bible for many years; I am much indebted to the author for dealing so comprehensively and insightfully with colour measurement and its application to reproduction.

Subsequently, when working for EMI at the time of the introduction of their NTSC encoder, I was grateful to its designer, David Kent, for his patient detailed explanations of the workings of each of the several unique circuits of which it was comprised.

I joined ABC Television in the United Kingdom at a time when broadcasters were experimenting with colour television and Michael Cox was leading the work there on the SECAM system. Eventually it became clear that the PAL system developed under the leadership of Dr. Bruch in Germany was to be the likely choice of system for the United Kingdom and at that time I began to work for Mike who taught me a lot as I 'bread-boarded' an experimental PAL encoder whilst he tackled the more difficult decoder.

In preparation for the introduction of a colour television service during this period, I, together with my colleague and friend Ray Knight, attended a post-graduate colour course under Professor Wright at Imperial College London, which turned out to be a milestone event. W. David Wright was one of the two colour scientists who had undertaken the work which led to the CIE standards on the measurement of colour. We were privileged to be able to derive our personal colour matching functions on the Wright colorimeter that had been used to obtain the data for those standards. David Wright's tremendous knowledge of the subject and his enthusiasm for sharing it with his students helped cement in me a lifelong interest in the subject.

The effect on my friend Ray Knight was very similar and since then we have kept in constant touch, sharing our interest in colour from slightly different perspectives. Nevertheless, we have continued to exchange ideas and, as the result of many, many discussions over the years, my knowledge of the subject has been further broadened by his enthusiasm and determination to ensure that we shared a common understanding of the more obscure aspects of the subject. Ray has also written on colour, particularly from the perception of an artist, on picture-matching and colour-mixing. As a result he has produced many excellent diagrams and I am very grateful for his generosity in allowing me to use several of them in this book. It follows from our close association with colour that he was the natural choice to review much of the manuscript as it was written and I am indebted to him for the exhaustive comments he has provided and the discussions that have led to the original material being much improved.

A little later Dr. Boris Townsend, a leading researcher and author on colour television at that time, who was much in demand for membership of international committees and as an invited college lecturer, joined the ABC colour team. His management style was exemplary and he encouraged my interest in colour by insisting that I stood in for him on a number of occasions, both in lecturing and attending European Broadcasting Union (EBU) technical meetings to agree on the chromaticities of primaries to be used in the first European colour television system.

More recently as the idea of writing a manuscript for a book took shape and it became necessary to discuss the finer points of some of the more obscure aspects of colour and to establish the direction in which international media committees were moving in laying the foundations of new system specifications, I called upon the support of several colleagues in the field.

Michael Pointer, a colour scientist internationally recognised for his work on the extent of the gamut of surface colours, has been invaluable with regard to CIE matters. In particular, referring me to papers on the history and technical detail of pertinent decisions in the evolution of colour measurement standard procedures, making CIE data available, sharing extensive worksheets on colour rendering indices, and taking the trouble to enter into email discussions on a range of topics. By sheer coincidence, he was also appointed as a reviewer of this book and I am much indebted for his suggestions both in clarifying my explanations and drawing to my attention lapses in English grammar.

David Bancroft has been particularly helpful in keeping me up-to-date with current thinking in the international committees which formulate specifications on such topics as the chromaticity of primaries for future television systems, the means of dealing with colour gamut mapping, and the parameters associated with the characterisation of standard monitors for picture matching in television vision control rooms.

Alan Roberts and John Emmett, who have recently jointly undertaken much work on colour rendering indices, which led to a new EBU recommendation on the topic, were very generous in allowing me free access to their work and giving permission to use it in this book. Alan also provided me with a wide range of measured spectral power distributions of light sources, which generously supplements the critical 'Illuminants' data sheet in the accompanying Colour Reproduction Workbook. John Emmett also kindly undertook the measurement of the spectral reflectance of a range of colour surfaces used as examples in the book.

Ionnasis N. Galidakis has an extensive website accommodating a comprehensive range of charts illustrating the spectral power distributions of various elements in their excited state and has given permission for these to be used freely within the book.

Seo Young Choi, who was prepared to enter into an email discussion with a total stranger on the finer points of specifying the scaling factors for colour difference signals in the constant luminance system variant of the UHDTV system, helped me resolve why a particular approach had been taken.

Mike Reed kindly provided me with access to his photographic workstation in order to extend the range of inkjet printer samples available for the chapter describing printers.

Scott Matthews also provided me with access to his professional photographic workstation and took part in detailed discussion and experiments on establishing the perceived contrast ranges of rendered images under different lighting and monitor set-up conditions.

Mick Vincent, a colourist at The Mill, gave me much of his time and expertise in describing the reality of using colour management procedures in a post-production grading environment.

Daniele Siragusano of Filmlight showed me the practicality of using a wide range of colour transform options in current post production systems and we discussed at length the pros and cons of adopting various specific procedures in the grading operation.

My friend, Frank Bateson, acted as a non-specialist reader on much of the manuscript and offered invaluable advice on text which I had thought was self-evident in its explanation but which he showed clearly was not the case.

Special thanks are due to my friend and colleague Bert White, who joined the ABC Television team on the same day as I, and has reviewed both the manuscript, chapter by chapter, and the worksheets included in the workbook which partners this book. He provided a wealth of feedback to improve the readability of the text and encouraged me to complete the work when other issues were pressing upon my time.

Finally of course, special thanks are also due to my wife Pamela who, for what seems an age, has borne with much patience a husband who, whilst buried in the work of the manuscript for hours on end, has, as she has put it, 'switched off' from all her attempts at conversation. In addition, she has had to endure many lonely periods when couples would normally be sharing a relaxing pastime whilst I caught up with a stage in the manuscript.

Michael Tooms
February 2015

About the Companion Website

This book is accompanied by a companion website:

www.wiley.com/go/toomscolour

This website includes:

- The Colour Reproduction Workbook of some 50 worksheets, including:
  - colour data sheets
  - derivation of chromaticity and camera spectral sensitivity plots from selectable primary chromaticities
  - calculation of illuminant rendering indices from spectral power distributions
  - calculation of matrix coefficients for transforming signal values to those of different system primaries
  - calculation of fidelity of performance of particular colour reproduction systems
- Guide to the Colour Reproduction Workbook (also available in this book as Appendix J)
- JPEG files containing images pertaining to references in the book, including:
  - colour bars
  - grey scale charts to establish perceived contrast range under different environmental conditions
- Guide to using the JPEG files

Introductions

The Book

This book is aimed at both the serious practitioners in the fields of photography, television engineering and cinematography, and those amateurs who have the enthusiasm to learn more of the reproduction medium which they enjoy as a hobby.

In essence the requirements of colour image reproduction may be simply stated, we need to:

- First understand what colour is,
- Determine a means of measuring it,
- Capture and measure the value of the colours in a scene,
- Establish the most efficient and precise way we can transfer those measurements to the device which creates and displays the reproduced image,
- Control the image creating device with signal components levels representing the measurements we originally made.

As the original structure laid out for this book took shape, it became clear that the subject matter associated with each chapter fell naturally into groups broadly reflecting the requirements listed above. These groups became the five parts of the book.

It also became evident that each part would benefit from its own introduction; thus rather than writing a general introduction at the beginning of the book to cover all the material in one step, it was decided to emphasise this natural grouping by writing an introduction to each part. These introductions describe the range of the topics covered and the approach adopted in addressing the material it contains.

Thus the five parts of the book are:

Part 1 Colour – Perception, Characteristics and Definition

Part 2 The Measurement and Generation of Colour

Part 3 The Concepts of Colour Reproduction

Part 4 The Fundamentals of Colour Reproduction

Part 5 The Practicalities of Colour Reproduction in:

Part 5A Television,

Part 5B Photography and

Part 5C Digital Cinematography

A more detailed indication of the material contained in each part is provided by reference to the Contents page, where the titles of the chapters and the sections comprising each chapter are listed. As an alternative, turning to the page containing the part heading will provide access to the introduction to that part.

This book is specifically about colour and its reproduction in photography, television and cinematography using electronics rather than film as the technology of implementation. Each of these media areas is highly technical in its own right, and there are good books available which describe every aspect of these individual technologies; thus, the material in this book is of necessity restricted specifically to the colour technology required of each of these media, which often tends to be only casually dealt with in more generic books. Only where an understanding of the associated technology would be helpful in clarifying the colour concepts in a particular area is the technology then described in very general terms.

The aim of this book is to ensure that it is easily read and understood by the widest range of readers, from those with only a passing understanding of physics and mathematics to those with a more specialised knowledge in these areas who wish for a deeper understanding of the subject. In consequence, I have relied on appendices to provide the depth the latter may require in order that these detailed explanations do not get in the way of the flow of the material.

In order to support the text and the numerous charts which appear in the book, a good deal of calculation was required in the form of worksheets and it seemed that it would greatly extend the usefulness of the book to make these worksheets available to those readers wishing to understand the underlying mathematics. Thus, 'The Colour Reproduction Workbook' is introduced in more detail below.

In researching material to support this book, it became evident that amongst much erroneous material on the World Wide Web there is also much which is excellent, and occasionally, in order to avoid 'reinventing the wheel', I have, where appropriate, included such material with due reference as to its source. I have also received much support from friends and colleagues who are active in the field of colour and reproduction and have welcomed the opportunity to note their contributions with appropriate references.

The Colour Reproduction Workbook

With very few exceptions, all the charts and supporting calculations appearing in the book are derived from worksheets produced by the author. These worksheets have been compiled into an Excel workbook and provide an invaluable resource to those readers who have the need or the interest to explore further the examples provided in the narrative of the book. Many of the worksheets contain icons controlling macros, which when activated will replace one set of data in a calculation with a different set, enabling a wide range of 'what if' questions which may arise in the mind of the reader to be answered. The dedicated data worksheets, which contain a very wide range of basic colorimetric data, provide the reader with the option of copying specific data into the example worksheet to meet their needs, or indeed enter new data, direct into the worksheets.

Each worksheet follows the numbering sequence of its associated chapter, more than one sheet being provided when necessary with an (a) or (b) suffix in order to avoid as far as possible the production of unwieldy spreadsheets. At the top left of the worksheet is a brief description of its functionality.

As the theme of colour reproduction is developed in the book, the supporting calculations become more extensive until the point is reached where a number of worksheets have evolved which effectively become simple mathematical models with accompanying performance charts which describe the whole or large sections of the reproduction process. These worksheets are particularly invaluable in exploring how different parameters can affect the performance of the colour reproduction process.

The workbook is accompanied by the 'Guide to the Colour Reproduction Workbook' which describes its structure and, for the more extensive worksheets, supplements the brief description provided at the top of each worksheet with a section which provides a description of its layout and how to operate the macros.

The Colour Reproduction Workbook and its associated 'Guide to the Colour Reproduction Workbook' may be downloaded from the companion website to this book at www.wiley.com/go/toomscolour.

In addition the 'Guide to the Reproduction Workbook' appears as Appendix J to this book.

Part 1

Colour – Perception, Characteristics and Definition

Introduction

Before the subject of colour reproduction can be addressed, it is important to first arrive at a common understanding of colour. Thus Part 1 is dedicated to explaining what colour is, how it is perceived, how it is characterised and how it is defined.

For the author of a book of this nature, which deals with a topic we all know *something* about but which inevitably extends into more advanced areas, it is important to know where to start. After much thought based upon the experience of many discussions about colour with a wide range of people from family and friends to those in the business of colour reproduction, it was decided to start at the very beginning, since unfortunately misconceptions about colour are often taught even within schools. Thus Chapter 1 commences with the basics of the perception of colour and colour naming (a minefield (Berlin & Kay, 1969)) in order to ensure that as the more advanced concepts are introduced all readers are at ease with the basics and the language used to describe the various parameters of colours in this field.

This chapter then goes on to introduce how the eye–brain complex perceives light and colour both in terms of its contrast range and how the overall spectral response of the eye at normal lighting levels is comprised of three different types of receptor, which respond to light in different parts of the spectrum.

In Chapter 2, the spectral responses of the eye are investigated further to explain how by using three primary colours a large proportion of the colours that can be perceived by the eye can be simulated by a mix of appropriate amounts of these primaries. The positioning of these primaries in the spectrum to optimise the size and position of the gamut of colours which

Colour Reproduction in Electronic Imaging Systems: Photography, Television, Cinematography, First Edition. Michael S Tooms.
© 2016 John Wiley & Sons, Ltd. Published 2016 by John Wiley & Sons, Ltd.
Companion Website: www.wiley.com/go/toomscolour

can be simulated is explored and the perceived ambiguity relating to which colours are the primaries for both light sources and pigments is addressed and hopefully eliminated.

Grassman's law regarding the behaviour of the eye to linearly add the components of broad spectral bands of colours is used as the crucial basis for explaining the rules for mixing colours, which is fundamental to the process of colour reproduction. Examples of the exploitation of this rule are given in several illustrations from the two-dimensional colour triangle to the three-dimensional colour space.

The preferred terms for describing the various parameters of colour are defined together with the terms in common use and reference is made to the manner in which some of these terms are used incorrectly and thus ambiguously in common speech.

The requirement to define colours in terms of categorising them in a manner which enables workers in colour to exchange information about colour in an unambiguous fashion is explained by reference to an example system of cataloguing a broad range of colours which explores virtually the full gamut of colours the eye can perceive.

Finally an abbreviated reference is made to the effect of the quality of illumination on the perception of colours in a scene, a topic which is explored in more depth in later chapters.

An in-depth understanding of the topics in Part 1 prepares the reader for addressing the material in Parts 2 and 3.

1

The Perception of Colour

1.1 Introduction

Before addressing the reproduction of colour it is essential to have firmly based ideas about what colour is, its spectral characteristics, the way it is described differently by different people and the importance of a common naming nomenclature. Thus this chapter describes how colour is perceived, and how it is unambiguously characterised, both in terms of the quantitative and qualitative responses it evokes in the eye.

The sensitive elements of the eye at normal levels of illumination are identified and characterised both in terms of their overall sensitivity and in terms of their spectral responses.

1.2 Setting the Scene

To introduce colour as a subject for study immediately presents one with a problem. In contrast to other subjects we may decide to investigate, we all have preconceptions as to what colour is. We think we already know much about colour, we have experienced it from early childhood, colour names crop up in speech on a regular basis, we have probably been taught at school how to mix colours to obtain a wider range than those available in the paint box and almost certainly at some stage we have been introduced to the concept of primary colours as the basis of obtaining a wide range of colours from the mixture in varying amounts of just three distinctly different primary hues. We will be formally defining the parameters which are used to describe the various aspects of colour later. For the present, however, the *hue* of a colour describes whether it is, for example, green, yellow or violet.

However the manner in which we perceive colour, though at an overview level not particularly complex, is just complex enough to require a level of attention beyond that which many of us have been prepared to give on a casual basis. Experience has shown that as a result there is widespread confusion about how colour is perceived.

One of the problems associated with initial considerations of colour perception is the naming of colours and the manner in which we differentiate colours of various hues through the spectrum. The following four paragraphs on *unitary hues* adapted from the work of Ray Knight provide a sound basis on which to commence consideration of this topic.

Colour Reproduction in Electronic Imaging Systems: Photography, Television, Cinematography, First Edition. Michael S Tooms.
© 2016 John Wiley & Sons, Ltd. Published 2016 by John Wiley & Sons, Ltd.
Companion Website: www.wiley.com/go/toomscolour

As we step through the visible spectrum from red to violet we pass a considerable number of quite distinct hues. A listing might read as: red, orange, yellow, green, cyan or turquoise, blue and violet. (Purple and magenta hues do not appear in the spectrum.) Because these seven colours continuously blend into each other we perceive many more than these seven hues, and certainly more hues than we have distinct colour names to identify them with.

Of the seven fundamental colours named above, only four are truly distinct and to these we can add black and white which together make up a group called *psychological colours* or, when hues only are being referred to, as *unitary hues,* attributed to Ewald Hering (1834–1918), which are red, yellow, green and blue. These four important colours, share the distinction that each one can be described without reference to the other three, or any other colour. Consider yellow, for example. To find a pure yellow without a hint of either adjacent colour means we look for a yellow with an absence of green and an absence of red – perhaps chrome yellow. Such a hue can be found, but not so with orange or purple and some other hues. Orange has within it an element of yellowness and redness, and purple has elements of blueness and redness.

This unambiguous isolation of hue only happens with the four unitary hues. It is quite fundamental that this occurs and is a matter of course without any teaching or learning.

The fact that some colours can be described by reference to their adjacent colours in the colour wheel, such as blue-green, means that the colour – just described – is *not* one of the four unitary hues of red, yellow, green or blue; if we mix blue and green the mixture is called blue-green, turquoise or cyan, and to confuse the situation, sometimes just a blue or green. Thus cyan is a hue with a blueness and a greenness about it which creates a colour naming problem, quite apart from the fact that this hue is not differentiated from blue or green in some cultures. So it is also for purple between red and blue; orange between red and yellow, and lime or yellow-green between yellow and green.

Thus one of the causes of the confusion alluded to above, is the predilection of many people, often men, to describe colours as variants of these unitary hues, red, yellow, blue and green, without differentiating them even into some of the other principal hues we experience such as orange, turquoise, violet and magenta, for example. One serious outcome of this casualness in describing principal hues has led to a common misconception that the so-called primary colours are red, yellow and blue. The author has had to face the almost impossible task, on more than one occasion, of persuading a young relative that their teacher was wrong in using these colours to describe the primary hues.

Returning to the complexities of colour perception, unless one understands the underlying rationale of what is going on, it is all too easy to become confused when faced with the prospect of one set of primaries for the mixing of coloured lights and a different complementary set for the mixing of pigments of various types. However neither set is the red, yellow and blue set referred to above. We will be looking further into what we mean by primary colours in Chapter 2.

All around us are examples of incorrect colour naming; take for example Robin Redbreast, so described, it would seem, to achieve an alliteration; in fact even a casual glance at a robin shows it to have an orange breast.

One might reasonably ask at this stage how we can be sure that we all perceive a particular colour in the same way. Early experiments, which we will be looking at in some detail later, showed quite clearly that those with normal vision do perceive colour in broadly the same way. There are of course people with defective colour vision; this does not mean they see in monochrome but usually one of the three colour receptors in the eye is defective to a degree, which leads to them being unable to differentiate between colours within a certain range of colours in the spectrum. In broad terms, some 8% of males are colour defective to a degree whilst for females the figure is only in the order of 1%. Total colour blindness is exceedingly rare.

Even for those prepared to differentiate between colours of vaguely similar hue and saturation there can be differences of opinion as to the naming of colours. Saturation is the term used to describe the intensity of a colour, adding white to a colour makes it increasingly desaturated. The situation is complicated further when we take the lightness and darkness of colours into account, many different words are used to describe the various parameters of a colour, and sometimes the words used are inappropriate and often are just wrong. Frequently for example, 'shade' is used to describe the hue of a colour which is only slightly different from the hue of another colour with which it is being compared. We will formally define the various terms used to describe colour a little later when some of the fundamentals of the manner in which we perceive it have been reviewed.

Though many readers who have picked up this book will not be amongst those used in the examples above, nevertheless in the author's experience there are sufficient readers who would benefit from a review of the way we perceive and describe colour before moving on to the more formal approach of the manner in which we measure colour in Chapter 3.

1.2.1 The Historic Developments Leading to an Understanding of Colour Perception

Already we take much for granted; it is difficult to put oneself in the position of someone attempting to grasp colour in the era before Newton. It was he who first showed that using a

Figure 1.1 Splitting and recombining white light with a prism and a lens respectively.

small hole cut into his window blind to enable a shaft of sunlight to fall onto a prism, white light could not only be dispersed into the colours of the spectrum but also by using a convex lens, the spectrum colours could be recombined to reproduce the white light once more as illustrated diagrammatically in Figure 1.1.

1.2.1.1 Naming the Spectrum Colours

The traditional way of looking at a spectrum is to use a prism and a narrow slit with a source of white light, either the sun or a tungsten lamp, to display the spectrum on to a sheet of white paper. More conveniently in this era of computer disks, one can hold a CD at an appropriate angle to a bright tungsten light source to see the spectrum directly.

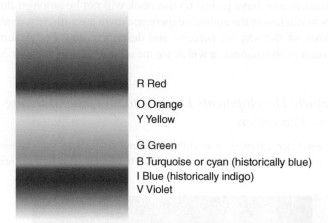

R Red

O Orange
Y Yellow

G Green
B Turquoise or cyan (historically blue)
I Blue (historically indigo)
V Violet

Figure 1.2   Colour naming the spectrum.

In doing so one obtains a range of colours which may be named as shown in Figure 1.2

The capital letters represent those of that old mnemonic to remember the spectrum colours: Richard Of York Gained Battles In Vain. The names of the colours as we are presently more inclined to use them have been added, together with their historical names for reference. Already one begins to see the opportunities for ambiguity.

Generally there is little ambiguity in naming the colours between red and green and these colours appear subjectively pretty much as one would expect. However between green and violet the situation is subjectively not so clear cut.

The colour most of us would perceive as turquoise, or cyan as it tends to be called in colour discussions, that is a colour subjectively perceived as half way between green and blue, occupies only a very narrow band of the spectrum between green and blue and is difficult to pick out within a continuous spectrum. In spectral terms most of that area of the band is taken up with a colour which many of us would describe as light blue. True blue, that is, the blue we describe as 'primary blue', occurs only briefly between cyan and violet. The spacing of the colours in the spectrum depends on whether the spectrum is generated by a prism or a diffraction grating.

Indigo is really an old name for a colour which is close to primary or 'true' blue. Unfortunately the use of current colour names prevents us falling back on that old mnemonic to remember the spectrum colours. The old names are included for reference. It is interesting to note that several well-recognised colours such as brown, pink, purple and magenta are not represented in the spectrum.

In discussing *colour* in terms of reproduction we generally take its most comprehensive meaning which embraces all colours of the same or similar hue, including those with ever diminishing levels of saturation as one approaches the neutral or grey tones between black and white. (A neutral white or grey surface colour is one which reflects equally at all wavelengths of light.) True there are exceptions, particular amongst those colours between red and violet where different levels of saturation lead to different names, the most common example being desaturated red which is universally known as pink.

1.2.2 Surface Colours

So, being aware that white light is actually a combination of all of the colours in the spectrum it becomes easier to appreciate that when it falls upon a surface the resulting colour we see is a mixture of all the colours reflected by the surface. If all colours are reflected we see a white surface but if some colours are absorbed, we see a colour which results from the *mixture of the colours of the spectrum which are reflected.*

Standard tiles with specified spectral reflection characteristics are available and samples from the range of Lucideon[1] standard tiles are illustrated in Figure 1.3.

Figure 1.3 Samples from the Lucideon range of standard tiles.

---

[1] Tiles are very colour fast and therefore are a very useful media for producing ranges of test colours. Lucideon is a company previously named CERAM, which specialises in producing tiles with specific reflection characteristics for use as standards within the industry. http://www.ceram.com/materials-development/colour-standards. The tiles are supplied by Avian Technologies in the United States; see http://www.aviantechnologies.com/products/standards/reflect.php#ceram.

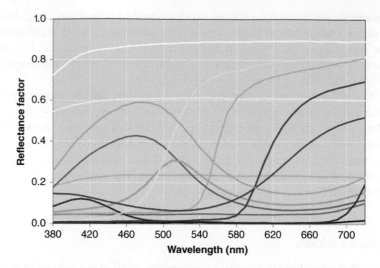

Figure 1.4   Reflectance characteristics of samples from the Lucideon CERAM range of test tiles.

In Figure 1.4 the spectral reflectance characteristics of a set of ceramic test tiles from the Lucideon range are illustrated. The colours of the curves are an approximate indication of the colour of the surfaces they represent. Note how the white tile reflects nearly 90% of the light across most of the whole of the colour spectrum, whilst the colour of the yellow tile is comprised of the light of the spectrum colours, green, yellow, orange and red.

The reader may wish to return to this graph once the information in the remainder of this chapter has been noted in order to review how the absorption of the light in certain spectral bands dictates the perceived colour of the surface. Ceramic test tiles are extremely colour stable and we shall be using the characteristics of this Lucideon range in subsequent chapters of this book.

We are now getting into the detail of what colours we see when certain parts of the spectrum are missing. We know that when the eye is exposed to a spectrum comprising broadly equal amounts of light from violet to red, we perceive the colour white; but in order to be able to predict what we see when a combination of elements of the spectrum are present, we need to investigate how the eye–brain complex responds to mixtures of elements of the spectrum. To do this we need to characterise in some detail how the eye responds to light of differing levels and to light of various frequencies or wavelengths within the spectrum.

1.3   Characterising the Responses of the Eye to Light

Colour is the term we use to describe how the eye perceives light of varying strength at different wavelengths, and light may be defined as the energy in that segment of the electromagnetic spectrum to which the eye responds. The electromagnetic spectrum in its entirety is extremely broad and comprises with increasing frequency: radio, infrared, light, ultraviolet, x-rays and gamma rays. In many branches of science and engineering electromagnetic energy is discussed in terms of frequency whilst in others it is in terms of wavelength. Wavelength and frequency of light are inversely related by the speed of light such that the wavelength 'λ' (lambda) equals the speed of light 'c' divided by the frequency 'f' in cycles per second.[2]

$$\lambda = c/f$$

where $c = 2.99792458 \times 10^8$ m/s or very nearly 3×10^8 m/s.

In treating the subject of light and colour the general practice is to refer to light of a given wavelength rather than to its frequency and to a band of wavelengths as a spectrum.

The eye perceives colour as a characteristic of light. Light is formally that very narrow segment of the electromagnetic energy spectrum occupying wavelengths of approximately 380–720 nm. (A nanometre or nm is one thousand millionth of a metre or 10^{-9} m). It is interesting to speculate on how we evolved such that our eyes are sensitive to just that part of the electromagnetic spectrum where the constituent molecules of surfaces are of such a range of dimensions that their interaction with electromagnetic energy allows light to be differentially absorbed or reflected across the visible spectrum.

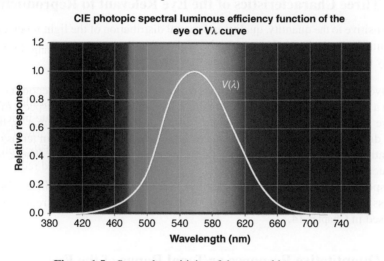

Figure 1.5 Spectral sensitivity of the normal human eye.

In Figure 1.5, the varying sensitivity of the eye over the segment of the electromagnetic energy spectrum to which it is sensitive is illustrated and is known formally as the CIE[3] Photopic Spectral Luminous Efficiency Function but often referred to in its abbreviated form as the 'luminous efficiency function' or in shorthand as the $V(\lambda)$ curve (pronounced the 'V lambda' curve). In the figure it is superimposed over a faint version of the spectrum to give an indication of the relationship between the spectrum colours, the wavelengths of light and the relative response of the eye. This curve, often also referred to historically as the photopic response of the eye or the luminosity function, is the average response of a large number of people with normal colour vision and correspondingly with very similar response curves. We will often use the '$V(\lambda)$ curve' as a recognised short hand in references to this function throughout this book. More recent work has shown the response to be very slightly uplifted on what is illustrated between about 380 nm and 450 nm[4] but nevertheless this is a CIE

---

[2] The unit of a cycle per second is the Hertz, named after the German physicist Heinrich Hertz who proved the existence of the electromagnetic waves theorised by James Clark Maxwell.

[3] Commission Internationale de l'Eclairage or International Commission on Illumination

[4] http://en.wikipedia.org/wiki/Luminosity_function

standardised response and as such is still used for all day-to-day luminous intensity and colour measurement work. You will note that the response is limited to roughly 400–700 nm and peaks at 555 nm in the yellow-green area of the spectrum.

The sensitive cells in the retina of the eye responsible for producing a sensation from the stimulus of light are comprised of two principal types to provide the ability to respond to a very wide range of the level of light perceived. The cones are responsible for *photopic* vision under normal levels of illumination, from bright sunlight down to low levels at dusk, and the rods are responsible for *scotopic* vision at very low levels of illumination, represented by moonlight, for example. Scotopic vision is monochromatic, in that colours cannot be determined; however, there is an overlap range of low-level illumination where both types of receptor are effective, which is described as *mesopic* vision. The remainder of this book will, unless specifically indicated otherwise, always relate to photopic vision.

1.4   The Three Characteristics of the Eye Relevant to Reproduction

The eye is sensitive to the quantity, quality and spatial distribution of the light it perceives. It is convenient initially to deal with the quantitative and qualitative responses of the eye separately before finally considering both of these aspects together.

- Quantitative response. How the eye responds to the amount of light. The accommodation of the eye to a wide range of levels of illumination; the lightness and tonal aspects of colours.
- Qualitative response. How the eye responds to the quality of the light, that is, how light energy of differing spectral content influences how we perceive the light in terms of its hue and saturation, or when these terms are taken together, its chromaticity. These terms are defined later in this chapter.
- Spatial response or acuity of the eye. The ability of the eye to resolve detail differs for changes in lightness and in colour. The relevance of these differences in acuity to reproduction will be addressed in Chapter 14.

1.5   The Quantitative Response or Tonal Range of the Eye

The ability of the eye to operate over a wide range of illumination is truly remarkable. In bright sunlight the illumination level may be between about 50,000 and 100,000 lx, whilst moonlight produces a peak of only about 10 mlx (millilux or 10^{-3} lx). Lux is a measure of the intensity of illumination and one lux is formally defined as equal to one lumen per square metre. The luminance of a surface reflecting light is measured in terms of nits[5] or cd/m^2. The derivation and use of photometric units together with their relationship to physical units is addressed in Appendix A.

This range encompasses both photopic and scotopic vision. Very broadly the vision ranges may be categorised as follows:

- Photopic vision covers the range of luminance greater than 10 nits or cd/m^2
- Mesopic vision covers the range of luminance between 10 mnit and 10 nits
- Scotopic vision covers the range of luminance less than 10 mnit

---

[5] See Appendix A. Using nits rather than cd/m^2 is more efficient, in the same way as we use amps or A to describe electric current rather than using coulomb per second or C/s.

However, we are unable to embrace this very large range of luminance in a single scene; the eye rapidly adapts to the lightest surface of significant image area in a scene; where 'rapidly' is a comparative description. We barely notice day to day changes of illumination when moving from an internal to external environment, even though the level of illumination may change in the order of 100:1 but we are probably familiar with wartime stories where the observers on ships, when moving from a dimly lit interior, required some 20 minutes to fully adapt to the outside dark conditions at night.

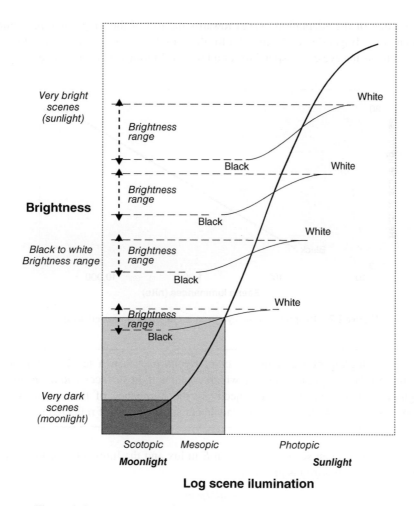

Figure 1.6   The response of the eye to increasing levels of illumination.

In Figure 1.6 the adaptation of the eye to a wide range of scene illumination is shown, indicating the relatively small range of perceived black to white sensation, or contrast range, that occurs at every level of adaptation.

This informative curve was drawn by Ray Knight to illustrate the data derived by Marshall and Talbot (1942).

In all, the eye has a response range of about a billion to one. However, it cannot of course see this enormous range at the same time. The eye adapts to the brightness of the scene and for any given brightness the visible contrast range is limited – as shown by the reduced contrast range curves crossing the main curve at various levels of illumination. It is interesting to note that not only does the contrast range increase with increasing levels of illumination but also the steepness of the curves increases, indicating a greater perceived change in brightness with change in illumination. This is one of the primary reasons scenes 'look better' at higher levels of illumination.

If we take one of the adaptation curves towards the top of the range as representing an outdoor scene on a bright day and expand it to fill a graph we obtain a representation of the range of brightness the eye can respond to for a level of illumination represented by sunlight.

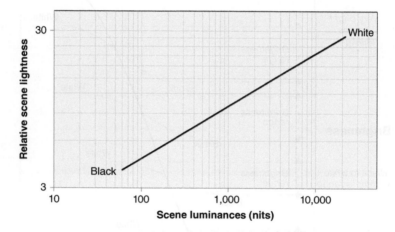

Figure 1.7　Response of the eye to an averagely illuminated scene.

Figure 1.7 is a log/log plot and in this case, we have chosen a level of illumination representing bright sunlight and a scene which contains light surfaces and deep shadows. The brightness of a scene, or more objectively the luminance of the various surfaces comprising a scene, is plotted in nits or candela per square metre of reflected light.

The relationship between scene illumination E in lux and the luminance of the scene L in nits (nt) or candela/m^2 (cd/m^2) is given by:

$$L = \rho E / \pi \; nt$$

where ρ is the reflection factor of a surface in the scene.

Thus taking a typical outdoor scene illuminated by the sun, the level of illumination may be about 75,000 lx, and the brightest surfaces may have a reflection factor of 0.90. Thus the luminance of the brightest surface in the scene, which will normally correspond to white, will be

$$L = 0.90 \times 75,000/3.14 = 21.5 \, knt$$

A log/log plot is chosen because the response of the eye in simplistic terms tends towards being logarithmic.[6] For the mathematically inclined, as we shall see in more detail later, the subjective response is roughly proportional to the cube root of the luminance of a surface, that is, L to the power of one-third. Such responses are produced as a straight line on log/log graph paper, as above. This subjective response to the relative level of light reflected from a surface, in comparison to the white of that surface, is referred to as the *lightness* of a surface.

The y axis gives the response of the eye in terms of the lightness of various elements of the scene or the tones in the scene from black through various less dark shades such as the dark greys, the browns and the blues to the lighter tones or tints such as the pale greys, the yellows and the pinks, for example.

The important factor to note is that the eye is much more responsive to small changes in the dark areas of a scene than similar changes in the lighter elements of the scene. Specifically it does not perceive a series of equal steps in luminance as equal changes of lightness; however, equal *percentage* changes in luminance of two samples with widely different luminances will produce a roughly equivalent equal percentage change in lightness. Work undertaken by Fechner and Weber indicated that in broad terms, depending upon the surround conditions and the adaptation of the eye, one can just perceive a 1% change in scene brightness over the adapted contrast range of the eye. This ratio of $\Delta L/L$ equal to a constant is now universally known as Weber's law.

Furthermore, it can be seen that in this example the scene contrast range is limited to a ratio of about 20,000 nits to about 60 nits or about 350:1. However, the actual contrast range will depend very much upon the type of scene, a darkish scene containing only a limited area of high brightness will evoke a greater range of perception because the low average level of illumination will cause the eye to adapt to that level whilst still accommodating the brighter elements of the scene. It is generally assumed for average scenes that the contrast range of the eye is limited to about 100:1.

The brightness of a scene is directly related to the level of illumination of the scene but the surface of an object within the scene may appear at a different 'lightness' depending upon its relative luminance compared with the average luminance of the whole scene and in particular the luminance of its immediate surroundings. An object of a particular luminance will appear to have a higher level of lightness when surrounded by objects of generally lower luminance and a lower level when surrounded by objects of a generally higher level of luminance.

We will expand further on tonal response and how it is affected by viewing conditions when we come to discuss the tonal response of the reproduction system in Chapter 13.

1.6   The Qualitative Response of the Eye

In Figure 1.8 is the same response of the eye we saw earlier and again including the hues that the different wavelengths of light evoke in the eye. Generally of course a surface will reflect light across a significant segment of the light spectrum. When all the light from an even broad spectrum source is reflected then the eye perceives white or a neutral grey, so it is useful to

---

[6] Also using logarithmic scales enables one to illustrate a much wider range of data than would be practical with a linear scale and furthermore one becomes familiar with the concept that a straight line on a logarithmic plot indicates a simple power law relationship between the parameters portrayed.

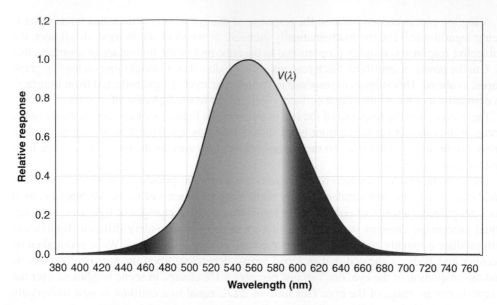

Figure 1.8   The response of the eye at different wavelengths.

consider that hues other than white appear only when the surface absorbs some of the incident white light. As noted earlier the response at the violet wavelengths has been shown to be slightly greater than illustrated here.

The spectrum starts with violet at below 400 nm, peaks at yellowish green at 555 nm and fades away again in the far reds above 700 nm.

If white light falls upon a prism, then we are all reasonably familiar with the coloured spectrum which is produced on a white surface placed to intercept the light leaving the prism, giving the spectrum colours shown earlier. As we have seen it is more useful to describe these colours using modern terminology to avoid the confusion which sometimes occurs when the colours are named by the names of the pigments which artists used centuries ago or even those names used by Newton who first created a spectrum with a prism.

Experiment shows that the addition of two lights of differing hues will evoke the response of a third colour in the eye. This gives a clue as to the mechanism by which the eye produces such an extraordinary range of colours in the brain. Early experiment indicated that the cones in the eye, which are the receptors responsible for vision at normal levels of illumination, were comprised of three types of receptors with very broad responses.

Unfortunately there is of course no direct way the spectral responses of the three types of receptors can be ascertained. Much work has been undertaken over the last 90 years or so by several workers based upon a number of different methods to establish the shape of these responses, including the work of Thomson and Wright (1947); Stiles (1978) and Estévez (1979). The results from each of these studies were similar enough to indicate they were at least representative of the actual responses. As we shall see later, knowing the actual shape of the curves is not critical to ensuring good colour reproduction since the method of measuring colour is not based upon a knowledge of these response shapes.

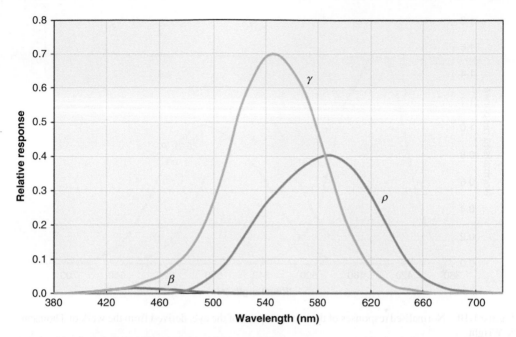

Figure 1.9 Spectral responses of the three cone receptors of the eye, derived from the work of Thomson & Wright.

It is now generally accepted that the responses of the cones are similar to those illustrated in Figure 1.9, peaking at wavelengths corresponding to the blue-cyan, yellow-green and red-orange bands within the spectrum. These three cone response functions are designated the *beta* (β), *gamma* (γ) and *rho* (ρ) curves, respectively; also sometimes referred to in the literature as the S, M and L responses for short, medium and long wavelengths, respectively.

Note the very low level of response of the beta receptor.

An indication that these three responses truly reflect the responses of the three types of cones may be ascertained by checking that the combined response of the three cone receptors equates to the luminous efficiency function of the eye.

In colour work, the shape of the curves is usually more important than their relative sensitivities and the area under the curves of Figure 1.9 are each normalised to 100% in Figure 1.10 in order to enable the shape of the curves to be better appreciated. By normalised, we mean that the area under each of the three curves are made equal.

It should be noted that the precise shape of these curves is not known but since the accuracy in colour work is dependent upon the accuracy of the *measured* colour matching functions which derive from these curves, as shown in Chapter 2, this is of little importance to us.

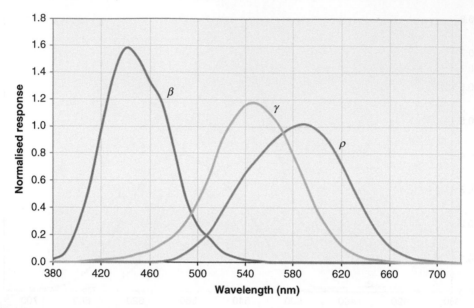

Figure 1.10 Normalised responses of the cone receptors of the eye, derived from the work of Thomson & Wright.

However, knowing the general shape of these curves is helpful in understanding the results obtained from appraising various aspects of colour.

The beta curve in particular has an extremely low comparative response and is shown here increased by a very large factor. (One can see that the peak of the beta curve at 445 nm relates to a response of the eye of only about 1% on the $V(\lambda)$ curve in Figure 1.8.) Although the beta receptor contributes very little to the luminance response of the eye, in colouring power terms it is of equal importance as the other two receptors.

As we shall see in later chapters, sufficient work has been undertaken to specify three colour matching functions relating to the measurement of colour and it follows from the method used to derive these functions that the data relating to the responses of the three cone receptors of the eye is contained within the data used to produce the standard colour matching functions. Thus the CIE, the international body responsible for standardising the colour matching functions, used these data to derive directly in both their 1997 and 2002 Colour Appearance Models (CIECAM97 and CIECAM02) (Hunt, 2004) the best match to the three receptors of the eye based upon the results of the workers listed above and several others responsible for more recent work. These calculations are undertaken in Worksheet 1 and illustrated in Figure 1.11.

As noted earlier, since the derivation of the $V(\lambda)$ curve, more recent work has indicated that the response between 380 nm and 500 nm is slightly higher than that indicated in Figure 1.8 and the evidence of violet at the extreme of the spectrum seems to point to this being due to the rho response falling to a minimum at around 460 nm but then recovering a little at shorter wavelengths. However, in order to ensure that full compatibility is retained between the $V(\lambda)$ curve and the sum of the cone response curves, for standardisation purposes the cone response curves continue to be derived with reference to the $V(\lambda)$ curve.

The eye–brain complex uses only the ratio of the levels of the signals from these three cone responses to evoke a specific visualised chromaticity, where chromaticity describes the

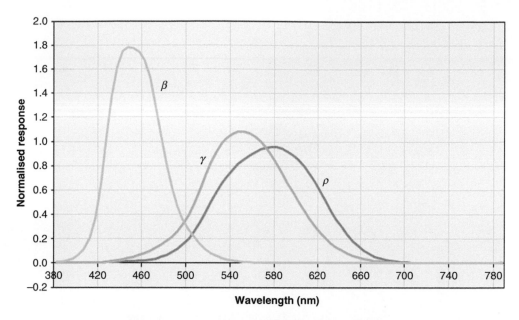

Figure 1.11   The CIECAM97 cone responses of the eye.

hue and saturation of a colour, usually shown plotted on a chromaticity triangle, circle or specific diagram (see Chapter 3). However, for a given level of adaptation of the eye, a colour with a defined spectral distribution may be described as orange for example at one level of illumination and brown at a lower level, even though the ratios of the responses in each of the receptors are identical. This explains how samples that may appear to be of a different colour may have identical chromaticities.

The above statement is so fundamental to the understanding both of what to expect from the mixing of colours and, as we shall see later, the fundamentals of colour reproduction that it is repeated to ensure that its importance has been fully appreciated:

*The eye–brain complex uses only the **ratios of the levels** of the signals from these three responses to evoke a specific visualised chromaticity.*

To be fully accurate, there are conditions where the visualised colour is also affected by other factors, such as colour adaptation but for colour reproduction this statement holds firm.

Note that all three responses are very broad and overlap and that the gamma and rho curves are relatively close together. If one were to produce optical filters with these characteristics and use them to view white light, then the beta light would be bluish; the gamma light yellowish green and the rho light an orangey red.

Figure 1.11 The CIE XYZ cone responses of the eye

hue and saturation of a colour, usually shown plotted on a chromaticity triangle or specific diagram (see Chapter 5). However, for a given level of adaptation of the eye, a colour with a defined spectral distribution may be described as orange, for example, at one level of illumination and brown at a lower level, even though the ratios of the response in each of the receptors are identical. This explains how samples that may appear to be of a different colour may have identical chromaticities.

The above statement is so fundamental to the understanding both of what to expect from the mixing of colours and, as we shall see later, the fundamentals of colour reproduction that it is repeated to ensure that its importance has been fully appreciated:

The eye–brain complex uses only the ratios of the levels of the signals from three receptors to assess the type of stimulus that is visualised.

It is truly remarkable, given the conditions where the luminant colours is also affected by other factors, such as chromatic adaptation but of colour combinations can vary so widely. Note that the three responses are very broad and overlap and that the peaks of the curves are intricately close together. If one were to produce optical filters with these characteristics and use them to view white light, then the light itself would be unsure are mainly blue, yellowish green and the red light an average red.

2

Mapping, Mixing and Categorising Colours

2.1 Primary Colours

We are now in a position to review which colours are the primary colours. But first we need to specify what we mean by the 'primary colours'. Experience indicates that a mix of two distinctly different colours leads to the perception of a different third colour and that mixing three distinctly different colours in various proportions enables a wide range of colours to be produced. It is generally acknowledged that the three colours which enable the widest range or gamut of colours to be perceived are *the primary* colours.

Much of the confusion surrounding just which colours are the primaries is as a result of not first indicating whether we are alluding to 'additive' primaries or 'subtractive' primaries. The more fundamental of these are the additive primaries which are used when we are adding lights of specific colour together; these may be lights from sources such as the pixels of a particular colour which form a computer screen display or lights which are selectively reflected from a surface. Subtractive primaries relate to pigments of one form or another which absorb the light of particular colours and reflect the remainder, thus *subtracting* from the incident light those colours that are not reflected or in the case of a transparency not transmitted.

2.1.1 Additive Primaries

As we have seen, a mixture of two colour stimuli will in the appropriate ratios be capable of producing any colour that lies on a line between them on a colour chart formed from linear primary values, effectively a two-dimensional range of colours. By adding a third colour, then if the three colours are reasonably well saturated, that is, of relatively narrow band in spectral terms, and evenly spaced across the spectrum, we have seen that we can produce a range of colours, or a colour gamut, which may be represented by a Maxwell triangle.

Colour Reproduction in Electronic Imaging Systems: Photography, Television, Cinematography, First Edition. Michael S Tooms.
© 2016 John Wiley & Sons, Ltd. Published 2016 by John Wiley & Sons, Ltd.
Companion Website: www.wiley.com/go/toomscolour

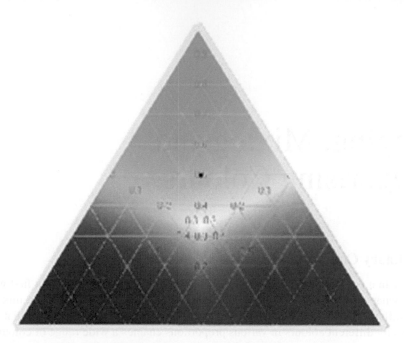

Figure 2.1   The Maxwell triangle.

As Figure 2.1 illustrates, this is an equilateral triangle which has three primaries located at its points and any mix of the colours being illustrated by a point within the triangle related to the relative amounts of the original colour. If the three colours have been appropriately chosen then roughly equal amounts of each of them will produce a neutral colour, that is, one that is represented on the greyscale between black and white.

Any mix of three colours will produce a gamut of colours between them but when we allude to primary colours we indicate that they have been chosen to provide the widest gamut of colours possible within certain constraints of the particular situation. Generally the closer the spectral power distributions (SPD) of the primaries approach a single (monochromatic) or very narrow band of wavelengths, the larger the gamut will be.

Now if there were wavelength positions in the spectrum where these three colour stimuli could be located in order that they each stimulated a different receptor in the eye, this would be the ideal choice for the location of the three 'primary' colours, since in varying ratios they could reproduce all the colours the eye–brain complex would have been capable of producing. This ideal situation is illustrated in Figure 2.2 which shows three *hypothetical* idealised eye responses and the associated ideal primaries.

In this idealised hypothetical example there are positions in the spectrum where each of the primaries stimulates only one receptor. In consequence it would be possible, when emulating the stimuli of a coloured sample comprised of a range of wavelengths, with the appropriate proportions of the three primaries, to emulate the colour of the sample precisely.

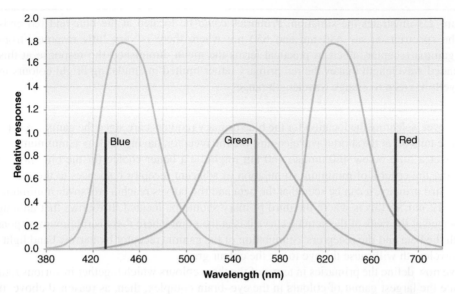

Figure 2.2 An idealised hypothetical set of eye responses with 3 possible single sensor primaries.

The reality however, is the situation illustrated in Figure 2.3 where there is a broad overlap of the eye's cone responses, particularly of the gamma and rho curves. We can still apply our criteria for selecting wavelengths for our primaries where the responses from adjacent curves are at a minimum but nevertheless, we are compromised in that each primary, to a greater or lesser degree, will stimulate more than one receptor in the eye making it impossible to emulate those narrow band colours which appear in the spectrum where the responses of the eye overlap and are also some distance away from the location of the chosen primaries.

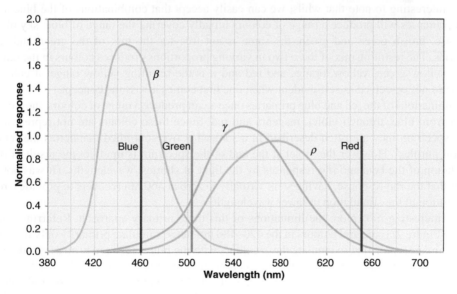

Figure 2.3 A set of probable eye responses with likely location of primaries for a large gamut of colours.

Figure 2.3 illustrates the situation. Primaries *could* be located at the blue and red ends of the spectrum at say 440 nm and 650 nm where there is very little response from the gamma receptor, albeit in practical terms the much diminished rho response at this extended wavelength makes the red primary rather limited in emulating bright colours in the yellow-green to orange wavelength range.

However, selecting the location for the blue primary requires care since the gamma receptor response tails off at the shorter wavelengths, the rho receptor, having sunk to a minimum starts to recover a little below 460 nm, so 460 nm might be a better choice for the blue primary. Applying the criteria of minimum stimulation of adjacent receptor responses to the location of the third primary, it can be seen that the beta and rho curves reach a crossover minimum at about 512 nm, so this is where the third primary should be located. It follows that although not located at the peak of the gamma curve it is at the point where there is minimum response in both the beta and rho receptors. Stimulation of the gamma receptor by narrow band light at this wavelength will cause the eye to see the colour green.

If we now define the primaries in terms of the three colours which together in various ratios produce the largest gamut of colours in the eye–brain complex, then, as reasoned above, the primary colours are red, green and blue. It should be emphasised that these are the additive primaries for lights as would be used in a display where the signals derived from a camera controlled the intensity of the red, green and blue light sources, as in, for example, a plasma or LCD display (see Section 8.3). Of course it is not necessary to locate our primaries precisely at the wavelengths nominated above, any reasonable trio of colours broadly described as red, green and blue will produce a large gamut of colours, albeit that the further the primaries are located from the ideal wavelengths, blue at 460 nm, green at about 512 nm and red at 650 nm, the more compromised will be the size of the gamut produced. The pigment subtractive primaries used for painting and printing are by their nature different as we shall see later.

It is interesting to note that whilst we can easily accept that combinations of the blue and green primaries will produce a range of colours broadly covering the range of blue, sky blue, turquoise, bluish green and green, the same is not true for the mix of the green and red primaries. The resulting mix of these two in varying proportions produces colours in the range green, yellow green, yellow, orange and red and it is the resulting yellow range of colours which are not perceived as self-evident when one first comes across the concept.

Combinations of the red and blue primaries once again produce a range of colours one would expect from blue, through violet, magenta and red. Since these colours are not represented in the spectrum they are often described in a somewhat oversimplified manner as the 'non-spectral purples'. However the violet at the end of the spectrum is in effect the result of the stimulation of the beta and rho responses by the light of shortest wavelengths. (It was noted earlier that the rho response, having sunk virtually to zero at 460 nm recovers to provide a red contribution to the stimulation at lower wavelengths.)

It is interesting to explore the limitations of this three primary approach. Referring once more to Figure 2.3, spectral colours in the 420–480 nm range stimulate primarily only the beta and gamma receptors but to emulate these colours with the blue and green primaries will lead the green primary to also stimulate the rho receptor significantly, effectively adding a little red into the response in the eye which will have the effect of making the emulated spectral colour appear less saturated than the original. (On the basis that an even mix of red, green and blue produce a grey or white and therefore any combination of three primaries are bound to be less saturated than two alone.) Similarly spectral colours appearing in the 560–660 nm range produce a range of highly saturated yellow-green to reddish-orange colours with no activation of the beta receptors; however, in using the green primary to emulate these colours it can be seen that inadvertently the beta receptor is also stimulated for the whole of this range which will once again cause a desaturation of these spectral colours.

In nature there is a preponderance of naturally occurring saturated colours in the yellow green to red range; the reason for which can be seen from an inspection of the curves in Figure 2.3. A colour formed of a broad spectral band of wavelengths located anywhere in the range of 550–700 nm will only stimulate the gamma and rho receptors leading to a fully saturated colour as perceived by the eye; however, a similar broad spectral band of wavelengths located anywhere in the range of 480–550 nm will stimulate all three receptors and will therefore take on a relatively desaturated appearance.

Having highlighted that with these primaries an extensive range of saturated colours in the yellow-green to reddish-orange range, which are comparatively commonly found in nature, are unable to be reproduced at their full saturation, it is worth investigating whether a better compromise could be found for the spectral position of our primaries.

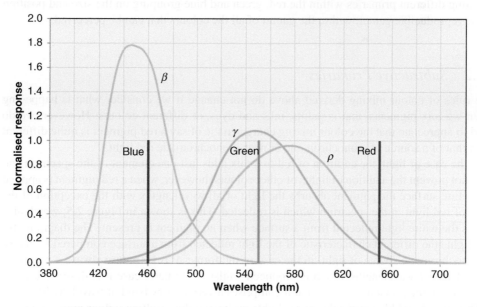

Figure 2.4 Likely locations of primaries for best compromise gamut.

Inspection of Figure 2.4 shows that if the green primary were relocated nearer the wavelength where the beta curve approaches zero at about 550 nm there would be no contribution from the beta receptor when using the new green primary over the whole of the yellow-green to reddish-orange range, thus leading to these colours being reproduced at their full saturation. This compromise would of course reflect adversely on the spectral blue-green colours, the cyans, between 420 nm and 560 nm but these are relatively rare in most scenes.

One might question at this stage whether increasing the number of primaries would enable a wider gamut of colours to be reproduced and a further inspection of Figure 2.4 illustrates that this is indeed the case. Using the three primaries already identified together with the new green at 550 nm would enable the limitations outlined above to be overcome. By adding a fifth primary at the point where the curve of the gamma cone response approaches zero at about 430 nm would extend the range of saturated blues in the gamut. However, it is a case of diminishing returns; each additional primary addresses those colours which are respectively increasingly rare in nature and therefore generally contributing nothing to the average colourful scene, so for good colour reproduction based upon light-sourced displays it is deemed that three primaries are sufficient.

In summary therefore, for the widest colour gamut, the primary lights should be located at about 460 nm for the blue, 512 nm for the green and about 650 nm for the red primary. However for a better compromise, which takes into account the more commonly occurring saturated yellows as opposed to the less frequently occurring saturated cyans, the green primary should be located at about 550 nm, albeit that this trio would produce a smaller gamut of colours. In reality factors other than the size of the gamut also affect the choice of primary colours as will be seen later. These conclusions will be re-affirmed perhaps more clearly in the next chapter where the chromaticity diagrams which are introduced clearly illustrate the effect of selecting different primaries within the red, green and blue grouping on the size and position of the reproducible gamut, within the gamut of all the colours that can be perceived.

2.1.2   Subtractive Primaries

The rules of colour mixing derived above do not change if we consider what is happening when we mix pigments, that is paints, inks and dyes, of different colours. However, we do need to appreciate that the colour mixing characteristic of say a red pigment is quite different from that of a source of monochromatic or near-monochromatic red light.

In the case of light, the addition of red to a mix adds just one element of the spectrum and does not prevent the addition of lights of other colours; however, when a red pigment is applied to a white surface the pigment absorbs the light of all wavelengths with the exception of the band of red light in the spectrum which is reflected. This is shown in Figure 2.5, which also shows the white light reflected from a surface when no pigment is present. The diagram also highlights the different characteristic of the light reflected from a surface compared to that of a monochromatic source of light, in the case of the former it is by nature always a relatively broad band of wavelengths which is produced whilst in the latter case it is desirable, as we have seen in the previous section, that it comprises a very narrow band of wavelengths.

Thus if red and blue monochromatic lights are mixed the resulting colour is magenta, but if a blue pigment is mixed with a red pigment then the blue pigment absorbs all colours with the exception of blue. However, there is no blue light reflected because the red pigment has already absorbed it and the blue pigment now absorbs the red light so the result is that no light is reflected. With ideal pigments the result would be black but pigments are not ideal, they do not absorb all the light at other parts of the spectrum so the result is often a dark muddy colour. Since both pigments generally absorb the green light this colour may receive double the absorption of the other colours in which case a 'black', possibly with a slight magenta cast may result.

It follows therefore that in selecting suitable pigments for primaries, we need colours that provide non-overlapping absorption bands, so that when mixed there are still reflecting

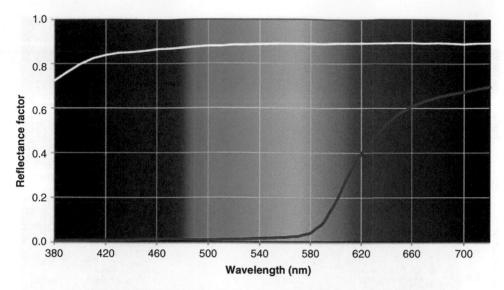

Figure 2.5 Light absorbed and reflected by white and red surfaces.

colours from both pigments to contribute to the reflected mix of light. Since these pigments are characterised by absorption, that is, they subtract colours from the incident light, they are known as subtractive primaries. Put at a more fundamental level, we need subtractive primaries which each stimulate two different receptors of the eye whilst providing maximum absorption at wavelengths corresponding to the sensitivity curve of the remaining receptor. The corollary of this statement is that the subtractive primaries in isolation will always reflect two of the additive primaries; thus if considered in this manner the normal rules for the mixing of coloured lights developed in the previous section will remain in force.

Figures 2.6 illustrates the characteristics of a set of ideal subtractive primaries, sometimes referred to as block primaries. In (a), the pigment reflects the green and red bands but absorbs in the blue and thus appears yellow; in (b), the pigment reflects the blue and green bands but absorbs in the red and thus appears cyan, whilst in (c), the pigment reflects the red and blue bands but absorbs in the green and thus appears as magenta. In (d), the characteristics of the three subtractive primaries are overlaid to emphasise the mutual crossover points of the block reflection characteristics.

As the figure above illustrates, when the yellow and magenta pigments are added together the green and the blue light is absorbed leaving only the red light. Similarly adding magenta and cyan pigments will produce blue, and adding yellow and cyan pigments will produce green.

A practical example of these results is illustrated in Figure 2.7 which is a scan of the artwork of the mixes of the three primaries appearing in the large circles. It shows the effect of mixing real pigments, in this case from the Daler Rowney System 3 Acrylic range. (It should be noted that the magenta primary is significantly nearer red than the ideal for the reasons discussed later.)

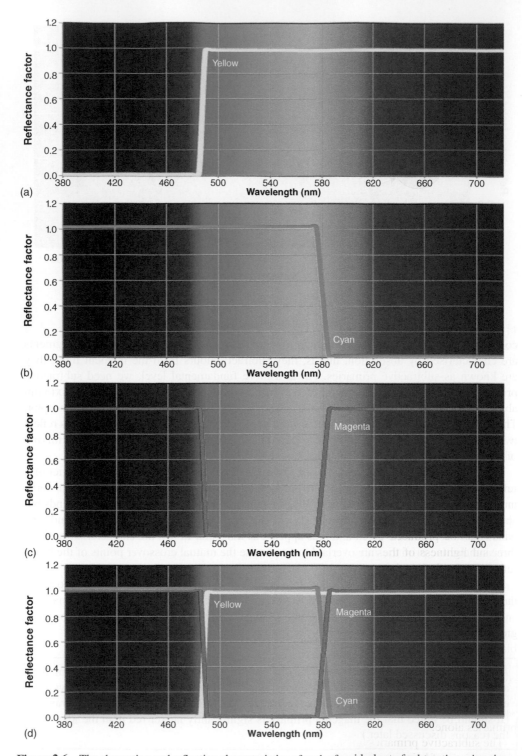

Figure 2.6 The absorption and reflection characteristics of each of an ideal set of subtractive primaries.

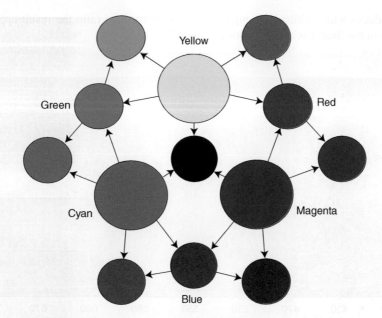

Figure 2.7 A practical mixture of the subtractive primary pigments. (Adapted from an original idea of Ray Knight.)

The arrows indicate the pigments contributing to the mixes in the smaller circles. The results illustrate the mixes to be reasonably saturated, however because the cyan pigment used is of a high saturation and therefore of a relatively low luminance, it was first mixed with a little white pigment to produce the cyan tint illustrated in the large circle.

The direct mix of any of the two primaries located in the large circles produces the colours green, blue and red, respectively, whilst further mixes of the primaries and the red, green and blue colour mixes produce the colours in the outer range of circles, including orange, lime green, turquoise, light blue and purple. The diagram illustrates the wide range of colours which can be produced by the yellow, cyan and magenta primaries, albeit that the saturation and lightness of the mixes is somewhat less than that which can be achieved by single pigments designed for a particular colour. When all three primaries are mixed together the incident light is mostly absorbed at all wavelengths and the result is the black of the centre circle.

When critically comparing the diagram on the computer screen with the original artwork, the on-screen yellow is very slightly more reddish, the purple is slightly desaturated and the black is less black; otherwise the colours are an excellent match to the original artwork.

Generally it will be noted that since all the pigments with the exception of white absorb some light, then invariably adding a further pigment to a mix will always result in a colour of lower luminance. For this reason mixing an amount of white pigment to a dark primary, whilst sacrificing a degree of saturation, will greatly extend the range of colours available from the primaries alone.

The subtractive primaries are sometimes referred to as the complementary primaries. This characteristic of being complementary carries through to the mix of the three primaries; with light this produces white whilst with pigments in an appropriate ratio the result approaches black as seen in the final mix at the centre of Figure 2.7.

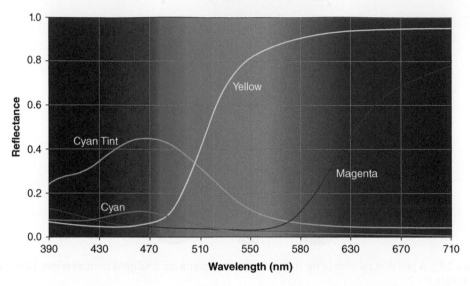

Figure 2.8   Spectral reflectance curves of the pigments used for Figure 2.7.

> Figure 2.8 is indicative of the actual reflection characteristics of the acrylic range of primaries used for the pigment mixes in Figure 2.7. The cyan tint is the result of mixing a little white pigment with the cyan pigment.

It can be seen that at wavelengths above about 500 nm the shape of the curves bear a good practical resemblance to the ideal curves portrayed in Figure 2.6; however, below this wavelength, for all three pigments the light reflected is only a very small percentage of the ideal. The result is that the yellow pigment is very close to the ideal of that shown in Figure 2.6(a). The 'red' component of magenta is also a reasonable match to Figure 2.6(b) but the 'blue' component peaks towards violet rather than blue and then only reaches about 15% of the ideal. The cyan pigment has an appropriate curve shape but only reaches about 12% of the ideal curve of Figure 2.6(b). Thus these pigments are relatively poor representatives of their corresponding ideal subtractive primary pigments.

The magenta pigment matches the ideal well in the red portion of the spectrum which explains why it appears midway between magenta and red rather than true magenta. The cyan pigment, although reaching a peak reflection in the blue portion of the spectrum falls away sharply in the green area leading to a hue which is on the blue side of true cyan; in addition, its very low reflectance explains its very dark appearance. It is somewhat surprising and gratifying to see that adding white pigment to cyan to produce the 'cyan tint' has greatly improved its lightness and curve shape with only very little loss in saturation caused by a small enhanced reflection of light over the yellow to red bands of the spectrum.

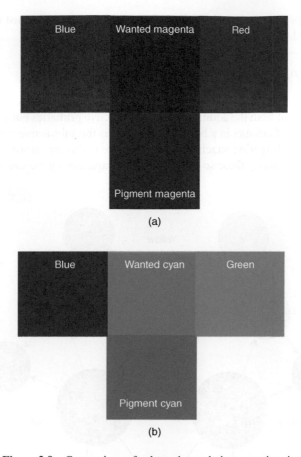

Figure 2.9 Comparison of subtractive and pigment primaries.

Figure 2.9 attempts to portray the difference in hue between the magenta and cyan subtractive primaries and the corresponding pigment primaries. The results are indicative only as the inks used in the printing of this book are likely to cause further distortions of the hues.

The subtractive primary is located between its two adjacent additive primaries for reference and the associated pigment primary is located beneath the subtractive primary to highlight the difference in hue.

In order to provide a realistic comparative match, the two rows of three computer-generated additive and subtractive primaries in (a) and (b) respectively have been markedly reduced in lightness to match the lightness of the pigment primaries.

Although the magenta pigment has an appearance closer to the red additive primary than to subtractive magenta, nevertheless, its spectral reflectance is effective in enabling it to provide a wide range of colours from a mix of primaries as illustrated in Figure 2.7.

The mixtures of subtractive primaries illustrated in Figure 2.7 are useful for showing what can be achieved with a set of pigments used for painting a picture. The success of the subtractive primaries, yellow, cyan and magenta can be seen when they are used in printing and photography to produce generally very satisfactory results with the vast range of colours we have become used to seeing in books, magazines and photographs.

2.1.3   The Non-Primaries

In Chapter 1, it was noted that frequently the primary colours are described as red, yellow and blue, effectively a mix of both the additive and the subtractive primaries but used it is believed to describe the mixing of colours in a box of paints, that is the subtractive primaries.

In order to dispel any lingering attachment there may be to this misinformation, Figure 2.10 illustrates the results of mixing these so-called primaries, again using the same range of acrylic pigments.

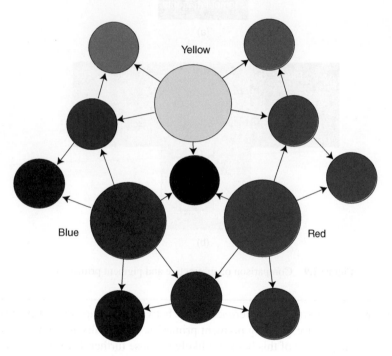

Figure 2.10   Illustration of the limited range of colours obtained by mixing red, yellow and blue pigments.

One can see that although the red and yellow make a good orange, the other mixes of the 'primaries' produce very desaturated and very low luminance samples, effectively with 'perfect' pigments producing black. The reasoning for this result is that, for example, the red primary has absorbed all colours but red and similarly the blue primary has absorbed all colour but blue, thus a mix of these two pigments leaves no light to reflect. For the blue and yellow primaries the result is even more striking; intuitively one might consider that adding the relatively bright colour yellow to the blue would lighten the resulting colour; however, the blue pigment absorbs all colours except blue and the yellow pigment absorbs the blue light, thus once again there are no colours left to reflect and the result is black.

This is conclusive evidence that the set of red, yellow and blue colours are no substitute for the yellow, cyan and magenta set of subtractive primaries as the basis for obtaining a broad gamut of colours from mixes of the three in various proportions.

So how is it that this misunderstanding, of which colours are the subtractive pigment primaries, is so relatively widespread and entrenched? One of the colours, yellow is correct, so the problem is related to those colours described as 'red' and 'blue'. It is believed the problem lies both with the situation addressed in the early paragraphs of the previous chapter, namely the casual naming of colours by categorising vaguely similar looking colours into one of the colours of the unitary colour group of red, yellow, blue and green and also because the reflection characteristics of these two primary pigment colours are far from the ideal. As a result, in terms of their hues they are not perceptually half way between the colours produced by the combination of the additive primaries red and blue, and blue and green, respectively but are somewhat closer to the red and blue additive primaries, respectively as illustrated in Figure 2.9. So pigment magenta is closer to red than to blue and pigment cyan does look subjectively nearer blue than green in colour, rather than cyan or turquoise. (Though the inks used for printing are much closer to the ideal than are the pigments used for the colour mixing experiment, nevertheless, they are not ideal and the cyan ink used in Figure 2.9 may still be perceived as closer to blue than green, rather than subjectively mid-way between them.)

Thus as a result of casual naming, any colour from orange through red to magenta may be called 'red' by some and similarly any colour from violet-blue through blue, sky blue and cyan may be called 'blue'. Since the pigment primaries are not as distinctively magenta and cyan as they ideally should be it is not surprising that ambiguity exists. The problem is highlighted in some areas of the print industry where the subtractive primaries are sometimes referred to as yellow, 'printers' red' and 'printers' blue'.

2.1.4 Primaries in Reproduction

As will be seen later, the concept of additive primaries based upon light sources and the concept of subtractive primaries, based upon pigments and inks, are both fundamental to the practical reproduction of colour images.

2.2 Colour Mixing

We can build upon the simple mixes used in the preceding paragraphs to illustrate the derivation of primaries by exploring the gamut of colours the eye is capable of seeing, or more accurately the eye/brain complex is capable of perceiving.

By arranging the spectrum colours around two-thirds of a ring and the non-spectral purples around the other third, a continuous ring can be produced which illustrates all the hues the eye is capable of recognising, as represented in Figure 2.11.

The two inner rings of sample colours are arranged to be at 180 degrees to each other in order that complementary colour pairs are adjacent. The fact that these colour pairs are truly complementary is illustrated by the fact that their mix results in the neutral grey circles which lie between them.

The additive RGB and subtractive CMY primaries are illustrated in the centre of the circle.

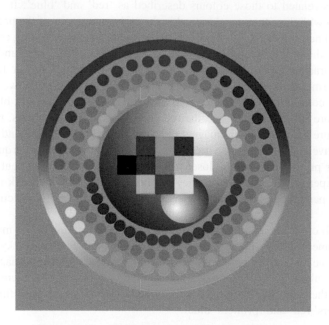

Figure 2.11   Colour mixture circle by Ray Knight.

As we have seen in the review of subtractive primaries, because the pigments are real and compromised compared to the block dyes, then the diagram in Figure 2.11, having been produced by printer's inks will also be compromised both in terms of the range of individual hues produced and the general level of saturation of the hues, compared to that which could be produced by the additive light primaries identified previously. Nevertheless, it serves to illustrate the general principle of the range of hues the eye can perceive. Further insights into colour mixing can be found in the works of Ray Knight (2014) and Gilbert and Haeberli (2007).

2.2.1   Grassman's Law

Herman Grassman (1853), a nineteenth-century scientist, established[1] that if two simple colours are mixed together, they give rise to the colour sensation which may be represented by a colour in the spectrum lying between them when that colour is mixed with an amount of white light. Later work showed that this law is a manifestation of a basic characteristic of the

---

[1] A description of this paper in English can be found in the book by Hyder (Hyder, 2009) and also an extract by searching the web on the title of Grassman's paper.

colour properties of the eye/brain complex; that is, colours add in a linear fashion, as we shall see in the next chapter.

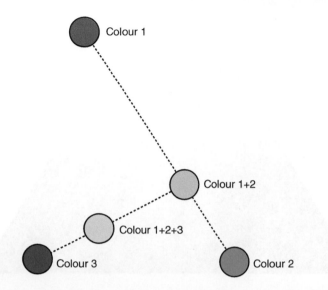

Figure 2.12   Illustrating the addition of a third colour to a mix of two colours.

In the context of the hue ring this law can be used to show that the mix of any two hues will produce a third hue which lays on a line connecting them at distances between them relating to the relative strength or saturation of the two hues.

Adding a third hue to the mix will produce a sample on a straight line between the new hue and the original sample produced by the two previous hues. The construction required is illustrated in Figure 2.12.

If we continue to assume we are using spectral hues as our primaries and then taking this approach to the limit, by for example, using a 1% change in each of three widely separate spectral hues in turn, we will establish $100 \times 100 \times 100$ different colour stimuli, which is one million colours, assuming for the moment that the eye is capable of differentiating these small changes in colour.

If we locate these primaries at three equidistant points, the colour triangle which results allows us to explore in simplified terms the concept of a flat plain representing all the hues and saturations of the colours perceived by the eye. This is the Maxwell triangle which is repeated here for convenience as Figure 2.13.

If the three spectral hues are suitably chosen, then as one approaches the centre of the triangle from the perimeter the saturation of the colours will reduce until white is obtained at the centre.

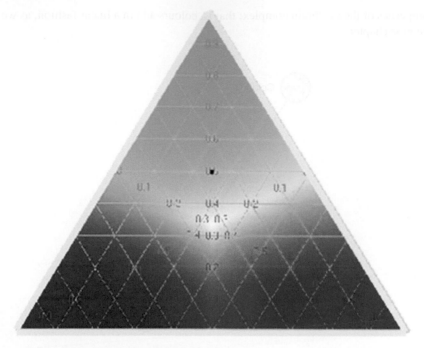

Figure 2.13   The Maxwell triangle. (A repeat of Figure 2.1.)

Because of Grassman's law relating to the result of a mix of colours appearing on a straight line between them, then any two hues opposite one another on the perimeter, when mixed in equal quantities, will produce white; that is, they are complementary colours.

Saturation of a sample is defined in terms of the distance between the white centre point and the sample expressed as a percentage of the distance between white and the hue of the sample at the perimeter of the triangle.

2.3   Colour in Three Dimensions

2.3.1   The Simple Three-Dimensional Colour Space

It may have been noted that some colours, brown for example, do not appear in the colour circle. This is because in this section we have been limiting our thoughts to only the qualitative aspects of light and ignoring the quantitative aspects. The colour brown, is predominantly a low-level luminance yellow or orange which reminds us that we have ignored the lightness of a colour. Thus taking on board this additional parameter it can be seen that a colour may be described in terms of its luminance factor, hue and saturation.

Since we have already established that the eye has three different receptors it is not surprising that colour is a three-dimensional property of the eye and we can alternatively use a classic three-dimensional model to illustrate this point as shown in Figure 2.14, where the red, green and blue primaries are shown as vectors at right angles to one another. (A vector is a directional arrow in space).

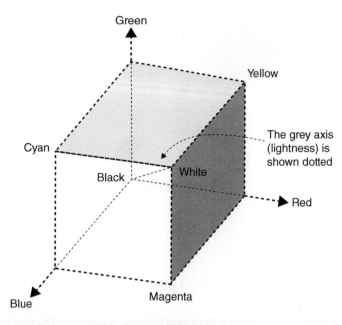

Figure 2.14   A cubic colour solid.

All colours capable of being produced by a mixture of the three additive primaries appear within the colour cube, the subtractive or complimentary primaries being vector additions of the primaries. Nevertheless, this simple diagram does not convey the actual shape of the colour solid representing all the colours that can be perceived as we shall see later.

Commencing from black, where each of the primaries is zero, and using equal increasing amounts of each primary, a neutral lightness vector is produced in which all the greys appear and which terminates at white.

Most colour solids are portrayed with this neutral lightness vector orientated in an upright vertical direction, which can be achieved by swivelling the cube on its black point.

2.3.2   The Lightness Axis

Figure 2.15, which portrays taking the lightness vector from the previous diagram and turning it upright, takes us one step nearer to appreciating the shape of the actual colour solid. One can imagine that at black we have no colour, as the RGB vectors increase in magnitude differentially we see each of the primaries and the colours between them creating a circle of constant lightness; further increase of the vectors will take us to the point where each reaches a value of 100% at white.

Figure 2.15   The result of turning Figure 2.14 on to its lightness vector base. (Drawing by Ray Knight.)

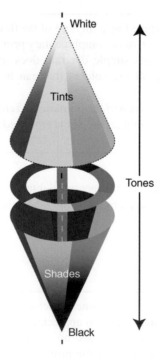

Figure 2.16   A useful geometric representation of the colour solid. (Drawing by Ray Knight.)

2.3.3   The Tone Scale

> Finally, until we are in a position to appreciate more fully the actual shape, we can imagine that the colour solid may be roughly considered as two cones connected at their bases as illustrated in Figure 2.16, with the vertical tone axis split into shades below the centre and tints above.
>
> In very broad terms the surfaces of the cones represent the maximum saturation at the appropriate tone level with the saturation diminishing to zero towards the axis of the diagram.
>
> I am indebted to Ray Knight for illustrating this concept and the execution of the associated graphics in this sequence of diagrams.

Unfortunately, as indicated earlier, the terms tints and shades and to a lesser extent tones, have been widely misused and are used by many different people to mean different things. Nevertheless, it is useful to have defined terms to describe these characteristics of colour.

2.4   Colour Terminology

Some of the terms used and defined earlier are collected together here with others that are used in a more objective manner in the study of colour.

Colours are usually fully described in terms of three subjective parameters variously interpreted by different people. Some of these descriptors are collected together in Table 2.1 under the headings of their currently preferred names.

Our general experience provides us with the basis of recognising the meaning of lightness and hue.

White and the various shades of grey are produced when the complete spectrum is present, the compensatory action of the eye ensuring that, given a few seconds to react, any broad complete spectrum which subtends a significant angle at the eye, will appear as white, irrespective of the shape of its SPD with wavelength. White and the greys have a hue and saturation of zero and a lightness that varies between zero and unity.

Hue describes the name of a colour, that is, whether it is blue, orange or yellow, for example.

Saturation, and a fourth derived term 'chroma', often have different meanings in different traditional colour measuring systems and are therefore terms worthy of some explanation. Saturation refers to the purity of a colour irrespective of its lightness, pink for example, is a low saturation red; chroma on the other hand is an indication of both saturation and lightness,

Table 2.1   Preferred and other terms used to describe a colour

Lightness	Saturation	Hue
Brightness	Intensity	Colour
Value	Purity	Dominant wavelength
Luminance	Chromaticity	
Chroma		

as in the above example where a particular brown might be a low chroma version of a yellow or orange having the same saturation.

Two other terms which are often used in general conversation to mean different things at different times, are 'tint' and 'shade' which are comparative terms. The preferred use of 'tint' is to describe a particular version of a colour which is moving increasingly towards white; whilst conversely 'shade' should be used to describe a version of a colour which is moving increasingly towards black. Unfortunately these terms, particularly the term 'shade' is often used incorrectly to also describe a version of a colour which is of a slightly different hue, so one must be careful in the interpretation of the use of the word.

2.5   Categorising Colours

For those in industries reliant on colour it is essential, if they are to operate effectively, there be a means of defining or categorising colours unambiguously, in order that characteristics of a colour can be communicated to others at different times and different places in such a manner that the colour can be replicated.

In the period before colour measurement procedures stabilised there were many attempts to categorise colours in terms of the colour-descriptive parameters outlined above. Several of the more successful attempts survive in useful forms today; one example of which has evolved into a fully specified system which has gained broad acceptance. Early in the twentieth century, Munsell (1912) realised that the efforts to place the full range of surface colours into a regular shape such as a globe, a cube or a pyramid were, in perceptual terms, unrealistic and recognised the requirement to describe surface colours subjectively but precisely in the terms of the time as hue, value and chroma. These legacy terms have been retained and relate to the hue, lightness and saturation defined in Section 2.4.

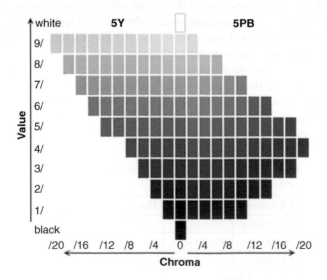

Figure 2.17   Two full ranges of colour samples of two complementary hues.

He found that if a full range of colour samples of a particular hue, each of a just perceptible difference in terms of value or chroma from adjacent samples, were arranged in value order on a vertical axis and chroma order on a horizontal axis, they produced very different encompassing shapes for widely different hues, as Figure 2.17 illustrates.

The figure shows the result for two complementary hues which share the same neutral value axis.

Munsell then went on to formalise his findings in the development of a new colour system for categorising colour samples.

2.5.1 The Munsell Colour System

The Munsell colour system is effectively based upon a cylindrical shape with the Munsell value represented by the vertical axis, the Munsell chroma by the distance from the value axis in a horizontal direction and the Munsell hue by the horizontal angle around the value axis, as depicted by Figure 2.18. In the following description of the Munsell system the Munsell parameters are described as value, chroma and hue.

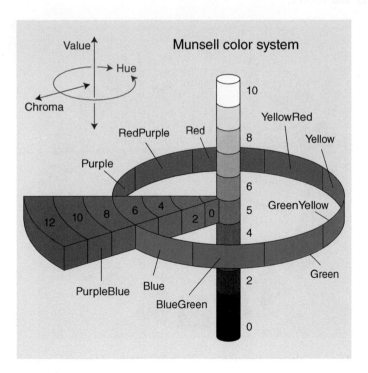

Figure 2.18 The Munsell system of colour. (From Wikipedia – Munsell Color System.)

> *The chips located in the Munsell colour space are always arranged in perceptibly uniform*
> *steps in terms of value (lightness), hue and chroma (saturation).*
>
> The value scale is divided into 10 steps between black and white. The hue scale is
> divided into 100 steps and the number of chroma steps at each hue is dependent upon the
> characteristics of the eye and varies significantly around the hue circle.

About 100 hues at a constant lightness and chroma come close to the limit of discernibility
and the ability to arrange 100 such chips in the correct hue order is often used as a critical
colour test.[2] In practice the implementation of the system in terms of a colour atlas normally
uses only 40 hue pages at a hue separation of 2.5 rather than 1.0.

Pages located on diametrically opposite sides of the circle are made to be complementary
in hue, that is, a mixture of any two symmetrically placed samples will produce a neutral hue
matching the central axis at that value level.

The hue notation of the Munsell colour circle is based upon five principal hues: Red, Yellow,
Green, Blue and Purple, and five intermediate hues: YellowRed, GreenYellow, BlueGreen,
PurpleBlue and RedPurple; each of these ten hues is described by its capital initial letter(s)
and the figure 10, for example, 10Y for Yellow. All other hues between these principal hues
are described in terms of their mixture, with nine integer steps for the 100 hue version and
three 2.5 unit steps for the 40 hue version. For example, in the 40 hue system the three steps
between 10YR and 10Y are 2.5Y, 5Y and 7.5Y, and between 10Y and 10GY the three steps
are 2.5GY, 5GY and 7.5 GY.

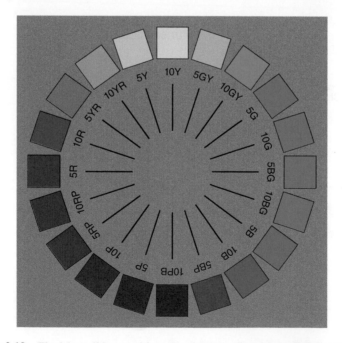

Figure 2.19 The Munsell hue notation. (From Wikipedia – Munsell Color System.)

[2] http://www.colormunki.com/game/huetest_kiosk

Figure 2.19 illustrates the notation described above for an abridged 20 hue version of the Munsell colour system. Each hue is represented at maximum chroma rather than at a constant lightness and chroma level.

Figure 2.17 is a combination of two abridged ranges of complementary pages from the Munsell atlas and clearly indicates the lack of symmetry of chroma and lightness values between widely different hue samples.

Figure 2.20   An artist's impression of a view of the Munsell Book of Color.

A limited 20-page artistic overview of the Munsell Book of Color is illustrated in Figure 2.20, where each page represents a particular hue. Rows of colours are arranged in levels of value or lightness with each row having samples with increasing levels of chroma in order across the page. The samples in each column of the page represent increasing levels of lightness for the same value of chroma.

Since Munsell originated his system over a hundred years ago, it has been adopted commercially and though the original concepts have been retained it has been refined, first in 1929 and then subsequently in 1943, the latter being known as the Munsell 1943 Renotation System comprising 2734 colour samples. The colour samples are made from very stable compounds and great care is taken to ensure equal perceptible steps between samples in hue, chroma and lightness values. All books are guaranteed to match one another in order that it is only necessary to quote the sample number to ensure that two workers with normal colour vision, each armed with a Munsell Book of Color in different locations under the same lighting will perceive precisely the same colour. The book is available in both glossy and matt editions.

Thus a system of colour categorisation has been developed based upon sound perceptible principles which enables colour workers to be precise about the definition of colours without reference to any objective measurement. Nevertheless, such an approach does not preclude the requirement to establish a system of objectively measuring colour and the next two chapters describe how such a system of colour measurement was evolved.

2.6   The Effects of Illumination on the Perception of Colour

So far, when considering the colour of pigment surfaces we have assumed that the illumination has been even across the spectrum. In fact this is rarely the case; the energy at each wavelength changes throughout the day as the radiation of the sun is absorbed differentially by the atmosphere, producing a distinct reddish hue when the sun is located near the horizon and approaching what we would term white throughout the day. In addition the illumination from a clear blue sky is quite different to that from a mixture of cloud and sky or from a cloud-filled sky.

The SPD of various phases of daylight illumination are significantly different as is shown in Chapter 6, with the effect being even more marked between daylight and artificial light from a tungsten source, for example. Clearly the light reflected from surfaces will be markedly influenced by these different sources of light, yet unless the effect is extreme we do not usually notice this to be so. When first switching on tungsten room lighting at dusk we may be disturbed for a second or two at the change of colour of the items in the room but very quickly the eye adapts to the new spectral distribution. This colour adaptation is a very critical characteristic of the eye which enables us to accommodate to light sources of widely different spectral distribution, subject to them being either relatively broad band or of a range of wavelengths which evoke a similar response in each of the receptors. In these circumstances, after sufficient time for accommodation, we are inclined to refer to all such sources as white. The extraordinary ability of the eye to accommodate these changes can be easily demonstrated when a camera set for daylight is exposed to a scene illuminated by tungsten light; the resulting image has a distinctly orange cast or alternately if the camera is set for tungsten the result when used in daylight is an image with a distinctly blue cast.

However the adaptation is not perfect, particularly when attempting to match colours; generally speaking we are at our most colour sensitive when viewing under an illumination which approximates to an even distribution of energy across the spectrum. This is why when attempting to critically match different items of clothing for example, some people will take them from the store into the street where a bright day with a few clouds will produce an illumination approximating to the ideal.

Many artificial light sources do not comprise solely of a smooth broad spectral distribution and in these cases it is often difficult if not impossible to match colours, particularly if they are reasonably well saturated. An extreme example of poor lighting is that provided by low-pressure sodium street lamps whose spectral distribution comprises a virtually single spectral line, making colour appraisal impossible since all colours will be perceived as various levels of orange.

We will return to the topic of illumination in Chapters 7 and 9 to consider its effect on colour reproduction and image appraisal in a more objective manner.

Part 2
The Measurement and Generation of Colour

Introduction

In Part 1, we provided the clues as to how the reproduction of colour could be made to work in a practical manner by exploring the mixing of colours using various levels of additive primary colours. In Part 2 these clues are extended to indicate that by measuring the level of these primaries we can accurately specify a colour in terms of the levels of these particular primaries. Once we can measure a colour in the scene we are well on the way to defining how that colour can be simulated in the reproduced image. The measurement of colour is therefore a fundamental first step to its reproduction in the rendered image.

Thus in Chapter 3 the fundamentals of colour measurement in terms of the values of the red, green and blue primaries, and the derivation of two-dimensional chromaticity diagrams, to portray the position of any chromaticity within the gamut of the complete range of chromaticities as perceived by the human eye, are developed from a historic perspective, a perspective which is still relevant today.

This is followed in Chapter 4 with a description of the work of the CIE in evolving the fundamental work of Chapter 3 into an internationally agreed method of measuring colour. This evolution is indicative of the foresight and brilliance of the early workers in this field and contains concepts which are often challenging to the casual reader. Of those not directly involved in the subject, the initially strange appearance of its several different chromaticity diagrams, which have often been used indiscriminately in the literature in the intervening years, leave many with an incomplete, if not confused, understanding of the interpretation of the results they portray. With this in mind, the evolution of the basic chromaticity diagram, evolved by the early workers in the field in Chapter 3 into the CIE system of colour measurement

Colour Reproduction in Electronic Imaging Systems: Photography, Television, Cinematography, First Edition. Michael S Tooms.
© 2016 John Wiley & Sons, Ltd. Published 2016 by John Wiley & Sons, Ltd.
Companion Website: www.wiley.com/go/toomscolour

in Chapter 4, is dealt with unapologetically in some detail to ensure readers have a good understanding of the remaining material in this book.

Ideally an objective system of colour measurement would express its defining parameters in perceptible terms and so be compatible with the subjective system of categorising colours described in Chapter 2. Unfortunately this is not the case and thus in the final sections of Chapter 4 the parameters associated with the original CIE method of measuring colour, using the values of the primaries, are mathematically transformed to a new set of subjectively related CIE parameters which provide a reasonably close match to the value, chroma and hue parameters of the Munsell system of categorisation.

In order to derive the colour characteristics of the eye in the formal manner described in the chapters on colour measurement it was necessary to strictly define the conditions under which the observers were adapted to ensure that the ability of the eye to change its characteristics, depending upon the lighting environment in which the measurements were carried out, did not influence the results. In Chapter 5 the attempts which are being made to reconcile the procedures developed earlier for colour measurement with the actual perception of what the eye–brain complex perceives when viewing colours under varying conditions of adaptation are reviewed.

The treatment of the basic theory of light and colour in Chapter 1 was of necessity somewhat curtailed as there was the need to first develop the basis for specifying the characteristics of colour before addressing the more fundamental aspects of the generation of coloured light. This is particularly true of the physics of generating light and reviewing its colour characteristics, a topic we need to explore if we are to understand the limitations of what can be achieved in the use of light in the various stages of colour reproduction: that is, in illuminating the scene; in generating the primaries for the display devices; and in providing the environmental lighting for appraising the reproduced image. Chapter 6 therefore provides some basic theory on the generation of light as it pertains to the requirements of the reproduction and perception of colour scenes. As indicated in the introduction to Chapter 6, it is not necessary for an understanding of the remainder of this book to retain the detail of the various means by which light is generated but reading this chapter will help in providing a broad understanding of the colour characteristics of the various light sources available.

3

A Practical Approach to the Measurement of Colour

3.1 The Fundamentals of Colour Measurement

In the previous chapter we provided the clues as to how the reproduction of colour could be made to work in a practical manner. If colour reproduction was reliant on reproducing precisely the same spectral distributions of the colours in the original scene, we would be truly up against it – the complexity of the system would be impractical with our current technology. However, we have seen that the majority of the colours commonly occurring in a scene may be emulated by an appropriate mixture of three primary colours, the missing link is in knowing in what quantities we need the three primaries in order to produce an accurate representation of the original colour.

Of course colour reproduction was not the only aspect of colour that required a method for its measurement. The measurement of colour was a universal problem during the first few decades of the previous century. In 1924 a standardised means of measuring the quantity of light had already been derived under the auspices of the CIE,[1] the international body responsible for recommending the standards for illumination and colour measurement, and it was known from experiments in the mixing of colours that, though not at that time identified, the three different cone receptors in the eye were responsible for evoking the perception of colour.

As we have seen in Chapter 2, the full gamut of colours the eye–brain complex perceives is related only to the relative levels from the three receptors. It is possible to simulate a colour in the spectrum by a mix of appropriate amounts of three primaries. Furthermore since we have also seen through Grassman's law that the response of the receptors are linear, then a combination of a number of adjacent spectrum colours may be simulated by the summation of the values of the primaries required to match each of those individual spectrum colours.

---

[1] Commission Internationale de l'Eclairage or International Commission on Illumination

---

Colour Reproduction in Electronic Imaging Systems: Photography, Television, Cinematography, First Edition. Michael S Tooms.
© 2016 John Wiley & Sons, Ltd. Published 2016 by John Wiley & Sons, Ltd.
Companion Website: www.wiley.com/go/toomscolour

Towards the end of the 1920s, J. Guild, an NPL scientist and W.D. Wright, at that time a PhD student, set out separately to establish an incontrovertible way of measuring colour. To do this they needed to establish, for the average of a number of observers with good colour vision, the values of the primary colours which match the colour of each wavelength of light through the visual spectrum. They used different apparatus, different primaries and a different group of observers to undertake their measurements, the results from which were subsequently found to be consistent between the two groups, thus indicating that they were free of experimental error.

3.1.1   Establishing a Method for the Measurement of Colour

The task of establishing a method for the measurement of colour may be broken down into the following list of activities:

- Select the specific colours of the primaries to be used.
- Build a colorimeter capable of: generating the spectrum; providing a source of controlled levels of suitable primaries and a means of matching the selected spectrum colour with the mix of the primaries.
- Establish a group of observers with normal colour vision.
- Measure the values of the primary colours to match the spectrum at closely spaced intervals of wavelength through the visible spectrum for each observer.
- Average the results from all observers and produce a set of three standardised colour matching functions (CMFs) which give the value of the primaries to effect a match at discreet wavelengths through the visible spectrum.

3.2   Colour Matching Functions

This fundamental work was carried out in the late 1920s independently by both Guild (1931) and Wright (1929). The author was fortunate enough to attend a postgraduate course on colour arranged by Professor Wright and experience at first hand the equipment constructed by him at Imperial College, London to produce the data for the standard CMFs. Thus we will, as an example of the method, describe his work which contributed so significantly to a standardised system of colour measurement.

3.2.1   Selecting the Primaries

Since the shape of the cone response functions of the eye, that is the rho, gamma and beta functions, are fundamental to selecting the wavelength of the primaries for evolving a method of measurement they are repeated here in Figure 3.1 for convenience.

It should be appreciated however, that it is not necessary to know these response curves in order to construct a method of measuring colour as will be seen later. However they are helpful in deciding where to locate the primaries in order to simplify subsequent measurements using the resulting CMFs.

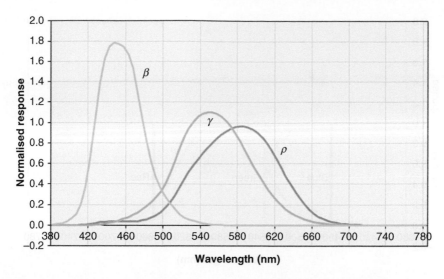

Figure 3.1 Normalised cone response functions of the eye, derived from CIECAM97 but with the rho response extended to lower wavelengths to accommodate the perception of violet colours.

Since all real colours are comprised of a range of wavelengths at varying powers within the colour spectrum (spectral power distributions (SPDs)), it follows that if our three primaries are used to match each of the discreet spectrum colours in turn, we will determine for each wavelength the amount of red (R), green (G) and blue (B) primary required to match each spectrum colour.

An inspection of Figure 3.1 shows that in order to select three primaries which ideally stimulate only one receptor each, they must be located in the red, green and blue portions of the spectrum. Nevertheless as shown in Chapter 2, it is evidently not possible to select a green primary which does not stimulate one or both of the other receptors.

The discussion on the selection of the 'ideal' primaries has been fully explored in the previous chapter and the findings are equally relevant to the choice of the primaries for the colorimeter to carry out the measurements for matching the spectrum. The process is summarised here for convenience.

The primaries should be located at points in the spectrum where as far as possible each only stimulates one of the three receptors of the eye. The blue primary is located where the rho receptor is at a minimum: note that this is at a point before the rho response just recovers at lower wavelengths; the red and green primaries are located at the points where the response of the gamma and beta receptors, respectively are minimally significant.

Figure 3.2 shows the location of the Wright blue, green and red primaries on the cone response functions of the eye at 460 nm, 530 nm and 650 nm, respectively. The effect of the primaries also exciting more than one receptor will be covered in the description of the matching procedure below.

3.2.2 The Colorimeter for Deriving the Colour Matching Functions

Wright built a colorimeter of the fundamental form shown in Figure 3.3 (Wright, 1928, 1939), where the components are laid out on a circular horizontal plane structure on two closely

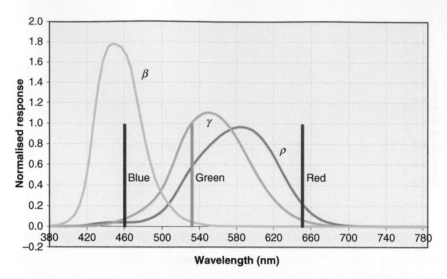

Figure 3.2 The location of the Wright primaries in the spectrum in relation to the CIECAM normalised cone response functions of the eye (modified).

adjacent layers one above the other. In the diagram, in order to simplify the illustration of the light paths, the two planes are actually shown side by side and what is a single set of optics to generate the spectrum and display the colour patches to be matched, is shown duplicated.

An intense source of white light passes through the path separating prism and is split into its spectral components by a further prism to form a wide spectrum of colours of a height to serve both layers. Two sets of narrow mirrors in the form of roof prisms are placed within this spectrum on each of the two layers; one set of three in the upper layer located at the selected

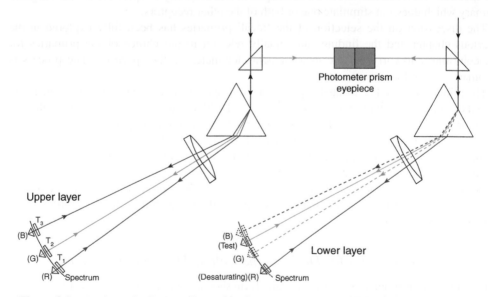

Figure 3.3 A schematic diagram illustrating the essential elements of the Wright colorimeter.

primary wavelengths reflects light back through calibrated variable density filters (T_1–T_3) to the left side of the rectangular field formed in the eyepiece where they are additively mixed. The other set comprising two further narrow mirrors are located in the lower layer, one of which may be moved by a calibrated adjustment through the spectrum to provide the test colour, which reflects light back to the right hand side of the rectangular field where the eye is able to make a critical comparison of the test colour and the colour formed by the addition of the primaries. The other reflecting prism of this set is placed in turn at the same position in the spectrum as the three primaries and is provided with the same level adjustment arrangements as are provided for the primaries. This calibrated primary light, referred to as the desaturating primary, is mixed with the selected test colour in the prism system appearing in the right hand side of the eyepiece. The prism eyepiece enables the two fields to be presented to the eye with no discernible separation between them, which allows a very critical match to be obtained. The eyepiece presents a field of 2 degrees to the eye. (Later work was carried out for a 10 degree field which produced marginally different results.)

Following a calibration of the system, the R,G,B variable density filters are then adjusted to provide a match of the mix of the primaries with the test colour at wavelength intervals throughout the spectrum.

The diagram illustrates the situation when a cyan from the spectrum is being matched. In this range of the spectrum a match can only be made when the spectrum test colour is desaturated with a small amount of the red primary, thus the desaturating primary mirror is placed in the spectrum position of the red primary.

3.2.3   The Observers

Wright established a group of 10 observers, each tested to have normal vision. Each of them completed the matching of the spectrum as described below.

3.2.4   Matching the Spectrum

As a match is made at each wavelength, whilst progressing through the spectrum from blue to red, it is generally found that it impossible to make a complete match. The colour produced by the mix of reference stimuli matches the spectrum colour in terms of lightness and hue but is insufficiently saturated. Inspection of Figure 3.2 indicates the problem. At the blue end of the spectrum where the beta receptor starts to make a contribution, the blue and red primaries used for the match are also stimulating the gamma receptor to a small degree, whilst the spectral test colour is not; thus an 'erroneous' signal is appearing in the gamma receptor. As one progresses through the spectrum, the same problem occurs with the other reference stimuli, first stimulating the rho receptor disproportionately once the wavelength of the blue primary is reached and then the beta receptor once the green primary is reached.

Clearly it is not possible to make a direct match of the test spectral colours with the reference stimuli since over a wide band of the spectrum the reference stimuli are stimulating receptors which the test colour is not. However, one way of achieving a match is to dilute the spectrum colour over the spectrum range where an unwanted receptor is being stimulated by the desaturating primary which matches the unwanted stimuli. Thus between 420 and 460 nm the desaturating primary mirror is located at the position of the green primary at 530 nm; between 460 and 530 nm the mirror is located at the position of the red primary at 650 nm; and beyond 530 nm the mirror is located at the position of the blue primary at 460 nm. Since the desaturating primary is contributing to the spectral colour rather than to the mix of primaries the filter value for any spectrum match is given a negative sign.

Thus if the three reference stimuli are represented by (R), (G) and (B) and the filter values by R,G and B, then a match may be written as follows:

$$C_1 \equiv R(R) + G(G) + B(B) \text{ where R, G, and B are the amounts of the primaries.}$$

The $\equiv$ sign should be read as 'is matched by'.

However, when a spectrum colour C_2 is located in the part of the spectrum where, for example, the rho receptor responds to the green primary but the spectrum colour does so less strongly, that is between 460 and 530 nm, then the desaturating red primary is used to effect the match:

$$C_2 + R_2(R) \equiv G_2(G) + B_2(B) \text{ and thus } C_2 \equiv -R_2(R) + G_2(G) + B_2(B)$$

The level of each primary is plotted at each wavelength, producing a set of three *CMFs*. These are illustrated in Figure 3.4. Clearly if different reference colour stimuli were to be used the functions at any wavelength will take on different values. It will be noted that the complementary pair of functions to a particular primary function will fall to zero at its primary wavelength.

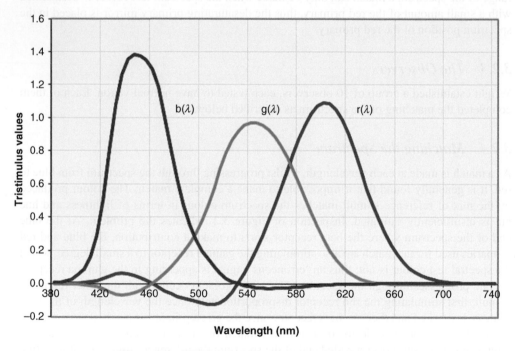

Figure 3.4   Wright primaries colour matching functions.

The values of (R), (G) and (B) stimuli required to match the spectrum colour at each wavelength through the spectrum are referred to as tristimulus values and the resulting curves, which are produced when the values are plotted against wavelength, are referred to as CMFs and assigned the symbols $\bar{r}(\lambda)$, $\bar{g}(\lambda)$ and $\bar{b}(\lambda)$, and referred to in shorthand as rbar, gbar and bbar curves.

As one would expect from inspection of the responses of the eye, starting at the red end of the spectrum, as the wavelengths to the left of the green primary are reached, the red curve goes negative and beyond the blue primary the green curves goes negative, but now the red curve goes positive as the spectrum turns towards violet.

3.2.5   Observer Results

The results from Wright's 10 observers were remarkably consistent and the average of the results of all observers was used to produce the CMFs shown in Figure 3.4. Nevertheless there are small differences between observers with normal colour vision and in certain circumstances as we shall see in Chapter 9, these differences can lead to differences in the perception of reproduced images by different observers. Furthermore, it should be recognised that the spread of results from only 10 observers may not be representative of the spread from a wider sample of the world's population.

3.3   Measuring Colour with the CMFs

If now we have a specimen colour, for example, the tile 'Arctic Blue', represented by an SPD shown in cyan in Figure 3.5, we can calculate the amount of our colour primaries needed to match that colour by in turn multiplying together each of the CMFs by the SPD of the colour, wavelength by wavelength throughout the spectrum and summing the results at each wavelength to give the amount of each primary needed. Usually the values are taken at 5 or 10 nm intervals through the spectrum.

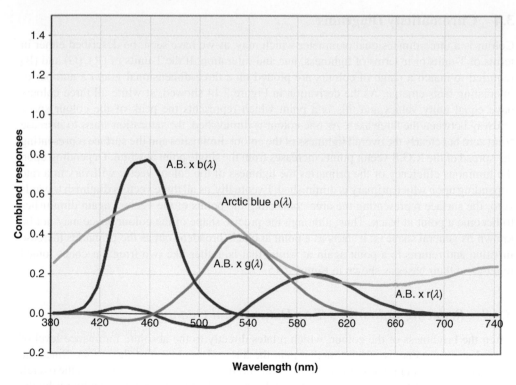

Figure 3.5   Illustrating the multiplication of the Arctic Blue (AB) response with each of the colour matching functions.

This may be carried out algebraically or graphically as shown in Figure 3.5. The units of the primaries are usually referred to as trichromatic units or simply as T-units. Should the summation lead to a negative value, then clearly the colour cannot be matched by real amounts of the primaries.

From Worksheet 3, which was used to develop Figure 3.5, by convolving the Arctic Blue response $\rho(\lambda)$ in turn with each of the CMFs, that is, summing the values at each 5 nm wavelength through the spectrum, the number of T-units of each primary may be obtained as follows:

For the Arctic Blue tile: $R = \sum \bar{r}(\lambda)\rho(\lambda) = 2.816$ T-units

$$G = \sum \bar{g}(\lambda)\rho(\lambda) = 5.683 \text{ T-units}$$

$$B = \sum \bar{b}(\lambda)\rho(\lambda) = 8.816 \text{ T-units}$$

It is important to appreciate that once a set of colour matching functions has been derived the use of a colorimeter for measuring colour is not necessary, all subsequent measurements may be made in terms of spectral power distributions and the amount of the primaries required to affect a match can be calculated as described above. However, the use of negative areas of the curves in calculation can easily cause miscalculation and the procedure was somewhat unwieldy to use before the advent of computers.

3.4   Chromaticity Diagrams

Colour is a three-dimensional parameter which may, as we have seen, be described either in terms of T-units or in terms of lightness, hue and saturation. If the T-units of (R), (G) and (B) required to match a range of colours are plotted on a three-dimensional graph a number of interesting facts emerge. As the derivation in Figure 2.14 showed, at white all three colours have equal unity values and this is a point which represents the peak of the colour space midway between the three axes. As one colour is diminished, the saturation starts to increase from zero but clearly the overall lightness of the colour diminishes and the surface representing the spread of the R,G,B vector points increases from the single point at white. Depending upon the luminous efficiency of the primaries the lightness of the colour vector will vary at a rate depending upon which primary is diminished. Eventually, as all three vectors diminish towards zero, the surface representing the area connecting the points of the vectors again diminishes to become a point at black. Thus, although the precise shape of the colour space may not be known its general shape is; it starts as a point at black, broadens out as the primaries increase in value and returns to a point again at white, that is, rather like two irregular cones joined together at their bases as shown in Figure 2.15.

3.4.1   Reducing Colour to a Two-Dimensional Quantity

Often the brightness of the colour, which relates directly to the absolute luminance level of the colour in a scene, is of less interest than the hue and saturation of the colour. Since the brightness of a colour relates directly to the total number of T-units of each primary, the overall brightness level of the colour may be removed from the T-unit equation by dividing it by the overall amount. Thus if a colour C is represented by R T-units of (R), G T-units of (G) and B T-units of (B) then dividing the equation by the total number of T-units, we obtain:

$$C/(R + G + B) \equiv R(R)/(R + G + B) + G(G)/(R + G + B) + B(B)/(R + G + B) = 1 \text{ T-unit}$$

Thus this *normalizing* process always results in a colour described by one T-unit and is usually written:

$$c = r(R) + g(G) + b(B)$$

Since the values of r,g,b always sum to unity, it is only necessary to quote two of these values since the third can always be derived. In effect by normalising the equation we have made 'c' a two-dimensional quantity. Furthermore, by dividing the equation by the overall brightness (R+G+B) we have eliminated the brightness and are left with quantities which represent the two remaining parameters, that is, the hue and the saturation. These two parameters when taken together refer to the chromaticity of the colour.

3.4.2   Three Steps to Producing a Chromaticity Diagram

Understanding the steps leading to the formation of a chromaticity diagram is crucial and is often dealt with in a somewhat cursory fashion. We will define the approach in three straightforward steps.

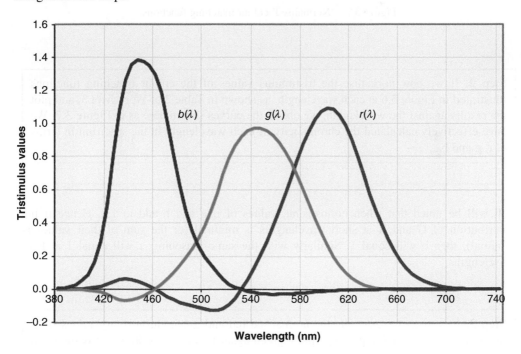

Figure 3.6   Wright primaries colour matching functions.

Step 1. We start with the tristimulus values derived from matching the spectrum colours and plotted as the CMFs as shown here, which is Figure 3.4 repeated for convenience.

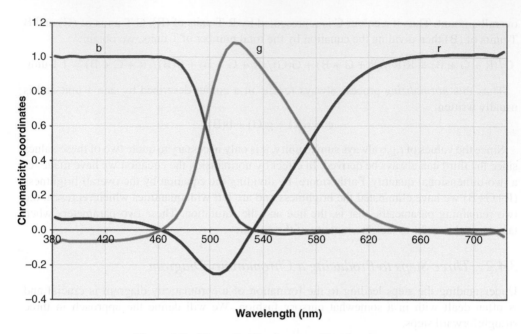

Figure 3.7   'Normalised' colour matching functions.

Step 2. If we now normalise the tristimulus values of the colour matching functions illustrated in Figure 3.6 at each wavelength, as shown in Table 2 of Worksheet 3, and plot the results against the wavelength, we obtain the curves shown here as in Figure 3.7. We have effectively calculated the chromaticity of each wavelength in the spectrum in terms of r, g and b.

It will be noted that when normalising, values of r, g and b add to 1.0. Hence if the contribution of G and R at short wavelengths is minimal (or the sum of their values is minimal), then b will equal 1. Similarly with the same reasoning, r will equal 1 at long wavelengths.

Step 3. If we use, for example, the normalised red and green chromaticity values from the graph in Figure 3.7 to plot a graph of green against red, the result, which is shown in Figure 3.8, is a chromaticity diagram.

The spectrum colours are located on the loci of the diagram and by using Grassman's law, which indicates that colours mix linearly, (as we saw in Chapter 2), we can by mixing the spectrum colours in varying degree fill in the colour space with the full range of chromaticities the eye can perceive.

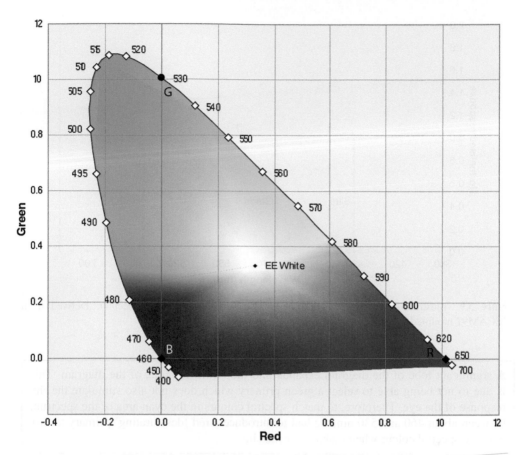

Figure 3.8 Wright primaries r, g chromaticity diagram.

The chromaticity diagram effectively represents a projection from the colour space described in simpler terms earlier. The colours used in the diagram are of course only representative, being limited by the gamut of the printing inks.

3.4.3 Characteristics of the Chromaticity Diagram

Since the three primaries in this case were spectrum colours then the spectrum locus, that is, the plot of chromaticity coordinates of the spectrum at each 5 nm, will pass through each of the primary points, where for the red primary $r = 1$, $g = 0$, $b = 0$, for the green primary $r = 0$, $g = 1$, $b = 0$ and for blue primary, which is at the axis of the diagram, $r = 0$, $g = 0$ and $b = 1$.

This diagram assumes that before commencing the measurements to derive the tristimulus values the three primaries of the colorimeter were adjusted to be equal on a system white equal to equal energy (EE) white. That is a colour which has an SPD which is flat across the spectrum and which as an illuminant is characterised by the CIE as Illuminant S_E. Thus the white point is represented by EE in the diagram and located at $r = 0.3333$, $g = 0.3333$ (and $b = 0.3333$).

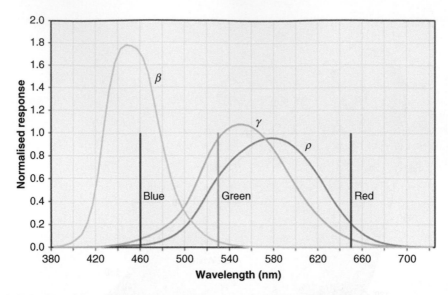

Figure 3.9 Illustrating how the green primary also stimulates the rho receptor. Derived from CIECAM97 modified. (A repeat of Figure 3.2 for convenience.)

A significant lobe of the diagram expands into the minus red area of the diagram. This is due to not being able to select a green primary which does not also stimulate the rho response of the eye. Therefore, to match spectral colours in the cyan area of the spectrum between about 460 and 530 nm, we had to introduce a red 'desaturating' primary to mix with the spectral colour when making our match.

Similarly there are minor lobes extending into the minus green and minus blue areas of the diagram. The green primary also stimulates the beta receptor to a small extent, whereas spectral colours between 380 and 420 nm do not, so a desaturating green primary must be added to the spectral colour side of the matching view over this range of wavelengths. Correspondingly the blue primary also stimulates the green primary to a small extent, whereas spectral colours between 560 and 660 nm do not, so a desaturating blue primary must be added to the spectral colour side of the matching view for these wavelengths. (A straight line drawn between the green and red primary on the chromaticity diagram represents the zero blue line and it will be noted that a very small lobe of the yellow and yellow green spectrum colours are located to the right of this line.)

Thus all colours which can be matched by the primaries alone are contained within the triangle specified by the primaries on the diagram.

As a consequence of locating the green primary at a location where the beta response has fallen to a low figure, the spectrum locus between the green and red primary is tending towards a straight line. Whereas if the green primary had been located at 512 nm, which in Chapter 2 we indicated would be the optimum position giving minimum interaction between the stimuli in the receptors, the cyan negative lobe would have been reduced in size but there would have been a corresponding increase in the yellow lobe beyond where the straight line connecting the green and red primary occurs in the current diagram; that is, for negative blue values of chromaticity coordinates.

As the representative colour in the diagram illustrates, the highly saturated colours are located close to the spectrum locus and as the saturation of the colours diminish they are located ever closer to the white point.

The chromaticity diagram is fundamental to the display of colour information; it is the one diagram to which virtually all colour measurement activities are brought to compare results. Familiarity with the derivation and use of this diagram is essential for fully understanding the remaining material in this book.

3.4.4   Plotting Colours on the Chromaticity Diagram

The values derived in Section 3.3 for the colour of 'Arctic Blue' in terms of T-units may be used to derive the chromaticity coordinates for the colour as follows:

$$\text{Arctic Blue C} = 2.816(R) + 5.683(G) + 8.816(B)\,\text{T-units} = 17.315\,\text{T-units}$$

So normalising the equation by dividing throughout by 17.315, the total number of T-units:

$$\text{Arctic Blue c} = 0.163\,r + 0.328\,g + 0.509\,b$$

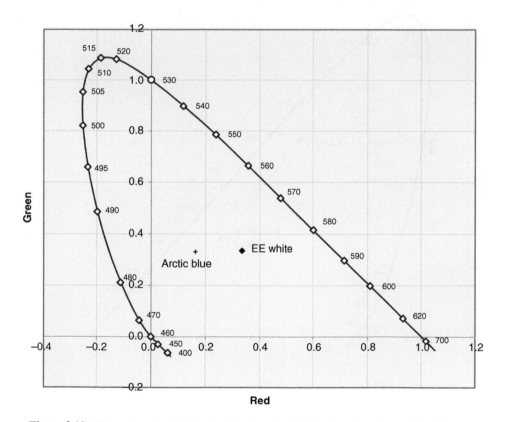

Figure 3.10   The colour 'Arctic Blue' plotted on the Wright primaries chromaticity diagram.

Using the r and g values, the colour 'Arctic Blue' can now be plotted directly onto the chromaticity diagram as shown in Figure 3.10.

One important property of the chromaticity diagram is that where two colours c_1 and c_2 are represented on the chart, a mixture of the colours will lie on a straight line connecting the two colours. The precise position on the straight line will depend upon the relative strength of the two colours but if the number of R,G,B T-units comprising each colour is known then the combination can be calculated.

If

$$C_1 = R_1(R) + G_1(G) + B_1(B) \quad \text{and} \quad C_2 = R_2(R) + G_2(G) + B_2(B)$$

then

$$C_3 = C_1 + C_2 = (R_1 + R_2)(R) + (G_1 + G_2)(G) + (B_1 + B_2)(B) \quad \text{and}$$

$$c_3 = (R_1 + R_2)/(R_1 + R_2 + G_1 + G_2 + B_1 + B_2) + (G_1 + G_2)/(R_1 + R_2 + G_1 + G_2 + B_1 + B_2)$$
$$+ (B_1 + B_2)/(R_1 + R_2 + G_1 + G_2 + B_1 + B_2)$$

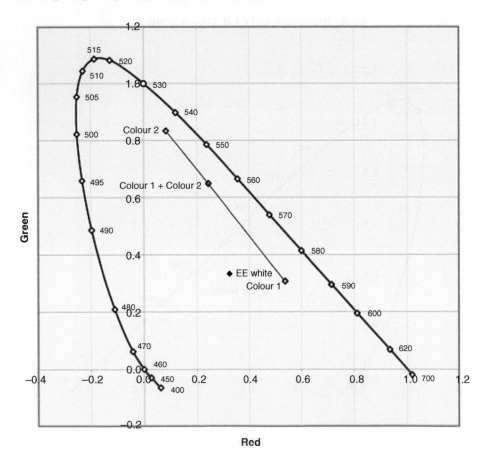

Figure 3.11   Result of the mix of two colours on the chromaticity diagram.

Figure 3.11 illustrates the mix of two arbitrary colours, as defined in Worksheet 3, when plotted on the chromaticity diagram.

It will be noted that the resulting mix lies on a straight line joining the two colours. Its position on the line is dependent upon the relative amounts of the two colours in terms of their number of 'T' units.

Although the system of colour measurement outlined in this chapter works, and is intuitively comfortable to work with, it has a number of drawbacks as it stands and requires refinement before the fundamental results gathered by Wright could be used in a standard manner. The descriptions of these refinements form the basis of the following chapter.

4

Colour Measurement Standardisation – The CIE System of Colour Measurement

4.1 Limitations of the Fundamental Approach to Colour Measurement

The method of measurement used by Guild and Wright and outlined in the previous chapter was not in principle unique. The concept had been recognised for a decade or so, though the care taken and the number of observers used to obtain the colour matching functions (CMFs) did set the results apart from those obtained in earlier work. The problem recognised by Wright, Guild and their other colleagues in the Commission Internationale de l'Eclairage (CIE) was that there was no standard procedure for measuring colour. Different groups in different parts of the world used methods based upon observer data that was unspecific; different sets of primaries and different system white chromaticities made it very difficult to compare results or simulate specific colours accurately.

In addition, it was recognised by this time that by adopting certain mathematical techniques it would be possible to establish a system with a number of advantages over the fundamental approach described in the last chapter, including reducing the amount of calculation required and eliminating the use of negative tristimulus values, thus greatly reducing the risk of error in summing the response readings across the spectrum to obtain the number of T-units.

4.2 The CIE

The CIE is the international body responsible for establishing standards in terms of illumination and colour. At the time Guild and Wright were undertaking their work to definitively establish the colour response of the average non-colour deficient human observer, the CIE was endeavouring to establish what was required to define an objective means of colour

Colour Reproduction in Electronic Imaging Systems: Photography, Television, Cinematography, First Edition. Michael S Tooms.
© 2016 John Wiley & Sons, Ltd. Published 2016 by John Wiley & Sons, Ltd.
Companion Website: www.wiley.com/go/toomscolour

measurement to enable results to be exchanged across the world without ambiguity. To achieve this end it was determined that two tasks needed to be undertaken, which would lead to:

- the definition of a 'standard' observer whose colour vision would represent any person with normal colour vision; and
- a method of colour measurement which is independent of human observation and which could be embodied into the work already undertaken to establish the sensitivity response of the eye, as manifest by the $V(\lambda)$ curve.

4.3   The CIE 1931 Standard Observer

Guild and Wright used different forms of colorimeter, different primaries and different observers in order to be sure that if successful, after appropriate processing of the data to a common set of primaries, their data would be mutually consistent.

The working primaries used by Guild (1932) were obtained by passing the light from an opal-bulb gas-filled lamp through red, green and blue gelatine filters. Guild then processed both his results and those produced by Wright to a common set of spectral primaries, which were blue at 435.8 nm, green at 546.1 nm and red at 700 nm. The blue and green wavelengths were chosen to enable calibration against known line spectra of mercury, the red primary was placed at the long wavelength end of the spectrum where there is very little change of chromaticity with wavelength, due to the cone response of the beta and rho receptors having fallen to zero at wavelengths above 650 nm, and thus the precise positioning in the spectrum is less critical than for the blue and green primaries.

The CIE, being a standards body, agreed that if there was good correlation between the two sets of results, the standard observer CMFs would be based upon the mean of the Guild and Wright results using the primaries adopted by Guild for bringing both sets of results to a common colour space. In fact the results showed a high degree of correlation and the mean values from both sets of data were adopted by the CIE as representing the CIE 1931 Standard Colorimetric Observer.

Figure 4.1 illustrates the CMFs of the CIE Standard Colorimetric Observer. The values of the relative contribution each of these CMFs make toward the luminance or brightness of a colour are defined as the luminosity coefficients and are of importance in developing the themes appearing later in this chapter.

In order to ensure a consistency between the established standards for photometry as exemplified by the luminous efficiency function, $V(\lambda)$, and the new standards for colorimetry, the CIE completed the Standard Colorimetric Observer standard by specifying that the summation of the CMFs shown in Figure 4.1, when suitably weighted by their luminosity coefficients, would equate to the CIE luminous efficiency function or $V(\lambda)$ curve shown in Figure 1.4. Thus if $V_{[R]}$, $V_{[G]}$, and $V_{[B]}$ are the appropriate luminosity coefficients for one T-unit of the (R), (G) and (B) primaries, respectively, then:

$$k \left(r(\lambda) \cdot V_{[R]} + g(\lambda) \cdot V_{[G]} + b(\lambda) \cdot V_{[B]} \right) = V(\lambda)$$

The values of $V_{[R]}$, $V_{[G]}$ and $V_{[B]}$ may either be found experimentally by photometry or by calculation. The CIE adopted the values calculated by the least square fit of the linear

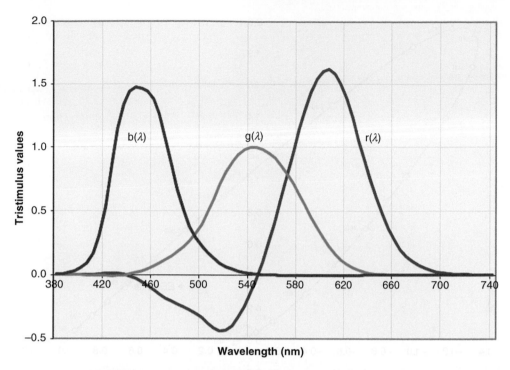

Figure 4.1   Colour matching functions of the CIE Standard Observer.

combination of the CMFs to the $V(\lambda)$ curve (Judd, 1925),[1] which resulted in the ratios of the values corresponding to $V_{[R]} = 1.00$, $V_{[G]} = 4.5907$ and $V_{[B]} = 0.0601$. In Worksheet 4(a) it may be seen that the addition of the CMFs in these ratios does in fact exactly match the $V(\lambda)$ curve.

The same approach used to obtain the chromaticity coordinates for the Wright primaries is used to obtain the normalised chromaticity coordinates for the standard observer primaries and from these the associated chromaticity diagram is derived and illustrated in Figure 4.2.[2]

It is worth noting that the large negative lobe to the left of the green axis is the result of using the line-spectra-calibrated CIE RGB primaries. In particular the green primary is located at a position in the spectrum nearer to the peak of the eye's rho receptor which, whilst avoiding stimulating the beta receptor, does produce a larger 'negative' stimulus in the rho receptor.

Similarly the blue primary is located beyond the point where the gamma receptor is active so there is no negative lobe in the minus green area of the diagram as there was with the Wright primaries and in consequence the spectrum locus is closed by the red axis.

Thus these primaries produce only one large negative red lobe, the negative blue and negative green lobes apparent in the Wright chromaticity diagram have been eliminated.

---

[1] The figures were recalculated using matrix inversion methods in 1996 in a very interesting paper 'How the CIE 1931 Color-Matching Functions were derived from Wright-Guild Data' (Fairman et al., 1997).
[2] See Worksheet 4(a).

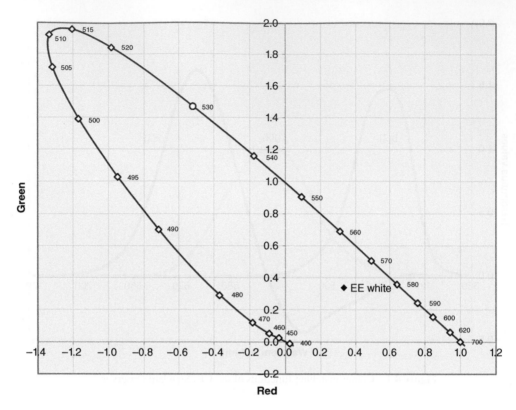

Figure 4.2 The CIE Standard Observer Primaries r,g chromaticity diagram from the normalised chromaticity coordinates of the CIE Standard Observer CMFs.

As we shall see, this diagram is never used for displaying the chromaticities of colours; it is only a tool to act as an intermediate step in developing the CIE system of colour measurement.

The CIE r,g chromaticity diagram is used as the basis for deriving a further diagram using the so-called 'imaginary' primaries which are placed outside of the spectrum locus.

4.4 The CIE 1931 X, Y, Z System of Colour Measurement

To overcome the difficulty of the 'negative' colours in the r,g chromaticity diagrams and to ensure that all realisable colours are always positive, the CIE r,g chromaticity diagram is used to derive a further diagram using 'imaginary' primaries placed outside the spectrum locus. This approach overcomes the difficulties outlined in Section 4.1 since all realisable colours now fall within the triangle connecting the three imaginary primaries and their chromaticity coordinates are therefore always positive.

The criteria used in selecting the new (X), (Y) and (Z) primaries are as follows:

- The primaries should lay outside the gamut of real colours such that all colours may be described by positive amounts of the (X), (Y), (Z) primaries.

- The straight line section of the original gamut between the red and green primaries should be used as one side of the new chromaticity triangle in order to ensure that the third primary falls to zero over this range of spectrum colours.
- The luminance of two of the primaries should be made equal to zero in order that the third primary also describe the relative luminance of the colour.

4.4.1   Ensuring that One Primary Carries All the Luminance Information

As we saw in the last section, the relative luminance of the RGB CMFs is given by

$$(r)(\lambda) \cdot V_{[R]} + (g)(\lambda) \cdot V_{[G]} + (b)(\lambda) \cdot V_{[B]} = (V)(\lambda)$$

where $V_{[R]} = 1.00$, $V_{[G]} = 4.5907$ and $V_{[B]} = 0.0601$.

Thus the relative luminance L of the primaries is given by

$$L = 1.0(R) + 4.5907(G) + 0.0601(B)$$

We need to establish the locus of the points on the r,g chromaticity diagram where the value of $L = 0$, we can then establish on this locus the location of the two new primaries.

$$\text{Thus } 0 = 1.0(R) + 4.5907(G) + 0.0601(B)$$

Normalising, that is sum the three values and divide each value by the sum, and remembering that $b = 1 - r - g$

$$0 = 0.177r + 0.8124\,g + 0.0106(1 - r - g)$$
$$\text{and finally } g = -0.2075\,r - 0.0132$$
$$\text{and when } \quad r = 1, g = -0.2075 - 0.0132 = -0.1943$$

In Figure 4.3 this expression is plotted on the chromaticity diagram for values of g against r and is the line of zero luminance called the alychne.

The straight line joining the R, G primaries is extrapolated in both directions; the intersection with the alychne gives the position of the first new primary, termed the X primary. The remaining two primaries are established by constructing the third line of the triangle to almost graze the spectrum locus; the intersect of this line with the alychne gives the second zero luminance primary termed the Z primary, whilst the intersect with the (R)(G) straight line locates the third primary termed the Y primary.

Since the X and Z primaries carry no luminance information then the Y primary value corresponds to the relative luminance value of the colour.

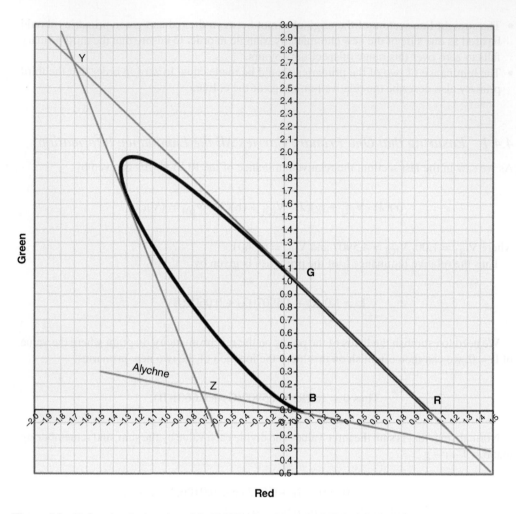

Figure 4.3   Estimating the location of the X,Y,Z primaries on the CIE Standard Observer r,g chromaticity diagram.

4.4.2   Transforming the R,G,B Diagram to the X,Y, Z Diagram

In colorimetry it is often necessary to change the axis of a chromaticity diagram so that the axes of the new primaries which have been established on the original diagram at non right angles are represented at right angles on the new diagram. There are two approaches to this problem, the first, which is more self-evident, is to envisage the two planes on which the two diagrams will exist at an appropriate angle to each other, such that the projection of the original diagram onto the new plane produces the required diagram projection. Though complex, one can envisage that with a suitable application of geometry it would be possible to calculate the relationship between the two diagrams to achieve these results.

The second approach is to use algebraic methods in solving the simultaneous equations in three unknowns which result from specifying the new primaries in terms of the originals. This approach is relatively complex and quite extensive, though made easier by the availability of the matrix functions built into computer worksheets which make the solution of these equations relatively straightforward.

However, even with the use of an appendix to remove the more complex mathematics, the following paragraphs can be overwhelming to those unfamiliar with the techniques. They are included here for those who consider it desirable to understand the process. However, it is suggested that once the principle of transforming chromaticity diagrams via a geometric projection is appreciated in essence if not in detail, it is not necessary to wade through the following mathematics and one can skim Section 4.4.3.

4.4.3   The Transformation Process

From the diagram in Figure 4.3 the coordinates of the (X), (Y), (Z) primaries can be approximately established in terms of their r,g chromaticity coordinates:

$$(X) = 1.28\,r - 0.28\,g$$
$$(Y) = -1.74\,r + 2.77\,g \tag{4.1}$$
$$(Z) = -0.74\,r + 0.14\,g$$

In colorimetry it is often convenient to transpose or *transform* such a set of equations in order to establish the alternate set of primaries as the base system; a task which is essentially a matter of solving three linear simultaneous equations in three unknowns. Since the technique is used so frequently in colorimetry but adds little to the theme of this section, the approach is illustrated in Appendix B where the relationships between the RGB and XYZ primaries are determined. The approach is also detailed in Worksheet 4(b) where the actual calculations are performed.

From Appendix B, equation (4.1) is used to derive an expression for the RGB primaries in terms of the XYZ primaries as follows:

$$(R) = 0.49000\,(X) + 0.17697\,(Y) + 0.00000\,(Z)$$
$$(G) = 0.31000\,(X) + 0.81240\,(Y) + 0.01000\,(Z) \tag{4.2}$$
$$(B) = 0.20000\,(X) + 0.01063\,(Y) + 0.99000\,(Z)$$

Note that the coefficients of Y are equivalent in ratio to the ratio of the luminosity coefficients of the RGB primaries. This is the inevitable result of placing the X and Z primaries on the alychne.

In Appendix B it is shown that since the $r(\lambda)$, $g(\lambda)$, $b(\lambda)$ CMFs represent a set of spectrum colours matched in the rgb system, then they may be used to derive the $x(\lambda)$, $y(\lambda)$, $z(\lambda)$ CMFs.

Thus transposing equation (4.2):

$$(x)(\lambda) = 0.490r(\lambda) + 0.310g(\lambda) + 0.200b(\lambda)$$
$$(y)(\lambda) = 0.177r(\lambda) + 0.812g(\lambda) + 0.011b(\lambda) \tag{4.3}$$
$$(z)(\lambda) = 0.000r(\lambda) + 0.010g(\lambda) + 0.990b(\lambda)$$

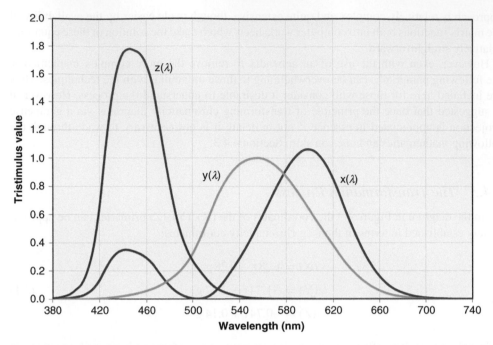

Figure 4.4 The CIE 1931 Two Degree Standard Observer x(λ), y(λ), z(λ) colour matching functions.

4.4.4   The X,Y,Z CMFs

Equation (4.3) gives the xyz CMFs in terms of our familiar rgb CMFs as illustrated in Figure 4.1. The x(λ), y(λ), z(λ) CMFs are derived in Worksheet 4(b) and plotted in Figure 4.4. These CMFs represent the basic tools of modern colorimetry. The published CIE tables for these functions are listed in Appendix I and in the worksheet entitled 'CIE'.

Colorimetrists traditionally used tables giving the values of $\bar{x}(\lambda)$, $\bar{y}(\lambda)$, $\bar{z}(\lambda)$ at 5 nm intervals throughout the spectrum to measure colour as described below. (See the tables in Appendix I). Note that the curves are all positive and that because we located the X and Z primaries on the alychne, the line of zero luminance, the y(λ) curve, which carries all the luminance information, is identical to the luminous efficiency function, V(λ). It is also interesting to note that the shape of a CMF is a function of the chromaticities of the other two primaries, not as one might imagine of its own chromaticity. Thus in the figure above, the shape of the y(λ) CMF is fixed by locating the X and Z primaries on the alychne and will remain the same wherever the Y primary is positioned. In fact, as long as the X and Z primaries remain on the alychne then also their position does not affect the shape of the $\bar{y}(\lambda)$ CMF.

4.4.5   The 1931 CIE Chromaticity Diagram

As with the $\bar{r}(\lambda)$, $\bar{g}(\lambda)$, $\bar{b}(\lambda)$ CMFs, the $\bar{x}(\lambda)$, $\bar{y}(\lambda)$, $\bar{z}(\lambda)$ CMFs may be normalised to produce chromaticity coordinates which in turn may be used to plot the spectrum locus on a chromaticity diagram. Such a diagram (derived in Worksheet 4(c)) is illustrated in Figure 4.5 together with the chromaticities of the original CIE RGB primaries from which it was derived, superimposed.

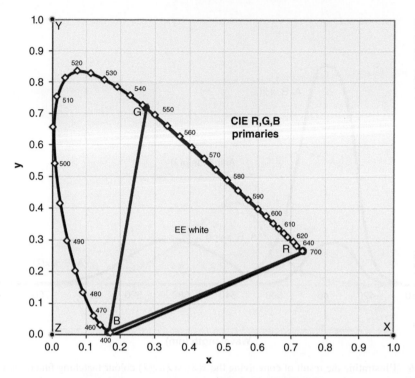

Figure 4.5   The CIE 1931 x,y chromaticity diagram.

Note that the alychne is now represented by the x axis which runs between the Z primary at zero and the X primary at x = 1. The line of zero values of Z is the diagonal which lies between the X and Y primaries and to which the colour gamut triangle is at a tangent, which as intended, leads to all the saturated red-to-yellow colours having zero values of Z.

The ubiquitous CIE x,y chromaticity diagram is the most frequently used diagram to illustrate the results of chromaticity measurements.

Figure 4.5 also illustrates the gamut of colours achievable with the CIE RGB primaries. Though it should be noted that with two of these primaries being so close to the limits of the V(λ) curve, their luminance would be impractical for any reproduction system since they would evoke too little brightness response. The possibility of extending the gamut by shifting the green primary further along the spectrum locus will be explored in later chapters.

4.4.6   Colour Measurement Using the $\bar{x}(\lambda), \bar{y}(\lambda), \bar{z}(\lambda)$ CMFs

The approach to measuring a colour with the $\bar{x}(\lambda), \bar{y}(\lambda), \bar{z}(\lambda)$ CMFs is fundamentally the same as that described in Section 3.3 for the Wright $\bar{r}(\lambda), \bar{g}(\lambda), \bar{b}(\lambda)$ CMFs.

A spectrophotometer is used to ascertain the spectral reflectance of the specimen tile, given by $\rho(\lambda)$, and the level at each measured wavelength is multiplied in turn (convolved) by each of the $\bar{x}(\lambda), \bar{y}(\lambda), \bar{z}(\lambda)$ CMFs to obtain the three sets of responses across the spectrum as shown in Figure 4.6 for a specimen tile referred to as 'Arctic Blue'.

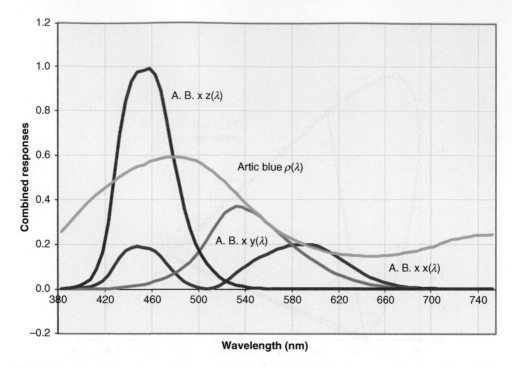

Figure 4.6   Illustrating the result of convolving the $\bar{x}(\lambda), \bar{y}(\lambda), \bar{z}(\lambda)$ colour matching functions with the spectral reflection functions of the Arctic Blue tile.

The area under each curve is then found by summing the values at each 5 nm wavelength through the spectrum to give the number of T-units. In order to compare the results with the T-units of a perfect white surface, reflecting all wavelengths through the spectrum at 100%, the tile T-units are normalised to the perfect white by dividing them with the value of the summation of the values of the $V(\lambda)$ curve.

From Worksheet 4(c) the T-unit values for the Arctic Blue tile are:

$$X = \sum x(\lambda)\, \rho(\lambda) = 5.6292$$

$$Y = \sum y(\lambda)\, \rho(\lambda) = 6.7329$$

$$Z = \sum z(\lambda)\, \rho(\lambda) = 11.6386$$

A reference white tile, reflecting an equal energy white source at 100% through the spectrum would produce a value of T-units:

$$Y_W = \sum y(\lambda)EE(\lambda)$$

As EE has a value of 1.0 through the spectrum

$$Y_W = \sum (y)(\lambda) = 21.3713$$

Thus the tile values 'normalised' to equal energy white are:

$$X_n = \sum(x)(\lambda)\rho(\lambda)/\sum(y)(\lambda) = 0.2634$$
$$Y_n = \sum(y)(\lambda)\rho(\lambda)/\sum(y)(\lambda) = 0.3150$$
$$Z_n = \sum(z)(\lambda)\rho(\lambda)/\sum(y)(\lambda) = 0.5446$$

Note that the luminance factor of the Arctic Blue tile is given by the value of Y_n, that is, 31.5%.

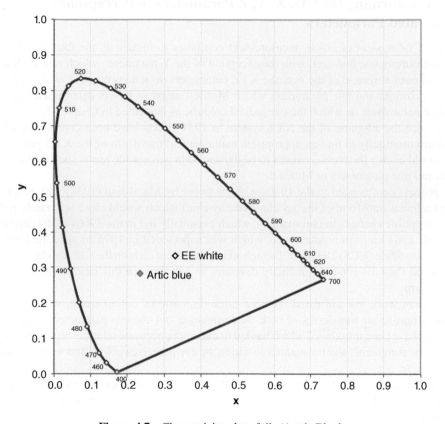

Figure 4.7   Chromaticity plot of tile 'Arctic Blue'.

Normalizing the T-units by dividing each value with the sum of the XYZ values to obtain the chromaticity coordinates gives:

$$x = 0.2345 \quad \text{and} \quad y = 0.2805$$

These values are plotted on the chart in Figure 4.7.

The procedure outlined above was the standard approach to measuring colour for several decades prior to the advances in digital instrumentation and computer processing. Now spectrophotometers are available which carry out these measurements and the associated processing automatically, providing the luminance value Y, and the chromaticity values x,y directly and much more besides.

It should be recognised that the XYZ primaries are not totally devoid of real colour meaning; the X primary is substantially a red primary, the Y primary is substantially a green primary and the Z primary is substantially a blue primary.

4.5   Transforming the CIE X, Y, Z Parameters to Perceptually Related Parameters

The CIE XYZ system of colour measurement continues to remain as the foundation of all colour measurements; however, with the exception of the Y parameter, which relates directly to the luminous response of the eye, the XYZ parameters do not correspond at all well with the value, chroma and hue parameters which Munsell amongst others determined were the subjective parameters on which the eye judged colours, as described in Chapter 2.

Thus since the adoption of the XYZ system in 1931 efforts have been continuous within the colour community to find an appropriate mathematical transform of the XYZ parameters which would enable the measurements to be expressed in perceptible terms such as the value, chroma and hue parameters of Munsell.

The process commenced in the 1930s with the paper by MacAdam (1937) with proposals to adopt a linear transform of the x,y chromaticity chart which would more accurately reflect equal perceptible changes in chromaticity, which eventually led in the 1960s to the definition of the CIE 1964 uniform colour space, which was superseded in 1976 by the CIE Uniform Chromaticity Scale (UCS) Diagram. Though an improvement on both the CIE x,y chromaticity diagram and the CIE 1964 chromaticity diagram, as will be seen this latter chart is still far from uniform.

Subsequent work has addressed the x,y,z three-dimensional colour space, with the introduction of non-linear transforms of both the luminance and chroma parameters. Successive transforms have been introduced which has led to ever more accurate relationships between the objective measurements and the manner in which the eye perceives colour and work continues in this area.

The following sections describe the work undertaken so far, defines the currently accepted transforms and outlines how the remainder of this book utilises them to measure the accuracy of colour reproduction.

4.6   The CIE 1976 UCS Diagram

4.6.1   Subjective Limitations of the CIE 1931 Chromaticity Diagram

Despite the almost universal use of the CIE 1931 Chromaticity diagram in books and magazine articles pertaining to the reproduction of colour, it does however have a serious disadvantage, in that perceived equal colour difference steps at different positions on the diagram are represented by vectors of greatly different lengths.

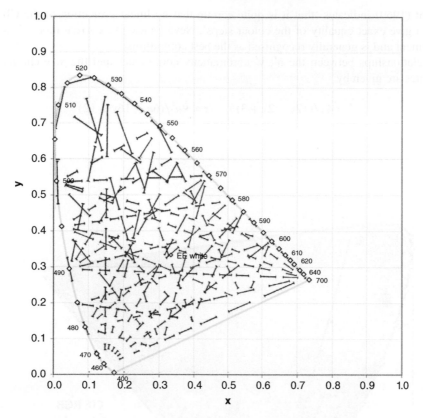

Figure 4.8 Subjectively equal chromaticity steps plotted in the CIE 1931 chromaticity chart. (After Wright, 1969.)

Wright's later work showed that plotting equal chromaticity steps of three 'just noticeable differences' or JNDs, in different areas of the diagram, led to lines of very different lengths, as shown in Figure 4.8, where the ratio of the long lines in the green area to the short ones in the violet area is in the order of 20 to 1.

In subjective terms the x,y diagram is therefore limited in its usefulness as a means of illustrating colour variations.

4.6.2   The UCS Diagram

Over the years colour scientists endeavoured through experiment to establish a different pair of axes which, by operating linearly on the x,y axes of the current 1931 chart, would portray equal subjective chromaticity steps as lines of equal length on the diagram.

Using the results obtained from this work the CIE addressed this problem by evolving chromaticity diagrams which are linear transforms of the x,y diagram. In 1960 the u,v uniform chromaticity diagram was introduced, which was itself superseded by the u′,v′ CIE 1976 UCS diagram, often referred to as the u′,v′ (pronounced u dashed, v dashed), nearly uniform chromaticity diagram.

Wright (1969) indicates 'that it is quite certain that no linear projection of the CIE diagram can give exact equality of the colour steps'. Nevertheless the current model is a great improvement and is generally recognised as the best compromise.

The relationships between the u', v' chromaticity coordinates and the x, y chromaticity coordinates are given by:

$$u' = 4x/(12y - 2x + 3) \qquad x = 9u'/(6u - 16v + 12)$$
$$v' = 9y/(12y - 2x + 3) \qquad y = 4v'/(6u - 16v + 12)$$

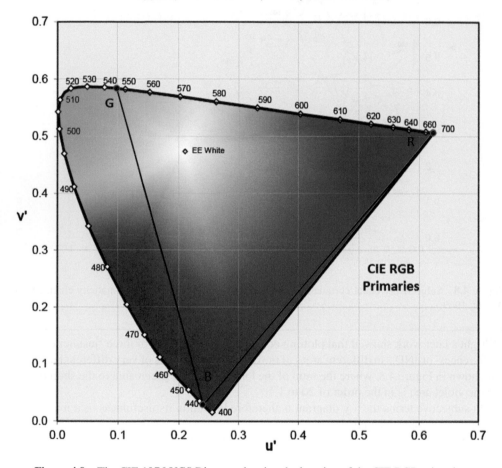

Figure 4.9   The CIE 1976 UCS Diagram showing the location of the CIE RGB primaries.

A plot of the u',v' chromaticity diagram is shown in Figure 4.9 where it can be seen that the white point has been significantly shifted towards the green area of the diagram. This observation highlights the danger of using the x,y diagram for subjective appraisal; an inspection of the x,y diagram would lead one to assume that the eye perceived a considerably wider range of chromaticities in broadly the green area of the gamut, whereas a similar inspection of the u',v' diagram would in fact show that the reverse was true.

The u′ axis at the reference white point represents reddish colours in the positive direction and greenish to cyanish colours in the negative direction, whilst positive directions of the v′ axis represent yellowish greens and in the negative direction the blues and violets.

Some of the lines representing three JNDs which appeared on the x,y chart in Figure 4.8 are plotted on the u′,v′ chart in Figure 4.10. In this case the ratio of the longest to the shortest lines is about four to one, a five to one improvement over the 1931 diagram but clearly still quite a compromise.

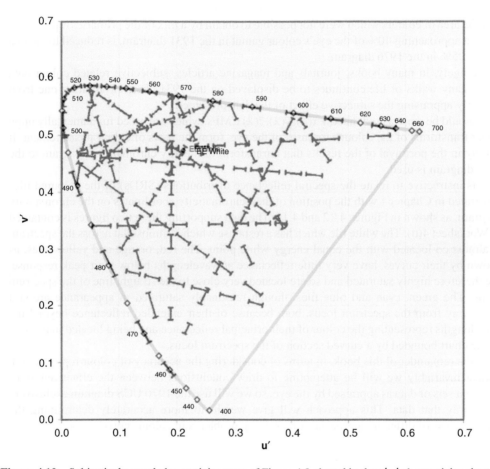

Figure 4.10 Subjectively equal chromaticity steps of Figure 4.8 plotted in the u′,v′ chromaticity chart.

As can be seen, though by no means perfect, the 1976 u′, v′ chart is a great improvement on the 1931 chart and has been adopted worldwide for the portrayal and comparative evaluation of subjective chromaticity data.

Both of the chromaticity JND diagrams are reproduced from Hunt's excellent book 'The Reproduction of Colour' (2004).

So we end up with two diagrams, one suitable for displaying objective results and the other which should be used whenever one wishes to show results which are subjectively important.

4.6.3   Comparing the Two CIE Chromaticity Diagrams

In the two diagrams in Figure 4.11, which are both showing triangles representing the gamut of colours which may be obtained from the mixture of the CIE RGB primaries, you will note that the area of colours which were not possible to obtain by a mix of the primaries represented an area approaching 40% of the eye's colour gamut in the 1931 diagram, is reduced to an area nearer 25% in the 1976 diagram.

Strangely, in many books, journals and magazine articles, subjective related colour data from many walks of life continues to be displayed on the 1931 chart, prohibiting one from properly appraising the subjective effect of the data presented.

It should be emphasised that the $\bar{x}(\lambda)$, $\bar{y}(\lambda)$, $\bar{z}(\lambda)$ CMFs, which are based fundamentally upon linear transforms of the colour responses of the eye, form the basis of colour measurement. It is only in the portrayal of the results that a transform of the x, y chromaticity diagram to the u′,v′ diagram is used.

It is instructive to relate the spectral reflectance distributions (SRDs) of the coloured tiles illustrated in Chapter 1 with the position of their chromaticity coordinates on the chromaticity diagram, as shown in Figures 4.12 and 4.13. The data supporting these two figures is contained in Worksheet 4(d). The white tile which has a response which is almost flat across the spectrum is almost co-located with the equal energy white point. The red, orange and yellow tiles, as shown by their curves, have very little reflectance at wavelengths below their peak response, are therefore highly saturated and so are located very close to the straight line of the spectrum locus. The green, cyan and blue tiles, though reasonably saturated in appearance are still well away from the spectrum locus, both because of their extended reflectance beyond the wavelengths representing the colour of their principal reflectance and being located in the area of the chart bounded by a curved section of the spectrum locus.

In the remainder of this book, in terms of considering the accuracy of colour reproduction, almost invariably we will be attempting to draw conclusions between the effectiveness of different sets of data as appraised by the eye, so we will use the 1976 UCS diagram exclusively to portray that data. This approach will give weight to more accurately determining the effectiveness of different approaches to resolving problems in colour reproduction.

4.7   The CIE 1976 (L*, u*, v*) Colour Space

4.7.1   Establishing a Perceptively Uniform Colour Space

In Chapter 2 we noted that a colour could be specified in three dimensions using three vectors at right angles to one another; either in amounts of the RGB primaries or using luminance, hue and saturation. Our aim here is to build on this concept to produce an objective means of measuring colour but to express the results in perceptively meaningful terms.

The chromaticity of a colour has been defined in a manner which, using only linear transforms, comes as close as possible to the subjective terms defined in Section 2.4, with the adoption of the u′,v′ chromaticity diagram, albeit the diagram not being as uniform as one

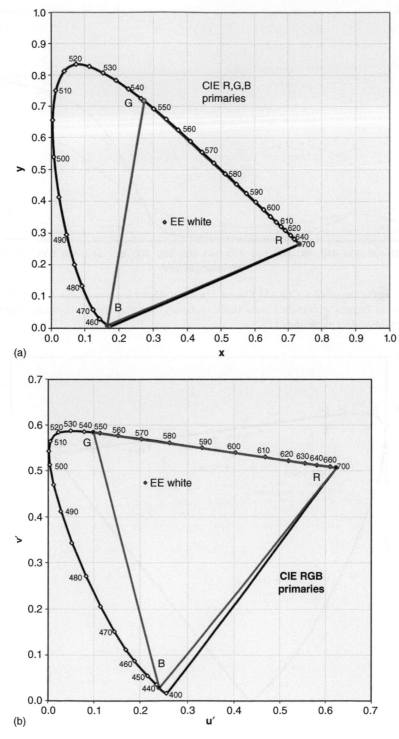

Figure 4.11   Comparison of the two CIE chromaticity diagrams both illustrating the CIE RGB primaries colour gamut.

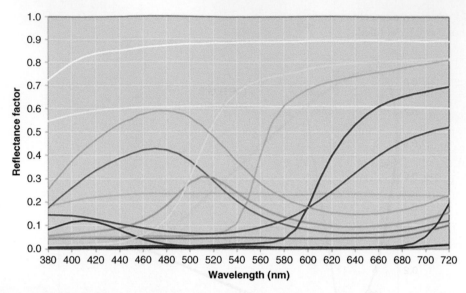

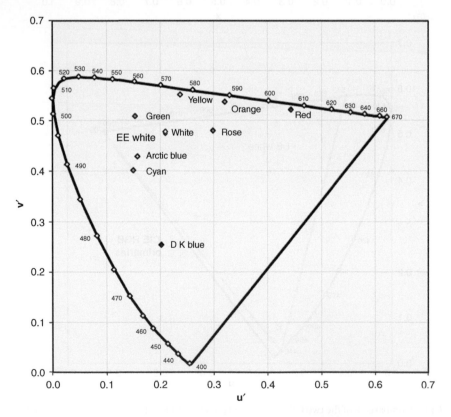

Figure 4.12 Reflectances of a CERAM Colour Standards 45/0 tile set available from Lucideon in Europe and Avian Technologies in the United States. (Adapted from http://www.aviantechnologies.com/products/standards/reflect.php#ceram.)

Figure 4.13 Chromaticity coordinates of the CERAM tile set.

would wish; however, although the subjective term *lightness* was described in Section 2.4, it was not formally defined.

4.7.2   Specifying the Lightness Characteristic

The material in this and the previous chapter has given much emphasis to measuring those two aspects of colour, hue and saturation, which together we have defined as the chromaticity of the colour; however as we saw in Chapter 2, colour is a three-dimensional quantity with the lightness or brightness of the colour being equally important in perception, measurement and reproduction, respectively.

In adopting the u', v' chromaticity diagram as our basis for getting closer to measuring as accurately as possible the subjective relevance of the chromaticity, we need to be equally as consistent in taking the same approach to measuring the lightness of the colour.

We saw in Chapter 1 that the intensity response of the eye tends towards being logarithmic and that a good approximation to its response may be achieved by adopting a power law function of the form which indicates the response of the eye is proportional to the cube root of the luminance in a scene. As we have seen in this chapter, the luminance factor is given by the tristimulus value of the Y primary, so lightness L is proportional to $(Y/Y_n)^{1/3}$, where Y_n is the tristimulus value of the reference white being used.

The CIE have specified a lightness parameter called the CIE 1976 lightness L*, which is defined in terms of (Y/Y_n) as follows:

$$\text{for } (Y/Y_n) > 0.008856 \qquad L^* = 116(Y/Y_n)^{1/3} - 16$$
$$\text{for } (Y/Y_n) < 0.008856 \qquad L^* = 903.3(Y/Y_n)$$

Note the scaling of the luminance and lightness parameters; white has a luminance factor value of 1.00 and a lightness value of 100.

The reason for this approach, where a different expression is used dependent upon the level of (Y/Y_n), is because the slope of the cube root expression at very low luminance levels tends increasingly towards infinity at zero, a situation which is impractical to implement; thus at very low luminance levels the cube root expression is replaced with a function of constant slope. This slope has a value fundamentally of 9.033 but because the relationship between maximum reflectance in terms of luminance and lightness is 1:100 the figure in the formula becomes 903.3.

For reasons which appear somewhat obscure, the CIE later expressed the lightness parameter using these expressions:

$$L^* = 116f(Y/Y_n) - 16$$
$$\text{where} \quad f(Y/Y_n) = (Y/Y_n)^{1/3} \qquad\qquad\qquad \text{if } (Y/Y_n) > (24/116)^3$$
$$f(Y/Y_n) = (841/108)(Y/Y_n) + 16/116 \quad \text{if } (Y/Y_n) \le (24/116)^3$$

Only the form of the expressions has changed, the constants equate to those in the original format.

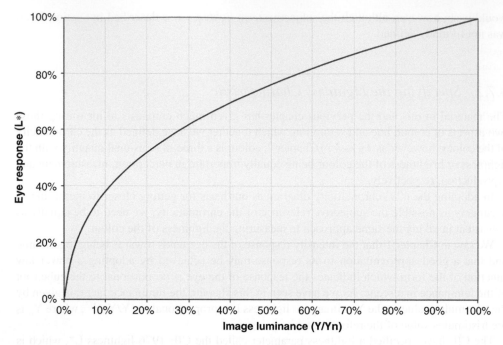

Figure 4.14   Graph of the CIE 1976 Lightness Response of the eye.

The particular arrangement of the formula, where the number 116 is included, is adopted to ensure both that a reference white surface will equal 100 and that the transition point on the curve between the two expressions is smoothly continuous as shown in Figure 4.14. As we shall see, this approach to specifying non-linear responses is much used in tonal reproduction and the supporting derivational mathematics is detailed in Appendix H, which supports the subject of non-linear processing in Chapter 13 and Worksheet 13(b).

The rather awkward constants appear to have been chosen to produce a break point between the linear and power law elements of the curves where the lightness value is precisely 8.00, or in terms of maximum lightness of white, 8%.

4.7.3   Constructing the Perceptibly Uniform Colour Space

With the u′,v′ chromaticity parameters and the lightness L* parameter defined above, all of which are based upon subjective visual effect, we appear to have the three parameters we require to form a three-dimensional colour space. However, we need to convert the chromaticity coordinates into quantities which represent hue and saturation before attempting to build our uniform colour space. The aim is to produce a colour space which notionally, and initially, is cylindrical in shape with lightness as its vertical axis; hue representing the angle of colours around the lightness axis and saturation on a horizontal plane where the degree of saturation is represented by the distance from the vertical axis.

On this basis the lightness axis will represent the colours of zero saturation, that is, the grey tones which occur between black and white. To convert the chromaticity coordinates to represent the hue and saturation we need to establish on a horizontal plane the distance and angle of the colour from the vertical axis, the latter represented by a tone of neutral equal energy grey on the chromaticity diagram at u'_n and v'_n. Thus the distance vector for a chromaticity represented by u' and v' in terms of rectangular coordinates of saturation will be $u'_s = u' - u'_n$ and $v'_s = v' - v'_n$.

However, the subjective relationship between lightness and saturation has so far not been defined: that is, since the relationship between luminance and lightness was defined in isolation it would be surprising if equal percentage changes in lightness and saturation as measured by L^* and u'_s, v'_s were perceived as equal changes and indeed this is not the case. It is found that a factor of 13 needs to be applied to the changes in saturation to make them subjectively similar to the changes in lightness at a particular level of lightness.

$$\text{Thus:} \quad s_u = 13u'_s \quad \text{and} \quad s_v = 13v'_s$$

and converting to polar coordinates we obtain the expression for CIE 1976 u', v' saturation

$$s_{uv} = \left[(s_u)^2 + (s_v)^2 \right]^{1/2}$$

However, early work noted that similar changes in saturation produced greater changes in the perception of the vividness of the colour as the lightness level increased, that is the changes in perceptibility of saturation in a colour space is roughly proportional to L^*. To accommodate these newly defined parameters the CIE have defined:

$$u^* = L^* s_u = 13L^* (u' - u'_n)$$
$$v^* = L^* s_v = 13L^* (v' - v'_n)$$

and converting to polar coordinates, the new colour parameters, formally called C^*, the CIE 1976 *chroma*, and h_{uv}, the CIE 1976 hue angle, are given by

$$C^*_{uv} = [(u^*)^2 + (v^*)^2]^{1/2} = L^* s_{uv} \quad \text{and} \quad h_{uv} = \arctan(v^*/u^*)$$

from which we can see that chroma is dependent upon both the lightness level and the saturation of a colour. The hue angle h_{uv} gives the hue value in degrees in a positive direction from the u^* axis.

To formally establish the position of a colour in the perceptibly uniform colour space we need to carry out the vector addition of the three contributing vectors, L^*, u^*, v^*. Thus a colour sample C_1 would have a position in the colour space given by a vector of length:

$$C^*_1 = [(L^*)^2 + (u^*)^2 + (v^*)^2]^{1/2} = \text{and angle of } h_{uv} \text{ to the } u^* \text{ axis}$$

Before illustrating the colour on a diagram it would be useful to be aware of the maximum value of C^*. An inspection of Figure 4.13 indicates that the maximum value of u'_S lies on the spectrum locus close to red and for v'_S close to blue; however, these values are amended by the value of L^* which prevents a simple estimation of the maximum value. In the next section this theme is developed further in the investigation of *optimal* colours where the associated Worksheet 4(e) shows that the maximum value of C* is about 196 on reddish colours.

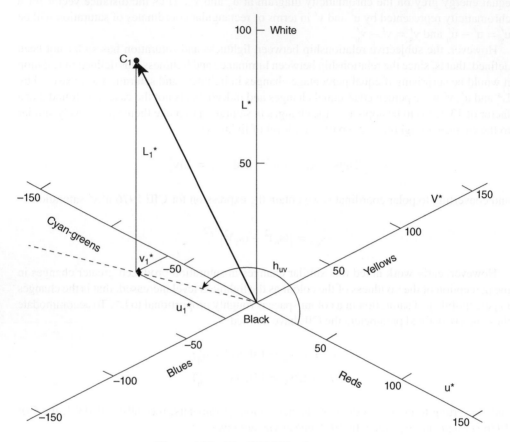

Figure 4.15   The CIELUV colour space.

The diagram in Figure 4.15 illustrates the situation for the perceptibility vectors, L*, u*, v* each at 90 degrees to each other. This colour space is referred to as the CIE 1976 (L*, u*, v*) colour space or more commonly the LUV colour space. This method of measurement is well established as one of the main methods of fully defining a colour.

The construction for the colour C_1 is also illustrated and indicates the angle h_{uv} makes with the positive u* axis.

At this point it may be noted that we have, to a first degree, defined a colour space with spatial parameters which are defined in the same terms as the value, chroma and hue terms of the Munsell colour space. In the next section we will endeavour to illustrate the relationship between these two colour spaces, one based upon measurement and the other based upon perception.

4.7.4 Measuring Colour Difference

In measuring the accuracy of colour reproduction one of the basic tools available is to measure the difference between the colour in a scene and the reproduced version of that colour. The LUV method of measurement provides the means of achieving this.

In Figure 4.16 the LUV colour space is illustrated in three-dimensional form with the lightness parameter L* on the vertical axis and the two u* and v* chroma parameters on the horizontal axis at right angles to each other. Two similar cyanish colours are shown in the resulting colour space as C_1 and C_2 with construction vectors included to aid in the interpretation of the diagram. The construction box which illustrates the difference between the two colours is enlarged and shown separately in Figure 4.17.

The difference in colour is represented by the length of the vector which spans the space between the two colours. As the figure shows, if the difference between the L*, u*, v* values of the two vectors is ΔL*, Δu*, Δv*, respectively, then this length is

$$\Delta E_{uv}^* = \left[(\Delta L^*)^2 + (\Delta u^*)^2 + (\Delta v^*)^2\right]^{1/2}$$

This quantity, ΔE_{uv}^*, is called the CIE 1976 (L*, u*, v*) colour difference or CIELUV colour difference.

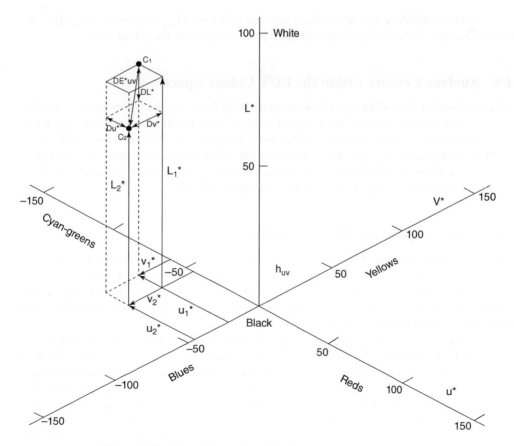

Figure 4.16 Two colours, C_1 and C_2, in CIELUV colour space.

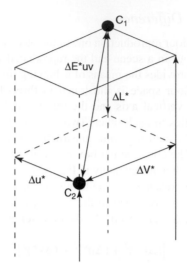

Figure 4.17   Colour difference vector.

One unit of Euclidean distance in the colour space, that is, $\Delta E^*_{uv} = 1$, is very roughly equal to 1 JND, dependent to a degree upon the position it occupies in the colour space.

4.8   Surface Colours within the LUV Colour Space

Taken in isolation the colour space defined above lacks form; to give it substance we need to envisage how surface colours will appear within the colour space and particularly to appreciate that the maximum vector length for each hue and lightness level is the maximum chroma level, which in turn specifies a point on the surface of the colour space; so perhaps the first task is to define the shape of the colour space surface, which is the three-dimensional loci of all the maximum chroma vector points for all hues and lightness levels.

4.8.1   The Shape of the LUV Colour Space

When first considering the shape of a colour solid in Section 2.3, based upon a hue circle and lightness levels from black to white we arrived at a simple shape based on two cones joined at the base, as illustrated in Figure 2.16. However, we are now in a position to amend this simple concept by replacing the hue circle with the C^*_{uv} chroma radial plot derived above and the lightness levels by the L* values.

In envisaging the shape of the resulting colour space we need to be aware of how the spectrum locus of the u',v' chromaticity diagram affects the brightness of fully saturated colours. As we have seen, because of the straight line section of the diagram, a fully saturated yellow may be achieved either somewhat dimly by a single wavelength of light at about 575 nm or by a broad spectrum of wavelengths from about 550 to 700 nm; the integration of energy over this band of the spectrum producing a relatively very bright, though still a very highly saturated, colour. In contrast, where the spectrum locus of the chromaticity diagram is convex, integration of the energy over a significant band of wavelengths around a selected hue will produce a desaturated colour and thus saturated colours in this area of the spectrum, being of limited spectral spread, are fundamentally of low levels of brightness. In addition, the response of the luminous efficiency function ensures that red and blue saturated colours at the edge of the luminance response are fundamentally of diminished brightness. Thus, the changing brightness which occurs with changing hues at maximum saturation leads to widely different values of maximum chroma at different lightness levels, which in turn leads to an irregular surface shape and makes it difficult to envisage how the shape of the CIELUV colour space will appear.

The surface boundary of the colour space for any particular hue angle may be found by calculating the maximum chroma value for an appropriate number of lightness values for that hue over the lightness range from black to white. The loci of these calculations will define the boundary shape for that hue and the colours which result from adopting this approach are named *optimal* colours.

4.8.2   Optimal Colours

In spectral terms optimal colours may be considered as block segments of the spectrum. For any hue of a particular wavelength at maximum saturation the size of the block will commence as a single wavelength, $\Delta\lambda$, with minimum lightness and, as the width of the block expands to incorporate a wider band of wavelengths, whilst ensuring the hue angle h_{uv} of the colour does not change, the lightness value will increase with increasing block width. Initially the chroma will also increase because of the dependency of chroma level on lightness level but at some point of increasing lightness the chroma will begin to diminish as the broadness of the band of the spectrum incorporated significantly reduces the saturation of the colour.

In Figure 4.18 the result of taking this approach for the hue based upon a yellow wavelength of 571 nm, which lies very close to the v* axis at 91.4 degrees, is illustrated. The position of the single wavelength of $\Delta\lambda = 1$ nm has a lightness value of 8.0 as shown in the upper most chart; as the width of the block encompasses more wavelengths so the lightness value increases as shown for the range of lightness values up to a value of 97.3 in the lower chart. Naturally, a block which encompassed the total spectrum would be the colour white at a lightness value of 100.

For non-spectral colours the block segments are arranged at either end of the spectrum as is illustrated in Figure 4.19 for a colour which may be described as magenta-ish red whose hue falls on the u* axis at 0 degrees.

The characteristics of optimal colours were explored by MacAdam (1935a, 1935b) in the 1930s, who defined optimal colours with the following theorem: *the maximum attainable purity for a material, from a specific given visual efficiency and wavelength, can be obtained if the spectrophotometric curve has as possible values zero or one only, with solely two transitions between these two values in all the visible spectrum.* As a consequence of this fundamental work, the loci of the optimal colours are sometimes referred to as the MacAdam limits.

The tables of the spectra of optimal colours for the four hues whose angles in the colour solid are close to the positive and negative u* and v* axis are listed in Worksheet 4(e). Establishing the spectra for a particular hue for each increase in lightness is a matter of trial and error; as the spectra block is increased in size on either side of the commencing wavelength, the calculated hue angle is continually monitored to ensure it remains constant. For non-spectral hues the procedure is to commence at white and the complementary wavelength and increasingly reduce the size of the white block around the complementary wavelength.

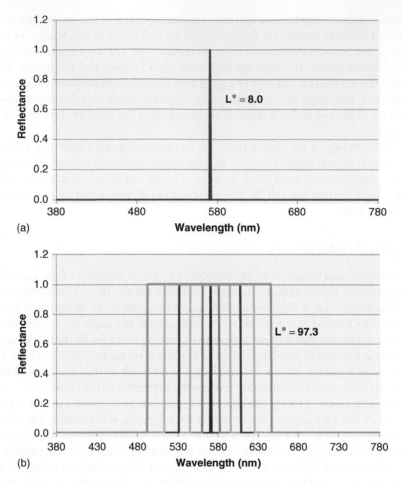

Figure 4.18 Illustrating the block spectral responses and corresponding lightness levels of the optimal colours with a hue aligned close to the v* axis.

For surface colours the optimal colours are in practice modified by taking account of the minimum reflection characteristics of surfaces, which is usually regarded as a level of not less than 0.50% and is referred to as surface correction.

4.8.2.1 Calculating the Loci of Optimal Colours

Once the spectra of the optimal colours have been established for the chosen hue angle it is a simple matter to calculate the values of L*, C* and the angle h_{uv} and hence plot the loci of the optimal colours to establish the shape of the colour solid surface for that particular hue angle. This is done in Worksheet 4(e) for the four colours which lay on the +/– u* and +/– v* axes, respectively and which are illuminated by equal energy white; the results are shown in Figures 4.20 and 4.21.

The relationship between the wavelength in nanometres and the hue angle h_{uv} in degrees changes rapidly so it is necessary to establish by interpolation in the worksheet the values of

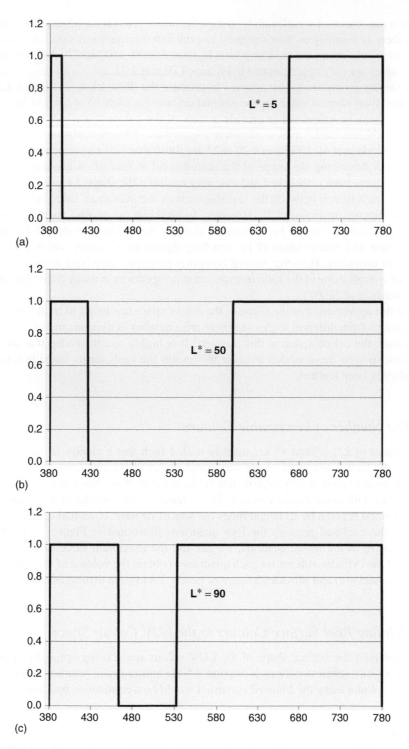

Figure 4.19 Spectral responses and lightness values for optimal colours with a non-spectral hue aligned to the u* axis.

x,y,z for each nanometre of wavelength from the CIE values given at each 5 nm of wavelength. Even then there is insufficient discrimination to establish the fractional value of wavelength required to precisely obtain the wavelengths at h_{uv} equal to 90, 180 and 270 degrees, which is why these values are only approximated in Figures 4.20 and 4.21.

Using a similar procedure to that adopted to produce the illustrations in Figures 4.20 and 4.21, the maximum chroma value for the optimal colours for each 10 degrees of hue around the u*,v* plot may be calculated to provide a view of the colour space looking down on it centrally, directly into the L* axis, as shown in Figure 4.22. These calculations are contained in Table 3 of Worksheet 4(e). Figures 4.20–4.22 are illustrated to the same scale.

Thus the loci describing the shape of the colour solid at four of its quadrants and from directly above have been established and one may envisage the shape for hues between any two quadrants as a morph between the two shapes with the maximum value being given by the outline illustrated in Figure 4.22. However, to establish the precise shape requires the above described calculations to be undertaken at every nanometre wavelength through the visible spectrum and manipulation of the resulting figures in a manner which a worksheet is incapable of providing. However, several computer programs have been written to provide images of the overall shape of the solid from different perspectives in a wire frame presentation (Martinez-Verdu et al., 2007).

To gain a full appreciation of the shape of the colour space one needs to be able to see it in three dimensions from different angles and there are a number of dynamic models on the web which illustrate the colour space in this manner.[3] It is highly recommended that the reader takes the time to view these models in order to confirm the basis for its shape in relation to what has already been learned.

4.8.3 The Number of Perceivable Colours

Since the values of L*, u* and v* are roughly scaled such that a change of value of 1.0 is equal to 1 JND, then if one makes some very rough and ready assumptions, the LUV colour space can be used to predict very approximately the number of colours that can be perceived by the observer with normal colour vision. The volume of the cylinder in which the colour space is contained is given by its height times the area of its base. If we make the assumption that each of the enclosed areas in the four quadrants illustrated in Figures 4.20 and 4.21 are representative of the whole quadrant, we can use the maximum value of C^*_{uv} times the percentage of the cylinder volume for each quadrant to obtain the volume of the solid. This is done in Worksheet 4(e) and provides a figure of nearly 3.8 million distinguishable colours.

4.8.4 Relating Real Surface Colours to the LUV Colour Space

Having established the surface shape of the LUV colour space using optimal colours, and described in strictly subjective terms in Section 2.5 the colour space resulting from the categorisation of colours using the Munsell system, it would be a confidence-building exercise in

---

[3] Bruce Lindbloom.com. See under 'Calc/Munsell Display Calculator' at: http://www.brucelindbloom.com/ index.html?LabGamutDisplay.html (It may be necessary to enter this address: http://www.brucelindbloom.com/ MunsellCalculator.html into the Exception List of the Java Control Panel to allow this Java applet to run.)

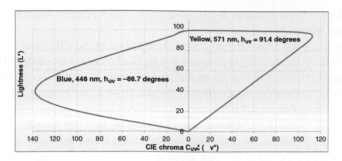

Figure 4.20 Loci of optimal colours for blue and yellow hues.

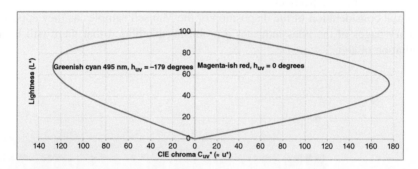

Figure 4.21 Loci of optimal colours for greenish-cyan and magenta-ish red hues.

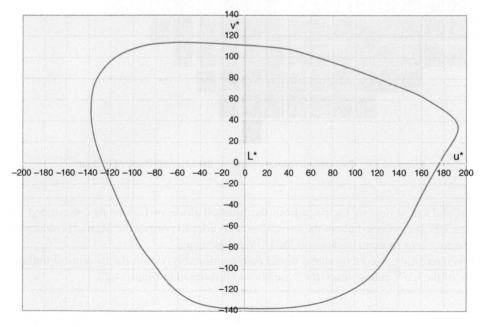

Figure 4.22 Outline of the L*, u*, v* colour space, maximum C* against angle h. looking down the L* axis from directly above.

the methodology adopted if we were able to reconcile these objective and subjective approaches to defining the colour spaces of real colours.

It may be recalled that each Munsell chip colour is described in the CIE terms of value, hue and chroma, terms which are directly synonymous with the L*, h and c* in the LUV system of measurement. Thus in plotting the values of the Munsell chips directly into the LUV colour space one should see a correlation. Since in the Munsell system the primary criteria for the selection of chip colour is that each chip is spaced one JND away from its adjacent chip in terms of value, hue and saturation, then one would anticipate that the samples appearing in the LUV colour space will be evenly distributed and confined to the positions relating to the page samples in the Munsell catalogue. Thus appropriate plots of the Munsell sample chips, in terms of their L*, u*, v* values, should fall within and fit comfortably within the shape of the optimal colour limits of the various views of the colour space as calculated and illustrated in Figures 4.20–4.22, respectively.

An initial consideration of the disposition of the Munsell samples as they appear on complementary pages of the atlas indicates a promising likelihood that there will be at least a reconciliation of sorts.

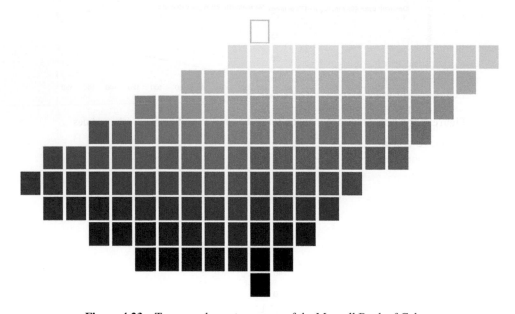

Figure 4.23 Two complementary pages of the Munsell Book of Colour.

Figure 4.23 is a view of the page from the Munsell Book of Colour first illustrated in Figure 2.17 but reflected about the vertical axis in order to portray the Munsell colours in the same colour space orientation as the LUV colour space.

It is interesting to note its shape would map comfortably within the theoretical outline shape of the LUV colour space for these hues as illustrated in Figure 4.20.

In comparing the shape of the profiles of the four hues identified above in Figure 4.20 and 4.21, respectively with the profiles of the corresponding hues in the Munsell Book of

Colour illustrated in Figure 2.20, it is satisfying to see how closely the shapes relating to the application of the derived parameters and those from actual samples correspond.

4.8.5   Reconciling the LUV Colour Space and the Munsell Book of Colour

The work of calculating the L*, u*, v* plots of the Munsell sample chips has been undertaken by a number of colour scientists and Bruce Lindbloom in particular has made his work freely available in graphical form on the web. His website, already cited, is not only useful in enabling the distribution of the plots to be captured for various views of the colour space but being dynamic and three dimensional, it is also educational in enabling one to view and grasp the shape of the open 'solid' representing the plots of all the Munsell colours from any direction in the colour space.

Bruce Lindbloom has kindly given permission for images captured in the required directions from his dynamic model to be used to compare with the outline shapes of the optimal colours shown in Figures 4.20–4.22. Figures 4.24–4.26 illustrate that indeed the plots of the Munsell samples fit well within the appropriate shape of the corresponding optimal colours.

The illustrations are to the same scale and their axes have been aligned to enable a meaningful comparison to be undertaken. In Figures 4.24 and 4.25, the nine lightness levels of the Munsell system can be clearly seen. The slight separation of the colours away from those at lightness level 50 is due to the three-dimensional aspect of the view. In Figure 4.26, radials of constant hue are also clearly apparent; however, there does appear to be a slight offset of the samples compared with the optimal colour outline, which is probably due to the Munsell sample plots being based upon Illuminant C rather than the equal energy white illuminant used to calculate the optimal colour outline.

From the above considerations it is reasonable to assume that the CIELUV colour space is a realistic and practical approach to specifying and measuring colours in a perceptibly meaningful manner.

4.9   Limitations of the LUV Colour Space as an Accurate Colour Appearance Model

At this point it must be emphasised that the LUV colour space is only an approximation to a perceptibly even colour space as was illustrated by the uneven lengths of the JNDs on the u′v′ chromaticity chart of Figure 4.10. Notwithstanding the points made in the preceding paragraphs it is apparent that, taking into account the work that has been undertaken to ensure the perceptibly even distribution of samples in the Munsell system, the portrayal of their distribution in the LUV colour space is not entirely even. This is not surprising since we have already seen the limitations in the evenness of the portrayal of equal JNDs in different areas of the u′,v′ chromaticity diagram which forms the basis of two of the dimensions of the LUV colour space. Despite the lightness parameter being significantly sensitive to the level of adaptation of the eye, which itself is dependent upon the distribution of luminance across the field of view, there is a very good correlation between the Munsell *Value* parameter and the lightness parameter in the LUV colour space, as shown by the evenness of the layers in Figures 4.24 and 4.25.

If the LUV colour space was truly uniform in all directions then the hue radials for each lightness level would overlay one another; they would be aligned with the 40 9-degree radials

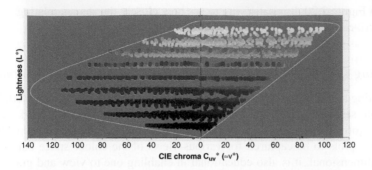

Figure 4.24   Munsell colours in the LUV space from the +u* axis looking towards the L* = 50 axes.

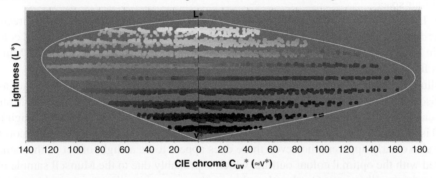

Figure 4.25   Munsell colours in the LUV space from the −v* axis looking towards the L* = 50 axis.

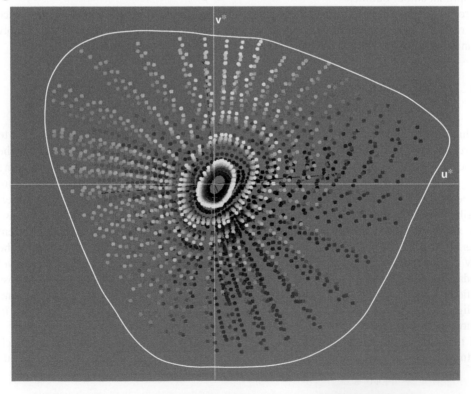

Figure 4.26   Munsell colours in the LUV space from the L* axis at L* = 100.

around the polar diagram; and they would be straight. Similarly the vaguely ovoidal loci of identical values of chroma would be equi-spaced circles.

Bruce Lindbloom has investigated this lack of conformity and his website enables plots to be produced of the distribution of the Munsell colour samples overlaid on charts of radials and circles, respectively, as captured in Figures 4.27 and 4.28, respectively.

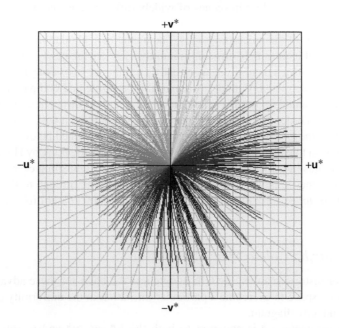

Figure 4.27   Plot of Munsell colours of constant hue. (Courtesy of Bruce Lindbloom.)

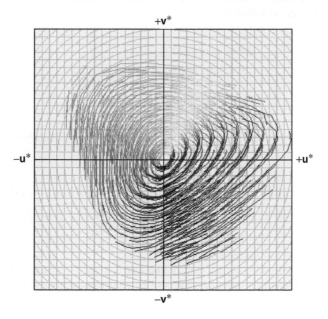

Figure 4.28   Plot of Munsell colours of constant chroma. (Courtesy of Bruce Lindbloom.)

In Figure 4.27 it can be seen that for most hues the radials for different value levels do not overlay and the spread is at least sufficient to encroach on the area of adjacent hues. In addition, for some hues the radials, particularly for colours of high chroma, are not straight lines and the spread can be up to about 5 degrees. Figure 4.28 also clearly indicates that the shape of the constant chroma loci is far from circular.

Colour scientists continue to establish with greater accuracy the manner in which the eye–brain complex responds to colour in scenes of widely different content and these results are regularly assessed by the CIE, from which new colour models appear which are refined and ever more complex versions of the proceeding ones. The mathematical relationships of these more refined models make it increasingly difficult to intuitively relate the newly derived parameters to the u'v' parameters, which themselves are already a number of steps away from fundamental RGB values. Thus since the LUV model is reasonably accurate and comparatively intuitive, it will generally be retained for portraying and measuring colour throughout the remainder of this book. However, in order to embrace relevant published material it will, where appropriate, be supplemented by the L*a*b* measurement methodology described below.

For those who wish to study this field further I would point to the book by Hunt and Pointer (2011), which explains in some detail the work of the CIE in establishing ever more complex and accurate colour metric models, from the CIECAM97 metric through to the CIECAM02 versions of a *colour appearance model* and also the CIEDE2000 colour difference equation.

4.9.1   The L*,a*,b* Colour Space

Unfortunately, for historic reasons, the CIE adopted as the basis for these advanced models the L*,a*,b* colour space to overcome a particular limitation in the uniformity of perception in the u',v' chromaticity diagram.

The lightness parameter L* is common to both the L*, u*, v* and L*a*b* systems of measurement and the a* and b* parameters are non-linear representations of the X,Y tristimulus values, which are defined as follows:

$$L^* = 116f(Y/Y_n) - 16$$
$$a^* = 500[f(X/X_n) - f(Y/Y_n)]$$
$$b^* = 200[f(Y/Y_n) - f(Z/Z_n)]$$

where X_n, Y_n, and Z_n are the X,Y,Z values for the reference white being used and the functions f have the same values as those for L*, namely for (Y/Y_n)

$$f(Y/Y_n) = (Y/Y_n)^{1/3} \qquad\qquad \text{if} \quad (Y/Y_n) > (24/116)^3$$
$$f(Y/Y_n) = (841/108)(Y/Y_n) + 16/116 \quad \text{if} \quad (Y/Y_n) \le (24/116)^3$$

and the same for $f(X/X_n)$ and $f(Z/Z_n)$.

The a*,b* colour axes broadly correspond in hue terms to the u*,v* axes, that is the a* axis represents colours from red to green (–a*) and the b* axis represents colours from yellow to blue (–b*).

The corresponding colour difference formula is given by:

$$\Delta E_{ab}^* = \left[(\Delta L^*)^2 + (\Delta a^*)^2 + (\Delta b^*)^2\right]^{1/2}$$

The Bruce Lindbloom website, already cited, also provides radial and polar diagrams for the Munsell colours plotted on the a* and b* axes chart. The spread of the Munsell samples on the hue radial diagram is similar if not more extensive in magnitude to those appearing on the u* and v* axes chart, though occurring at different hues. However, the plot of lines of constant chroma on the a* and b* axes polar chart leads to shapes which are significantly closer to circles than the same shapes on the u* and v* axes chart.

L*,a*,b* is an alternative colour space originally championed by the dye industry but since it is not related to a linear chromaticity diagram it is considered less relevant to the subject of colour reproduction; in fact recent work[4] has indicated that in colour reproduction, where significant differences in colour are being measured, in terms of changes of perceptibility the CIELUV colour metric gives more accurate results than the CIELAB colour metric. Thus unless standards are being considered which relate directly to CIELAB, the CIELUV colour metric will be used for illustrating colour differences in the remainder of this book.

Notwithstanding the limitation on the extent to which it is intended to explore colour appearance models, we shall, in the next chapter and also in Chapter 9, consider in some depth the influence the adaptation of the eye has on its ability to appraise reproduced scenes under different conditions of ambient lighting.

4.9.2   The CIEDE2000 Colour Difference Equation

The current, CIEDE2000 colour equation, produces through a formula based upon L*a*b* values, values of colour difference, ΔE_{00}^* which are apparently shown to be perceptually closer[5] to JNDs than any other metric; therefore, where colour differences are being measured we will continue to illustrate the magnitude and direction of the differences on u′,v′ diagrams but where appropriate also provide figures for colour differences in terms of ΔE_{00}^*.

The derivation of ΔE_{00}^* from L*a*b* values is not straightforward and requires a number of relatively complex calculations to be carried out in sequence. Fortunately these calculations have been undertaken and made available in spreadsheet format (Sharma et al., 2004) via the website in the citation, a copy of which is inserted into Worksheet 4(e) for reference. This calculator will be used in a practical manner in Part 5.

---

[4] EBU TECH 3354 'COMPARISON OF CIE COLOUR METRICS FOR USE IN THE TELEVISION LIGHTING CONSISTENCY INDEX (TLCI-2012)'

[5] However, the reader is warned that the author has found that in using the three different colour metrics for ΔE discussed in this section the values for ΔE_{uv}^* and ΔE_{ab}^*, though understandably different, are of the same order; whereas, the values for ΔE_{00}^* are between two and four times smaller for a comprehensive range of colours, which causes concern; at these levels of difference they cannot both be measuring quantities which relate closely to JNDs.

5

Colour Measurement and Perception

5.1 Chromatic Adaptation

The manner in which the eye adapts to changes in the spectral distribution and the level of illumination is complex and whilst the fundamentals of colour measurement have been well understood and defined for many decades the same is less true for the understanding of chromatic adaptation, which continues to be improved as more research is undertaken. In consequence there is much ongoing work in evolving methods to best emulate its effects in colour reproduction and in the supporting literature on the subject. This chapter is therefore little more than a very brief introduction to the subject in order to provide a basis of why it is important to recognise its influence in colour reproduction.

In Section 2.6 the effects of large changes in the spectrum of lighting on the change in the perception of illuminated colour surfaces were briefly reviewed and it was noted that to a large degree the ability of the eye–brain complex to adapt to these changes and perceive colours, broadly similar under very different spectral illumination, is quite remarkable. However, in colour measurement terms, using the $\bar{x}(\lambda), \bar{y}(\lambda), \bar{z}(\lambda)$ CMFs to measure the colour of a surface under these changes of condition would lead to very different colours being recorded.

To make sense of this situation in a colour reproduction system, these two disparate results need to be reconciled if the reproduction is to be perceived as similar to the perception of the original scene. In order to do this we need to develop a scheme that predicts how a set of colour measurements in the form of trichromatic units, for one example of spectral illumination, will be modified to provide the same perception of the colour under a different example of illumination.

As we noted above, the eye–brain complex sees the colours amended by the change in illumination as 'broadly similar' but not identical to the original. Perhaps one of the most common examples where the limitations of adaptation becomes obvious is the comparison of darkish blues and browns, in good daylight and in low levels of tungsten lighting, where in the former the differences are all too apparent whilst in the latter it can become very difficult to differentiate between them, particularly if they are of a similar luminance.

Colour Reproduction in Electronic Imaging Systems: Photography, Television, Cinematography, First Edition. Michael S Tooms.
© 2016 John Wiley & Sons, Ltd. Published 2016 by John Wiley & Sons, Ltd.
Companion Website: www.wiley.com/go/toomscolour

There are two main aspects to take into account in considering the effect of adaptation when there is a change in the spectral distribution of the illumination. Primarily it is the balance of the weighting of the energy across the spectrum of the illumination which is so successfully compensated for by adaptation; again taking the extremes of daylight and tungsten lighting which have a bias towards the blue and red end of the spectrum respectively, the adaptation effect works well. However, if the surface reflection characteristics of the colour in the scene is irregular, particularly in terms of having a number of peaks and troughs in the response across the spectrum, then the perception of the colour under different conditions of illumination can change significantly.

5.2   Metermerism

In Chapter 2, the example of a problem with illuminants of different characteristics was given in which the procedure for matching colours in a shop, perhaps two different fabrics where it is important to obtain as close a match as possible, was described. The experienced customer will first check if there is a reasonable match under the interior lighting of the shop but to ensure there will not be a problem later, both sets of material are then taken outside to make the comparison in daylight. Frequently it is found that a match in the shop is an unacceptable mismatch in daylight and vice versa.

In order to explain how this can happen we need to look once more at how the eye–brain complex perceives colour, commencing with the cone responses of the eye. As described in Section 1.6, these ρ, γ and β cone responses, derived from the $\bar{x}(\lambda), \bar{y}(\lambda), \bar{z}(\lambda)$ colour matching functions, are the closest match to the measured cone responses (Estévez, 1979) and the equations which define this relationship are laid out below.

$$\rho = 0.38971X + 0.68898Y - 0.07868Z$$
$$\gamma = -0.22981X + 1.18340Y + 0.04641Z.$$
$$\beta = 1.00000Z$$

The inverse relationship is:

$$X = 1.91019\rho - 1.11214\gamma + 0.20195\beta$$
$$Y = 0.37095\rho + 0.62905\gamma + 0.00000\beta.$$
$$Z = 1.00000\beta$$

The resulting cone responses are shown in Figure 5.1 (This is the relationship which was used to derive the ρ, γ and β cone responses referred to in Section 1.6 and Figure 1.11.).

It was noted in Section 1.6 that the colour the eye perceives is related only to the levels of the responses of the three receptors.

Although we cannot measure directly the signals generated in the eye, accepting that these curves are a meaningful indication of the responses of the eye, we can indicate what the tristimulus values of the colour signals will be when the eye is responding to light reflected from a surface illuminated by light of a specific spectral power distribution or SPD.

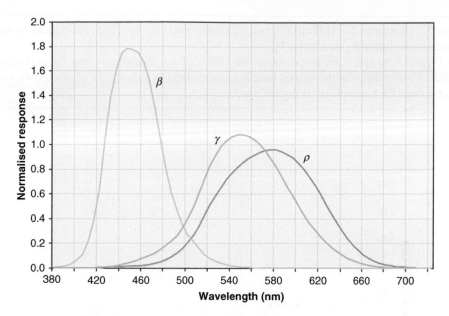

Figure 5.1 The normalised cone response curves of the eye.

So for example within the shop, the two materials, fabric F_A and fabric F_B, are successively illuminated by Illuminant A (SA), that is tungsten lighting, and daylight which may be characterised as one of the CIE phases of daylight, D65; both illuminants being defined in Part 3. Since they match under illuminant A, we can assume that the β, γ and ρ cone responses are equivalent for both illuminants, irrespective of their different reflection characteristics.

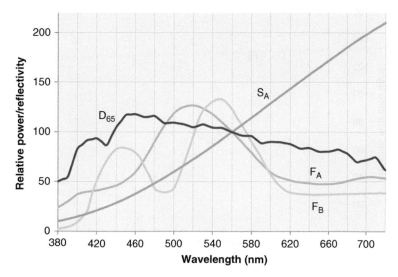

Figure 5.2 Illuminant SPDs and spectral reflection functions, respectively of the illuminants and two matching fabrics under Illuminant A.

Figure 5.2 illustrates the SPDs of the two illuminants and the spectral reflectances of the two materials, fabric F_A and fabric F_B.

In Figure 5.3a and 5.3b, respectively, the SPDs of the illuminants and the reflective spectral characteristics of the fabrics have been convolved to indicate the characteristics of the light from each sample reaching the eye.

(a)

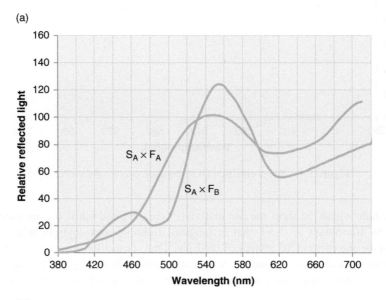

(b)

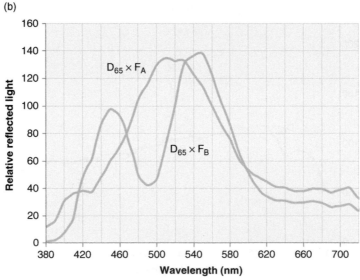

Figure 5.3   (a) Reflected light from fabrics with Illuminant A. (b) Reflected light from fabrics with Illuminant D65.

Table 5.1 Eye responses for the metameric match

| | S_A | | D65 | |
	F_A	F_B	F_A	F_B
ρ	1,855	1,856	1,802	1,784
γ	1,873	1,873	2,112	2,028
β	527	527	1,439	1,602

If now each of the curves in Figure 5.3 is convolved in turn with each of the β, γ and ρ cone responses of the eye, then the area under the resulting curves represents the responses of the eye to these two colours under each of the illuminants. The levels of the responses represented by the areas under the curves are calculated in Worksheet 5 and are shown in Table 5.1.

When two surfaces with different spectral reflectance visually match, as for the SA illuminant in Table 5.1, they are described as being *metameric* pairs or metamers. Thus metamerism occurs because each type of cone responds identically to the cumulative energy from a broad range of wavelengths, so that different combinations of light across all wavelengths can produce an equivalent receptor response.

It is not surprising, given the very large differences in the spectral distributions of the two illuminants that the appearance of the colours, based on the figures in Table 5.1 will be significantly different under the two illuminants. However, to a large degree, the ability of the eye to adapt to different illuminants will compensate for most of the differences given a few seconds to accommodate but what the eye cannot do in these circumstances is compensate for the different signals from the two fabrics and the mismatch will be all too obvious.

In the foregoing the acknowledged responses of the three receptors of the eye were used to establish a metameric match which is both intuitive and fundamental. However, these responses are derived rather than being directly measured and are therefore regarded as inadequate for formally defining metamerism. Thus metamerism is defined in terms of a match using tristimulus values derived from the $\bar{x}(\lambda), \bar{y}(\lambda), \bar{z}(\lambda)$ CMFs. However, since the eye responses used for this calculation were themselves derived from the x,y,z barred colour matching functions, it is not surprising that a match using one set of curves will also produce a match (with different values) using the alternate set, as is illustrated in Worksheet 5, the results from which are listed in Table 5.2.

Table 5.2 Tristimulus values for the metameric match

| | S_A | | D65 | |
	F_A	F_B	F_A	F_B
X	1567	1569	1384	1476
Y	1866	1866	1997	1937
Z	527	527	1439	1602

5.2.1 An Index of Metamerism

It was noted earlier that metamerism is most likely to occur when either or both the spectral characteristics of the illumination or reflective surface is irregular and once recognised, it is intuitive that the more extreme the peaks and troughs of the characteristic, the larger will be the difference in the perceived colours of the mismatch.

The CIE recommend that the degree of metamerism for changes in illumination be defined by an Illuminant Metamerism Index (Hunt & Pointer, 2011), based upon the difference in colour of two samples when measured under a standard illuminant and a test illuminant. As the models for the behaviour of the eye have evolved, so the methods of measurement of the colour of the samples have changed and therefore the method used should be stated.

5.3   Quantifying Chromatic Adaptation

The effects of chromatic adaptation are such that when two stimuli are viewed in different conditions of illumination and yet appear to match under both, they are defined as *corresponding colours*. In order to predict the actual colour perceived as a match under a different illuminant a model of chromatic adaptation is required. Thus, this model when applied to a set of tristimulus values of the colour under the first illuminant will provide the trichromatic values of the colour which will be perceived as the same colour when the eye is adapted to the second illuminant.

Attempts at quantifying the basis for adaptation were carried out in the early days of colour science and it was evident to von Kries, a German physiological psychologist, that, at least to a first degree of approximation, the behaviour of the eye was such that the individual β, γ and ρ cone receptors reacted to the average illumination by effectively adjusting the gain of each receptor in inverse proportion to its stimuli so that the response produced from a neutral white reflecting surface was equal from each receptor irrespective of the chromaticity of the illumination.

In mathematical terms this hypothesis may be described as follows:

$$\rho_{\text{adapted}} = \rho / \rho_{\text{white}}$$

$$\gamma_{\text{adapted}} = \gamma / \gamma_{\text{white}}$$

$$\beta_{\text{adapted}} = \beta / \beta_{\text{white}}$$

where ρ, γ and β are the levels of the original cone responses and ρ_{white}, γ_{white} and β_{white} are the responses from the white in the scene.

This observation led later colour scientists to define a *chromatic adaptation transform* or CAT based on this hypothesis and is known as the von Kries transform. Since this is effectively a normalisation to the white in the scene the von Kries adaptation is referred to as a *white point normalisation*.

As so much of the calculation in colour measurement is in terms of matrices it is often convenient to express the above relationships in terms of a matrix as follows:

$$\begin{bmatrix} \rho_{\text{adapted}} \\ \gamma_{\text{adapted}} \\ \beta_{\text{adapted}} \end{bmatrix} = \begin{bmatrix} \frac{1}{\rho_{\text{white}}} & 0 & 0 \\ 0 & \frac{1}{\gamma_{\text{white}}} & 0 \\ 0 & 0 & \frac{1}{\beta_{\text{white}}} \end{bmatrix} \begin{bmatrix} \rho \\ \gamma \\ \beta \end{bmatrix}. \tag{5.1}$$

One of the earlier forms of chromatic adaptation transforms adopted by the CIE was based upon using a geometrically derived formula to modify the u,v chromaticity values of a test illuminant in such a manner that they became equal to the chromaticity values of the reference illuminant. This relationship could then be used to apply a correcting formula to the chromaticity values of a series of colour samples illuminated by the test illuminant to indicate how they would appear once the eye was adapted to the test illuminant. Although this method is now obsolete, it is still used in the formal CIE specification for deriving the colour rendering index of a test illuminant, as will be addressed in Section 7.2.

Generally, chromatic adaptation transforms have been based more directly on the hypothesis of the von Kries transform whereby the aim of the transform is to produce an equal response from the cone receptors of the eye for neutrals under different illuminants. However, since chromaticity data are usually in the form of X,Y,Z coordinates the transforms are based upon converting the X,Y,Z values of samples under the reference illuminant to the X,Y,Z values under the adapted illuminant.

As noted in the preceding paragraphs the ρ, γ, and β cone responses are derived responses from the $\bar{x}(\lambda)$, $\bar{y}(\lambda)$, $\bar{z}(\lambda)$ CMFs and it is therefore usual to use the inverse of these relationships to express the von Kries transform in terms of the CIE CMFs.

Thus the procedure for applying a basic von Kries transform would comprise a number of steps. First converting the X,Y,Z values to R,G,B values using the appropriate XYZ to $\rho\gamma\beta$ cone matrix. Then applying the correction values obtained by making the values of R,G,B equal for a neutral under the adapting illuminant, as shown in equation 5.1, and finally applying the inverse of the original cone matrix to obtain the X,Y,Z values representing how the sample would appear under the adapting illuminant.

Since the 1980s much experimental work has been undertaken to improve the results obtained using the relatively simple von Kries adaptation transform. However, in essence the various transforms which have evolved are all based upon the fundamental assumptions of that transform and use the same approach of a sequence of operations using the cone matrix, a correcting matrix and an inverse cone matrix. Albeit that the values of the figures in the cone matrix may differ from the cone responses derived from the $\bar{x}(\lambda)$, $\bar{y}(\lambda)$, $\bar{z}(\lambda)$ CMFs. Additional adjustments are defined to take into account the complex adaptation processes of the eye–brain complex when the levels of the illuminants differ considerably and also when the field of view does not embrace a scene entirely illuminated by the adapting illuminant.

The Bradford transform (Luo et al., 1998), so named since it resulted from work originally carried out at the University of Bradford by Dr. Clement Lam and subsequently taken up by Luo at the University of Leeds, has become accepted as a sound basis for predicting the effects of adaptation to a reasonable degree of accuracy. A modified version of this transform known as the CIE CAT97 transform forms the basis of the CIECAM97 colour appearance model and a further simplified version forms the basis of the CIECAM02 colour appearance model. In various forms it has been much used to support a number of interchange specifications within the colour reproduction field, as will be addressed in the appropriate later sections of this book.

In its comprehensive form the Bradford transform comprises cone matrices which are modified forms of the CIE cone response curves as shown in Figure 5.4 where the two sets of curves are compared. In addition the derived β value in the resulting RGB set takes on a marginally non-linear form. However, frequently a simplified version of the transform is used which does not incorporate the non-linear term; it is this simplified form which is used in CAT02.

The Bradford transform is frequently alluded to in the literature and since it forms the basis of much of the current work in the field of adaptation it would be remiss not to provide a realistic indication of its use. However, it is considered the above explanation will satisfy

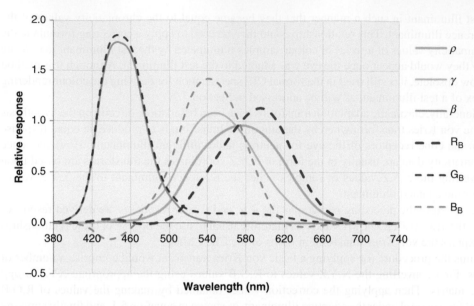

Figure 5.4   The cone and Bradford transform response curves.

the majority of readers and in consequence the more detailed description of the transform in mathematical terms is consigned to Appendix C.

As indicated in the introduction to this topic, chromatic adaptation has become a very extensive and complex subject which tends to be subsumed in the study of *colour appearance models* a subject which is beyond the scope of this book. Readers who require more information in this area are recommended to the book by Hunt and Pointer (2011) which contains much original material and extensive references on this topic.

6

Generating Coloured Light

6.1 Introduction

In colour reproduction light is required at the scene to be shot for illumination, at the display for producing the colour primary lights and in the viewing area for both illuminating the prints and providing ambient illumination.

This chapter has been provided as background information and for those particularly interested in how light is generated in luminaires for scene lighting and in generating the primaries of the display device. The characteristics of these devices are described in some detail in Chapters 7 and 8. This chapter is confined to describing in minimally mathematical terms the physics of light production. Thus it is not necessary to fully understand the detail of the material in this chapter in order to comprehend the material in the remainder of this book. However, it is recommended that as a minimum 'skimming' the material here will leave the reader with sufficient information to determine whether or not to investigate further, whilst at the same time providing an overview of the mechanisms involved. Furthermore it will provide an insight into the reasons why the practical gamut of colours which can be faithfully reproduced is usually somewhat less than that which could ideally be achieved from a three colour system.

6.2 The Physics of Light Generation

Since the mechanisms for generating sources of light are often common to both illumination and displays we will not differentiate between them at this fundamental level. As we noted in Chapter 1, light is a form of electro-magnetic energy. So for generating light it must be produced from a source of energy; that energy can be in the form of heat; electrical, in terms of energetic electrons; or indeed, in the form of more energetic forms of electromagnetic energy outside the visible spectrum such as ultraviolet (UV) light.

The conversion of these forms of energy into light at a detailed level is beyond the scope of this book. However, it may be helpful to describe the conversion concepts in general terms as a means of bringing a greater understanding to the characteristics of the light generated by various sources, including daylight, the various lamps experienced in illuminating a scene and in the production of primary colour lights for various display devices.

Colour Reproduction in Electronic Imaging Systems: Photography, Television, Cinematography, First Edition. Michael S Tooms.
© 2016 John Wiley & Sons, Ltd. Published 2016 by John Wiley & Sons, Ltd.
Companion Website: www.wiley.com/go/toomscolour

In Chapter 1, bowing to general usage, we determined to describe light in terms of its wavelength. However, in physical terms it is more intuitive to use frequency rather than wavelength to describe the conversion processes; for example, graphs using a frequency scale will show energy content increasing from left to right, whereas on a wavelength scale, energy would diminish from left to right. It is fortunate that, if the units of wavelength are in nanometres and those of frequency are in terahertz, then a mirror image of a very similar range of numbers occurs for describing the visible spectrum. Thus light in the spectrum between violet wavelengths of 380 nm and red of 780 nm occupy a frequency band of approximately red at 380 THz and violet at 790 THz, where a THz is equal to 10^{12} Hz. For the sake of consistency we will revert to the use of wavelengths at the completion of this chapter.

The process of generating light from heat is described as incandescence whilst generating light by various processes at room temperature is termed luminescence.

6.3   Incandescence: Light from Heat – Blackbody or Planckian Radiation

At the turn of the nineteenth century into the twentieth, one of the most outstanding problems in physics was to explain in mathematical terms the intensity and spectral distribution of electromagnetic energy radiated from a warm body, which as the body reaches a temperature of about 730 degrees centigrade, or 1,000 degrees Kelvin, just begins to radiate in the visible spectrum, a phenomena known as incandescence.

This is a continuous spectrum which commences below radio frequencies but terminates relatively abruptly at some higher frequency corresponding to the temperature of the heated material. It was Max Planck, a German physicist, who conceived the idea of quantising the bundles of thermal energy, or phonons, in a warm body in order to explain the cut-off maximum frequency obtained for a particular temperature; in this respect he is considered the founder of what became known as the physics of quantum mechanics; albeit that at that time he was unaware of the broader ramifications of his proposal, which was only adopted to make the equation he derived fit the observations.

The application of heat to a body causes the atoms or molecules of the body to vibrate with increasing amplitude as the temperature is increased from absolute zero causing the atoms to each lose an electron. This energy is in the form of discreet quanta called phonons; the average increase in energy per phonon being equal to kT where T is in degrees Kelvin and k is Boltzmann's constant, equal to $1.3806504 \times 10^{-23}$ joules/K. (Absolute zero on the centigrade scale is at about minus 273 degrees, so room temperature is at roughly 293–300 K.) In material at temperature T, the phonons will have a statistical distribution of energy centred on a level of kT joules.

In the special case of a perfect 'black body' radiator, that is a body which absorbs all electromagnetic energy incident upon it, the material comprises atoms with electrons which become freely available when absorbing energy. Soot is a compound which comes close to behaving as a perfect black body radiator at all temperatures, other bodies such as metals, for example, only approach the conditions of a perfect black body at much higher temperatures.

The oscillations of the charged bodies in the material give rise to electromagnetic radiation; the frequency of radiation being proportional to the level of energy absorbed from the thermal quanta or phonon which instigated the movement. Statistically the phonons will transfer energy around a range of levels equal to kT depending upon the efficiency of the transfer and whether at any instant more than one quanta acts instantaneously on the charged particle. The likelihood of increasing numbers of phonons acting together on a charged particle diminishes rapidly, thus setting a maximum to the energy available for transfer to electromagnetic radiation.

Observation had shown that increasing temperature led to not only an increase in the intensity of radiation but also an increasing maximum frequency. Planck established the relationship between the energy E and the frequency f of electromagnetic energy by introducing a constant such that $E = hf$, where h is Planck's constant equal to 6.626×10^{-34} Joule-seconds or J.s. Some years later Einstein also showed that electromagnetic energy is in the form of packets of energy equal to hf; the term photon being adopted in the 1920s to describe these packets of electromagnetic energy.

Thus phonons of thermal energy kT joules are converted into packets of photons of energy hf joules, so if the conversion process was on a one-to-one basis the frequency of radiation would be given by $f = kT/h$ Hz. However, as described above this is a statistical process based upon a distribution of phonon energies which in turn produces a spectrum of frequencies centred on this value but peaking at a value equal to approximately $2.82\ kT/h$.

Planck used statistical and probability functions to develop a precise formula for the frequency spectrum of black body radiation which introduces the natural logarithm function e. The derivation is beyond the scope of this book but can be found in Wikipedia[1]. The formula derived there gives the intensity of radiation I_f per unit frequency for a temperature T in this form:

$$I_f(T) = \frac{2h}{c^2} \frac{f^3}{e^{\frac{hf}{kT}} - 1} \quad \text{J/sr/m}^2\text{/s/Hz} \quad \text{or} \quad \text{W/sr/m}^2\text{/Hz}$$

where c is the velocity of light at 299,792,458 m/s and the other constants are defined above. The equation in this form gives the energy distribution in the form of joules per steradian per square metre per second per hertz or watts per steradian per square metre per hertz.

In the accompanying Worksheet 6(a) this formula is plotted for a number of temperatures from room temperature up to 10,000 K on a log–log plot to show the shape of the resulting curve against frequency and is reproduced here as Figure 6.1. This graph spans several orders of magnitude of both frequency and intensity of radiation in order to illustrate the power law rise in energy with frequency – until the point is reached where f_{max} is approximately equal to $2.82\ kT/h$ Hz, whereupon there are diminishing numbers of phonons of the required energy to produce photons of higher frequency and the curve starts to deviate from the straight line and shortly afterwards the radiation diminishes exponentially. Note that the vertical scale of the graph is also a log scale so that the fall in intensity with frequency is very rapid to relatively very low numbers of photons, each of increasing energy.

The useful part of the spectrum for producing light is in the frequency range of about 400–790 THz, which in Figure 6.1 is illustrated by the light blue band. This segment of the spectrum is normally illustrated on a linear graph which changes the shape of the curves as shown in Figure 6.2.

---

[1] http://en.wikipedia.org/wiki/Planck%27s_law

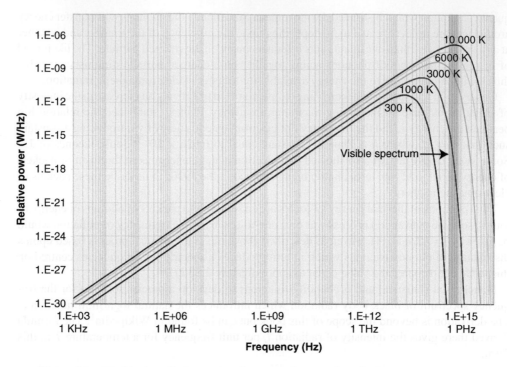

Figure 6.1   Blackbody radiation against frequency for a number of relevant temperatures.

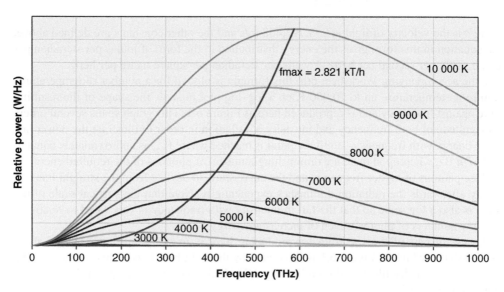

Figure 6.2   Planckian radiation in the spectrum which includes the infrared, visible and ultraviolet radiation.

Note that the frequency at the peak of the curve is directly proportional to temperature.

The red line is the locus of the frequency of maximum energy per Hertz for each value of temperature K.

In interpreting this curve it should be noted that the full radio (10 KHz–1 THz) and lower infrared part of the spectrum (1 THz–10 THz) is confined to only about 10% of the first 100 THz of the graph.

Having reviewed the physical basis of light production from thermal energy in terms of the more intuitive parameter of frequency we will now revert to the more familiar parameter of wavelength.

The derivation of the formula for $I_f(T)$ above is based upon an integral of an expression in f and if f is replaced by c/λ prior to integration then the following expression for $I_\lambda(T)$ is obtained:

$I_\lambda(T) = 2hc^2 \times \lambda^{-5}/(e^{hc/\lambda kT} - 1)$ J/sr/m^2/s/ metre of wavelength or W/sr/m^2/metre of wavelength.

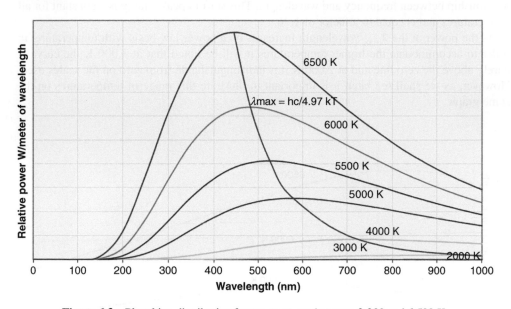

Figure 6.3   Planckian distribution for temperatures between 2,000 and 6,500 K.

This formula is graphed in Figure 6.3 over a range of wavelengths from UV to infrared for a number of temperatures which relate to those of interest to us.

The locus of the peak of the temperature curves is the red line and follows the relationship:

$$\lambda_{max} = hc/4.97\,kT.$$

It will be noted from Figures 6.2 and 6.3, that the frequency and wavelength, respectively relating to the peak of a curve for a particular temperature do not follow the usual relationship relating f and λ, that is $f_{max} \neq c/\lambda_{max}$. Fundamentally this is because the units of energy in the two formulas are different, that is, energy per hertz is not equal to energy per wavelength. In mathematical terms, prior to integration, the small change in frequency Δf over which the energy is integrated is equal to $f_2 - f_1$, but if wavelength is substituted for frequency at this point then:

$$\Delta f = f_2 - f_1 = c/\lambda_2 - c/\lambda_1 = c\left(\frac{\lambda_1 - \lambda_2}{\lambda_1\lambda_2}\right)$$

and as $\lambda_2 \rightarrow \lambda_1$, $\lambda_1\lambda_2 = \lambda^2$ thus $df = c\,d\lambda/\lambda^2$.

Thus substituting for df in the equation from which $I_f(T)$ was derived when integrating, leads to the formula for $I_\lambda(T)$ above. In the paper by Soffer and Lynch (1999)[2] it is pointed out that at a fundamental level this apparent paradox comes about because the Planck function is a density distribution function and is defined differentially.

As an example, the f_{max} at a temperature of 6,000 K is 353 THz which relates to a wavelength of 849 nm; however, the λ_{max} at this temperature is 483 nm. Thus the peak wavelength has shifted to a considerably lower wavelength than one would anticipate using the simple relationship between frequency and wavelength. This shift in peak energy is a constant for all temperatures and is equal to a factor of 1.76.

As the power at the λ_{max} wavelength increases on a power law basis with temperature, in order to accommodate the higher temperatures it will be noted that at 3,000 K the curve is barely above the zero line and at 2,000 K it is indistinguishable from zero on the scales used. However, as we shall see later, it is important to illustrate this range of temperatures on the same graph.

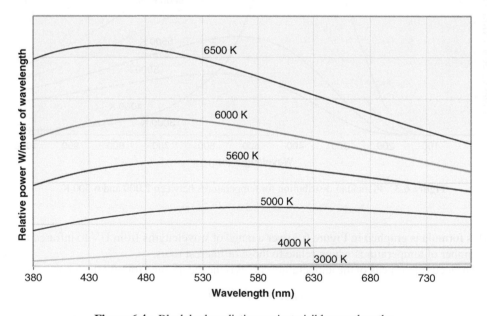

Figure 6.4   Black body radiation against visible wavelengths.

[2] http://escholarship.org/uc/item/8q007697

The formula is also graphed against the visible wavelengths in the worksheet to produce Figure 6.4.

This highlights the problem of scaling the curves in a relative power manner; what is of more interest to us, once a sufficient amount of light is available, is the colour of the source.

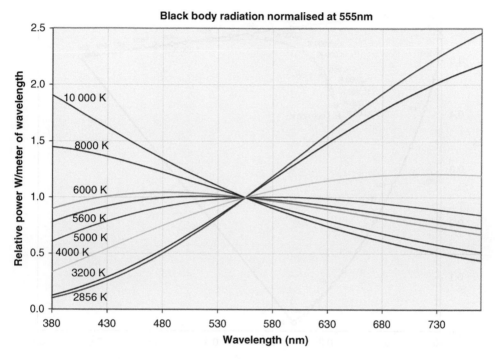

Figure 6.5   Visible blackbody radiation normalised to a wavelength of 555 nm, the wavelength at the peak of the luminous efficiency function.

The data for the curves in Figure 6.4 are therefore normalised to a wavelength of 555 nm and the resulting curves are illustrated in Figure 6.5.

From these curves it is apparent that at 3,200 K the light is biased towards the red end of the spectrum whilst at 10,000 K the light is biased to the blue end of the spectrum.

At 5,600 K the curve approaches that of the straight-line curve of equal energy white (EEW).

6.4   Colour Temperature

Historically black body radiators such as the sun and the early artificial lamps used throughout the last century have been of prime importance in illuminating scenes for reproduction. In consequence the colour of that radiation is critical and thus the plot of the locus of black body radiators over the visible temperature range on the chromaticity diagram has led to it having become one of the mainstays for describing the colour of illuminants of all types.

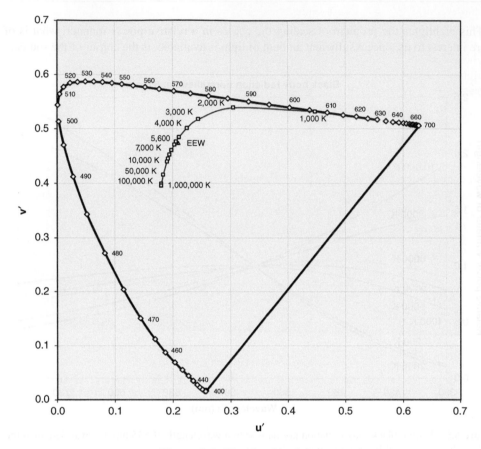

Figure 6.6   The Planckian locus.

The plot of chromaticity against temperature of a blackbody is illustrated in Figure 6.6 and is known as the Planckian locus.

Although it commences at lower temperatures than the 1,000 K illustrated, in the far red end of the spectrum, the level of radiation is below that which the eye can see when accommodated to normal light levels. As the temperature increases the colour moves from red through orange, pale yellow, white and on to pale blue.

As the temperature increases beyond 10,000 K the peak of the curve of radiation level against frequency, illustrated in Figure 6.1, extends beyond visible frequencies into the UV leaving a straight line of constant slope but increasing amplitude in the visible band, thus in the limit further increases in temperature increase the intensity of the light but do not change its chromaticity since the ratio of the energies by frequency does not change.

Also included in the chromaticity diagram is a plot of EEW which is useful in indicating the range of temperatures which come closest to matching white. Broadly any temperature between about 5,000 K and 6,500 K will appear as a good white with only a minor requirement for the eye to adapt; the closest temperature to EEW is at about 5,400 K. Reference back to Figure 6.5 will indicate why this is so; this is the range of temperatures where the response across the visible frequency range is at its flattest.

The Planckian locus became the accepted basis for comparing the chromaticity of sources of illumination, such that in shorthand the source might be described as having a colour temperature of 2,800 K.

6.4.1   Correlated Colour Temperature

As new non-blackbody sources of illumination were developed their chromaticities were also referred to the Planckian locus. Since it was acknowledged that their spectrum might be widely different from a black body, the term 'correlated colour temperature', often abbreviated to CCT, is used to describe their chromaticity. The theme of correlated colour temperature is explored further in Chapter 7.

6.5   Luminescence

At the atomic level, when energetic particles such as a moving electron or a photon are in collision with an atom of matter, the energy of the collision can be absorbed by the atom on a temporary basis and in certain circumstances may then be released by the emission of a photon of light of specific frequency. Any excess difference in energy between that of the collision and the emitted photon is absorbed in terms of heat, in the case of a solid that is in terms of the vibration of the lattice. Except in certain circumstances, the emission occurs in an extremely short time scale measured in nanoseconds which, from the perspective of many of the sources of light used in image reproduction, is instantaneous. The exception to this general rule is phosphorescence, a mechanism related to the absorption of photons, when the mechanism of photon release may be delayed for relatively long periods, on a time scale between milliseconds and many seconds. Although historically the term *phosphorescence* was originated to describe the glow of phosphorus when first exposed to oxygen, it now has a more generic connotation, being used to describe what should be called a *chemiluminescent* effect in a large range of materials.

In general terms the mechanism of absorption and emission is directly related to the energy state of the atom in the collision; this energy state is associated with the level of the orbital position of the electrons orbiting the nucleus of the atom, which in turn may only occur at fixed discreet quantum levels. In the ground or zero-energy-level state, all of the electrons are at their lowest orbital positions; depending upon the characteristics of the material, as the atom absorbs more energy one or more electrons are raised to higher energy bands and ultimately, if sufficient energy is available, an orbital electron may be knocked from the atom completely, whereupon the atom is described as being *ionized.*

Each quantum level or shell may contain a maximum number of electrons, the number increasing by a simple formula the higher the energy level. If the shells are numbered from the nucleus outwards the formula has the form $n = 2 \times s^2$ where s is the shell number and n is the maximum number of electrons the shell can contain. Thus the maximum number of electrons in shells 1–3 are 2, 8 and 18, respectively. The shells are usually referred to by a series of letters from the alphabet commencing with K. Hydrogen, helium and lithium have 1, 2 and 3 electrons, respectively. In the case of helium the inner K shell is full and since a full shell is very stable helium is inert. Whereas for lithium there is only one electron in the L shell which has seven vacancies and therefore the element is chemically very active.

A free electron in an electrical field will be accelerated towards the positive source of the potential gaining speed and therefore energy as it is accelerated. The energy of the electron due to its velocity may be conveniently measured in terms of electron volts (eV), where one electron volt is the energy gained by an electron when accelerated through an electric field of one volt. The charge of an electron is 1.602×10^{-19} Coulomb and thus the relationship between the electron volt and the standard for energy measurement, the Joule, is that one electron volt is equal to 1.602×10^{-19} Joules.

When an electron strikes an atom, depending upon its energy it may raise an electron up to a higher shell at one, two or more levels above the shell the electron resides in when unenergised.

Since the electrons within an atom may exist in only discreet energy bands, as they return to a lower energy band they emit a photon with energy precisely equal to the energy difference between the bands. The frequency of the photon is related to the change of energy state by the formula derived by Planck as described in Section 6.3, that is, $f = E/h$. In general, since atoms have a number of energy bands, then many materials will be capable of emitting at a number of discrete frequencies across the electromagnetic spectrum, often encompassing the infrared, visible and UV spectra. These spectra are referred to as line spectra in recognition of the discreet frequencies of which they comprise. Different materials will absorb and emit in a range of specific frequencies which has led to the growth of the science of spectroscopy which can identify specific materials from the characteristics of their line spectra.

Depending upon the element, whether it is in an elemental or molecular form; gas or solid; exists in isolation or in a complex compound or in a doped form in close proximity to other doped material, the structure of the spectra may comprise of single lines, line pairs in close proximity, a narrow band of frequencies based upon a particular characteristic frequency or a relatively broad band across much of the visible spectrum.

When used for illumination or generation of colour primaries for displays, the actual characteristics of these sources can have a profound effect on the quality of reproduction. Therefore it is essential to be aware, at least in general terms, of the nature of these characteristics for the various forms of light generation outlined above. In the following the mechanism of light generation in each of the sources relevant to those found in colour reproduction is briefly described.

6.6   Electroluminescence

Electroluminescence is the description given to the generation of light derived from the energy of electrons in an electric current striking an atom in a vacuum, gas or solid. The characteristics of the light generated vary somewhat in each of these three situations and are therefore described separately in the following.

6.6.1   Cathodoluminescence

Cathodoluminescence is the generic title given to the process which occurs when a beam of electrons impacts upon a luminescent material. Earlier, a beam of electrons was generated in a tube fitted with an anode and a cathode; the cathode being coated with a material which when heated by a separate filament released electrons into the surrounding area. When a voltage was applied between the anode and the cathode, the free electrons formed a beam of electrons flowing towards the anode. Thus such a beam of electrons became known as the cathode ray.

6.6.1.1   Cathode Rays Impinging on a Target

Cathode rays in vacuum are not light generators in their own right but are the providers of energy in the interaction of the electron beam and its target material. In reproduction, this is invariably in the context of a cathode ray tube (CRT), which comprises an electron gun, mounted facing a glass screen coated with a phosphor at a very high positive voltage with respect to the cathode and thus acts as the anode. When the electron beam strikes the phosphor it fluoresces. Fluorescence is described in Section 6.7.

6.6.1.2   Cathode Rays in a Gas Discharge

When two electrodes are placed within a gaseous environment in a sealed glass tube and the negative electrode, the cathode, is heated by means of a separate electrical filament it will start to emit electrons. If an electric potential is now applied across the electrodes the electrons will respond to the electric field between the electrodes and will be accelerated towards the anode causing a small electric current to flow. These electrons will collide with the atoms of gas and if the applied voltage is high enough to impart sufficient energy to the electrons the collision will cause valence electrons to migrate to higher-level orbits, which on return to lower orbits will produce line spectra. At higher voltages the collisions will eventually become energetic enough to ionise the gas. At this point the gas will comprise of electrons, ions and impurities and is referred to as plasma, after its similarity in terms of a mixture of content in blood plasma.

The following descriptions of the spectra of gas discharges are intended to give only an indication of what is achievable from the principal elements used in gas discharges in support of illumination technology. The author has drawn freely from the research work of Ioannis Galidakis[3] in preparing the material for this section and is grateful for his permission to include the tables in Figures 6.7–6.10 which are amended versions of those appearing on his website. Each of the graphs shows the power emitted normalised to a maximum value of 100 in the visible spectrum and the chromaticity coordinates are recorded to an accuracy of five decimal places as is done in the originals.

Mercury Discharge

The spectra produced are dependent upon the pressure and density of the gas. Because of its ease of vaporisation and its useful spectra, mercury is often used as the gas in practical discharge tubes and has a spectrum with two distinct distributions of energy depending upon

---

[3] Since initially citing this website it has experienced problems. Galidakis suggests the following approach to gain access to the relevant area. I found it only works when using the Google Chrome browser:

1. Disable javascript in Google Chrome.
2. Go to www.archive.org
3. Enter string and search http://ioannis.virtualcomposer2000.com/spectroscope/elements.html:
4. Click on 2012 6th June.

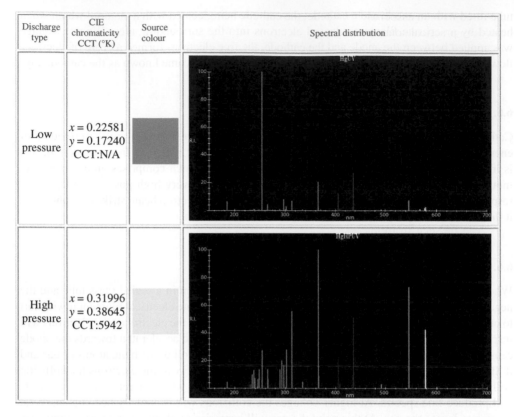

Discharge type	CIE chromaticity CCT (°K)	Source colour	Spectral distribution
Low pressure	x = 0.22581 y = 0.17240 CCT:N/A		
High pressure	x = 0.31996 y = 0.38645 CCT:5942		

Figure 6.7 The spectra of mercury plasma at different pressures in nm. (By permission of I.N. Galidakis.)

whether the gas is at low pressure at small fractions of atmospheric pressure or at a pressure of several atmospheres.

In Figure 6.7, the primary emission is located at the resonance line of 253.7 nm, well into the UV band, with only limited blue and to a much lesser extent green emissions in the visible band at 404.7 nm, 435.8 nm and 546.1 nm. Transitions between an energised state and the ground state are usually the most frequent at low pressures and the resulting lines are termed resonant lines. In consequence low-pressure mercury discharges in isolation are luminously inefficient and produce a cyanish light of poor colour rendering due to the lack of any emission in the red sector of the spectrum. In high-pressure discharge the voltage and current are increased which imparts more energy to the gas such that more transitions occur between the more closely associated higher-energy orbits, enabling lower-energy photons in the visible spectrum to be produced. The UV resonant emission is greatly reduced, the marginally dominant line is now at 365 nm and the green and the dual orange lines in the visible spectrum at 546.1 nm and 578.2 nm are greatly enhanced. The result is an increase in luminous efficacy from about 10–50 lm/W and a much improved colour rendering, as indicated by the colour patches in the figure which represent the colours of the discharges.

Although at first sight it might appear that the low-pressure mercury discharge has little practical value compared to the high-pressure version, as we shall see later, the UV lines in both discharges in association with the effect of fluorescence described in Section 6.7, makes them both practical sources of light.

6.6.1.3 Sodium Discharge

The other element frequently used in gaseous discharges is sodium which has a spectral distribution illustrated in Figure 6.8 and which is also derived from the Ioannis website.

At low pressures sodium produces effectively only two lines very close together in the visible spectrum at 589 and 589.6 nm and, since this is a wavelength close to the peak of the luminous efficiency function, the discharge is very efficient at about 200 lm/W. Other lines have relative emission energy of less than 1%. However, because the emission is restricted to effectively only one wavelength, though orange in colour, the source is effectively monochromatic and all colour in the scene is eliminated.

As can be seen from the lower section of the figure, at high pressure, like mercury, the low-level lines are greatly enhanced providing some ability to differentiate colours in the scene; however, the relatively poor distribution of spectral lines of similar amplitude still leads to very poor colour rendering of the illuminated scene. At high pressure the original resonant line is subject to thermal and pressure broadening and self-absorption which leads to the relatively broad peak of energy around the original resonant lines. The luminous efficiency at high pressure reduces to about 100 lm/W.

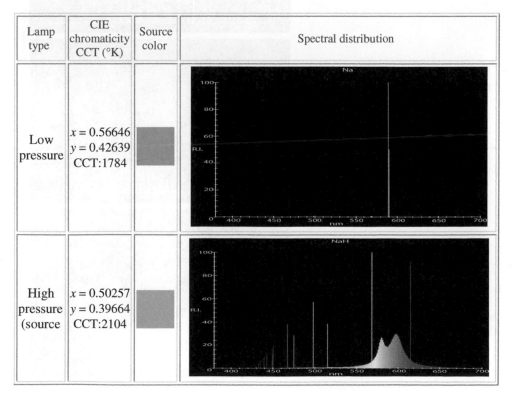

Lamp type	CIE chromaticity CCT (°K)	Source color	Spectral distribution
Low pressure	x = 0.56646 y = 0.42639 CCT:1784		
High pressure (source	x = 0.50257 y = 0.39664 CCT:2104		

Figure 6.8 Sodium discharge spectrum for low- and high-pressure gas. (By permission of I.N. Galidakis.)

Xenon Discharge

Xenon gas in a discharge situation produces many more spectral lines than mercury or sodium in the visible part of the spectrum and, as the following three table-graphs from the Galidakis website illustrate, as the pressure is increased the energy spread in the visible spectrum improves even further.

At low pressures, despite the relatively broad spectrum, there is nevertheless a preponderance of energy in the long-wavelength end of the spectrum and in consequence the discharge produces a warm colour as indicated by the colour patch in the top table of Figure 6.9. At

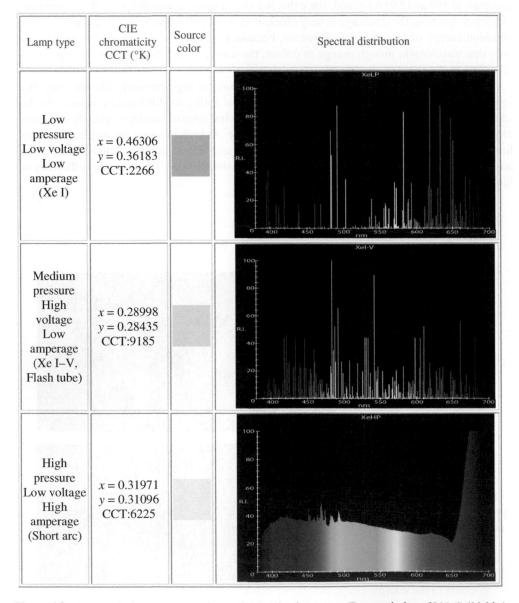

Figure 6.9   Xenon discharge spectra at increasing levels of pressure. (By permission of I.N. Galidakis.)

medium pressures the number of lines produced increases very significantly and the balance of lines across the spectrum is much improved, resulting in a slightly bluish colour. Finally at high pressure levels of some 30–100 atmospheres, there is such an increase in lines that a continuous spectrum of energy across the visible spectrum is effectively produced giving an excellent comparison with daylight, with correlated colour temperatures in the range 5,600–6,300 K. Inspection of the third table in Figure 6.9 illustrates that the level of emission is rising rapidly into the infrared segment of the spectrum and although this does not detract from the excellent CCT produced, it does limit the luminous efficiency of these sources to about 40 lm/W.

Mercury Metal Halide Discharge

The metal halide discharge is a variant of the high-pressure mercury discharge, the variation being in the form of various metal halides being added to the mercury vapour.

Metal halides are compounds of halogen and the chemically defined metals. The halogen group of elements are those elements in group 17 of the periodic table, which are characterised by being strongly electronegative as a result of having only one electron missing from a full outer shell of the element atom; they comprise fluorine, chlorine, bromine, iodine and astatine. The metals with which they combine are complementary in terms of being electropositive and form groups 1–12 of the periodic table;

At normal temperatures the metal halides are usually solids or liquids. However, in the mercury gas plasma close to the centre of the electron beam, the temperature will rise to a level in the order of 700 degrees centigrade, which has the effect of breaking down the metal halide into its constituent elements. The metals are drawn into the beam to be struck by electrons and thus participate in the emission spectrum by adding spectral lines towards the centre and red end of the spectrum. The released halogen forms a vapour which effectively protects the glass envelope of the discharge tube from the highly reactive alkaline metals, combining with them to again form metal halide before they can react with the glass. The process of disassociation and recombination continues on a cyclic basis.

Development of metal halide technology commenced in the early 1960s with the aim of producing efficient sources of illumination with a broad distribution of energy across the visual spectrum. Very many compounds were investigated and used for several different purposes but a representative selection of those that are useful in the aim to emulate daylight to various degrees of success are illustrated in Figure 6.10 from the Ioannis source. (Ioannis notes that these spectra are representative and do not equate to measurements of particular compounds.)

6.6.2   Solid State Electroluminescence

6.6.2.1   The Active Semiconductor Junction

The semiconductor junction forms the basis of the light-emitting diode (LED), which is a variant of the semiconductor diode which was first manufactured in the late 1950s; its operation is briefly described in Appendix D. In the following a brief indication of the principles of light emission in an inorganic semiconductor junction is given. For a comprehensive coverage of the topic, Schubert's book on Light Emitting Diodes (2006) is recommended.

The p-n junction of the diode when forward biased enables the electrons and holes on either side of the junction to diffuse across the barrier and recombine; in doing so they dissipate energy equal to the energy gap of the junction plus any kinetic energy they have associated with the temperature of the material at the junction.

Lamp type	CIE chromaticity CCT (°K)	Source color	Spectral distribution
Sodium/ Thallium/ Indium metal halide (European)	$x = 0.37426$ $y = 0.41000$ CCT:4366		
Sodium/ Scandium metal halide (American)	$x = 0.35185$ $y = 0.32282$ CCT:4575		
Dysprosium/ Thallium/ Thulium/ Caesium "daylight" (European) metal halide	$x = 0.30179$ $y = 0.35347$ CCT:6855		

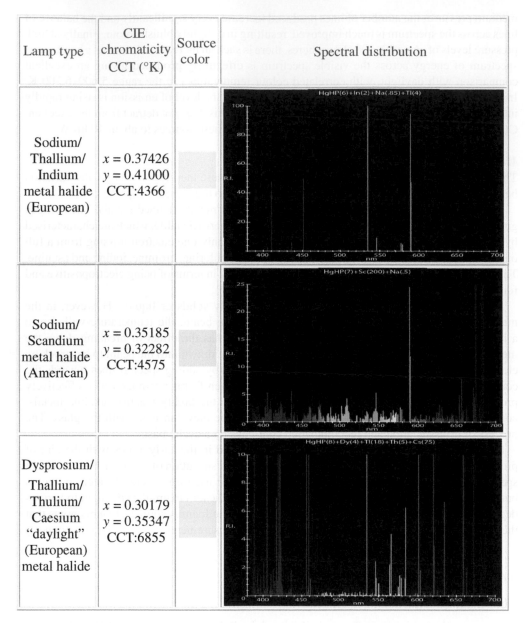

Figure 6.10 Representative spectra of various combinations of mercury/metal halide plasmas. (By permission of I.N. Galidakis.)

The electrons in the conduction band and the holes in the valence band are assumed to have parabolic energy dispersion relations as follows:

$$\text{for electrons} \quad E_e = E_C + \frac{k^2 T^2}{2m_e c^2}$$

$$\text{for holes} \quad E_h = E_V + \frac{k^2 T^2}{2m_h c^2}$$

where E_C and E_V are the conduction and valence band edges, respectively, m_e and m_h are the electron and hole effective masses, k is Boltzmann's constant, T is absolute temperature.

The second terms of these relations are those due to the kinetic energy of the particles.

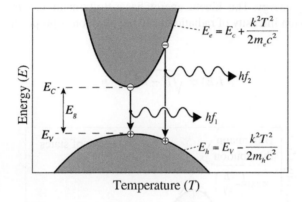

Temperature (T)

Figure 6.11 Parabolic electron and hole dispersion and recombination at the semiconductor junction. (Adapted from Schubert, 2006).

Figure 6.11 illustrates the distribution of energy across the junction. As can be seen the lowest energy exchange occurs at the point the electron has no kinetic energy when the band gap energy $E_g = E_C - E_V$, producing monochromatic light. Generally the gap energy will be higher, leading to a spread of the monochromatic light radiated at zero temperature into a band of frequencies commencing at the monochromatic frequency.

Depending upon the material, its purity, the freedom of the crystalline structure from faults, the doping concentration and other less significant factors, the energy released at recombination is in the form of either phonons or photons. The purer and fault free the material, the more the energy is released as photons. The energy of the phonons is released in vibrating the crystal lattice which leads to a rise in temperature.

Since the energy of the band gap is known for most semiconductor materials it is a simple matter to calculate the notional frequency of the emitted light. The frequency of the photon is related to the change of energy state by the formula derived by Planck as described in Section 6.3, that is, $f = E/h$, where E is in Joules and h is Planck's constant. The energy of the band gap is more conveniently described in terms of electron volts, eV, rather than joules, and as indicated in Section 6.5 one electron volt is equal to 1.602×10^{-19} Joules.

Thus by selecting a material with the appropriate band gap energy, suitable junctions can be fabricated for a wide range of frequencies. A common material for this purpose is gallium arsenide which has a band gap energy of 1.42 eV, thus at 0 degrees Kelvin:

$$E = E_g = E_C - E_V = hf$$

$$\text{and} \quad f = \frac{1.42 \times 1.602 \times 10^{-19}}{6.626 \times 10^{-34}} = 343\,\text{THz}$$

which corresponds to a wavelength λ of 873 nm in the infrared band.

At room temperature the electrons and holes have an average kinetic energy of kT which adds to the band gap energy. This kinetic energy has a Boltzmann distribution and Schubert shows that the emission intensity of radiation from the junction is proportional to the product of two terms:

$$I(E) \propto \sqrt{E - E_g} \;\; e^{-\frac{E}{kt}}$$

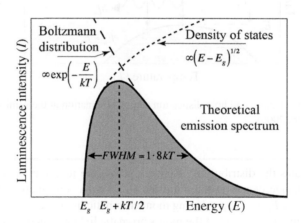

Figure 6.12 Theoretical emission spectrum of a semiconductor junction. (Based upon a diagram from Schubert, 2006).

The theoretical luminescence intensity of the junction against particle energy is illustrated in Figure 6.12 where the dotted lines show the contribution from the two terms.

The maximum emission intensity occurs at $E = E_g + \dfrac{kT}{2}$.

The full width at half maximum (FWHM) of the emission is

$$\Delta E = 1.8\,kT \quad \text{or} \quad \Delta f = \frac{1.8kT}{h}$$

Thus the FWHM at a room temperature of 300 K is:

$$\Delta E = 18 \times 1.38 \times 10^{-23} \times 300 = 7.457 \times 10^{-21} \text{ J}$$

$$\Delta f = \frac{E}{h} = \frac{7.457 \times 10^{-21}}{6.626 \times 10 - 34} = 11.254 \text{ GHz}$$

$$\text{and } \Delta \lambda = c \left(\frac{f_2 - f_1}{f_1 f_2} \right) \text{ if } f_2 \approx f_1 \text{ then } \Delta \lambda = \frac{c \Delta f}{f^2} = \frac{1.8kT\lambda^2}{hc} = 28.58 \text{ nm}$$

As indicated above the optical efficiency of the junction is related to the number of particles which recombine to produce photons rather than phonons and this relates primarily to the purity of the semiconductor material. In the 1960s the levels of purity achieved led to efficiency levels of fractions of 1%; however, due to several steady improvements in the purity of the semiconductor material since that time efficiencies now exceed 90% and in some cases 99%.

The simplified situation outlined in the diagrams and formula above has become very much more complicated by the developments of the past 50 years as new materials, compounds and techniques have been discovered to improve the efficiency and frequency range of the emitted radiation. As new materials were discovered, the frequency of the emissions increased, starting in the infrared and progressing over the years into the visible red, orange, yellow and green frequencies. For many years practical junctions emitting blue light were not available but this changed in the 1990s and in the early years of this century the range has been extended into UV frequencies.

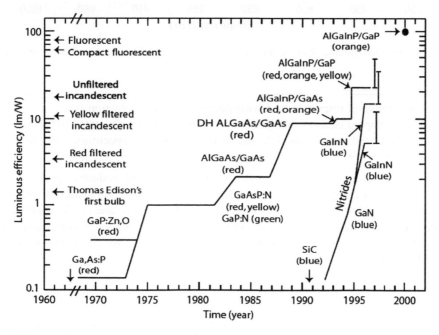

Figure 6.13 Semiconductor junction developments. (From Schubert, 2006, adapted from Craford, 1997).

Figure 6.13 illustrates the progress made since the 1960s and compares the efficiencies obtained with those of more traditional light sources. Remarkably the efficiency has improved by several hundred to one over the period. For a particular LED the efficacy falls off with increase in current.

The cost of production of these devices has seen similar large reductions over the same time period.

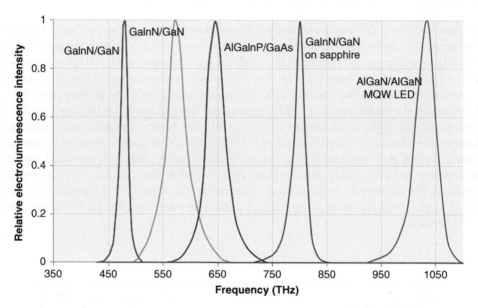

Figure 6.14 Junction spectral distributions at room temperature of various semiconductor junction types.

The spectral emission responses of a wide range of different junction types in terms of base materials, doping elements and junction configurations are illustrated in Figure 6.14, which include the red, green, blue, UV and far UV bands of the spectrum, respectively.

In recent years research has extended into using organic semiconductors as the junction materials. The use of compounds as semiconductors significantly complicates the physics of the process, nevertheless it is analogous in as much as the valence and conduction bands of inorganic compounds may be regarded as synonymous with the highest occupied and lowest unoccupied molecular orbitals of the organic compound. Devices which use this technology are termed organic light emitting diodes (OLEDs).

6.6.2.2 Laser

The term "laser" is an acronym for *light amplification by stimulated emission of radiation* and is a device which generates and emits a highly coherent, virtually single wavelength, beam of light.

The laser comprises a source of energy, often termed the pump source; an optical resonator and an optical amplifier. The pump source may be an electric current or light of shorter wavelength and thus of higher energy per photon than that required from the laser. The amplification mechanism comprises an optical resonator filled with the material that provides the gain as shown in Figure 6.15.

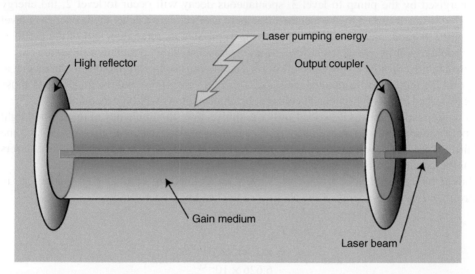

Figure 6.15   Laser principal components. (Adapted from http://en.wikipedia.org/wiki/Laser.)

The optical resonator comprises a cylinder terminated at either end by mirrors. At one end the mirror is highly reflective whilst at the other, it is semi-silvered to act as an output coupler. This allows some reflection of the generated beam back through the amplifying medium and the remainder to leave the resonator in a highly collimated beam of relatively high intensity.

The amplifying medium may be a gaseous, liquid or solid substance which is capable of being energised by either an electric current or a high-energy source of photons above its ground-level state and emitting photons upon return to a lower or ground state.

As with previous examples of energised materials investigated in this section, normally the energised particles will return to their ground state spontaneously and there will be no coherence between the photons. However, if prior to spontaneous emission the energised particle is perturbed by a passing photon generated by another particle of the same characteristics, it will return to the ground state by emitting a photon which is coherent with the photon that disturbed it. Thus two photons *of the same frequency and phase* are generated at the point of perturbation, an amplifying effect known as stimulated emission. This effect also occurs of course in the events previously described in this section, but unless specific conditions are met the coherence is lost by the absorption of one or both of the photons in the media material.

With the operation of the two-level amplifying medium material described above it is not possible to energise sufficient proportions of the particles making up the media to ensure that more stimulated emission pairs are generated than absorbed. This can be achieved by using a material with at least three levels of energy, including the ground state, under the right conditions.

Using level 1 to describe the ground state, level 2 the intermediate energised state and level 3 the highest energised state, the criteria required is that the decay time from level 3 to level 2 is significantly shorter than the spontaneous decay time from level 2 to level 1. As the particles are energised by the pump to level 3, spontaneous decay will occur to level 2, the energy generated being absorbed by the media. Since there is effectively a delay in the energy of level 2 being discharged, then eventually a high proportion, that is, more than half of the particles in the media, will be energised at level 2, a situation known as population inversion. Since the resonator ensures by continuous reflection a continuous source of photons in the media, at this point the number of stimulated emissions surviving absorption will be greater than those being absorbed and a high-intensity coherent beam of photons will be generated.

As with the mechanisms described earlier in this section, the frequency of the emitted light is dependent upon the difference in energy levels between the level with the slow decay time and the lower level to which the particle settles, if albeit briefly. In practice, three-level lasers are inefficient and lasers are usually of four or a higher number of levels.

Thus if the energy levels of these two layers are defined in terms of E_{slow} and E_{lower} in electron volts then the energy of the photons of the laser will be given by:

$$E = E_{slow} - E_{lower} \text{ eV}$$
$$\text{and} \quad f = \frac{E \times 1.602 \times 10^{-19}}{6.626 \times 10^{-34}} = \text{Hz}.$$

Unlike LEDs where the fundamental monochromatic radiation is broadened as a result of the temperature of the material, laser radiation is virtually monochromatic.

A more detailed description of the mechanism of stimulated emission and population inversion is given in Appendix E.

Lasers used for large screen displays have a luminous output in excess of 60,000 lumens at an efficiency of between 10% and 30%. The bandwidth of the laser is between 0.1 and 2 nm which is ideal for producing a large display gamut; however, the highly coherent laser beam can introduce a 'speckle' effect at the display surface. This is the result of interference patterns which occur when waves of the same frequency are reflected from an irregular surface, producing a range of waves of different phase which add in a manner producing different intensities at each adjacent location. The effect is ameliorated by using lasers of broader bandwidth.

6.7 Fluorescence

Fluorescence occurs when energetic particles, electrons or high-energy photons, strike a semiconductor material. Therefore, in terms of luminescent categorisation fluorescence can occur as a result of either cathodoluminescence or photoluminescence. The wide range of semiconductor materials used for this purpose are termed phosphors and as explained in Section 6.5, they have no relationship with the element phosphorus.

The mechanism of generating photons from the phosphors is as described in Section 6.5. Since the phosphors are generally compounds with trace amounts of dopant material called activators, the SPDs produced are usually complex and relatively broad band compared to the monochromatic lines produced by the elements.

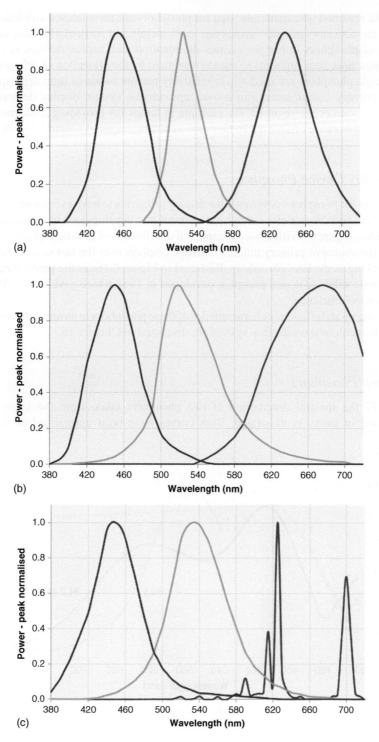

Figure 6.16 Various groups of display phosphors. (a) Silicate phosphors. (b) Sulphide phosphors. (c) Sulphide-based phosphors with the red phosphor doped with a rare earth element.

Some of the common host materials used for phosphors are the silicates, sulphides, oxides and nitrides; the activators include various rare earth metals and particularly europium.

The range of phosphors in use for various illumination and display devices is very large indeed and there have been attempts to standardise them by the allocation of 'P' numbers.[4]

Usually single phosphors are used where relatively narrow bands of light are required, for example, as primary colour sources in colour reproduction system display devices. Where broad spectrum sources are required, for example in lamps for providing illumination, then mixtures of phosphors are often used to facilitate matching-required spectral distributions.

6.7.1   Display Device Phosphors

The requirement for phosphors developed for display primaries to have ever more brighter and saturated colours with higher efficiencies has ensured that development has continued apace over the decades. Figure 6.16 illustrates the spectral distribution of groups of phosphors which have formed the source of primary colours in image displays over the last several decades.

What is striking in the later phosphors illustrated in Figure 6.16c is the peaky nature of the red yttrium oxide sulphide doped phosphor developed in 1964 which gave for the first time a true red with good efficiency.

The chromaticity and efficiency characteristics of these phosphors in terms of their suitability as primaries in a colour reproduction system are discussed in Chapter 16.

6.7.2   Lamp Phosphors

In Figure 6.17 the spectral distribution of two phosphors taken from the large range of phosphors used in lamps is illustrated. Both curves have been normalised at 500 nm to

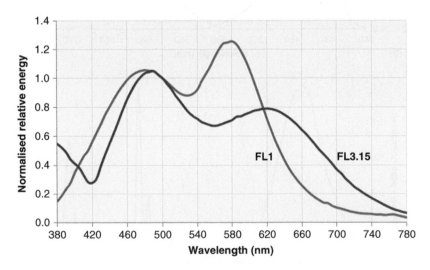

Figure 6.17   Lamp phosphors.

---

[4] http://en.wikipedia.org/wiki/Phosphor

emphasise the difference in the curve shapes. F1 is from the range of calcium halophosphate phosphors which consist of two semi-broadband emissions based upon activations of antimony and manganese producing a good efficiency and a satisfactory colour. F3.15 is from a range of phosphors which are each based upon a mix of a wide range of phosphors. This mix provides improved colour, in terms of matching daylight but at the cost of a corresponding reduced efficiency.

Part 3

The Concepts of Colour Reproduction

Introduction

Electronic colour reproduction, whether by television, photography or cinematography comprises a number of processes, the majority of which, in one form or another are common to all three media types. The exception to this general rule is photography where only the final element of the workflow, that is the production of the print, has no corresponding process in television or cinematography. Nevertheless, apart from this final stage the remainder of the photographic processes do correspond to similar processes in television and cinematography. Thus with the exception of the print process, which is separately described in Part 5, all the processes described in Parts 3 and 4 are equally applicable to television, photograph and cinematography. Specifically, reference to display screen characteristics in the following chapters may relate to a television screen, a computer monitor screen for previewing a print or a cinema screen.

The complete process of colour reproduction includes, at the start of the work flow, the illumination and capture of the scene, followed by the transfer of the red, green and blue (RGB) camera signals to the display, the generation of the reproduced image and the viewing and appraisal of the image in the viewing environment. For each of the three activities: scene illumination, image generation and viewing environment illumination, light needs to be generated and utilised in a manner appropriate to that activity. In Part 3 an overview is provided of the various types and characteristics of the light and lighting used for these operations, building on the physical processes described in Chapter 6, before describing the concepts of the system which captures the image, processes it and displays it.

Colour Reproduction in Electronic Imaging Systems: Photography, Television, Cinematography, First Edition. Michael S Tooms.
© 2016 John Wiley & Sons, Ltd. Published 2016 by John Wiley & Sons, Ltd.
Companion Website: www.wiley.com/go/toomscolour

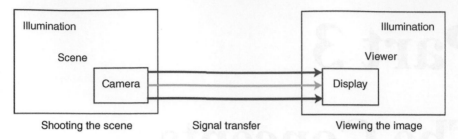

Figure P3.1   Signal chain or work flow of a conceptual colour reproduction system.

Thus Part 3 commences with a review of the characteristics of the light required both for the illumination of the scene, in order to ensure colour accuracy in the reproduction, and the illumination of the viewing environment, where the characteristics of the ambient lighting will influence the adaptation of the eye and therefore the perception of the rendered image. Ideally illumination would have been dealt with in Part 1 where it was briefly introduced but without the knowledge and the colour measurement tools developed in Part 2 many readers may have found that some of the concepts would have been difficult to comprehend.

The material covered in Part 2 on colour measurement may now be used to establish the critical factors which control the colorimetric aspects of a colour reproduction system.

We have seen how a system of colorimetry is built upon the foundation of the property of the eye to emulate the colour of a spectrally complex reflecting surface by the simple addition of three primary colours in the correct proportions. It is a small step from here to visualise how such an approach could form the basis of a colour reproduction system.

In the following we will initially use the fundamental tristimulus approach to define in simplistic terms the colorimetric requirements of any colour reproduction system. Having determined the basic approach we will then use the CIE system of colorimetry to obtain in a classical manner the ideal colour spectral sensitivities of the camera of any reproduction system. These, as we shall see, are entirely dependent upon the chromaticity coordinates of the three primary colours of the reproduction display and the corresponding colour matching functions (CMFs). The basis for the adoption of these CMFs as camera spectral sensitivities will be explained in some detail.

The conceptual approach described here has historically formed the basis of early simple practical colour reproduction systems for television. So in this respect the concepts derived in Part 3 are capable of being implemented in real systems, which accepting the technical limitations of the period, delivered good colour pictures.

Finally the importance of the viewing environment in terms of the level, spectral distribution and positioning of the ambient illumination in the manner in which it affects the subjective appraisal of the reproduced image is described.

7

Sources of Illumination

7.1 Overview

In Section 2.6 the wide range of sources of illumination of a scene and the remarkable ability of the eye to adapt to them, both in terms of the level and the colour of the illumination, were broadly outlined. However, a simple colour camera is not capable of these excellent levels of adaptation so it is important that the colour characteristics of the scene illumination should match the appropriate illumination parameters used in defining the colour reproduction system. (Defining these parameters is the subject of the next chapter.) In the following, the use of the word 'illuminant' is used both as defined by the Oxford English Dictionary, that is 'a source of illumination', and as defined in formal CIE[1] terms, where an illuminant is a table of illuminant data; the context will indicate which meaning is appropriate.

Furthermore, in viewing the reproduced image it is important that the level and colour of the illumination of the viewing environment matches the intended viewing conditions specified in the reproduction system, as the eye will adapt to the generally larger area of view of the surroundings. Thus if there is no match between the colour of the environmental illumination and the colour balance of the reproduced image, the latter will not be perceived as intended and the results will be disappointing at best.

Having established objective methods for measuring colour and from these measurements calculating the luminance and chromaticity of colours in Part 2, we are now in a position to investigate in more detail the colour characteristics of sources of illumination.

Daylight of course is the most fundamental source of illumination, indeed since early historical times it was the only source of illumination. Technology came to the rescue where daylight was lacking in relatively recent times with various lamps based on black body radiators in the form of tungsten-based bulbs. These were of relatively poor efficacy, in as much as they were inefficient and did not produce light of an even spread across the spectrum. Various forms of gas discharge lighting followed of which the simplest and most efficient form were mercury-based discharge lamps. In more critical environments, florescent lighting is used, where the narrow band energy of the gas discharge is partially converted by a fluorescent coating into a broader band spectrum. Lately, the efficacy and level of illumination of light-emitting diodes

---

[1] Commission Internationale de l'Eclairage.

Colour Reproduction in Electronic Imaging Systems: Photography, Television, Cinematography, First Edition. Michael S Tooms.
© 2016 John Wiley & Sons, Ltd. Published 2016 by John Wiley & Sons, Ltd.
Companion Website: www.wiley.com/go/toomscolour

(LEDs) have reached the point where they are capable of providing relatively efficient indoor lighting with ever improving levels of illumination and colour characteristics.

Before the various sources of illumination available are reviewed, it is necessary to develop an objective method for determining the quality of the colour rendering of the illuminants, both in terms of that perceived by the eye and that perceived by the cameras so that their suitability for illuminating scenes for reproduction can be assessed.

7.2   Illuminant Colour Rendering Quality

In reviewing the characteristics of the sources of illumination used in colour reproduction we need to be aware of how the relationship between the spectral power distributions (SPDs) of the illuminants, the reflection characteristics of surfaces in the scene and the cone responses of the eye affect the capability of the illuminant to provide good colour rendering when compared with a broadband source such as daylight or tungsten light. Generally speaking, sources of even energy distribution across the spectrum with no energy peaks or troughs will, subject to adaptation, provide good colour rendition, whilst those with gaps in the spectrum and concentrations of energy around two or three wavelengths will provide poor rendering for at least some colours. The shortcomings of light sources need to be characterised so that a method can be developed of relating how one type of illuminant compares with another with regard to their ability to provide good colour rendering for satisfactory colour reproduction.

The general approach to formulating a method of measuring the ability of a light source to become a satisfactory source for illuminating a scene for colour reproduction is to test the source against a reference source of known even spectral distribution when both in turn illuminate a range of specified test colours. The reference illuminants used are theoretical spectral distributions based on either tungsten or daylight sources at a colour temperature matched to the correlated colour temperature (CCT) of the test source.

The methodology adopted is based on using the spectral distributions of the reference illuminant and test sources, the spectral reflectivity of the test colours and colour response curves of the cones or the camera to calculate the overall response to the stimuli in terms of the values of the red, green and blue signals generated for both the reference illuminant and the test source. In the case of a camera, the signal levels are converted first into XYZ values and subsequently into one of the CIE metrics for measuring colour difference such that an index based on the range of colour samples can be generated to reflect the suitability of the test source as an illuminant for colour evaluation or reproduction.

The current international standard for defining the suitability of light sources as the basis for illuminating a scene is the CIE defined *colour rendering index* (CRI) which is based on measuring the differences in colour of a number of coloured samples when illuminated by a reference illuminant and by a test illuminant.

The CRI was standardised in 1965 primarily for use in assessing the colour rendering quality of illuminants for industrial, commercial and show areas, where the criteria was the accuracy of colour rendering to the human eye. Since that time, as was described in Section 4.9, the metrics for measuring colour differences have evolved through several CIE standards and indeed continue to improve as indicated in Chapter 5. Furthermore, for those involved in colour reproduction the primary interest is how well the light source performs in providing good colour rendering as 'seen' by the colour camera as well as that perceived by the eye, and there are important differences between the two as will be seen.

These shortcomings were recognised many years ago, shortly after colour television was introduced into Europe but the proposals formulated at that time for an index which overcame these limitations never reached the stage of standardisation. However, an informal index based on these proposals is in use by several users who regard the CRI as unsatisfactory and is described below as the MCC[2] index.

Despite the shortcomings, the original CIE methodology of deriving the CRI is also common to its later proposals evolved in 1995. Furthermore, since it is the only international standard for the present, it is appropriate to describe its formulation since it is still used by many manufacturers of illuminants to indicate the suitability of their luminaires[3] for colour reproduction, and forms the basis for judging the metamerism of ambient light sources in ISO Standards for viewing conditions when appraising rendered images.

Subsequent to writing this section, the work of Alan Roberts, John Emmett and Per Böhler cited in the following paragraphs has been adopted by the European Broadcasting Union as the basis for Recommendation R 137 which recommends the use of a Television Lighting Consistency Index-2012 as a means of evaluating the suitability of luminaires for television and is described in Chapter 18.

The use of each of these three indices is described in the following.

7.2.1   The CIE CRI

The procedure for calculating the CRI of a lighting source is defined by the CIE publication: 13.3-1995[4] and uses the SPDs of a reference and a test source, their CCTs and the chromatic adaptation characteristics of the eye to measure, under both the reference and test sources, the colours of a range of eight test colour samples whose spectral reflectances are defined. The values of the differences in colour between the two illuminants are used to derive a CRI. The calculations, though mathematically straightforward, are quite complex and extensive and are not essential to the understanding of the process. In consequence, the approach is described here in outline form and the actual calculations required are assigned to Worksheet 7(a).

The CIE specify both a *special CRI* and a *general CRI*. The special CRI is the index obtained for each of the range of test colours whilst the general CRI is the mean of all the special CRIs.

The special CRI is given by:

$$R_i = 100 - 4.6d_i$$

where d_i is the distance between the points representing the chromaticity of the test colour in the CIE 1964 $U^*V^*W^*$ colour space when illuminated by the reference and test sources of illumination respectively. This is an obsolete colour space which was superseded by the CIE 1976 LUV colour space described in Section 4.6 and is one of the reasons why the CRI comes in for criticism. (The CIE 1964 $U^*V^*W^*$ colour space was based on the original u,v chromaticity coordinates, and as these had been superseded this colour space was not

---

[2] The index is based upon the Macbeth ColorChecker® Chart (MCC). However, production of the chart has now been taken over by X-rite and is known simply as the ColorChecker® Classic. See Section 7.2.2.

[3] A luminaire is a lamp source accommodated in a suitable housing for illuminating a scene.

[4] http://www.cie.co.at/index.php/index.php?i_ca_id=464

developed in Chapter 4; however, the parameters of the colour space are defined and utilised in Worksheet 7(a).)

The value of '4.6' in the formula was designed to give a value of $R_i = 50$ for a warm white fluorescent lamp. However, for sources with a large colour rendering divergence d_i, it is possible to produce an unrealistic negative index value, another cause of criticism of the index.

If the CCT of the test light source is below 5,000 K, then the reference light source should be that of a Planckian radiator at the same CCT as the test source, and if it is 5,000 K or above then its SPD should be the Illuminant 'D' SPD of the matching colour temperature. (The CIE daylight or 'D' illuminants are defined in Section 7.3.)

In order to take account of the chromatic adaptation characteristics of the eye, an adjustment is made to the measured test sample chromaticities under the test illuminant to make the test illuminant appear neutral, by using the CIE specified adaptation formulae, which is based on the von Kries adaptation hypothesis described in Chapter 5. This formula is detailed in Worksheet 7(a).

Eight test colour samples were drawn from an early Munsell Book of Colour (as illustrated in Figure 2.20) evenly divided around the axis of the colour space but comprised of samples of low saturation; a cause for further criticism of the index.

Table 7.1 CRI colour test samples. From Wikipedia.[5]

Name	Munsell Notation	Appearance under daylight	Swatch
TCS01	7,5 R 6/4	Light greyish red	
TCS02	5 Y 6/4	Dark greyish yellow	
TCS03	5 GY 6/8	Strong yellow green	
TCS04	2,5 G 6/6	Moderate yellowish green	
TCS05	10 BG 6/4	Light bluish green	
TCS06	5 PB 6/8	Light blue	
TCS07	2,5 P 6/8	Light violet	
TCS08	10 P 6/8	Light reddish purple	

The test colour samples are illustrated in Table 7.1; the samples being defined as those to be used for calculating the special indices, which by taking the mean of all the samples leads to the general CRI. The Munsell notations of the colour samples are also given together with indicative representations of their colours.

The spectral reflectances of the eight test colour samples are defined by the CIE and illustrated in Figure 7.1.

Once the eight special CRIs have been calculated the general CRI may be found by taking the mean of the values for the eight samples as follows:

$$R_a = 100 - (4.6/8)(d_1 + d_2 + d_3 + d_4 + d_5 + d_6 + d_7 + d_8)$$

By taking the mean figure for the samples, the effect of any one sample having a large deviation will be diminished which gives a better result than might well be experienced in

---

[5] http://en.wikipedia.org/wiki/Color_rendering_index

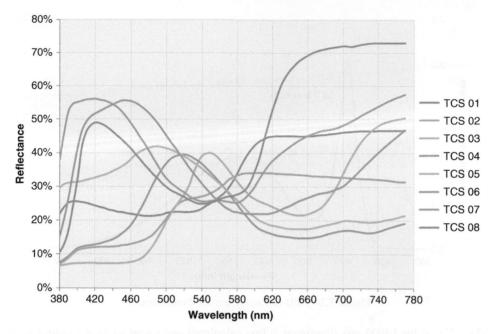

Figure 7.1 Spectral reflectance of the CIE CRI test colour samples.

practice. This is another point against the present form of the CRI. Experience indicates that this is probably the most important criticism, since with marginal illuminants significant negative values of R_i can be obtained, which on averaging the result actually improves the R_a value unrealistically.

In very general terms, values of CRI between 85 and 100 are considered to provide good colour rendition; however, where levels of illumination and efficacy are an important criteria, such as in the illumination of sports arenas, then a compromise may be reached and sources of illumination with CRIs in the range from 75 and above are likely to be used.

Calculations required for obtaining both the CCT of a test illuminant from its SPD and the R_a value of its CRI are laid out in Worksheet 7(a) with text box explanatory notes where possibly unfamiliar formula are used.

A large number of measured SPDs of various illuminants are contained in the 'Illuminants' worksheet, any of which may be copied across to Worksheet 7(a) as required in order to calculate both the CCT and the R_a of the test illuminant. Similarly any new lamp SPD values may be entered directly into the calculator. The calculator was used to establish the CCT and the R_a of all the SPDs outlined in the remainder of this chapter.

Figures 7.2 and 7.3 illustrate an example of the use of Worksheet 7(a) in calculating the change in chromaticities which occur with changes in scene illumination.

Figure 7.2 illustrates the SPD of a mid-range CRI LED test illuminant (LED4 in the 'Illuminants' worksheet), as currently in use in studio lighting, and the corresponding SPD of the reference illuminant of the same CCT as calculated in the worksheet.

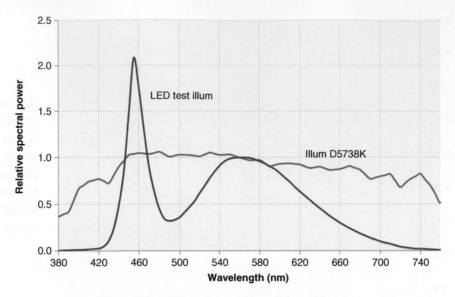

Figure 7.2 LED test and reference illuminants.

The CCT of the LED4 test illuminant is first calculated and found to have a colour temperature of 5,738 K. Using this colour temperature the SPD of the illuminant 'D' reference illuminant is then calculated using the method described in the Section 7.3 and detailed in Worksheet 7(a).

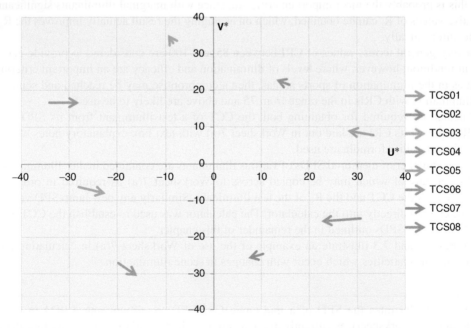

Figure 7.3 Plot of the change in chromaticities of eight CRI test colours with a typical LED lighting source.

Figure 7.3 illustrates the changes in chromaticity which occurs for the eight CRI test colours when illuminated in turn by the LED source of Figure 7.2 and the D5738 illuminant reference. The values of U* and V* are taken from the CIE W*U*V* colour space as defined in the CRI worksheet. Since a unity step on this chart is notionally equal to one just noticeable difference, then it can be seen that several of these colours have shifted between about 5 and 8 JNDs.

The value of the CRI calculated in the worksheet for this illuminant is $R_a = 76.7$ which might appear reasonable, but it can be seen that the lowest R_i is only 62.2 on colour TCS08, showing that by taking the mean of the R_i values there are occasions when this can lead to indicating a considerably better performance than the lowest R_i would indicate.

7.2.1.1 Addressing the Limitations of the CIE CRI

As new and improved metrics of colour spaces evolved from the time of the original specification of the CRI, its several limitations highlighted in the above paragraphs have been recognised; in addition, the requirement for an index based on the colour performance of a television system rather than the eye has also been addressed by Roberts et al. (2011). It would appear therefore that there is a requirement for two new indices, one for use when lighting is being appraised by the eye and the other for when the scene is being captured by a camera; ideally both with a common set of parameters and test colours where appropriate.

The CIE has addressed the need for improvements in the CRI on a number of occasions[6] and a number of improvements have been proposed; an early attempt extended the number of test colours to include more saturated colours as shown in Table 7.2.

Table 7.2   The extended range of CRI test colours

TCS09	4,5 R 4/13	Strong red
TCS10	5 Y 8/10	Strong yellow
TCS11	4,5 G 5/8	Strong green
TCS12	3 PB 3/11	Strong blue
TCS13	5 YR 8/4	Light yellowish pink
TCS14	5 GY 4/4	Moderate olive green

The additional six samples include four highly saturated samples which were introduced at a later stage to provide supplementary colour rendering information.

Figure 7.4 is based on the same parameters as those used for Figure 7.2 but with the additional test sample colours indicated in Table 7.2. In order to embrace the higher saturated colours the scale of this diagram has been extended and it is immediately evident that the shift in chromaticities of the saturated colours is very much larger than for the original pastel colours. The red sample in particular indicates a shift of about 24 JNDs. The resulting CRI

---

[6] http://en.wikipedia.org/wiki/Color_rendering_index

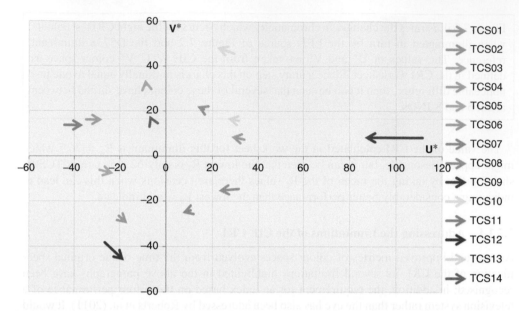

Figure 7.4   Plot of the change in chromaticities of 14 CRI test colours with a typical LED light source.

R_a value has been reduced from 76.7 to 67.5 and the worst case R_i has been reduced to –8.3, an example which clearly highlights the limitations of the original pastel test colours and of taking a mean value of the values of R_1 in order to represent the performance of marginally acceptable illuminants.

Another improvement in the 1990s addressed all the points raised earlier with regard to the limitations of the currently specified CRI. Unfortunately, although a new specification was produced, lack of agreement between the research and industry members of the appropriate CIE committee prevented a recommendation based on the specification being approved. This specification, known as the $R96_a$ method, will provide a result which is more soundly based on the rendition obtained but its lack of standardisation has prevented it being adopted within the lighting industry.

7.2.2   The MCC Index

The workbook contains a further worksheet, 7(b), which amends the structure of Worksheet 7(a) to incorporate current practices amongst some users of lighting equipment and provides a means of calculating this more realistic alternative index. It is emphasised however that the procedure outlined is an example only of several methods which are informally in use as an alternative to the CIE CRI and has no formal basis as a standard.

The MCC (Macbeth ColorChecker) index is based on the colour samples of colours from the ubiquitous ColorChecker colour rendition chart illustrated in Figure 7.5. It applies the CIE CAT02 colour adaptation transform to the test results and uses the CIELAB colour space for calculating the colour differences.

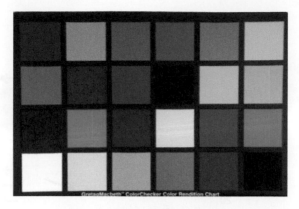

Figure 7.5 The ColorChecker[7] chart.

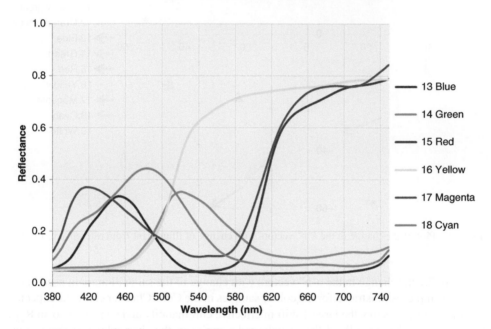

Figure 7.6 The spectral reflection characteristics of the ColorChecker primaries.

The spectral reflection characteristics of the additive and subtractive primary samples of the ColorChecker chart are illustrated in Figure 7.6.

---

[7] Also known by its previous names as the Macbeth ColorChecker Chart and subsequently as the Gretag-Macbeth ColorChecker chart when the companies merged and following further company mergers now known as the ColorChecker Classic, http://en.wikipedia.org/wiki/ColorChecker. It includes a range of colour and neutral chips to enable the performance of a colour reproduction system to be evaluated. The top two lines of patches represent common colours including human skin, green grass and blue sky; the third line the additive and subtractive primaries and the fourth line a range of neutral chips from black to white.

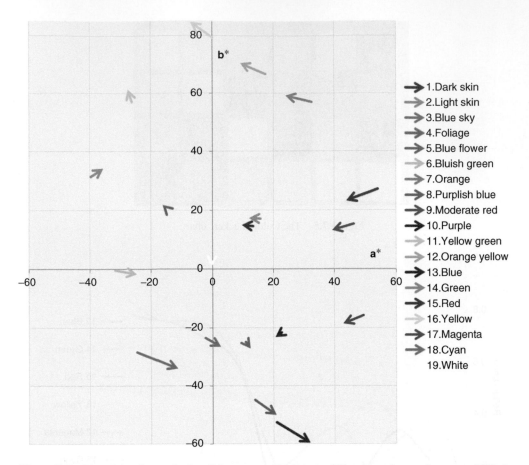

Figure 7.7   Plot of the change in the a*b* vectors with change of illuminant from reference to LED 4.

In Figure 7.7 the change in the a*b* vectors with change of the same illuminant used in the previous examples is illustrated for the colour samples in the ColorChecker chart. As expected the saturated colours show the largest shift in effective chromaticity and contribute to an R_{CC}, of 67.6 based on the mean R_i of the 18 coloured samples of the chart which, perhaps coincidentally, compares very well with the value obtained using the original CRI method on the extended range of CIE colour samples. However, the minimum R_i value is 32.4 compared with −8.3 for the CIE method, indicating a more realistic result. This is a fairly typical result for what appears initially to be a marginally acceptable illuminant based on the R_{CC} but indicates the range of R_i is likely to fall between about 30 and 98 for usable if somewhat compromised illuminants. Nevertheless, this approach is still capable of producing a negative R_i as substituting a low-pressure sodium source as the test illuminant in Worksheet 7(b) will show.

The results of using this approach to measuring the performance of illuminants may be broadly summarised as follows:

- A range of saturated samples from around the colour circle should be included
- The range of measurements of R_i is improved using current colour difference criteria

- Taking the mean value of R always leads to the real performance of the illuminant being artificially significantly enhanced.

Thus it is recommended that if disappointment in the colour rendition of the scene is to be avoided, then the full range of values of R_i should be considered in evaluating the performance of the test illuminant.

The impetus to develop more efficient lighting, in some cases without due regard to the quality of rendition, has led to renewed attempts by Roberts et al. for an agreed index for colour reproduction. Their work has led to proposals being formulated which have triggered the formation of a technical committee of the European Broadcasting Union to review the proposals and agree the parameters to be adopted to support them in a proposed EBU Recommendation for a 'Television/Film Lighting Consistency Index'. Often industry recommendations and standards of this type are submitted for adoption by the appropriate international standardisation committees. (Since completing this chapter the work of the EBU has now been completed. However, as it is based on the characteristics of a camera rather than that of the eye and the colour television reproduction system is not fully described until Part 4, the Television Lighting Consistency Index (TLCI) will be described in Chapter 18.)

7.3   Daylight

The primary source of daylight, particularly when the sky is clear, is the Sun.

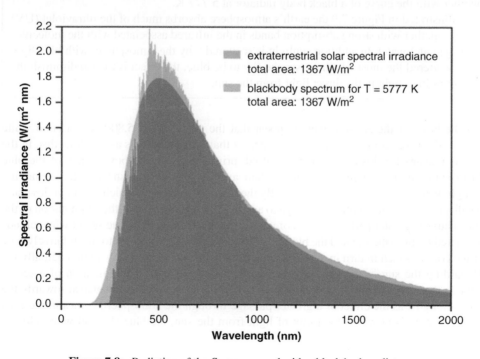

Figure 7.8   Radiation of the Sun compared with a black body radiator.

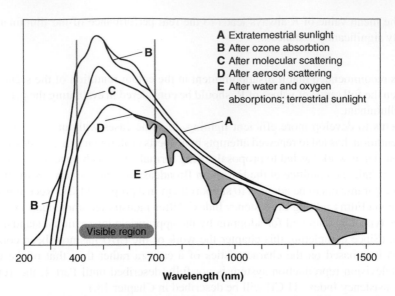

A Extratemestrial sunlight
B After ozone absorbtion
C After molecular scattering
D After aerosol scattering
E After water and oxygen
   absorptions; terrestrial sunlight

Figure 7.9   Successive processes affecting sunlight during its penetration of the atmosphere. (Henderson, 1970. Reproduced with permission of Elsevier.)

> The spectrum of the radiation from the surface of the sun is very close to that of a black body radiator at about 5,800 K. Figure 7.8 illustrates the measured radiation from the sun together with the curve of a black body radiator at 5,777 K.
>
> As illustrated in Figure 7.9 the earth's atmosphere absorbs much of the ultraviolet (UV) energy together with strong absorption bands in the infrared associated with the molecules of oxygen and water. In addition sunlight is scattered [8] by the atmosphere, with blue light being scattered the most which causes the sky to be blue, the effect is to also diminish the amount of blue light in the direct light from the sun.

On the basis of the above it would appear that the sun at about 5,800 K is a good white source of light, yet common experience indicates that it appears to us as a yellow disc in the sky. The reasons for this appear to be twofold: primarily as noted above, much of the blue light from the sun is scattered by the atmosphere and, as was indicated in Chapter 1, removing blue light from white light leaves principally the red and green light which add to yellow; and secondly, the adaptation ability of the eye to adjust to the average hue of the illumination of the scene. Thus the greater part of the field of view is provided either by a blue sky or white clouds illuminated by both the sun and the blue sky, thus the eye accommodates to this relatively blue average colour which in turn inclines to make the sun appear more yellow than it actually is.

Regarding the sun in the sky and its contribution to the overall illumination of a scene, except at the period around dawn and dusk, when there is a strong imbalance towards the red wavelengths as the effect of a greater passage through the atmosphere comes into effect, the illumination is provided by a mix of light from the sun, the blue sky and white clouds.

---

[8]Rayleigh scattering. Lord Rayleigh (1842–1919). http://en.wikipedia.org/wiki/Rayleigh_scattering.

The overall colour of the illumination will therefore change as the ratio of cloud to clear sky changes. Nevertheless, the SPD of daylight is reasonably balanced across the spectrum with an emphasis towards a bluer average in more northerly latitudes.

7.3.1   CIE Standard Daylight Illumination

As the ratio of the sources of illumination between the sun, the blue sky, the addition of various amounts of cloud and a completely overcast sky changes, the colour characteristic of the daylight in terms of its CCT changes substantially. It was considered important by the CIE to make available a means of specifying the SPD of daylight over a range of CCTs representing various phases of daylight in a standardised manner (Hunt & Pointer, 2011) and to this end three SPDs have been defined as $S_0(\lambda)$, $S_1(\lambda)$ and $S_2(\lambda)$. $S_0(\lambda)$ represent the mean obtained by a number of workers who have plotted the SPDs of daylight under a wide range of conditions and $S_1(\lambda)$ and $S_2(\lambda)$ provide variations to the mean plot depending upon the phase of daylight and corresponding CCT required.

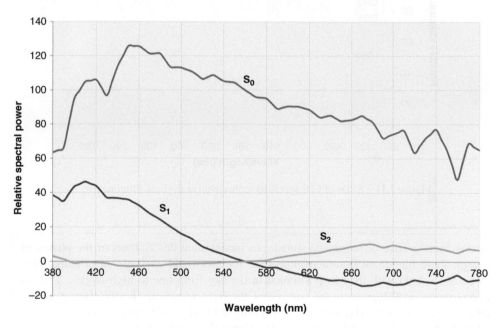

Figure 7.10   The CIE standard 'S' spectral plots.

Figure 7.10 illustrates the spectral distribution of the three standard CIE daylight 'S' curves which are summed in various proportions in order to obtain an SPD with a particular CCT.

Thus the SPD at a particular CCT is given by

$$S(\lambda) = S_0(\lambda) + M_1 S_1(\lambda) + M_2 S_2(\lambda)$$

where M_1 and M_2 are factors calculated from knowledge of the CCT required. The formulae for calculating these factors are detailed in the above reference and also in Worksheet 7(c).

From these standard SPDs a set of specified SPDs have been calculated in Worksheet 7(c) which represent commonly used phases of daylight. These are the CIE Illuminant D range of illuminants, D50, D55, D65 and D75, which are illustrated in Figure 7.11. The D numbers refer to their CCT in hundreds of degrees Kelvin. These are hypothetical illuminants in as much as there are no artificial sources available which will emulate these SPDs; however, being good representations of daylight they are useful illuminants for contributing towards the calculations required in various aspects of colour reproduction as will be seen.

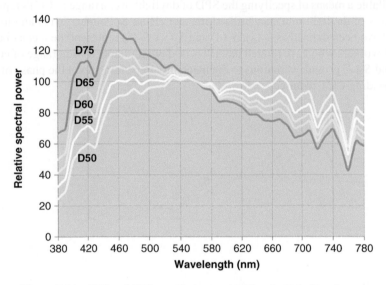

Figure 7.11   SPDs of CIE specified representative daylight illuminants.

Various sources have different and sometimes inconsistent descriptions of the phases of daylight these illuminant D sources represent. As one would envisage, in principal they represent daylight with the sun positioned in the sky from low to high angles. Thus in general terms, D50 represents the warmth of the light sometime shortly after or before dawn and dusk, respectively, when the sun is fully in the sky; D55 represents mid-morning and mid-afternoon light; D65 average daylight and D75 North sky light.

At the time these standards were adopted the now obsolescent u,v chromaticity diagram (as opposed to the u′,v′ diagram) was in use and remains the diagram on which the chromaticities of the standard illuminants are illustrated as shown in Figure 7.12, which is an enlarged section of the diagram showing the area encompassed by the Planckian locus.

This diagram shows the formally defined lines of constant CCTs which are defined to be always at 90 degrees to the Planckian locus. In addition to the standard D illuminants the CIE Standard Illuminant A (SA) at a CCT of 2,856 K and the obsolescent CIE Illuminant C (SC)

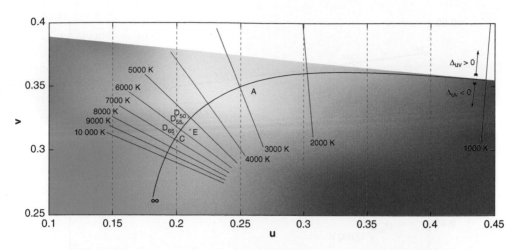

Figure 7.12 Chromaticities of CIE defined representative daylight illuminants.

at a CCT of 6,774 K are shown. Illuminant A was representative of the tungsten lamps almost universally in use at the time before the introduction of fluorescent lamps and Illuminant C was the forerunner of D65, representing northern latitude average daylight without the extended blue into UV content present in daylight but which is now considered necessary to accommodate the florescent effects of some surfaces. The theoretical CIE equal energy white (EEW) illuminant, SE, at about 5,400 K is also illustrated as 'E' on the diagram.

As noted above, the derivation of the D illuminants is mathematically based and explains why these illuminants form a locus, known as the CIE Daylight Locus, parallel to and offset from the Planckian locus as shown by the three plots in Figure 7.12.

As will be seen, three different D illuminants are adopted as the reference white in the three reproduction systems that will be described later.

For the sake of conformity with the material in the remainder of the book the chromaticity coordinates of a number of CIE Daylight Illuminants are also given in terms of x,y and u′,v′ in Table 7.3 and Figure 7.13.

Table 7.3 Chromaticities of CIE D Illuminants

	D50	D55	D60	D65	D75	D93
x	0.3457	0.3325	0.3217	0.3128	0.2991	0.2831
y	0.3586	0.3475	0.3377	0.3291	0.3149	0.2970
u′	0.2092	0.2045	0.2008	0.1978	0.1935	0.1888
v′	0.4881	0.4808	0.4742	0.4684	0.4586	0.4457

Figure 7.13 illustrates that EEW is closest to a temperature of 5,460 K on the Planckian locus and the nearest CIE Daylight illuminant to this point is D55.

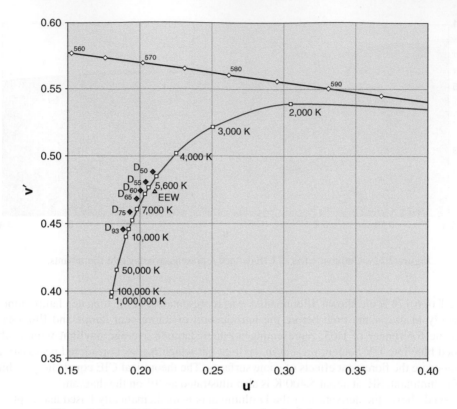

Figure 7.13 CIE Daylight illuminants on the u′, v′ chromaticity diagram.

The varying spectral distribution of daylight can cause problems in reproduction since the balance of the illuminating colour may vary with time or position of the shot; for example, on a sunny day the illumination in sunlight and in the shadows is completely different, the former being illuminated by the sun and the sky and the latter only by the relatively blue sky. The colour adaptation mechanism of the eye usually prevents us from noticing these changes but the reproduction system, depending upon its design characteristics, may not be able to produce a satisfactory result in all circumstances.

7.4   Incandescent-based Lamps

Studio lighting has traditionally been based on luminaires which use as their light source tungsten filament lamps of various powers. These have the advantage of being flexible in fitment arrangement and may be controlled in intensity by dimming the voltage that drives them. As we saw in Section 6.3 incandescent sources produce fundamentally black body or Planckian radiation with the characteristics therein described. However tungsten melts at about 3,700 degrees and no other material is available which has mechanical stability at the colour temperature of daylight, that is roughly in the range 5,000–7,000 K. Therefore professional luminaires are limited to a colour temperature of about 3,200 K. The CIE Illuminant A (SA) is defined as having the same relative spectral distribution as a black body or Planckian radiator at about 2,856 K. As reference to Figure 7.12 shows, at these colour temperatures tungsten lamp sources have a distinctly yellow cast. Nevertheless, they do satisfy the criteria of an even distribution of energy across the visible spectrum and the eye will accommodate to this colour so that generally it appears white in non-critical situations.

Although the lamps can be dimmed there is a strict limit to the reduction in voltage which can be applied since the colour temperature drops significantly at reduced levels making it difficult to obtain a good colour balance across a scene. Matching the lighting across a complex scene calls for significant experience and whoever is responsible for lighting will normally rely on the positioning of lamps to obtain the level of illumination required and only use the dimming function for minor trimming of levels. Both light and colour temperature meters are used to achieve satisfactory results.

One of the main problems with tungsten illumination is its very poor efficacy, that alone would be problem enough but much of the power which does not appear as light is in the form of infrared heat which causes problems in the studio. As Figures 6.3 and 6.5 illustrate, at 3,000 K the peak of the emission curve is in fact in the infrared region of the spectrum. The efficacy of incandescent lamps increases with increase in power varying from about 5 lm/W at 25 W to 20 lm/W at 250 W.

7.4.1   Tungsten Halogen Lamps

The limitations of tungsten lamps can be offset to a considerable degree by the introduction of a small amount of halogen, usually bromine, into the inert gas of the bulb which surrounds the filament. As the tungsten atoms from the filament eject into the gas they combine with the halogen, circulate and cool within the bulb before reappearing at the filament where its high temperature causes the molecules to dissemble and the tungsten atoms are deposited back onto the filament. The lifetime of the lamp may be increased by this approach from about 1,000 to 2,000 hours and it is operated at a higher temperature than a tungsten lamp to assist the recycling mechanism. Additional increases in lifetime and operating temperature may be brought about by the inclusion of multi-layered dichroic[9] filters on the surface of the bulb which reflect back some 60–70% of the infrared light onto the filament, whilst allowing visible light to radiate from the bulb. The combined effect of these various improvements enables the bulb to be operated up to 3,400 K with corresponding improvements in efficacy of up to 30 lm/W and an increase in lifetime of up to 5,000 hours.

7.4.2   Accommodating the Difference in Colour Temperature between Daylight and Tungsten-based Lamps

The wide difference in colour temperature between tungsten-based luminaires and daylight can cause problems in mixed lighting environments, most of which can be overcome with the use of suitably specified filters which can either amend the SPDs of daylight to provide an approximate match to tungsten or those of tungsten to daylight. Generally speaking, daylight

---

[9] Dichroic layers are layers of material deposited on glass filters and mirrors which are a quarter of a wavelength in thickness at the frequency at which it is required to filter or reflect the light. They are highly efficient, either reflecting or transmitting virtually all the incident energy.

is of a much higher intensity than tungsten illumination and since cameras were initially of limited sensitivity it is usual where practical to provide the former type of filter.

Historically it was sometimes a requirement for a television studio, in for example a live news environment where the immediacy of a downtown background scene is desirable, to have illuminated the studio with tungsten lighting yet show a window on the external daylight environment. Such a situation whilst not looking too offensive to the eye would ensure that the colour camera will show the outside scene apparently lit by a very blue source. Apart from being a dramatic example of how wonderful the eye is in accommodating the colour differences up to this point, the solution is to cover the window with appropriate correcting filter material to bring the daylight into balance with the studio lighting.

7.5   Electrical Discharge-based Lamps

In Chapter 6, a description was given of the spectra available from the low- and high-pressure electrical discharges of mercury, sodium and xenon. The same chapter also described the physics of fluorescence. These two physical processes are combined in the discharge lamps which are manufactured for the illumination of scenes to be captured for the reproduction of colour.

Our interest is generally limited to those sources of illumination which are pertinent to both illuminating a scene intended for capture and reproduction and for the viewing environment of the rendered image. Nevertheless it may be useful to highlight the limitations, in terms of image capture and reproduction, of other forms of illumination which are common in our surroundings.

Lighting for public areas, where efficacy of operation is more important than colour rendering of the scene, uses both high-pressure mercury and sodium lamps. High-pressure sodium lamps, which usually also contain mercury are often based on the spectrum of the discharge only and are used primarily for street lighting and produce an orange illumination. In high-pressure mercury lamps a number of variations in the technology are available from straightforward high-pressure mercury vapour discharge lamps which produce dominantly cyan illumination, to lamps with phosphor coatings on the inner glass walls, super high-pressure lamps and hybrid lamps which combine an incandescent filament. These variations are designed to improve the colour rendering of the emitted light, which they do but even the best are unlikely to have a CRI above 50.

Generally speaking, scenes illuminated by lamps of this type should be avoided in capturing images for colour reproduction.

The aim of the lamps described in the remainder of this chapter is to emulate the spectral distribution of daylight as far as possible and to achieve a level of efficacy in the lamps that exceeds the relatively poor efficacy of tungsten lighting. These two requirements are usually not compatible; as will be seen, a large jump in efficacy is easily achieved but at a cost of poor colour rendering and as the rendering is improved the efficacy initially achieved falls off significantly.

7.5.1   Xenon Discharge Lamps

As we saw in Section 6.6 the spectrum of a xenon discharge under medium to high pressure produces a rich spectra across the visible band and as the pressure is increased large numbers of new lines in the visible spectrum are introduced making the xenon discharge eminently suitable for emulating daylight or even an equal energy source across the spectrum.

One of the most common uses of xenon (prior to the developments which produced xenon car headlamps) is in the flash bulbs used in photography where the correlated colour temperature and spectral distribution is acceptably close to daylight at about 6,000 K.

At higher pressures, in the order of 30–100 atmospheres, the spectrum is virtually continuous across the visible spectrum and because at these pressures the arc between the electrodes is relatively short, the xenon bulb can become a small, very intense source suitable for constructing lamps with well-focussed beams. This makes them very suitable for use in spotlights for highlighting a small area within a scene.

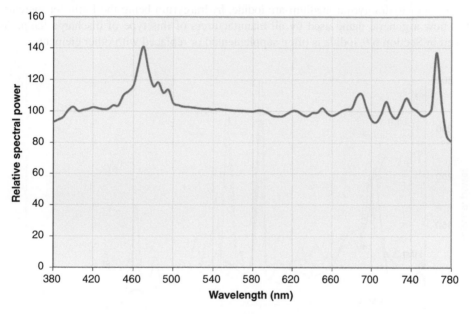

Figure 7.14   Typical spectrum of a xenon lamp.

The spectrum of a xenon lamp is illustrated in Figure 7.14 which indicates a good spread of energy across the spectrum. The peaks in the blue and red areas of the spectrum provide the xenon light with a faint violet appearance. The precise spectrum will depend upon the pressure of the gas in the bulb.

Xenon lamps may be operated at relatively high power and brightness, with power up to 15 kW being available, providing efficacies of 35–50 lm/W; whilst this is a relatively poor efficacy compared to fluorescent lamps, it is compensated for by the small size of the light source (an arc of only a few millimetres) and the high power available.

However, xenon lamps are relatively expensive and complicated to operate, requiring complex support ballasts to ensure a constant current supply under variable load conditions and because of the high pressure under which they operate, they require considerable safety precautions, particularly for the higher power variants.

The CCT of xenon lamps varies in the range 5,600–6,300 K which makes them an excellent match to daylight and their corresponding CRI R_a is in the range of 90–95. These figures for colour performance together with the high output powers available make them a first choice as projector lamps for large cinema screens.

7.5.2 High-pressure Vapour Discharge Lamps

The high-pressure vapour discharge lamps used for illuminating scenes for colour reproduction are usually mercury metal halide lamps which were developed specifically for the media industry, initially by Osram in Germany who trademarked them as HMI lamps. HMI is an abbreviation of hydrargyrum medium-arc iodide, hydrargyrum being the Latin for mercury. HMI is now a generic name used by all manufacturers of this type of discharge lamp. As described in Section 6.6, iodide is often supplemented or replaced with other elements.

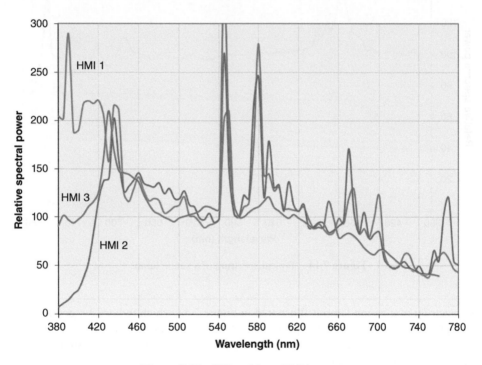

Figure 7.15   SPDs of three HMI lamps.

Figure 7.15 illustrates the SPDs of three typical HMI lamps, HMI 1,2 and 3. The precise shape of the SPD will depend both upon the pressure of the gas and which metals are used in combination with the halogen to produce the halide.

In all three of these lamps the lines produced by the mercury as described in Section 6.6 remain dominant but the spectrum between them is filled by the emissions from the metal in the plasma stream derived from the halide.

The CCTs of these lamps are usually arranged to be close to daylight between 5,000 K and 6,800 K; however, due to the reduction in the length of the electrodes as the bulb ages and therefore the corresponding increase in the length of the arc, the voltage required to maintain the arc increases with a resultant decrease in colour temperature. New bulbs usually require a period of several hours to be 'burnt in' since the initial CCT can be as high as 15,000 K. The lamp represented by HMI 3 in Figure 7.15 is somewhat atypical in that the peaks above 550 nm are not present which leads it to having a relatively superior performance with a CCT of 6,740 K and a CRI R_a of 87.2.

Bulbs are available over a large power range from 125 W up to 24 kW with efficacies in the range of 85–108 lm/W, which makes them suitable for complementing daylight and for shooting those events at night which require the high level of illumination and efficacy available from these lamps. Development continues to increase further the power and CRI of HMI lamps.

The CRI R_a of the HMI 2 bulb illustrated in Figure 7.15 is quoted at 70 (found to be 77.4 using the Worksheet 7(a) calculator) which makes it an acceptable compromise rather than an ideal solution for colour reproduction. Some manufacturers quote CRI figures up to 95 for their HMI lamps but rarely if ever publish the lamp SPD to support these claims.

7.5.3   Low-pressure Vapour Discharge Lamps – Fluorescent Lamps

The energy source of fluorescent lamps is based on a mercury low-pressure discharge. The phosphor deposited on the inside of the glass envelope is activated by the UV light at mercury's resonant frequency of 253.7 nm which in turn emits light in the visible spectrum at wavelengths and levels dependent upon the characteristics of the phosphor used. The secondary emission lines of the mercury discharge appear at a comparative level to the phosphor-generated emissions and contribute significantly to the composite visible spectrum emitted.

These lamps have a multitude of uses, from warm lights designed to emulate tungsten lamps, to highly efficient cold lights at CCTs of 9,000 K and those that emulate daylight to a lesser or greater degree – again depending upon the efficacy required. In general terms the higher the CRI required, the less efficient is the illumination. Kitsinelis (2010) lists nearly 50 different compounds used for phosphors with particular emission characteristics.

The CIE have defined a range of illuminants based on fluorescent lamps with various phosphor compounds; FL1-FL6 are so called 'standard' fluorescent lamps using calcium halophosphate phosphors with antimony and manganese activations; FL10–FL12 are lamps with narrow triband phosphors in the red, green and blue areas of the spectrum, respectively; and the lamps which are more relevant to the requirements of colour reproduction are in the FL7–FL9 range, which use multiple phosphors to obtain broadband spectra and higher CRIs.

Unfortunately, it appears that the various manufacturers of these lamps do not use the CIE 'FL' numbers as a means of providing a guide as to the SPDs of their lamps in their catalogues. Also there has been a trend over recent years where less and less technical information, particularly with regard to SPD of lamps, is generally available. Wikipedia[10] provides a list of the CIE FL range lamps and the corresponding common names used by some manufacturers, though there is no guarantee that different manufacturers will use the same name for lamps which meet a particular FL number specification.

---

[10]http://en.wikipedia.org/wiki/Standard_illuminant

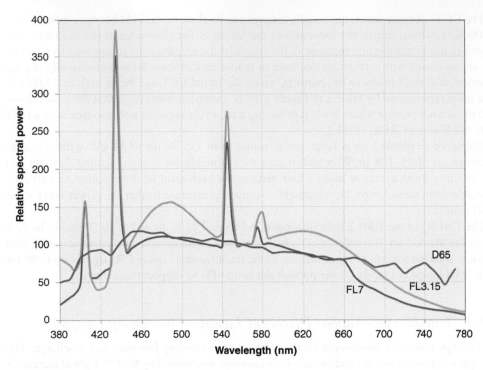

Figure 7.16   SPDs of fluorescent lamps and D65.

Wikipedia describes the FL7–FL9 range with the names 'D65 simulator', 'D50 simulator' and 'cool white deluxe', respectively. The CIE have defined a further set of 15 fluorescent lamps FL3.1–FL3.15 (Hunt & Pointer, 2011) of which FL3.15 is also described as a D65 simulator. The 'Illuminants' worksheet provides figures for the energy spectra of the full range of the FL and FL3 lamps.

In Figure 7.16, all SPDs have been normalised at 560 nm. The SPDs of the two 'D65 simulator' lamps defined above are plotted with D65 as a comparison. They are characterised by the same mercury discharge peaks at 435.8 nm and 546.1 nm but by a different distribution over the visual spectrum range. As measured using the CIE CRI spread sheet, the FL7 lamp has a CCT of 6,496 K and a CRI R_a of 90.2, whilst the FL3.15 has a CCT of 6,508 K and a R_a of 98.5. These are comparatively special lamps with much higher CRIs than are generally available from lamps of this type.

As indicated earlier, the efficacy of fluorescent lamps tends to vary inversely with the CRI and to achieve a CRI which is satisfactory for reproduction, high efficacies which can be achieved by this form of lighting drop to be within the range 60–80 lm/W. A very good match to daylight colours in the 5,500–6,500 K range can be achieved with chromaticities very close to the Planckian locus. A CRI R_a of 80–85 can be achieved with relatively good efficiency which makes them suitable for non-critical lighting of scenes and, as was indicated above, if required lamps with R_a values in the 90–98.5 range are available.

Efficiency considerations have driven a trend to the use of lamps of this type in new or refurbished studio situations as a replacement for tungsten halogen lighting, though often the latter is retained for critical productions where a CRI R_a of 100 and the flexibility of tungsten lighting is required.

For the viewing environment fluorescent lighting is ideal in providing both display surround lighting and subdued ambient lighting well matched to the standard white of the media to which the eye can adapt.

7.5.3.1 Cold Cathode Lamps

Cold cathode lamps are essentially a form of fluorescent lamp where the cathode is not separately heated to cause the emission of electrons but which instead uses an initial high voltage to induce secondary emission. The characteristics of the light emitted from a cold cathode lamp are essentially as described above for fluorescent lamps which make them ideal as a back light for liquid crystal displays (see Section 8.3). The SPD required for this purpose is less demanding than for general illumination, the criteria being to ensure there are similar amounts of light output at wavelengths corresponding to the selected primaries of the reproduction system (see Chapter 8).

7.6 LED Lamps

From the turn of the present century, the efficacy and SPDs of LED-based luminaires have continued to improve and have now (2012) reached the point where they can, under certain circumstances, be used for illuminating the scene and perhaps more appropriately, illuminating the viewing environment.

As we saw in Section 6.6, electroluminescent semi-conductor junctions form the basis of LEDs, which in simple configurations are capable of producing light emission of only relatively narrow spectral bandwidth, which in turn makes them useless for illuminating scenes or environments for colour reproduction.

Light with a broad spectrum is required for scene illumination and the early attempts at producing a satisfactory illuminant used a combination of red, green and blue LEDs to produce white light.

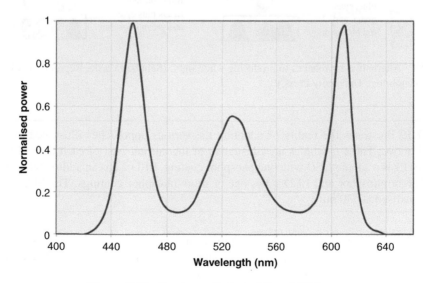

Figure 7.17   Spectrum of a typical three LED lamp.

However, the SPDs of such sources, as shown in Figure 7.17, still leave much of the spectrum with very little energy and thus for certain saturated colours the scene will appear quite different to that illuminated with a broad band source such as daylight or tungsten. Nevertheless, this combination has a CCT of 6,500 K, a claimed CRI of 84 and an efficiency of 32 lm/W.

Tetrachromatic and pentachromatic sources using four and five LEDs with SPDs across the spectrum can improve the CRI further at a cost.

An alternative approach is to use a combination of electroluminescence and a form of secondary emission wavelength conversion; the most common form of which is fluorescence, whereby the shorter wavelength, higher-energy photons from the junction strike a phosphor compound surrounding the junction causing the emission of longer wavelength energy to complement the original. In the simplest situation the junction wavelength is in the blue range and the phosphor energy is in the complementary yellow band, giving the effect of white light. However, the approach can be extended and a number of different solutions are available as illustrated in Figure 7.18.

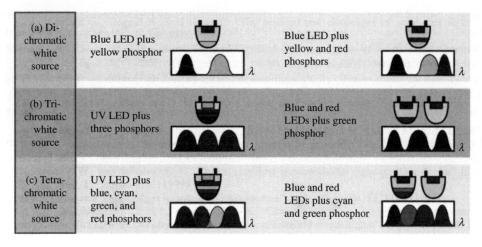

Figure 7.18   Alternative approaches to producing white light. (Schubert, 2006. Reproduced with permission of Cambridge University Press.)

Figure 7.19 illustrates the reality of adopting the various approaches illustrated in Figure 7.18 above. These examples are illustrative of the current best of each type of LED lamp. LED 1 is a simple LED with one phosphor coating, LED 2 has an additional coating of a different phosphor and LED 3 has one or more phosphor coatings. The SPDs have been normalised at 530 nm.

All the LEDs in Figure 7.19 provide light with a white appearance but as the dips in energy are filled by fluorescence from additional phosphors, so the CRI is improved. The CCT and CRI of these three LEDs as calculated in Worksheet 7(a) are shown in Table 7.4.

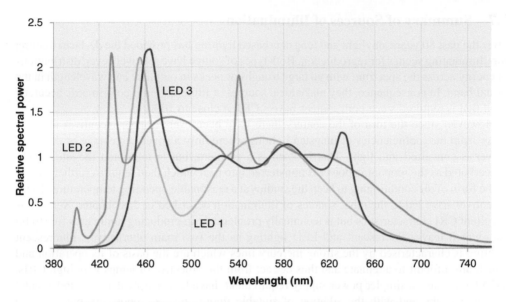

Figure 7.19   SPDs of a representative range of LED lamp technologies.

Other solutions using UV as the primary light source are available together with two or three phosphor types which give a good CRI. However, by using UV as the primary energy source the Stokes losses will be considerably greater which will lead to a loss in efficacy.

A further approach to secondary emission uses an additional semiconductor junction in the path of the original radiation and therefore goes under the name of photon-recycling semiconductor LEDs or PRS-LEDs. This approach is potentially capable of producing highly efficient LEDs as developments progress.

The promise of higher-efficiency LEDs has caused a huge investment in their development, with reports that in the laboratory efficacies in excess of 150 lm/W are being achieved. In the practical world where room temperature plus internal heating and the requirement to achieve high CRIs are the order of the day, typical efficacies are 60–90 lm/W with CRIs of 75–95.

Though the power outputs of LED lamps are somewhat limited, the availability of ever more sensitive cameras enables them to be used for limited studio lighting where large throws between the source and the subject are not required. Like fluorescent lamps they are ideal for environmental lighting in viewing the reproduced printed image and to provide the source of illumination for liquid crystal display devices (see Section 8.3).

Ultimately it would appear that LED sources have the potential to win out in the race to replace tungsten halogen lighting in the studio.

Table 7.4   Performance parameters of the LED lamps illustrated in Figure 7.19

	CCT	R_a	Worst (R_i)
LED 1	5,951 K	72.0	−42.3
LED 2	6,537 K	96.4	78.22
LED 3	5,594 K	86.6	65.4

7.7   Summary of Sources of Illumination

Over the past 80 years, daylight and tungsten-based lighting has provided the de-facto lighting for illuminating scenes for reproduction. Both types of source have relatively even distributions of energy across the spectrum with no large troughs or peaks in output at any wavelength in the visual band. In consequence, they make ideal sources of illumination if colorimetric accuracy is all that is required since both sources have a CRI of close to 100.

However, since the turn of the century there have been ever stronger incentives to move away from the inefficiencies of tungsten lighting where only a small percentage of the input power is converted into light. That in itself is a problem, but in television the situation is exacerbated as the remaining power is transferred into heat which then requires further energy in the form of air conditioning to keep the studios at a reasonable operating temperature. These incentives have led to the other sources of illumination described in this chapter. Xenon has excellent CRI characteristics but is technically problematic in producing luminaires for studio use, which leaves fluorescent and LED lighting as the two main alternatives. Fluorescent lighting is characterised by the strong mercury lines which are the basis of its operation and which are difficult to attenuate and thus detract from the objective of lamps with high CRIs. LED lamps have a simpler power supply requirement, have less sharply defined fundamental emission peaks and with the addition of suitable fluorescing phosphors are potentially a cheaper, more efficacious and better matched solution to higher CRI sources than discharge-based lamps. A comparison of the principal sources used for scene lighting is shown in Table 7.5.

In the enthusiastic move to ever greater efficiency, it is of some concern that colour rendition may be being compromised. In television, in general terms one is not able to compare the reproduction to the original and as long as flesh colour is being reproduced satisfactorily, the use of lighting with CRIs of less than say 90 may be acceptable; though such an approach is likely to be occasionally problematic in the production of commercials, where a particular pack colour is well known to many of the audience. In photography, where often one is able to compare the final image with the scene, versions of both fluorescent and LED lighting with poor CRIs are likely to lead to problems.

In Section 7.2, several limitations of the current CRI in accurately defining the rendition performance of illuminants was discussed and reference was made to ongoing work to establish

Table 7.5   Indicative comparison of illuminants designed for reproduction

Source	Max Output (lm)	Efficacy (lm/W)	CCT (K)	CRI (R_a)	Worst (R_i)
Daylight	Not applicable	Not applicable	5,000–7,000	100	100
Tungsten	5,000	20	~2,860	100	100
Tungsten Halogen	45,000	30	3,200–3,400	100	100
Xenon	700,000	35–50	5,600–6,300	90–95	86–92
HMI	2,400,000	85–108	5,600–6,000	80–90	75–82
Fluorescent	3,650	60–70	5,500–6,500	85–97	89–96
LED*	320 1,000	160 100	5,000–7,000	72–97	50–92

*Efficacy reduces with increasing power.

a new index[11] which more accurately reflects the rendition performance of illuminants for reproduction. It seems likely that when the new index is introduced it will show many of the current illuminants in less favourable light than their current values of R_a would indicate which in turn implies one should take a conservative view of currently quoted CRIs.

A safer means of establishing the rendering performance of an illuminant is to rely on the individual values of R_i which form the basis of CRI R_a. If only one figure is to accurately represent the performance of an illuminant then the lowest value of R_i will usually give a better indication than the R_a. However, it is found that in general, the lowest value of R_i of the CIE CRI is overly critical and that values of R_i based on the R_{MCC} index will give a result which more accurately describes the rendering performance of the illuminant.

---

[11] Since writing this chapter the EBU have issued a recommendation for a new Television Lighting Consistency Index-2012, Recommendation R137 which is described in Chapter 18.

8

The Essential Elements of Colour Reproduction

8.1 The Basic Reproduction System

In Chapter 3, we have seen how a set of three primaries may be used to match any colour sample in the spectrum and thus any mixtures of spectrum colours, with the proviso that negative quantities are allowed; that is, for specific segments of the spectrum it may be necessary to add a percentage of one of the matching primaries to the spectrum colour for a match to be achieved. Initially the amount of the primaries required to match the colour must be measured using their colour matching functions (CMFs) and subsequently these measured values are used to control the level of the primaries to produce a match to the sample colour.

In a colour reproduction system the same approach that is used for colour measurement and colour matching is adopted for reproduction. The camera measures the level of the primaries required to match the colour of an element of the scene and these measurements are used to control the level of the primaries of the corresponding element in the display device.

Thus in a simplified arrangement as used in the early colour television cameras, the optical system of the camera splits the light from the image of the scene produced by the lens into its red, green and blue components in a manner which is described in detail in Section 8.2. These three colour images of the scene are focussed onto the light-sensitive surface of image sensors, formally described as opto-electric image sensors[1], which are comprised of a large matrix of rows of picture elements or pixels, each of which produce red (R), green (G) and blue (B) electrical voltages, the levels of which correspond to the intensity of those colours in the scene. The voltage levels are then read off from each pixel in the image in sequence, an operation often referred to as scanning the image (Poynton, 2012). The sequence of voltages from each of the image sensors is referred to as either a *component,* in a static image situation such as photography, or a *signal,* in a dynamic situation such as television or cinematography; expressions which are commonly used when describing the workflow of reproduction. To avoid duplication, we will use the term *signal* to cover both situations in Parts 3 and 4 of the book but revert to the term *component* in Part 5B which deals with photography. These RGB

---

[1] Sometimes the form 'opto-electronic image sensors' is used.

Colour Reproduction in Electronic Imaging Systems: Photography, Television, Cinematography, First Edition. Michael S Tooms.
© 2016 John Wiley & Sons, Ltd. Published 2016 by John Wiley & Sons, Ltd.
Companion Website: www.wiley.com/go/toomscolour

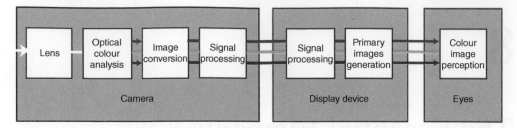

Figure 8.1   The signal path of the basic reproduction system.

signals control the intensity of the primaries of the corresponding pixels of the display device, which comprises the means of converting the signals back into red, green and blue images. The means of displaying these three images in a manner which enables the eyes to combine them in order to perceive a colour image of the scene depends on the form of the display device as described later in this chapter.

Both the viewing environment, in terms of the characteristics of the display and the ambient lighting, whether it be for viewing a print, a television screen or a cinema screen, and the parameters associated with the rendered image, in terms of the chromaticities of the colour primaries, tonal relationships, and pixel numbers, defines the standards for a particular reproduction system, as we shall see in more detail in Part 4. Thus the specification of the *camera parameters* reflects what is required to match the environment for the *display and viewing of the image.*

The optical path and signal flow of the system is illustrated in outline form in Figure 8.1. In this basic arrangement it is assumed that the characteristics of each stage of the workflow are linear. When this is not so, then additional signal processing is required as described in Chapter 12.

8.1.1   The Technological Approach to Colour Reproduction

As indicated in the Introduction, this book is not intended to convey a comprehensive description of the technology of the equipment used in media reproduction systems but to concentrate on the colour reproduction aspects. However, the following brief descriptions of cameras and display devices will hopefully provide a sufficient understanding of the functioning of a colour reproduction system on which to base the concepts described in the following chapters. At this fundamental level the signal processing blocks illustrated in both the camera and the display device may be assumed to be linear interfaces between the two devices.

8.2   The Camera

8.2.1   Camera Optical Colour Analysis

The optical systems of electronic cameras (Sproson, 1983) are designed to split the light from the scene into its red, green and blue components and, as will be described in more detail in Chapter 9, the characteristics of the splitting mechanism should be such that the spectral response associated with each of these components should match the characteristics of the CMFs associated with the primaries of the display device.

8.2.1.1 The Three Sensor Camera

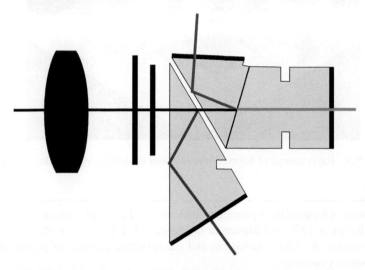

Figure 8.2 Illustrating the technique for splitting the light from the scene into its red, green and blue components.

Generally in professional cameras used for television and in some top-of-the-range cameras for other media, this is usually achieved by a system of prisms whose exit surfaces are coated with dichroic layers that form colour selective mirrors which route the appropriate bands of light for either reflection or transmission as shown in Figure 8.2.

The dichroic mirrors (Sproson, 1983) are very efficient, either reflecting or transmitting virtually all the light and absorbing very little.

The light exiting the prism system forms an image on each of three separate image sensors not shown in the diagram.

Optical trimming filters may be used to ensure that the convolution of the spectral characteristics of the scene lighting; the camera optics; the dichroic mirrors; the trimming filters and the image sensors, which together form the camera spectral sensitivities, combine to produce spectral responses which closely match the positive responses of the CMFs of the display primaries.

8.2.1.2 The Single Sensor Camera

Generally in consumer cameras the image sensors are integrated into one device with the light-sensitive surface split into a matrix of cells, each cell incorporating either a red, green or blue filter.

Figure 8.3   Bayer mosaic of filters. (From http://en.wikipedia.org/wiki/Bayer_filter.)

The most common form of filter pattern, or mosaic, used in single sensor cameras, is that patented by Bayer in 1976 and illustrated in Figure 8.3. Figure 8.4 is the corresponding diagram illustrating the filter mechanism and the resulting patterns of pixels relating to each of the primary colours.

It will be noted that for each 'composite' pixel, there are twice as many green pixels as there are red and blue pixels.

Interpolation is used to derive values for the red, green and blue signals at each pixel site. As indicated in Section 8.4, the number of composite pixels required would normally match the number of pixels in the display device. In photography, the minimum number of pixels required relates to the maximum size required of the resulting prints. In television and cinematography the number of pixels needed when viewing the resulting image in order to avoid a loss in perceived resolution is defined in Section 8.4.

It will also be recalled from an inspection of the CMFs in Chapter 3 that the green CMF is a reasonably close match to the V_λ curve which characterises the luminance response of the eye. In consequence, by doubling the number of green pixels, the virtual luminance response of the camera has twice the resolution and sensitivity of the red and blue responses. An important consideration as will be explained in Section 14.5.

In fact the 'green' spectral response may be made to be very close to the luminance response and the red and blue spectral responses may be made close to a *non-luminance response*. (For example the Y and the X and Z responses, respectively of the CIE system of colour measurement.) Different manufacturers use different algorithms in deriving appropriate values for the RGB signals from these raw Bayer signals. The implications of these approaches will be explored in Part 4.

It may be remembered that when describing the various characteristics of the eye in Section 1.4, reference was made to the spatial resolution of the eye. The relevance of this parameter will be discussed in some detail in Chapter 13. It is sufficient to indicate here that the spatial resolution characteristic of the eye to luminance data is significantly greater than for chromaticity data.

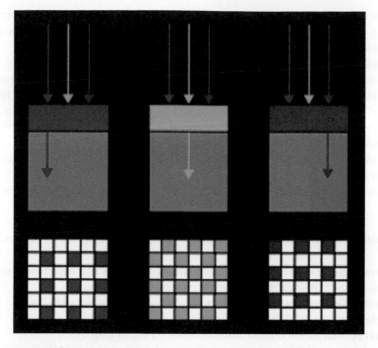

Figure 8.4 Illustrating the relationship between the filters and the pixels of a Bayer mosaic. (From http://en.wikipedia.org/wiki/Bayer_filter.)

8.2.1.3 Cameras and Lenses

In practical terms, in designing a camera it is not always as straightforward as indicated above to determine whether to use a beam splitting block or a single sensor with a matrix of red, green and blue pixels, since other criteria influence the dimension available between the lens and image sensors and therefore whether there is space available to locate the prism assembly. In historical cinematography, where film was used to capture the image, the lens to sensor distance is not long enough to incorporate a beam splitting system. Since there is a huge financial and emotional investment in the use of film camera lenses, then electronic cameras designed for shooting major productions must be capable of incorporating these lenses if they are to be commercially viable, which in turn has led to intensive development of high-resolution single image sensors for cameras used in a cinematographic environment.

8.3 Display Devices

8.3.1 Light Generation and Modulation in Display Devices

In ensuring a colour reproduction system is capable of rendering images with a large colour gamut, the critical elements of the display device are the generators of the primaries, both in terms of their chromaticity and the degree of control of the light level on a pixel by pixel basis. As we saw in Chapters 3 and 4 the nearer the primary colours are to the spectrum locus of the chromaticity diagram, the better is the system's ability to reproduce saturated colours and, the greater the degree of spectral separation of the primaries, the wider is the gamut of reproducible colours. In the previous chapter, we saw that for good reproduction, the spectral power distributions (SPDs) of scene illuminants should broadly span the visual spectrum, but for generating the primaries the opposite is true – the SPDs of illuminants should be as narrow as it is practical to achieve.

In Chapter 6, the characteristics of light generators were reviewed and broadly speaking, the mechanisms of generation fell into two categories: incandescence and luminescence. Illuminants falling into the former category are broad band generators whilst those in the latter are fundamentally narrow band generators.

Thus the candidates for primary light generators of display devices will be luminescent generators and specifically those that generate narrow band spectra. However, where intense sources of primaries are required to generate large displays then powerful broad band sources in association with red, green and blue optical filters are also used.

The generators which meet the narrow band criteria and the technologies which utilise them are:

- **Cathodoluminescence/fluorescence** – Three cathode ray tubes (CRTs), the shadow mask CRT
- **Cathodoluminescence/fluorescence/optical filters** – One CRT in association with a spinning disc comprised of sequential sectors of red, green and blue filters
- **Gas discharge fluorescent/optical filters** – Liquid crystal displays (LCDs) for flat panel and projection and digital light processing (DLP) for projection
- **Gas discharge fluorescence/photoluminescence** – Plasma displays
- **Electroluminescence/LEDs** – LCDs for flat panels and projection, DLP projectors and organic LEDs (OLEDs) for flat panel displays
- **Electroluminescence/Lasers** – DLP and other projection displays.

From the point of view of colour reproduction, interest is limited only to the means of generating the light sources; the various technologies adopted for display implementation for each type of primary source do not to a first degree influence the colour quality of reproduction. However, it must be acknowledged that the choice of the technology of implementation can influence the contrast range of the displayed image and therefore the quality of the rendition of the wider range of colours which comprise the colour space. Nevertheless, since without a broad understanding of the technological approaches the reader may be left with an incomplete understanding of colour reproduction, a very brief description of each of the technologies is included in the following paragraphs. Wikipedia on the web may be accessed by those wishing to study the appropriate technologies further.

There are two basic approaches to rendering images on colour displays. One approach is based upon individual red, green and blue images derived from the camera signals on each of three display devices and uses an optical system, similar to the camera analysis system in reverse, to combine them into a single image. Historically, for video displays in the 1950/60s, three CRTs displaying the red, green and blue images, respectively were used via a projection system to overlay the images onto a screen; modern cinema projectors are based upon the same basic approach, though using a combination of optically filtered light from a discharge lamp or lasers and a modulating liquid crystal arrangement to produce the images for projection.

The second approach uses the integrating ability of the eye–brain complex, in either temporal or spatial terms, to combine the red, green and blue images. The temporal approach, which is usually used as a cost-effective compromise, is to use a single monochrome display; a rotating filter wheel containing red, green and blue optical filters; and a method of switching the RGB signals to the display in sequential synchronism with the filter wheel to produce a sequence of red, green and blue images. If the rate of display of the images is sufficiently rapid, the eye–brain system will integrate the images into a full colour display. However, the system is prone to problems of movement both within the scene and by the viewer such that sometimes individual red, green and blue strobe effects may be seen. The chromaticities of the primaries are dictated by the characteristics of the red, green and blue optical filters and the SPD of the source light; the narrower the filter the nearer the spectrum locus will the chromaticity be located but the less bright will be the display.

In exploiting the spatial integration ability of the eye, the three images are produced in a single display device in which each display pixel is in turn comprised of three independent red, green and blue pixels. The image formed on the retina is such that the individual colour pixels are too small for the eye to resolve and a composite colour equal to the addition of the levels of the light of the three primary pixels is perceived by the eye.

The fundamental means of generating light using the above methods was described in Chapter 6 and a brief description of how the technology is used in display devices follows. Most of the above technologies are available in both simultaneous and sequential format displays; however, only the former approach is described in the following paragraphs.

8.3.1.1   Cathodoluminescence/fluorescence – CRTs

A CRT comprises an electron gun with a grid to which the signals are applied to control the intensity of the beam which is then deflected by scanning waveforms to produce a raster on the faceplate of the CRT. The faceplate is coated with phosphor and where the beam strikes the phosphor, electroluminescence occurs as described in Section 6.7. Depending upon the doping of the phosphor, light of the required primary colour is emitted.

It was evident that the bulky projection systems which resulted from this approach were never likely to be acceptable in the majority of homes and in the early days of colour television, RCA developed the shadow mask CRT which contained three electron guns in the neck of the tube and a mosaic of phosphor dots on the faceplate arranged in triangular groups of three, called triads, to represent each camera pixel. The beams from the guns were focussed to converge on to the triads from three different 120 degree directions and a *shadow mask* fitted close to the screen with one hole per triad ensured that only electrons from the appropriate gun passed through and onto the red, green or blue phosphor. In the passing decades between the developments of the shadow mask tube in the 1950s and the 1990s, several variations of this basic three-gun CRT were developed by different manufacturers and these displays were the mainstay of television and computer screens over this extended period. The chromaticity of the primaries depended upon the doping of the phosphors as shown by the SPDs illustrated in Figure 6.16.

8.3.1.2   Gas discharge fluorescent/optical filters – Cold Cathode
Backlight/LCD Panels

Though an improvement on the three CRT projection systems the shadow mask CRT itself is a comparatively bulky and heavy device and the requirement to develop flat panel displays became the aim of the industry. However, it was not until the 1980s that the first flat panel colour LCD was produced, initially for portable or laptop computers and later for television displays. The display has a faceplate comprised of a number of voltage-controlled variable density red, green and blue filters, back illuminated by a white source of light in the form of a cold cathode fluorescent lamp (see Section 8.2).

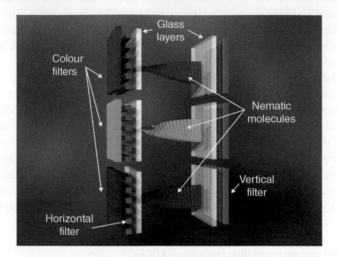

Figure 8.5   LCD faceplate operation.

Figure 8.5, which is derived from the web, illustrates how a voltage-controlled nematic polarising filter, sandwiched between two polarising filters, controls the level of light passing through the combination. The degree of rotation of the polarised light from the vertical filter is controlled by the voltage applied to the nematic crystal layer and the horizontally polarised exit filter will attenuate the light passing through in proportion to the angle of twist of the polarisation.

Thus applying the RGB signals to the appropriate cells in the matrix will produce a colour image on the front surface of the screen.

These LCDs started to displace the CRT display for television viewing at about the turn of the century and since the early 2000s have become the popular choice for computer and television screens. Early versions of the screen suffered from brightness and colour variations with viewing angle, though these effects have been reduced latterly. In addition, the voltage control of light level by the combination of polarising filters is imperfect, since even when the polarising filters are at 90 degrees to one another the light is not completely extinguished, making the display of black a compromise and thus limiting the contrast range of the display.

Liquid crystal filters are also used as the engine of large-screen projector displays, often in a reflective form where the source light once it has passed through the liquid crystal modulating assembly is reflected back through the filter sandwich, thus traversing the filter twice. Higher-power light sources are required for this application, usually xenon lamps.

Another technology which uses the gas discharge fluorescent/optical filter combination as a primary colour source is the DLP display. This display is based on a semi-conductor chip which is comprised of an array of micro mirrors arranged in a rectangular matrix with each mirror representing a display pixel. Each mirror is assembled on a pivot such that it can be tilted with the application of a voltage so that varying the time of the tilt will vary the intensity of the light reflected from the mirror. Depending upon the voltage applied, the reflected light is either directed through the lens onto the display screen or onto an absorbing black surface which acts as a light and heat sink.

As indicated above, the chromaticities of the primaries are thus dictated by the characteristics of the red, green and blue optical filters and the SPD of the source light.

8.3.1.3 LED/Optical Filters – LED Backlight/LCD Panels

In cold cathode fluorescent backlit LCD displays the saturation of the primaries is limited both by the characteristics of the source of the backlight and the coloured optical filters associated with the liquid crystal cells which are relatively broad spectrum in nature and therefore the chromaticities of the primaries are located some distance away from the spectrum locus and are therefore not ideal.

Static LED Panels
As described in Section 7.6, LED lamps started to appear in the early 2000s and more recently tailored versions, comprising a mixture of red, green and blue LEDs, started to replace the cold cathode lamps used as backlights for LCD displays, where they exhibited improved efficiency, contrast range and colorimetry. Nevertheless, they continue to retain the limitation of LCD displays in as much as the use of polarised filters, which control the level of light passing through the panel, does not enable the backlight to be completely filtered out on zero level signals, thus preventing the display of black.

Although fundamentally they remain LCD displays, in order to differentiate them from the cold cathode backlight technology, the manufacturers refer to them as LED displays, albeit that the LEDs are used as a back light source, not as a pixel source.

Dynamic LED Panels
In order to ameliorate the difficulty of producing black from an LCD display with an LED backlight arrangement, a dynamic illumination technology has been introduced whereby the level of illumination from the LEDs is modulated in level in accordance with an algorithm based upon the signal level and the dispersion pattern of the LEDs at the rear of the screen which are arranged in a coarse representation of the pixel arrangement. Thus when the signal level indicates an area of the image should be at a low level of luminance, the intensity of the LEDs is reduced appropriately, enabling black to be produced on the screen. Early versions of the technology for the consumer market were a compromise, in that the resolution and dynamic control of the LEDs was inadequate to the requirement, leading to the image often appearing as 'black clipped', that is, dark tones in the scene being produced as black over significant areas of the display beyond the scene dark area. Recent introductions of the technology (2014) into the professional market are based on an LED pattern of some 1,500 LED triads, the intensity

of which is controlled on a frame-by-frame basis leading to a much improved rendition of the image.

8.3.1.4   Gas discharge/fluorescence/photoluminescence – Plasma Displays

In the early days of LCD development there were limitations to the size of the flat panel display which could be successfully produced and plasma displays were developed to provide screens of the dimensions sought by those seeking a more inclusive experience. These are flat-panel displays where the panel comprises a matrix of red, green and blue subpixels arranged in vertical stripes. Each pixel is a glass cell which contains a rarefied mixture of noble gases and mercury. The phosphor-coated walls of the glass cells also contain a cathode and anode such that, when a voltage is applied across the cell the gas is ionised into a plasma and emits ultraviolet photons which in turn activate the phosphor causing the cell to emit light at a wavelength dependent upon the phosphor doping, as described in detail in Section 6.6. The intensity of the light from each subpixel is controlled by varying the duration of the voltage applied across each cell.

These plasma screens were introduced as professional television displays in the early 2000s and became the de facto domestic standard for the larger size of screen from about the year 2005. Latterly (2014) as LED/LCD displays have become available in larger screen sizes at less cost, production of plasma screens appears to have ceased.

8.3.1.5   Electroluminescence/LEDs – LED Displays

Potentially the use of LEDs as the pixel source of light in display devices is an attractive proposition since their chromaticities are generally located close to the spectrum locus of the chromaticity diagram thus potentially providing a large chromaticity gamut. However, historically their physical dimensions prevented them from serving as pixels in display devices. Nevertheless as the increase in both the range of colours and the level of brightness available occurred in the 1990s, as described in Section 6.6, these components were used for the large outdoor screens seen at sports events, where each pixel is comprised of a red, green and blue LED.

The ongoing development of OLEDs has led to the availability of active matrix organic light emitting diode (AMOLED) displays, the active matrix being the semi-conductor pixel switching system integrated into the material of the display.

Though initially available only as small screens of the type used in mobile phones and camera viewfinders, despite manufacturing difficulties, they have (2013) become available in television screen sizes, albeit currently at premium prices. This technology, which incorporates subpixel LEDs at full high-definition resolution, is able to fully exploit the potential of the LED in providing primaries located very close to the spectrum locus of the chromaticity diagram and with very high contrast ratios. The current technology limits the brightness of these displays and to date no data has been released by the manufacturers giving the chromaticity coordinates of the primaries of these devices.

8.3.1.6   Electroluminescence/Lasers – Laser Displays

Lasers are beginning to replace the xenon light source in LCD and DLP cinema projectors and since lasers produce monochromatic light, they are ideal for use as primary light sources in colour reproduction systems. The laser-illuminated projector (LIP) has the potential to produce brighter, higher contrast, wider gamut and more efficient displays than their predecessors as will be described in more detail in Chapter 32.

8.4 Reconciling Minimum Image Resolution with Maximum Perceivable Resolution

We saw in Section 8.2 that the optical system of the camera forms an image on each of the red, green and blue image sensors and that these converters comprise a number of pixels usually laid out in a rectangle of horizontal rows. The ratio of the width to the height of the display is termed the 'aspect ratio' and in the general case, where square pixels are used, also describes the ratio of the number of pixels in the horizontal rows to the number of pixels in the vertical columns.

In a well-designed system, the number of pixels present in each converter should be large enough to ensure that at 'normal' viewing distance the recognition acuity of the eye, which is its ability to discriminate between two adjacent objects, is not compromised in viewing the reproduced image.

The implications of this last statement relate entirely to the manner in which the reproduced image is viewed and in what follows, it will be assumed that the number of pixels in the image sensors of the camera is matched by the number of pixels in the display.

The closer one is to the displayed image the more likely one is able to discriminate the individual pixels. It becomes apparent that the critical parameter regarding the number of pixels required at the display to satisfy the resolution criteria of not compromising the acuity of the eye is the angle of view subtended at the eye by the spatial separation of the pixels. Such a parameter clearly takes into account the size of the screen, the viewing distance and the number of pixels as Figure 8.6 illustrates. Early work indicated that for most individuals with good eyesight the recognition acuity of the eye is about one minute of arc or about 300 micro radians.

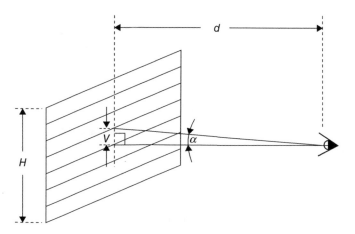

Figure 8.6   The relationship between the parameters associated with establishing the minimum number of pixels required not to compromise the acuity of the eye.

The distance of the viewer from the screen is d metres. The angle subtended at the eye by 1 pixel height is α.

The height of each pixel in terms of the angle subtended at the eye is:

$$V = d \sin \alpha \text{ metres}$$

Number of pixels per picture height $N_H = H/V$

Therefore $N_H = H/d \sin\alpha$

The number of pixels per picture width is dependent upon the aspect ratio A, and assuming square pixels:

Pixels per picture width $= N_W = N_H \times A$

and the total pixels per picture $N = N_H \times N_W$

$$N = N_H^2 \times A$$

In the above arrangement of parameters, the 'resolution' relationship between the acuity of the eye, the dimensions of the screen, the viewing distance and the number of pixels required to ensure they are not visible is determined. These results are used in Worksheet 8 to produce a relationship between the four parameters, which in turn enables any two to be fixed and the relationship between the other two to be graphed. As an example, when applying this relationship to a television screen in Figure 8.7 the aspect ratio and the viewing distance are fixed at 16:9 and for what might be assumed is an average viewing distance of 3 metres respectively. The minimum number of pixels per picture height required for a particular screen size in order that the perceived resolution of the system is not compromised may be read off from the graph.

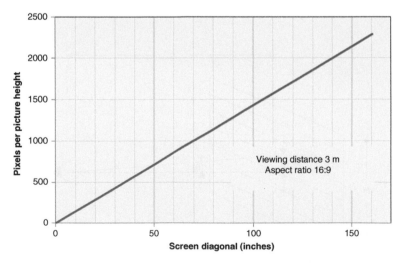

Figure 8.7 Graph illustrating the minimum number of pixels required for a specified screen size at a viewing distance of 3 metres.

Following tradition, the screen size is shown in terms of the diagonal dimension in inches and the pixels in terms of the number per picture height. The formula in the worksheet enables these parameters to be alternatively read off in terms of screen height in metres and the total number of pixels, for example. The minimum number of pixels required for a 50″ diagonal screen viewed at a distance of 3 metres is seen to be about 713 vertically, which corresponds to about 1,268 horizontally and about 904,700 in total.

By rearranging the formula in the worksheet other graphs which illustrate the relationship between viewing distance and screen size for a particular number of pixels can be drawn. In Chapter 14, this relationship is used to give examples for screens containing pixel numbers which relate to various system standards.

9

Colorimetry in Colour Reproduction

9.1 The Relationship between the Display Primaries and the Camera Spectral Sensitivities

The primaries of a simple colour reproduction system are the primaries of the display device and the corresponding range of tristimulus values necessary to match each colour through the spectrum using these primaries, that is the colour matching functions (CMFs), become the camera spectral sensitivities. Since image sensors respond to light of all wavelengths, they effectively carry out the integration of the scene spectral power distribution (SPD), surface by surface against the appropriate red, green and blue colour matching functions to produce the tristimulus values, that is the RGB voltage signals. The level of the output signals will correspond directly with the tristimulus values of a colorimetric system of measurement of the surfaces in the scene, pixel by pixel.

Colour Reproduction in Electronic Imaging Systems: Photography, Television, Cinematography, First Edition. Michael S Tooms.
© 2016 John Wiley & Sons, Ltd. Published 2016 by John Wiley & Sons, Ltd.
Companion Website: www.wiley.com/go/toomscolour

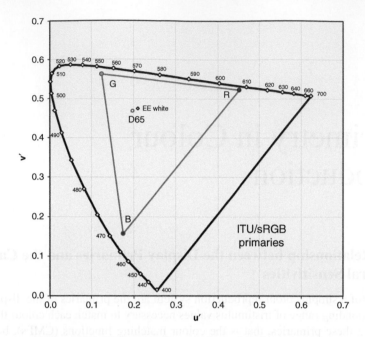

Figure 9.1 The gamut of a typical set of display primaries.

When the RGB signals are used to drive the display primaries, the colour gamut obtained will correspond to the area of the triangle formed by the primaries on the chromaticity diagram as illustrated by a practical set of display primaries in Figure 9.1.

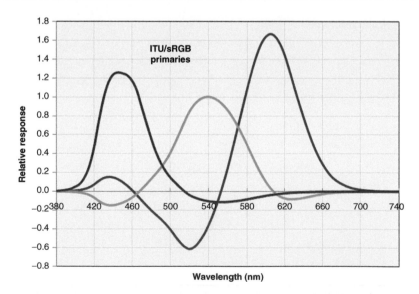

Figure 9.2 Idealised[1] camera spectral sensitivities matching the primaries of Figure 9.1.

---

[1] 'Idealised' in the context of this and the following camera spectral sensitivities charts alludes to the inclusion of the negative lobes of the characteristics which cannot be achieved directly at the output of the image sensors.

Every set of primaries therefore has a corresponding set of CMFs or matching camera spectral sensitivities. In Figure 9.2 the spectral sensitivities corresponding to the primaries in Figure 9.1 are illustrated. Their similarity to the colour matching functions derived in Chapter 3 will be noted.

Subject to the practical camera spectral sensitivities matching the positive lobes of the colour matching functions illustrated in Figure 9.2, colours in the scene which fall inside the gamut will be reproduced accurately whilst those outside of this gamut will in broad terms be produced with the same hue but with a saturation limited by the gamut of the triangle.

The inability of the image sensors to provide a negative output over the range of the spectrum where the curves dip into the negative response area will, when a colour in the scene has a spectrum with significant energy at wavelengths corresponding to the negative areas of a curve, generally lead to the signal associated with the appropriate negative response being higher than it would otherwise be, thus leading to desaturation of the displayed colour compared to the original. For example, a green colour represented by an SPD with power mainly in the 500–600 nm band will evoke a strong green response, a significant red response and a small blue response from the camera due to the red response curve between about 555 and 600 nm and the blue response curve between 500 and 515 nm being positive. However there will be no corresponding negative red or blue response from the camera over the 500–555 nm and 515–600 nm bands, respectively as there should be in accordance with the ideal spectral sensitivity characteristic. The red and blue signals will therefore be significantly higher than they would otherwise be causing a possible hue shift and a significant desaturation of the original colour.

Generally speaking, for broad band colours of a particular hue, the complimentary response levels will always be higher than they should be, thus leading to a general desaturation of the reproduced colour.

9.2 The Choice of Reproduction Display Primaries

Ideally, the colour reproduction system should be capable of reproducing all colours accurately. However, as highlighted in Figure 9.2, an inspection of the idealised camera spectral sensitivities indicates that in reality, since the curves dip into the negative region over portions of the spectrum, it is not possible to provide a perfect match as the sensors are incapable of providing a negative output. The means by which we overcome this inability to match the negative lobes of the colour matching functions with the camera spectral sensitivities is dealt with in Chapter 12.

From the above, it is apparent that at a fundamental level it is the chromaticity coordinates of the display primaries which set the design parameters for a colour reproduction system since the colour matching functions for a particular set of system primaries become the spectral sensitivities of the camera.

Clearly the ideal set of display primaries for any reproduction system should be that set which gives the widest gamut of *useful* colours. At first sight this criteria would appear to have been met if the primaries were to be located at the apexes of the u',v' 'triangle' as shown in Figure 9.3, with the red and blue primaries at the extremes of the spectrum and the green primary between 505 and 510 nm. However, this simplistic approach is far from the ideal for a number of reasons.

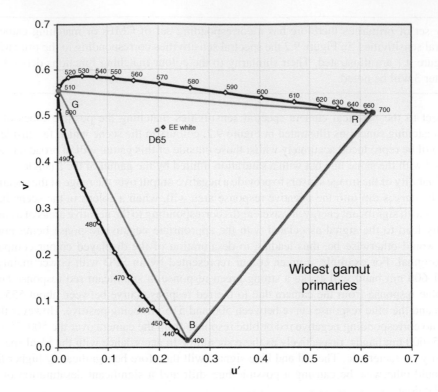

Figure 9.3 A simplistic approach to specifying a set of 'ideal' primaries.

First and foremost, it must be remembered that colour is a three-dimensional quantity and brightness is very important in regard to the reproduced image. If an image of impeccable colour rendition is produced at a brightness level which cannot be viewed comfortably in a lighted room, it is of little practical use.

It is important therefore that primaries with low luminance factors are not selected, such as the red and blue primaries proposed above which are located at the extremes of the spectrum locus, since they are also located at the extremes of the V_λ curve where the eye has very little response.

Another important factor is the shape of the colour gamut. We have already noted that the red to green section of the spectrum locus is virtually a straight line over much of its length. The implication of this is that even a colour comprising a broad band of energy with an SPD limited between red and green will still, by the straight line laws of addition, produce a colour located near the spectrum locus, that is, it will be close to 100% saturated, irrespective of its wide spectral distribution. In consequence, it is not unusual for surface colours to be found which have chromaticities which fall close to the straight line section of the spectrum locus between yellow-green and red. Thus in order to reproduce these not uncommon colours, the red and green primaries should ideally be located on the straight section of the spectrum locus.

Conversely, the convex nature of the curve of the spectrum locus between green and blue causes any colour with a broad band SPD over this portion of the spectrum to inevitably produce a mix of its components which is away from the spectrum locus and as a result, saturated cyan colours are relatively rare in nature. Thus when selecting the green primary it is a further reason to locate it nearer to the straight G to R line than to the point which would embrace saturated colours in the green to blue range; a point close to 525 nm would be a good compromise.

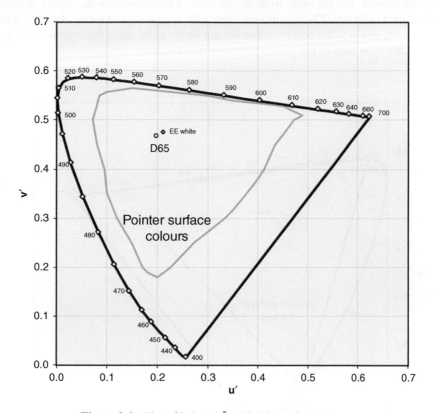

Figure 9.4 Plot of Pointer's[2] realisable surface colours.

This approach is supported by the plot of the extent of realisable surface colours shown in Figure 9.4, an approximation of the results achieved by Pointer (1980). Note how virtually 100% saturated red, orange and yellow colours are realisable. Clearly a gamut which embraces the realisable surface colours is highly desirable and may be considered by some as the only criteria. It seems desirable however, if primaries of sufficient luminous efficiency could be found, to broaden the reproduction gamut to cover as much of the overall gamut as possible, not least because the gamut of realisable surface colours may continue to increase as new dyes are developed.

---

[2] Pointer used Illuminant C to determine his colour plots and a recent paper by Li, Luo, Pointer and Green (Li, et al., 2013), notes that the maximum real surface colours gamut is a little larger than the Pointer original. The CIE has established a Technical Committee, TC1-73, Real Colour Gamuts, to address this issue.

Note that if the red primary is located at a compromise position between the edge of the gamut of surface colours and the extreme of the spectrum response at zero luminance, and the green primary is located at 525 nm, there is still a difficult choice to make for the blue primary. Too far towards the end of the spectrum causes an increasing area of saturated cyan colours to be missed; too far away from the spectrum end diminishes the ability to produce saturated magenta colours. The compromise normally accepted is to locate the blue primary as far as possible towards the spectrum end but not so far as to cause the reproduction gamut to cut across the cyan area of the gamut of real surface colours.

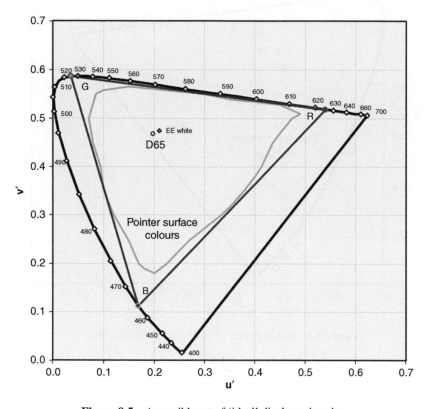

Figure 9.5   A possible set of 'ideal' display primaries.

Figure 9.5 illustrates what may be considered to be close to the ideal set of display primaries; these are monochromatic primaries located at 465 nm, 525 nm and 625 nm. However, no practical illuminants for use in a consumer environment have so far been found that will enable a picture of acceptable brightness to be produced with these chromaticities.

The chromaticity coordinates of these primaries together with an arbitrary selected system white are shown in Table 9.1.

Table 9.1   Chromaticities of a set of 'Ideal' display primaries

	x	y	u′	v′
Red	0.7007	0.2993	0.5400	0.5190
Green	0.1142	0.8262	0.0360	0.5861
Blue	0.1355	0.0399	0.1690	0.1119
White D65	0.3127	0.3290	0.1978	0.4683

9.3   Derivation of Colour Reproduction System Camera Spectral Sensitivities

As we saw in Section 4.4 on colorimetry, the standard system of colour measurement and specification is based upon the CIE non-real stimuli X,Y,Z and their corresponding colour matching functions the $\bar{x}(\lambda), \bar{y}(\lambda), \bar{z}(\lambda)$ curves.

It is usual therefore to specify a colour reproduction system in terms of these CIE internationally accepted parameters. This has the advantage that once this relationship between system primaries and camera spectral sensitivities has been derived, any set of colour system primaries and a system reference white may be selected and from a knowledge only of their chromaticity coordinates, the corresponding camera spectral sensitivities may then be calculated and expressed in terms of the $\bar{x}(\lambda), \bar{y}(\lambda), \bar{z}(\lambda)$ colour matching functions.

Creative adjustments apart, the aim of a colour reproduction system is generally to reproduce on the display the original colours in the scene. However, as we have seen in the previous chapter, the quality of illumination will clearly affect the value of RGB signals derived from the camera. It is important therefore in a reproduction system to standardise on the illuminant used for the scene and the complementary matching white used for the display. In reality it is not practical to standardise the illuminant for the wide range of scenes, both indoor and outdoor, which will be presented to the camera. Thus the solution is to design the camera for a standard illuminant and provide adjustment within the camera to compensate for any departure from the standard when capturing a scene with a different illuminant. The display white is the white produced when maximum equal level RGB signals drive the display. This white of the scene illumination and matching display is referred to as the system reference white.

The procedure for establishing the relationship between the RGB primaries of the system, the system reference white and the corresponding idealised camera spectral sensitivities in terms of the $\bar{x}(\lambda), \bar{y}(\lambda), \bar{z}(\lambda)$ colour matching functions is very similar to that used originally to derive the $\bar{x}(\lambda), \bar{y}(\lambda), \bar{z}(\lambda)$ colour matching functions themselves from the CIE RGB primaries. In Appendix F, a general set of equations are derived which express the relationship between the chromaticities of any set of RGB primaries and system white point and their corresponding camera spectral sensitivities (i.e. the $\bar{r}(\lambda), \bar{g}(\lambda), \bar{b}(\lambda)$ colour matching functions), in terms of appropriate values of the $\bar{x}(\lambda), \bar{y}(\lambda), \bar{z}(\lambda)$ colour matching functions.

By entering the values given in Table 9.1 into the relationships derived in Appendix F, we can use Worksheet 9 to calculate the coefficients of the XYZ colour matching functions for the 'ideal' display gamut primaries as follows:

$$\bar{r}(\lambda) = 1.5947\,\bar{x}(\lambda) - 0.2021\,\bar{y}(\lambda) - 0.2523\,\bar{z}(\lambda)$$

$$\bar{g}(\lambda) = -0.7183\,\bar{x}(\lambda) + 1.6816\,\bar{y}(\lambda) + 0.0367\,\bar{z}(\lambda)$$

$$\bar{b}(\lambda) = 0.0326\,\bar{x}(\lambda) - 0.0764\,\bar{y}(\lambda) + 0.9956\,\bar{z}(\lambda)$$

In the worksheet, a weighting factor is applied to the table of curve values in order to bring the peak of the green curve equal to 1.00 for comparative purposes.

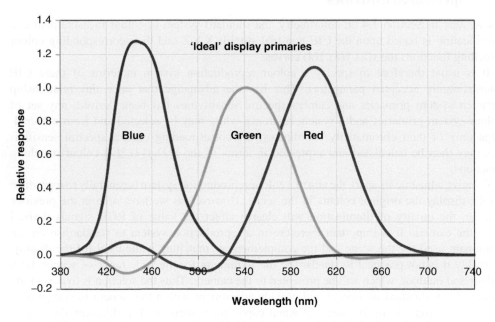

Figure 9.6   Camera spectral sensitivities for the set of 'Ideal' display primaries.

The ideal camera spectral sensitivities which result from plotting these relationships are shown in Figure 9.6. These are the characteristics which complement the primaries illustrated in Figure 9.5.

Note that as a result of locating the primaries on the spectrum locus and using an extended gamut, the negative lobes of these curves are very much less than those related to a set of current practical primaries as illustrated in Figure 9.2.

One of the criteria set in defining the relationships between the display chromaticities and the system white was that a neutral reflecting surface in the scene, illuminated by a white source whose SPD matched the system white, would produce R,G and B signals of equal value. In Worksheet 8(b) the ideal camera spectral sensitivities are each convolved with the system white SPD (D65 in this case) and integrated to show that the resulting values of the RGB signals are identical.

Each different set of primaries will require a different matched set of camera spectral sensitivities which is reasonable for a closed system but where it is useful or necessary for a common signal to drive displays with primaries of different chromaticity values, then problems can occur unless steps are taken to define the overall system appropriately, as is described in Part 4.

The advantages of selecting spectral primaries located close to the spectrum locus of the chromaticity diagram have been described. However, there is also an advantage to having primaries with a slightly broader spectral distribution when the relatively minor differences in observer responses described in Section 4.3 are considered. A set of broader spectrum primaries will tend to mask the differences between observers but as primary SPDs approach spectral lines, so the differences between observers will become more apparent, in terms of reproducing colours which will be perceived differently by those observers whose responses differ slightly from those of the average CIE observer.

10

Appraising the Reproduced Image

10.1 Introduction

Let us assume for the moment that we have produced an image which is as close to a technically correct reproduction of the original scene that it is practically possible to achieve with current technology. How do we appraise the image in perception terms, being aware that when viewing the image the eye–brain complex will accommodate for both the brightness level and the colour of the lighting environment in which the reproduced image is situated? Section 1.4 and Chapter 5 reviewed the degree to which the eye will accommodate in these circumstances.

Depending upon the medium we are appraising, it is likely that the conditions under which the image is being viewed will vary considerably. Taking the viewing of television as an example, the screen will generally be located in a domestic environment where the viewing conditions might be a proportion of daylight filtered through the windows of the room or artificial light of a warm colour temperature. If daylight, the intensity of illumination could vary over a very wide range, whilst the screen is likely to have a fixed peak level of luminance.

The scene the eye perceives in these cases will include not only the screen but also the surrounding area, both of which are illuminated by the environmental lighting; thus the eye will accommodate to the average brightness and colour of the combination of the screen together with the surrounding surfaces which in turn could cause the perception of the reproduced image to be adversely affected.

The degree to which the perception of the image will be affected by the environmental lighting is related to the ratio of the area of the image to the area of the total field of view of the eye. The greater the area of the display to the field of view the less will be the effect of the surrounding environmental illuminated surfaces.

Thus generally speaking, the appraisal of an image will produce the most accurate perception of the required results if the peak luminance of the surrounding environment is considerably less than the peak luminance of the image and if the illumination is of the same colour temperature as the white reference point of the screen. Furthermore, the greater the proportion of the field of view of the eye occupied by the screen, the more accurately will the rendered image be perceived.

Colour Reproduction in Electronic Imaging Systems: Photography, Television, Cinematography, First Edition. Michael S Tooms.
© 2016 John Wiley & Sons, Ltd. Published 2016 by John Wiley & Sons, Ltd.
Companion Website: www.wiley.com/go/toomscolour

Having circumscribed the factors affecting the perception of the reproduced image we will now investigate each of them in more detail and address the steps which can be taken to ensure the image is viewed in an environment which is conducive to perceiving the image as intended.

10.2   The Environmental Lighting

As noted above both the intensity and the colour temperature of the environmental lighting is critical to the perception of the reproduced image.

10.2.1   Intensity of Environmental Lighting

As was described in Section 1.3, as the intensity of illumination increases so the eye compensates by reducing its sensitivity, using both the iris of the eye and the automatic gain compensation effects in the eye–brain complex to ensure the scene is perceived satisfactorily.

As the sensitivity of the eye is reduced, so the reproduced screen image will be perceived as a less bright or dull image. In perception terms, dull images are increasingly less pleasing than bright images over the adapted contrast range of the eye. Furthermore, the less bright the image the more the perceived chroma of the image reduces, further reducing its impact.

Naturally, when viewing a photograph under environmental lighting, the reproduced image brightness will increase in the same ratio as the increase in the level of the environmental lighting; so there will be no corresponding adverse effects on the perceived image. In fact, in general terms the brighter the environmental lighting, the better will the reproduced image appear. For cinema viewing the environmental lighting is usually kept to an absolute minimum consistent with local safety regulations and usually below a level which directly affects the perception of the reproduced image.

10.2.2   Colour Temperature of Environmental Lighting

The brighter the environmental lighting and the smaller the reproduced image in terms of the field of view of the eye, the greater the effect of the colour temperature of the environmental lighting will be on the perceived colour balance of the image.

The reference colour temperature for reproduced images in all media is usually related to a daylight correlated colour temperature of between 5,000 and 6,500 K, though not the same colour temperature for all media as will be seen later. However television, particularly after dark, is usually viewed in a lighting environment based on a colour temperature which matches tungsten illumination at about 3,000 K. Therefore there is the potential for a considerable mismatch if conditions are such that the eye adapts to the colour temperature of the environmental lighting.

Thus in an environment illuminated by relatively bright lighting of a low colour temperature, all the deleterious effects noted above will occur in addition to the image appearing to be 'cool' in comparison to the surroundings.

This effect will be noticeable in appraising both television and photographic images.

10.3   Reflections from the Display

The environmental lighting under which the image is viewed will also fall upon the image, which for a photograph is the primary illuminant by which it is perceived. However, for the remaining media any light reflected from the screen will add to the image and detract from it.

In the cinema environment that light falling on the screen which is not generated by the image is a combination of the safety lighting and any light produced by the projector when the incoming signal is at black. This small but significant level may be perceived as detracting from the depth of the blacks in the image, that is, the contrast ratio of the rendered image will be impaired.

For television and computer screens the light reflected from the screen may be considerable, particularly if the screen is mounted in a position where a bright surface in the adjacent environment or a window reflects directly from the screen into the viewer's line of vision. The reflected light normally has two components: that reflected from the front surface of the glass on which the image forming structure is mounted, and that reflected from the structure itself. Means of minimising these reflections have been in use for many years. Nevertheless it is surprising that from time to time marketing and fashion trends support the sale of shiny screens which naturally result in the highest level of reflection from the front surface of the screen and a dramatic drop in the image contrast ratio. Other means of minimising the reflections from the image structure material on the rear of the screen naturally lead to a dark appearance when the screen is not activated, which is generally perceived in a domestic environment as unaesthetic in appearance.

10.4   Image Size

Generally speaking, the larger the angle the image subtends at the eye the less the eye is distracted by the viewing environment. In the cinema, the trend is towards images which fill the angle of view of the eye, creating a more immersive experience. As the resolution of display devices have improved so there has also been a trend towards larger screens at home, both for the computer and particularly for the television screen. Large photographs, rather than snapshots are preferred as evidenced when visiting a photographic exhibition.

10.5   Managing the Viewing Environment

From the foregoing it begins to become evident that in order to gain the maximum benefit from viewing images it is important that recommendations be made with regard to the viewing environment. It is also clear that each reproduction media will require a different set of recommendations in order to take account of the different viewing conditions in each case. For example, though a cinema environment would lead to a more critical environment for viewing television and would be fine for a home theatre situation, it would not be acceptable for general viewing where other distractions are from time to time the order of the day.

A deeper review of the factors affecting the perceived contrast range is addressed in Chapter 13 and the recommended viewing conditions for each of the reproduction media are addressed in Part 5 where other factors which interact with the perceived image are also taken into account.

10.6   System Design Parameters

It has been noted that with the possible exception of outdoor screens it is impractical to reproduce the brightness range of an outdoor scene. Instead the system strives to reproduce the contrast range of the original scene, albeit that this aim is compromised by the viewing environment to varying degrees as described in the preceding paragraphs.

In particular, the perceived contrast law of the reproduced image in an environment of high illumination is distorted by adaptation and by reflections from the screen. To a degree this distortion can be partially compensated for by pre-distorting the contrast law in a complementary manner.

This chapter has been located at the end of Part 3 which describes the workflow of a colour reproduction system. However, it might be surprising to learn that it is the typical conditions of the viewing environment that are used to define environmental parameters, based upon the impairing factors described in this chapter, which in turn enable the fundamental system parameters to be specified for each of the media colour reproduction systems. These system design parameters will be addressed in Chapter 15.

Part 4

The Fundamentals of Colour Reproduction

Introduction

The conceptual elements of a colour reproduction system, in terms of the light and signal flow through the camera and the display were illustrated in Figure 8.1, where both the camera and the display contained an element labelled 'signal processing'. For the conceptual system it was assumed the signal processing elements were linear in operation and had no effect on the colour signals; often however, this is not the case. Furthermore, other elements in the signal flow have characteristics which are sometimes not ideal.

Thus as the early conceptual systems evolved into the systems of today the signal processing elements progressively expanded to encompass the correction of these various shortcomings as illustrated in Figure P4.1. Depending upon whether the reproduction process is for television, photography or cinematography, these additional processing elements are included not only in the camera but also form an additional unit external to the camera. This arrangement enables the option, post of shooting the scene, for a greater flexibility of settings and adjustments in dealing with the shortcomings of the conceptual camera system. In very general terms, in television the additional processing is usually carried out only in the camera system. In serious photography there is an option for this processing to be carried out externally and in cinematography there are invariably facilities for processing the signals subsequent to capturing the scene.

The 'post' (for post shooting or post production) processing is usually carried out on a computer-based system with appropriate hardware and software to enable the parameters of the workflow elements to be adjusted for optimum picture quality.

Colour Reproduction in Electronic Imaging Systems: Photography, Television, Cinematography, First Edition. Michael S Tooms.
© 2016 John Wiley & Sons, Ltd. Published 2016 by John Wiley & Sons, Ltd.
Companion Website: www.wiley.com/go/toomscolour

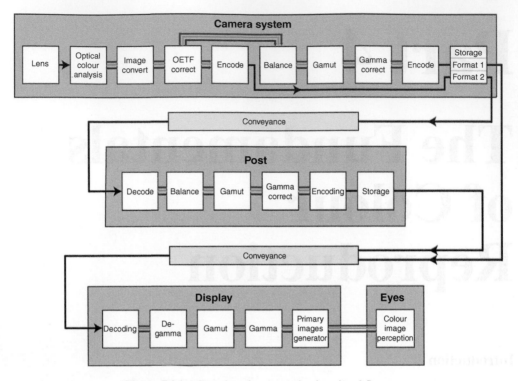

Figure P4.1   Generic colour reproduction signal flows.

Principally the four additional elements of post control the colour balance, colour gamut processing, gamma correction (transfer characteristic pre-correction) and colour encoding for storage and delivery. Irrespective of whether a post phase is included in the workflow these elements are always included in the camera in order to provide images for the viewfinder and to provide basic images for preview and non-critical display.

In addition, two additional functions may be required depending upon the transfer characteristics of the image sensor in the camera and the image generators in the display device. As will be seen it is important that colour balance and gamut processing are carried out on signals which are linear representations of the red, green and blue content of the original scene. Thus in the event that the camera image sensor's opto-electric transfer function (OETF or transfer characteristic) is non-linear, it would be necessary to correct the signals with a complementary function before further processing was undertaken. However, though historically image sensors were sometimes nonlinear, current devices are essentially linear in their response up to the point where the level of the incoming light overloads the device, and thus an OETF corrector is not usually required.

The processing in the display device will depend upon the vintage and sophistication of the device. In devices using a CRT, its classic power law electro-opto transfer function (EOTF) is broadly complementary to the characteristic of the gamma corrector in the camera. Thus in such historically simple configurations the de-gamma, colour gamut and EOTF correction processing elements are unnecessary and the signals are fed direct from the decoder to the CRT. In modern systems, where the image generator transfer function is unlikely to complement the camera gamma corrector, some or all of the additional processing elements illustrated will be required.

Part 4 addresses the practicality of processing, storing and transferring the native[1] RGB signals between the camera and the display or photograph, particularly with regard to the functional processing elements identified above, where in each chapter, the function and shortcomings are described and the correction processes are detailed.

---

[1] 'Native' is commonly used to describe signals as initially generated before any processing takes place.

11

System White and White Balance

11.1 System Reference White

In a practical colour reproduction system, the colour of the scene illumination may not be that for which the camera is designed, so adjustments have to be made to compensate for the resulting change in the colour balance of the image. Also the requirement to serve different reproduction systems, which may have different reference whites, with a common camera source, may require the system reference white to be reset between the camera and the display.

In Chapter 9, system reference white was defined as the chromaticity of the white of the standardised scene illumination and the matching chromaticity of the white of the display. The system white of a reproduction system is an important parameter which can significantly influence the optimisation of the viewing of an image and it is thus worthwhile investigating further the background which led to the choice of system reference white for the various media reproduction systems.

In a general sense it is reasonably evident from what we have learned about equal proportions of the red, green and blue primaries summing to white that we would expect the camera to produce equal voltage levels for the red, green and blue signals when scanning a neutral surface, that is, a spectrally non-selective white or grey surface in the scene. Thus, it becomes a required condition that equal RGB signals applied to the display will produce white or a neutral grey. However, as we have seen in Chapter 6, on illumination, both the adaptation characteristics of the eye to illuminants of different colours and the variable colours of illuminants used to light a scene mean that we must be specific in defining the white of the scene and the white of the display or print.

As we saw in Chapter 10, when viewing a display or print which subtends a relatively small angle to the eye, the eye adapts to the illuminant of the general surrounding areas, and thus in critical viewing situations, it is usual to match the white of the illuminant used to light the surrounding viewing area with the system white in order to avoid any challenge to the adaptation characteristics of the eye. In the case of a photographic print, the white perceived by the eye is the white which results from the reflection of the environmental lighting incident upon the white of the photographic paper, which may or may not reflect equally at all wavelengths. The situation in the cinema is in this respect less critical; first,

Colour Reproduction in Electronic Imaging Systems: Photography, Television, Cinematography, First Edition. Michael S Tooms.
© 2016 John Wiley & Sons, Ltd. Published 2016 by John Wiley & Sons, Ltd.
Companion Website: www.wiley.com/go/toomscolour

the image subtends a much greater angle of view to the eye and, second, the surrounding illumination is generally at a very much lower relative level than for a television display or a photographic print; thus, unless the imbalance is extreme, the image itself will set the adaptation white point of the eye.

In colour measurement, the colour measuring procedure is completely objective, that is, since the colour mixture curves are based upon the standard observer and equal energy white, there are no subjective effects to take into account. However, colour reproduction is highly subjective for the reasons outlined in the above paragraphs and, in addition, practical sources of illumination with an equal energy white characteristic are not available.

With the exception of large-screen LED displays in public places, the viewing of the reproduced image is generally undertaken in either subdued daylight, artificial light or near darkness, depending upon both the type of media and the circumstances. In this context, artificial lighting, realistically meaning the choice for the majority of domestic environments, is tungsten lighting or its replacement at the same correlated colour temperature (CCT). In general terms therefore, the choice of a system white lies between one of the daylight standards and one with a CCT in the range of about 3,000 K. In reality however, although the eye is very successful at adapting to tungsten-based illumination, given the option, the preference for white illumination is always daylight; thus, it was generally agreed that the system white for the reproduction of display images and prints should be based upon one of the daylight standards. The problems this will cause when viewing the reproduced image in a tungsten-based environment was discussed further in Chapter 10. For the theatre or cinema display, the lack of a significant level of ambient illumination means that the white point of the 'master' can be set objectively neutral at equal energy white and the data then transformed (see Chapter 12) to match the white point of the projector as required.

One might consider that the daylight standard to be selected as the white point for displays and prints would be the one that came closest to matching equal energy white, particularly so since its CCT lies between the extremes of the range of daylight CCTs at about 5,500 K. Unfortunately this was not to be; historically, television was the first medium to set the standard and that was prior to the adoption of the current range of CIE daylight standards, leaving the only realistic choice the now obsolescent Illuminant C, which as we saw in Chapter 6 does not fully emulate daylight at the ultraviolet end of the spectrum. Thus, when the opportunity arose to reconsider the primary television standards, the nearest of the new illuminant specifications to Illuminant C, that is D65, was adopted as the system white point.

The situation has been exacerbated by the photographic industries later selecting a different white point of D50, which, because the computer industry has also adopted the D65 standard, means that the computer displays on which the prints are adjusted and previewed are at D65, whilst the print itself is specified to be viewed under D50 lighting, thus presenting a mismatch between screen and print when viewed in close proximity, albeit the recommendations for viewing prints specifically state that the display and print images should not be viewed in the same environment. Nevertheless, this difference in system white point causes much misunderstanding and confusion in those situations not fully professionally equipped.

These problems are explored in more detail in the relevant reproduction medium chapters in Part 5; suffice here to indicate that when specifying a colour reproduction system, it is essential to include the chromaticity of the system white point. Reference to Worksheet 8, which is used to calculate the camera colour analysis characteristics derived in Section 8.3, illustrates the use of the chromaticity coordinates of the chosen standard illuminant in these calculations.

Table 11.1 Chromaticity coordinates of defined illuminants

Illuminant	CCT	x	y	u'	v'
SA	2,856 K	0.4476	0.4074	0.2560	0.5243
SC	6,774 K	0.3101	0.3162	0.2009	0.4609
SE	5,460 K	0.3333	0.3333	0.2105	0.4737
D50	5,000 K	0.3457	0.3585	0.2092	0.4881
D55	5,500 K	0.3324	0.3474	0.2044	0.4807
D60	6,000 K	0.3217	0.3377	0.2008	0.3161
D65	6,500 K	0.3127	0.3290	0.1978	0.4683

The chromaticity coordinates of the illuminants commonly used in the theory and practice of reproduction are listed in Table 11.1. The corresponding spectral power distributions are illustrated in Chapter 6 and are detailed in the 'Illuminants SPDs' worksheet.

The adopted system white point for television and computer systems and for the appraisal of images prior to printing is D65, for viewing photographic prints D50 and for cinematography SE and D60.

11.2 White Balance

One of the criteria used in Appendix F for deriving the camera spectral sensitivities is that equal levels of the RGB signals from the camera, representing effectively the number of T-units, will produce a neutral white or grey on the display or print. The consequence of this requirement is that when a neutral surface in the scene is illuminated by light whose spectral distribution corresponds to the system standard illuminant, then the RGB signals from the camera will be equal in level.

In mathematical terms, the RGB signals are proportional to the summation of the products of the spectral power distribution (SPD) of the scene illuminant and the camera spectral sensitivities at each wavelength interval through the spectrum and for a neutral surface:

$$\mathrm{R} = \sum r(\lambda)i(\lambda) = \mathrm{G} = \sum g(\lambda)i(\lambda) = \mathrm{B} = \sum b(\lambda)i(\lambda)$$

where $r(\lambda)$, $g(\lambda)$ and $b(\lambda)$ represent the camera spectral sensitivities, and $i(\lambda)$ is the spectral distribution of the standard illuminant of the system.

In the white balance worksheet, Sheet 11, an example calculation is carried out for both the current television and the obsolete NTSC analysis characteristics against their respective standard reference white illuminant SPDs, which are D65 and SC, respectively. The results for both of these examples are that to two places of decimal R = G = B, a satisfying result.

In practice, the success in achieving in the camera a precise balance of the RGB signals on a white or neutral grey in the scene, even if it were possible to illuminate the scene by the system white, without adjustment is unlikely. This is because the required camera spectral sensitivities are a combination of the prism dichroic filters, any optical correction filters and the spectral response of the image sensors, and even if it were possible to achieve the desired match to the shape of the individual spectral sensitivity characteristic, it is inevitable that the relative responses of the characteristics will suffer different levels of attenuation. Thus, in the design of the camera, an amplification factor in each of the RGB channels must be inserted to compensate for the variable attenuation in each of the optical paths. Ideally, these would be fixed factors which would be set when the camera was commissioned.

However, in addition to the attenuation factors described above, the SPD of the lighting will only approximate to the standard illuminant of the system, as, for example, the characteristics of daylight change depending upon the time of day and the amount of cloud cover. The result of these practicalities is that without some adjustment to the gain of the RGB channels, the output level of the RGB signals will not be equal when the camera is scanning a neutral surface in the scene. The critical criterion here is to ensure that the shapes of the characteristics are as close as possible to the ideal; often this requires a sacrifice in the sensitivity of one or more of the optical paths; however, in colour fidelity terms, this can be fully compensated for by adjusting the gain of the amplifiers in the RGB signal paths to *colour balance* the camera, that is, to make the RGB signals equal to the standard voltage representing the maximum reflectance from the scene when the camera is scanning a standard white surface under the current illuminant.

On a day-to-day basis, the attenuation characteristics of the optical paths of the camera are unlikely to change by very much; however, the spectral distribution of the illumination, whether daylight or artificial light, will change, as likely also will the gain of the amplifiers in the signal path. Unless these imbalances are compensated for, the resulting reproduced image will show a colour cast related to the specific imbalance in the level of the RGB signals on white.

These considerations lead us to the two most important adjustments which are required as a first step to ensuring that the reproduced image contains no colour cast:

- The camera must be white balanced to produce equal levels of signal on neutral greys and white located in the scene which is illuminated as for the scenes to be captured.
- The display or printer must be adjusted such that when processing equal levels of RGB signals, the chromaticity coordinates of the image will match the chromaticity coordinates of the system white point. In the case of a print, this implies that the combination of the spectral characteristics of the viewing illuminant with both the white of the paper and greys in the image, together produce a chromaticity which matches the chromaticity of the system white.

The means of ensuring the display and the printer are properly adjusted are dealt with in Part 5B.

The procedure for achieving a white balance from the camera is dependent to a degree upon the media in which the camera is being used and the circumstances associated with the shooting of the scene. For critical colour matching, a manual procedure is usually the optimum approach, but for less critical operations or where a manual procedure is impractical, the camera electronic system will usually provide the option of an automatic white balance.

11.2.1   Manual White Balance

Manual white balance is invariably the chosen approach for television production situations, as unlike photography or theatre/cinema, there is often no intermediate 'post' processing which enables imbalances to be subsequently corrected. Nevertheless, despite the availability of adjustment of white balance in post-production, cameras for shooting productions for the cinema will also use a manual white balance procedure, as will many amateur and professional photographers who need to get the best from their cameras. Television studios, particularly for live productions, will contain a number of cameras, the outputs of which under the control of the vision mixer will be switched randomly, effectively directly to the television audience in the case of a live show or to a recorder for later transmission. The viewer will see the image from different cameras successively and since the eye will adapt to each image in turn, any small change of white balance between cameras will be very apparent.

The approach used in the studio or outside broadcast is to place a greyscale chart within the scene illumination and use a waveform monitor and the RGB gain controls on the camera to adjust the white chip of the greyscale to equal the standard peak white voltage for each of the colour outputs. The greyscale has a number of neutral grey chips between black and white, their reflectances being arranged in a logarithmic order so that after gamma correction (see Chapter 13), the chips appear on the display as roughly equal steps. Figure 11.1 illustrates a typical greyscale chart, and Figure 11.2 the chart as displayed on a waveform monitor.

Figure 11.1   A typical greyscale chart.

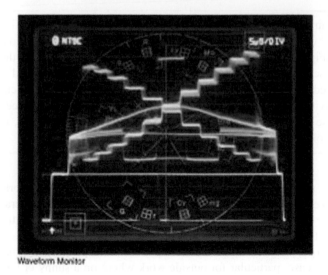

Figure 11.2   Waveform display of greyscale.

The use of a greyscale of the type illustrated in Figure 11.1 is to a degree traditional; in the early days of television, the transfer characteristic of the image sensors could and often did vary a little from sensor to sensor, the result being that despite being able to adjust the gains of the RGB channels to match the white chip, the law differences between the sensors prevented a match being achieved on all the chip levels simultaneously. The waveform monitor enabled the RGB displays to be overlaid so that the mismatch on the steps was accentuated and adjustments were available to vary the transfer characteristics to enable a match to be produced on all steps of the greyscale. Usually this was achieved by minor adjustments to the law of the gamma correctors (see Section 13.4). Developments over the decades have improved the basic technology very considerably to the point that the transfer characteristics of the image sensors generally match; nevertheless, the availability of a chart which enables the operator to confirm at a glance that the camera is balanced at all levels of scene luminance provides an indication that in colour balance terms all is well.

Modern 'greyscales' may take different forms, often only three or four surfaces are supplied between black and white. One of the problems is to provide a black of sufficient low reflectance to represent the black in a real scene which may be in shadow and thus produce a voltage well below the black of a critically illuminated greyscale. One solution adopts a historic technique of using a hollow cube with black internal surfaces and a hole cut in it to represent shadow black. The other surfaces of the cube have reflectances to represent the scene white and a number of intermediate greys.

It is generally assumed that scanning the white of a test chart in the scene will produce the maximum levels of R, G and B from the camera, enabling these values to be adjusted for the system maximum level. However. in certain circumstances, this is not always the case. An example of when this is not so is given in Worksheet 11, where the RGB values are calculated for the Lucideon test tiles (whose SPDs are illustrated in Figure 1.4). In this example, as the calculations in Worksheet 11 illustrate, compared with a perfectly reflecting scene white, the white tile produces RGB values of 0.893, 0.889 and 0.896, respectively, whilst the orange tile produces RGB values of 0.923, 0.218 and 0.016, respectively; that is, the value of the red signal on the orange tile exceeds the red signal on the white tile.

As an exercise in understanding at a deeper level why this should be so, compare the reflectance of the white and orange tiles with the red analysis curve related to the television red primary in Figure 11.3. Although at no point in the spectrum does the orange tile spectral reflectance reach the level of the white tile; nevertheless, the red signal on the orange tile exceeds the value of the red signal on the white tile. On inspection, it can be seen that convolving the camera red response and the white tile will produce a substantial negative response over the spectrum range between 460 nm and 550 nm, whilst for the orange tile, convolution produces only a positive response.

In photography, many of the more advanced cameras provide the opportunity to semi-manually undertake a white balance. This is achieved by setting the camera in a prepared mode to capture the reflectances of a white in the scene and adjust the gain settings in the RGB channels to equate the RGB signals to the standard white level.

The description of white balance in this section is fine for a controlled situation; however, often that is not the case, particular for outside work where on a sunny or partly sunny day,

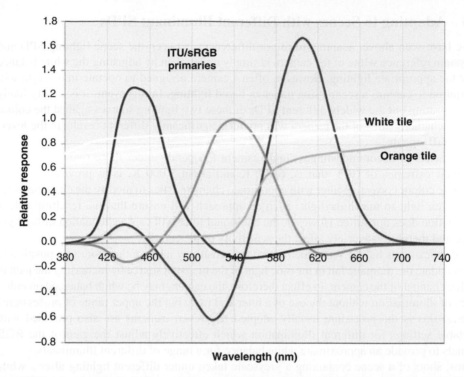

Figure 11.3   A comparison of the spectral response of the white and orange tiles against the idealised camera spectral sensitivities.

the shadows are illuminated by the blue sky whilst the remainder of the scene is sunlit. Even the situation of a passing cloud obscuring the sun will cause the spectral distribution of the illumination to change considerably; the adaptation of the eye prevents the untrained observer from noticing these effects at the scene, but in the controlled environment of the image display, the effects are subjectively enhanced.

11.2.2   Automatic White Balance

For television cameras used in rapidly changing situations such as a news shoot and for consumer-based photographic cameras used by the general public, there is no opportunity for a manual white balance and automatic white balance is the only answer. This is achieved by making the assumption that in general the sum of all the colours in a scene will equate to a roughly neutral grey and by automatically adjusting the gain of the RGB channels in the camera to ensure that this is so.

Often this can lead to disappointing results, such as when shooting a red sunset, when of course the sum of all the colours in a scene is far from neutral, leading to the red component being attenuated and the scene appearing as it would in the middle of the day. Most modern cameras of the 'aim and shoot' variety now have a range of scene selection modes which enable the camera to adjust the RGB gains to an average value related to the different lighting conditions.

11.3   Adapting to Scenes with Different Illuminant SPDs

As we have seen above, adapting to minor differences between the scene lighting SPD and the system reference white of the camera is simply undertaken by adjusting the white balance under the appropriate lighting. However, often a camera designed to operate in daylight will be required to capture a scene using tungsten-based lighting. In this event, it is clearly likely that integrating out the widely different SPDs of these two lighting sources against the colour analysis characteristics of the camera will produce significantly different results in the levels of the RGB signals.

One means of accommodating a single camera to operate under these two light sources with their extremes of CCT, that is, 6,500 K and about 3,000 K, is to provide a built-in selectable colour correcting filter with appropriate characteristics to modify the characteristics of tungsten light to match daylight. Such an approach will ensure that the resulting colour reproduction does not suffer. However, the additional filter will cause attenuation of the light source and therefore adversely affect the sensitivity of the camera.

Modern cameras have a very much improved sensitivity and are therefore in a position to accommodate the dramatic fall in the blue light of the tungsten source by increasing the gain in the blue channel of the camera. In effect therefore, the camera may be white balanced on either source of illumination without the use of a filter and in all but the upper range of professional cameras, this is the procedure usually adopted. However, cameras are also provided with selectable settings for different illumination which effectively adjust the gain of the RGB channels to provide an approximate white balance for a range of different illuminants.

Thus, shots of a scene containing a greyscale taken under different lighting after a white balance will show neutral greyscales; however, if one considers the integration of the SPDs of these very different light sources against each of the three camera spectral sensitivities, it is not surprising to find there are differences in response for the same scene colour from the two different illuminants.

In the 'White Balance' Worksheet 11, the convolution of the current television primaries analysis characteristics with the D65 and SA light sources is carried out. As anticipated, the D65 integration produces equal levels of RGB signals, whilst the SA integration produces the widely different values of R = 2636, G = 1182 and B = 334. The inverse of the ratio of these values is used with the integration of the camera spectral sensitivities and the SA SPD to obtain a mathematical white balance. If now the resulting values at each wavelength are convolved with the SPD of the tiles from the Lucideon range illustrated in Figure 11.3, we obtain the RGB values for each of the tiles. The worksheet illustrates the transform of the RGB values to u',v' values for plotting on the chromaticity chart.

These values can be compared with the values obtained when the calculations are carried out using the D65 illuminant, and as shown in the worksheet and illustrated in Figure 11.4, where the SA illuminant chromaticities are represented by the arrow heads. The results are similar but significantly different. In fact, for those saturated colours close to the red-to-green section of the spectrum locus, the colours produced are 'out of gamut', which implies the RGB signal levels will exceed their system maximum levels for these colours despite the white balance levels having been set at the system maximum level.

It is apparent therefore that for accurate reproduction, using white balance as the only means of using one camera for widely different illuminants will not produce results which match and which may also cause 'out of gamut' problems.

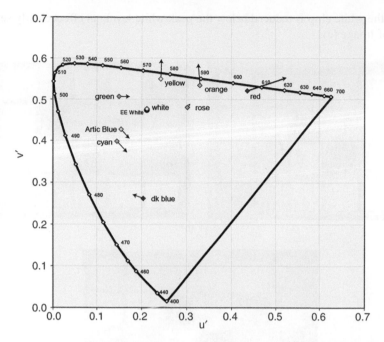

Figure 11.4   Chromaticity values of the Lucideon CERAM range of tiles under D65 and SA lighting.

In practice, is this likely to be a serious problem? To obtain some sort of measure of the extent of the problem, the ColorChecker chart was captured by a number of still cameras under different lighting conditions and the results were compared.

Though all cameras showed similar results, the three shots illustrated in Figure 11.5 were taken with a current (2012) camera from the top professional range then available. Being aware that a print is generally less able than a display to show differences in saturated colours, Figure 11.5 attempts to illustrate the difference in the colour chart shot under three different sources of illumination; in the shade under an exceptionally clear blue sky, under full early afternoon sun plus blue sky light and under a tungsten light source.

In each case, the level of the white chip was initially adjusted for white balance at peak white level. In visual terms, the most marked effect is to the lightness of the colours rather than their chromaticity. Taking Image (b) as that shot under conditions closest to the system reference white of D65 and therefore the norm, then when the illumination is biased towards blue, as in Image (a), the blue colours are too light and the red colours are too dark; whilst when the illumination is biased towards red, as in Image (c), the red colours are too light and the blue colours are too dark.

In measurement terms the results support those obtained from the worksheet calculations on the Lucideon tiles, the yellow chip returns a higher value for the red signal than does the white chip. This is the case for all three images but increasingly so for the warmer the source of illumination. Since the chip SPDs indicate this should not be so, it would appear that the camera or Camera Raw matrixing (see Chapter 24) is incorrect.

Thus, in adjusting the exposure in Camera Raw to avoid an overload of the red signal on the yellow chip in Image (c), whilst retaining colour balance on the white chip, inevitably, the level of the white chip is reduced below the peak white level, an effect clearly seen on an inspection of Image (c).

(a)   (b)

(c)

Figure 11.5   The ColorChecker chart captured under various lighting situations.

Image (a)

    Situation: In the shade.

    Ambient: Full sun in a cloudless exceptionally blue sky in early afternoon.

Image (b)

    Situation: Full sun at about 45 degrees plus blue sky.

    Ambient: Full sun in a cloudless exceptionally blue sky in early afternoon.

Image (c)

    Situation: Tungsten lighting about 10 degrees from the norm in an average domestic situation.

    Ambient: Tungsten lighting.

12

Colorimetric Processing

12.1 Introduction

As we saw in Chapter 9 for accurate colour fidelity, it is essential that the colour spectral sensitivities of the camera match the colour matching functions (CMFs) of the primaries of the display image generator; however, it is not always practical or desirable for the display device to use the primaries associated with the camera. Also for practical reasons, the 'native' colour spectral sensitivities of the camera are unlikely to precisely match the required CMFs of the system primaries. Thus, 'colour gamut' transform processing or colorimetric transform processing[1] is usually required both at the camera and, where appropriate, at the display device.

As noted in Section 9.3, in many cases, there is not a one-to-one correspondence between the camera and the display device; several different forms of display device may receive their signals from a common camera source. In this event, it is essential to signal to the display device which primaries relate to the spectral sensitivities of the camera in order that it can match the signals to the display primaries with appropriate colour transform processing.

The advantages of the concept of transfer primaries, independent of both the camera spectral sensitivity curves and of the display primaries, will be reviewed in order to provide the background to establishing international standards for these critical parameters in later chapters.

In Chapter 9, it was shown that a conceptual colour reproduction system is defined by the chromaticity coordinates of its primaries which dictate the shape of the resulting camera spectral sensitivities. It can be seen however that if this was the full story, it would be somewhat limiting; for example, presuming that at a particular point in time, a set of primaries were chosen for a system which reflected those that were the best of those then available, how would developments which produced primaries capable of larger and improved gamuts be incorporated into the system?

---

[1] Often referred to as 'matrixing', the process adopted for carrying out this function.

Colour Reproduction in Electronic Imaging Systems: Photography, Television, Cinematography, First Edition. Michael S Tooms.
© 2016 John Wiley & Sons, Ltd. Published 2016 by John Wiley & Sons, Ltd.
Companion Website: www.wiley.com/go/toomscolour

This example of the limitation in the fundamental relationship between display primaries and camera spectral sensitivities could be resolved if there was some way in which the camera RGB signals could be processed such that they could be made to represent signals from a camera which matched a different set of primaries. The clue to how this may be achieved is outlined in Chapter 4 where the linear RGB CMFs derived by Wright were matrixed to produce the CIE XYZ CMFs.

The same approach is adopted for processing the linear RGB signals derived from the image sensors of colour cameras. However, operating mathematically on electrical signals requires a technological approach which is difficult to achieve accurately on analogue signals, and it was not until the advent of the application of digital technology to the processing of image-generated signals in the 1990s that this technique could be accurately applied.

The means by which matrixing was achieved on analogue signals is now historic and is therefore limited to being addressed in Chapter 17, where a brief description is given of the pre-digital days of television, when analogue signals were processed to provide an approximate match of the camera spectral sensitivities to the CMFs of the CRT display primaries.

This colour space conversion is often referred to as colour gamut transformation and the means of achieving the transformation as a transform, the latter being adopted in what follows. However, it should be appreciated that strictly speaking, the transforms required to achieve the results discussed above are actually chromaticity transforms; colour space transforms require luminance, the third dimension of colour to be included in the calculations, and as we shall see, the signal representing this parameter is not always linear. Thus, in the wider use of the term, though the matrix is at the heart of the transform, it may be complemented on its input and output by processors to linearise and functionalise the signal, respectively. Nevertheless, for the remainder of this chapter, reference to 'colour gamut' or just 'gamut' will mean reference to a 'chromaticity gamut' and 'transform' will describe the process of matrixing the RGB signals.

In the example outlined above, the problem of transforming from a comparatively small gamut to a larger gamut was addressed, a relatively simple situation since any colour existing in the source gamut is theoretically capable of being displayed in the target gamut. But what of the complementary situation, where signals ostensibly derived from a larger gamut camera are required to be displayed on a device with a relatively small gamut? For those colours which are constrained to the gamut of the display device, there is no problem, but those that are outside of the gamut cannot be displayed at their original chromaticities. A strategy is therefore required for mapping these colours into the gamut of the display device or printer; the process of implementing this strategy is referred to as gamut mapping.

It is sometimes convenient, though not strictly accurate, to refer to the signals derived from a capture device, such as a camera or a scanner, as being based upon a particular colour gamut, when more accurately, we are implying that the signals have been captured and processed in accordance with the CMFs of a particular set of primaries, which do describe a colour gamut. As we shall see, depending upon the spectral sensitivities of the capture device, any subsequent matrixing, and the integrity of the processing of the resultant signals, that is, whether or not the signals have been clipped in either the positive or negative direction, the capture device itself does not have a colour gamut, but subject to all elements of the signal being retained, it may be characterised as having one. Thus, by imposing a description of a colour gamut on a capture device, we are not limiting its ability to describe colours outside of that gamut, that is, unless the signals have been clipped, either optically or electronically.

12.2   Manipulating the Colour Space – Chromaticity Gamut Transformation

12.2.1   The Requirement to Change the Chromaticity Gamut

In addition to the example given in the introduction, there are a number of other reasons why it becomes necessary to undertake gamut transformation.

It was also shown in Chapter 9 that the camera spectral sensitivities that complement the primaries which are currently available have large negative lobes which cannot be implemented in a camera, since it is incapable of producing negative signals from the image sensors. As was also shown, this leads to a significant loss of colour fidelity. Thus, although one might envisage that the selection of primaries close to the spectrum locus of the chromaticity diagram or even beyond it would lead to much reduced or zero negative lobes respectively, the resulting RGB signals would not then match the display primaries.

One approach to this problem would be to make the camera spectral sensitivities match the X,Y,Z CMFs[2] (see Figure 4.4) and transform the signals so derived to match them to whatever set of display primaries were in use. Such an approach has an immediate advantage that since these characteristics have no negative lobes, then assuming the curves can be accurately matched by the various camera optical filters, the signal produced by the camera would be free of chromatic distortion for those colours contained within the gamut of the display primaries, which certainly no camera attempting to use only the positive characteristics of the matching spectral sensitivities of the primaries would be able to achieve. Unfortunately as we shall see later, such an idealistic approach has significant practical ramifications which initially had prevented its widespread adoption.

In a one-to-many camera to display systems, such as television, for example, much stress has historically been placed on minimising the complexity of the display device – it is much easier and more cost-effective to undertake any processing required once at the source prior to distribution, rather than in each display device. So traditionally in television terms, the RGB signals derived from the image sensors within the camera are processed to match as closely as possible the spectral sensitivities complementary to a display device with 'standard' primaries. The same has been historically generally true for any electronic one-to-many systems.

Even in the situation where RGB signals are derived from a matching set of primaries and spectral sensitivities, negative signals will be produced on those highly saturated colours which fall outside of the primaries' gamut. The display cannot of course react to a demand to produce negative light, so for the information contained in these negative signals, the display will limit at the maximum saturation it is capable of and any variation of colour which occurred in the negative range of the signal will be lost. The appearance is of a general loss of detail in highly saturated areas of the picture due to the 'clipping' of the signal over that portion of the spectrum which generates a negative signal.

One may question at this point whether there is anything to be gained by transferring signals which may from time to time contain negative excursions. In a viewing environment where all display devices are limited to a standard where the gamut of the primaries does not fully embrace the gamut of surface colours in the scene, the answer is that there is not. However,

---

[2] A camera or image capture device whose spectral sensitivities or spectral responsivities can be expressed as linear combinations of the colour matching functions of the *CIE 1931 Standard Colorimetric Observer* is sometimes described as colorimetric.

if, as the result of technological advances, some percentage of the display population is more sophisticated, in that their display gamut is greater than that of the gamut of the primaries represented by the transfer signals, then, subject to that device containing an appropriate transform to process the incoming signal to that of the display, an enhanced image will result. In the ultimate, should the primaries have chromaticity coordinates close to the 'Ideal Display' set described in Table 9.1, perfect colour rendition is possible for all known surface colours.

As we have seen in the above, there are a number of reasons why it might become necessary to change the colour gamut of the RGB signals and these may be summarised as follows:

- To match the camera native gamut to the system standard gamut
- To match historically recorded media to a current media system with a different gamut
- Before the advent of world standard primary sets, to match material from one country to that of another
- To match the system gamut to a different display gamut
- To accommodate the various 'capture' gamuts available in photography to the current display gamut
- To match the various sources of media available to the computer display gamut

12.2.2   Deriving a Matrix for Gamut transformation

12.2.2.1   Source Gamut Smaller than the Display Gamut

The basic approach to deriving a matrix for gamut transformation is the same for all of the examples listed above, but as a practical example of two systems using quite different display gamuts, we will look in some detail at the means of converting media recorded in the consumer photographic sRGB gamut format (which is also the chromaticity gamut for the current high definition [HD] television system) to the Adobe RGB gamut, which is the gamut adopted by some current professional displays. The sRGB gamut is specified by the International Colour Consortium and by IEC 61966-2-1:1999[3]; the world HD television gamut is specified in an International Telecommunications Recommendation – ITU-R BT.709-5. For shorthand convenience in what follows, since colorimetrically they are the same, we will refer to this mutual chromaticity space as the sRGB colour space or gamut.

The chromaticity coordinates of these two gamuts are listed in Table 12.1 and the gamuts are illustrated on the chromaticity diagram in Figure 12.1.

Consideration of the matrix mathematics used in deriving the relationships between the CIE standard X,Y,Z data, the chromaticities of the display RGB primaries and the camera spectral sensitivities in Chapter 9 indicate that this approach may be extended to process any set of mutually compatible linear data relating to one set of primaries to that relating to an alternative set of primaries.

The generic approach is to commence with the signals from the source camera and to successively apply a number of matrix operations, firstly, to convert the camera RGB trichromatic values back to the original scene XYZ values under the source camera illuminant and then apply the system white point chromaticities to convert these values to scene XYZ values under equal-energy illumination. At this point, the values represent the XYZ values of the colour as it would appear in the original scene under an equal-energy illuminant and may therefore be

---

[3] See Section 24.2.1.

Table 12.1 System primaries chromaticities

	x	y	z	u'	v'
Rec709 and sRGB					
Red	0.6400	0.3300	0.0300	0.4507	0.5229
Green	0.3000	0.6000	0.1000	0.1250	0.5625
Blue	0.1500	0.0600	0.7900	0.1754	0.1579
White D65	0.3127	0.3290	0.3583	0.1978	0.4683
Adobe RGB					
Red	0.6400	0.3300	0.0300	0.4507	0.5229
Green	0.2100	0.7100	0.0800	0.0757	0.5757
Blue	0.1500	0.0600	0.7900	0.1754	0.1579
White D65	0.3127	0.3290	0.3583	0.1978	0.4683

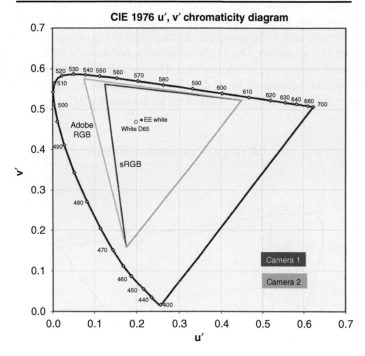

Figure 12.1 The chromaticity gamuts of the system primaries.

processed by matrices representing a camera with the new system primaries and white point, that is, with the target spectral sensitivities relating to the CMFs of the Adobe RGB primaries.

In practice, as shown in Appendix F and Worksheet 12(a) Cells C47:F50, the matrices may be concatenated into one matrix and lead to the result as follows for converting signals from a camera with spectral sensitivities associated with sRGB primaries to signals which appear to have been derived from a camera with Adobe RGB spectral sensitivities. (The 'a' subscript refers to Adobe RGB and the 's' subscript to sRGB.)

$$R_a = 0.7152R_s + 0.2849G_s - 0.0001B_s$$
$$G_a = 0.0000R_s + 1.00G_s + 0.0000B_s$$
$$B_a = 0.0000R_s + 0.0412G_s + 0.9588B_s$$

Note that the sum of the coefficients of RGB in each line sum to 1.0000, a necessary condition when the system white point is the same for both colour spaces in order to preserve white balance when the matrix is inserted into the signal path. Also, in this particular case, the green output from the matrix is unmodified by the red and blue signals; this is because the blue and red primaries chromaticities happen to be identical for the sRGB and Adobe RGB specifications.

As the system white for both sets of specifications is D65, the appearance of colours on the Adobe RGB display when supplied with the corrected signals will be identical to the same colours as seen on an sRGB display.

If however the signals from the sRGB camera had been applied direct to the Adobe RGB display without the matrix correction, then the displayed colours would have been in error. Figure 12.2, which is derived in Worksheet 12(b), illustrates the colour gamuts of the two systems and the shift in chromaticities which occurs on the Lucideon tiles when viewed on an Adobe RGB display without gamut correction; the arrowheads indicate the direction of shift of the chromaticities. In the case of these two gamuts, the red and the blue primaries have the same chromaticities, so in consequence, there is a diminishing shift of chromaticity as the red–blue axis is approached. The displayed colours, particularly in the green area of the chromaticity diagram, would appear too saturated. However, in the more general case, where the new gamut extends the triangle in all directions, the signals are in effect driving more saturated primaries than those they are matched to, so the result will be a general increase in

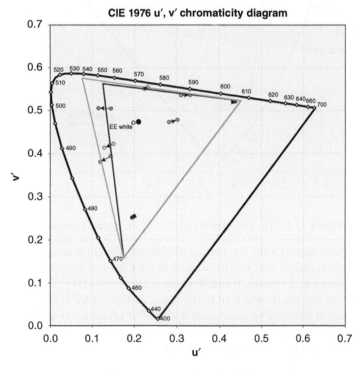

Figure 12.2 Chromaticity shifts resulting from applying uncorrected sRGB signals to an Adobe RGB display.

the saturation of the scene, that is the scene will be oversaturated. In the example reviewed here the effect when viewed on screen is very obvious.

Although it is convenient to use the chromaticity diagram to illustrate the change in chromaticity, it must be remembered that colour is a three-dimensional quantity and one should also take into account the change in luminance which may occur when non-matching signals are applied to a display. In Section 4.6, the CIELUV Colour Difference parameter was defined which measures the difference between two colours in the LUV colour space, where a value of 1 is approximately equivalent to a just noticeable difference (JND) in the colours. Table 12.2 shows the CIELUV colour difference values for the Lucideon tiles when signals derived from an sRGB camera are applied without correction to an Adobe RGB display.

Table 12.2   Colour difference values between correct and displayed colours when signals from an sRGB camera are viewed on an Adobe RGB display without correction

Tile	White	Orange	Cyan	L Green	D Blue	Yellow	Rose	Red
ΔE^*uv	0	13	21	22	2	9	8	16

These colour difference values at over 20 JNDs are subjectively very noticeable and what also is worth noting is that the red tile, which shows only a small chromaticity difference, nevertheless has a high colour difference value due to the large change in luminance which occurs.

12.2.2.2   Source Gamut Larger than the Display Gamut

In the above example of unmatched source and display gamuts, the chromaticity coordinates of the source gamut were contained within the gamut of the display primaries. In consequence, the transform matrix resulted in RGB levels which were smaller than the original levels and thus the display was able to accommodate them and display the colours accurately. However, if the opposite is true, in that the source gamut exceeds the display gamut, then following the correction matrix, any scene colours within the display gamut will result in positive RGB levels but those outside the display gamut will produce a negative value in one or other of the RGB levels.

Assuming for the moment that the reproduction system is able to accommodate the negative signals from the correction matrix, then these signals will be applied to the display device, which clearly will be unable to respond. In effect therefore, whenever out-of-display gamut colours are viewed by the camera, then one or other of the RGB signals will appear at zero level to the display, that is, the negative signals will be 'clipped' by the display.

In order to illustrate the effect of clipping, it is necessary to use colours which are outside the display gamut. Unfortunately, the reference colours in the ColorChecker chart and the Lucideon tiles are very nearly constrained within the sRGB gamut and are therefore unsuitable for this investigation. Nevertheless, many flowers in particular do present colour surfaces which are outside the sRGB gamut. Therefore, a set of hypothetical spectral reflectance distributions (SRDs) representing highly saturated additive and subtractive primaries within the Pointer gamut of surface colours have been generated in the 'Surfaces' worksheet for use in exploring the effect of out-of-gamut colours in reproduction.

In addition, in order to avoid ambiguity in resolving the cause of the effects which occur, we will assume in the following that the source camera has colour spectral sensitivities which match the 'Ideal' primaries defined in Section 9.3, albeit at this stage we have not yet dealt with how to produce the negative lobes in the spectral sensitivities of such a camera.

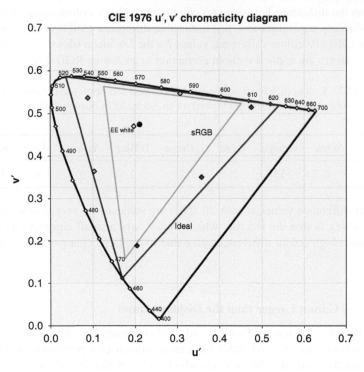

Figure 12.3 Illustrating the location of the hypothetical saturated sample colours between the gamuts of the camera and the display.

Thus the source camera gamut, the display sRGB gamut and the hypothetical saturated colour samples which are located within the Pointer gamut of real surface colours are illustrated in Figure 12.3.

By selecting the appropriate primaries in Worksheet 12(a), the conversion matrix required to modify the RGB signals from the 'Ideal' display gamut camera to emulate those that would originate from an sRGB display gamut camera may be obtained as follows:

$$R_s = 1.78R_I - 0.74G_I - 0.04B_I$$

$$G_s = -0.12R_I + 1.14G_I - 0.03B_I$$

$$B_s = -0.02R_I - 0.08G_I + 1.10B_I$$

This matrix may be used to calculate the values of the RGB signals from the emulated sRGB camera, whereupon the negative values obtained may be 'clipped' to zero to emulate the effect of applying negative signals to a real display.

In addition to the practicality of the situation where the display cannot respond to negative signals, the system constraints are usually such that signals greater in value than those obtained from the system white reference colour, that is a value of 1.00 for R, G and B in the above example, cannot be accommodated and therefore the positive values in excess of 1.0 are also clipped to a maximum value of 1.00.

Table 12.3   Camera RGB values obtained before and after matrixing

Camera signals		Red	Green	Blue	Yellow	Cyan	Magenta	ER White
1. Camera:	R	0.56	0.17	0.11	0.96	0.02	0.70	1.00
'Ideal'	G	0.03	0.86	0.03	0.41	0.46	0.03	1.00
primaries	B	0.02	0.25	0.98	0.02	1.02	0.71	1.00
2. Camera:	R	0.96	−0.35	0.14	1.41	−0.35	1.20	1.00
Emulated	G	−0.03	0.96	−0.01	0.36	0.49	−0.07	1.00
sRGB primaries	B	0.01	0.21	1.08	−0.03	1.08	0.77	1.00
3. As 2:	R	0.96	0.00	0.14	1.41	0.00	1.20	1.00
Negative values	G	0.00	0.96	0.00	0.36	0.49	0.00	1.00
to zero	B	0.01	0.21	1.08	0.00	1.08	0.77	1.00
4. As 3:	R	0.96	0.00	0.14	1.00	0.00	1.00	1.00
Values > 1	G	0.00	0.96	0.00	0.36	0.49	0.00	1.00
limited to 1	B	0.01	0.21	1.00	0.00	1.00	0.77	1.00

The RGB values from the camera for these four conditions are calculated in Worksheet 12(b) and detailed in Table 12.3. Points to note are that in the case of the original camera with the 'Ideal' camera spectral sensitivities, the RGB signals are all positive for all the colours, since they all have chromaticities within the gamut of the 'Ideal' display primaries. In the emulated sRGB camera values produced by the above matrix, each of the colour samples, with the exception of equal reflectance (ER) white, produces only one negative value in each group of RGB values, since the apexes of the colour gamut are on the spectrum locus. It will be noted that four of the colour samples also produce one of the signals at a greater value than the allowed maximum of 1.00. In the third and fourth set of RGB values, firstly, the negative values are clipped and then in the fourth set, both the negative values and those values above 1.00 are clipped to their limiting values.

In Worksheet 12(b), the RGB values in Table 12.3 are converted to u', v' values and the difference vectors between the Ideal camera and the two versions of the emulated cameras chromaticities are plotted on the chromaticity diagrams illustrated in Figure 12.4 and 12.5, respectively.

In Figure 12.4, it is apparent that the result of clipping the negative values of the matrixed RGB signals is to place their displayed chromaticities on the periphery of the sRGB gamut. Though we have approached this result as a consequence of matrixing, it is evident that an idealistic camera whose spectral sensitivities matched the sRGB primaries would of course produce the same results. As the change of chromaticity is towards the system white point of

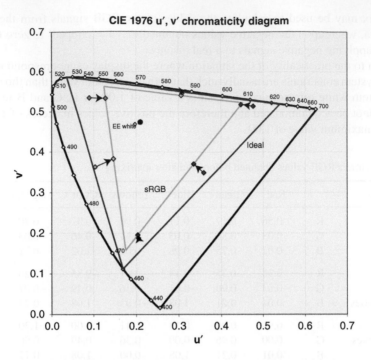

Figure 12.4 Effect of limited display gamut on saturated scene colours. RGB negative values clipped to zero.

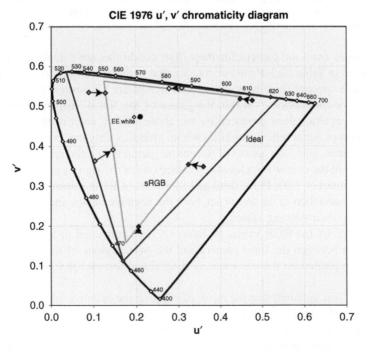

Figure 12.5 Negative values and values in excess of 1.00 clipped to zero and 1.00, respectively.

the diagram, to a first degree of approximation, the distortion is characterised by a reduction in the reproduced saturation of the colour samples.

The result of also clipping the positive values of RGB which exceed a value of 1.00 is illustrated in Figure 12.5, where it is apparent that the directions of some of the arrows have been skewed further away from pointing approximately at the system white chromaticity, indicating a significant change in hue as well as saturation. As noted previously, hue changes are very much more apparent than saturation changes.

Although the chromaticity charts in Figures 12.4 and 12.5 indicate a similar amount of chromatic distortion, in fact, when the CIELUV colour difference values are calculated, it can be seen from Table 12.4 that the yellow and magenta samples suffer considerably more as a result of the signal levels being limited to 100%. The yellow sample in particular is reproduced at a significantly lower level of luminance as the 40% overload of the R signal is clipped to 100%.

Table 12.4   CIELUV colour difference values for sRGB limited display primaries

Out-of-range values clipped	CIELUV	Red	Green	Blue	Yellow	Cyan	Magenta	ER White
Negative	ΔE^*uv	13	39	1	6	36	16	0
All	ΔE^*uv	13	39	5	37	39	27	0

These clipping distortions manifest themselves in the reproduction where subtle changes in highly saturated areas beyond the display gamut are eliminated by the clipping action and can appear at first sight to be a localised loss of focus. The effect is very apparent in close-up shots of flowers such as red roses and in the fabric of clothes.

12.2.2.3   Moving Successively Between Gamuts

Taking note that when manipulating colour gamuts in a comprehensive colour processing system, the move from a small gamut to a larger gamut which embraces the smaller gamut can be made without changing the chromaticities of the colours and when moving from a large gamut to a smaller gamut, any colours with chromaticities outside of the smaller gamut will be clipped. It is important to recognise therefore that in a processing system which does not retain negative RGB values, then:

- successively moving from a small to a large gamut and back again does not change the chromaticities of the colours, but
- successively moving from a large to a small gamut and back again will permanently loose the chromaticities of those colours outside the smaller gamut.

12.3   Gamut Mapping

Gamut mapping is the process whereby out-of-gamut colours are mapped to bring them within the display gamut in a manner which preserves as far as possible the perception of good colour rendering. Immediately, it can be seen that this is a subjective description of the process and, in consequence, there are a number of strategies for implementing gamut mapping depending upon both the range of colours out of gamut and the desired appearance of the result. These strategies are referred to as *rendering intents*.

Clearly the examples given in Section 12.2.2.2 whereby a range of saturated colours are effectively automatically clipped by both the system constraints and the inability of the display to respond to negative signals is a crude form of gamut mapping where the chromaticities of the out-of-gamut colours are relocated on the display gamut extremities in an uncontrolled fashion. For out of gamut colours this leads to: loss of colour differentiation for colours of the same hue but of different luminance or saturation; loss of saturation; and selective hue shifts. However, scene chromaticities located within the display gamut are reproduced accurately.

One may envisage a range of approaches to selecting a strategy for improving on the crude result of leaving the system to clip the values of the signals produced by the out-of-gamut colours. Perhaps the most obvious and simplest solution would be to bring the chromaticities of those out-of-gamut colours to the nearest in gamut point. A more sophisticated approach would be to ensure that in restraining the saturation, the new location would lie on the cross point of the gamut extremity and the path between the original chromaticity and the system white point, thus reducing the shift in perceived hue.[4] Figures 12.6 and 12.7 illustrate the effect of adopting these two approaches, respectively.

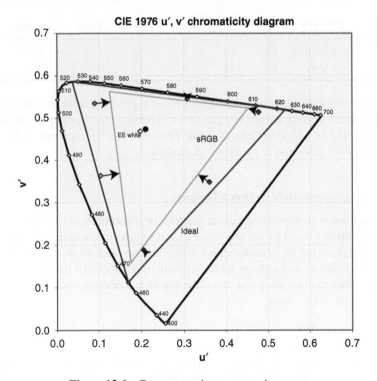

Figure 12.6   Gamut mapping – nearest in gamut.

Since in a particular scene a subject such as a rose of highly saturated colour is often likely to have a small range of variation of out-of-gamut chromaticities, a range which would be significantly reduced or eliminated by either of the two strategies outlined above; a different strategy is required if a variation in this range of chromaticities is to be preserved. The most obvious

---

[4] As Figure 4.27 illustrates, lines of constant hue on the u′,v′ chromaticity diagram are not always straight lines; thus, this strategy does not always lead to a preservation of the original hue.

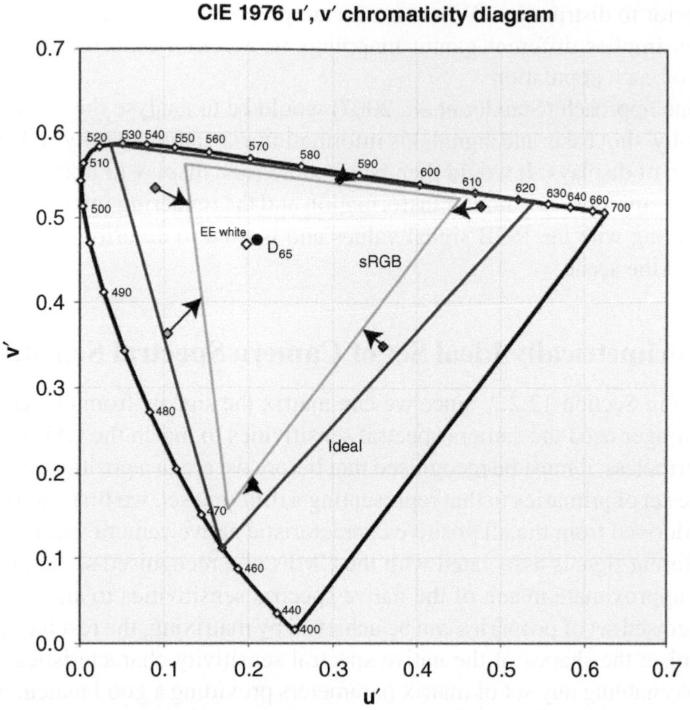

CIE 1976 u′, v′ chromaticity diagram

Figure 12.7 Gamut mapping – hue retention.

approach is to reduce the saturation of signals representing the scene until the most extreme of the out-of-gamut colours is just located on the gamut extremity. This approach preserves all the scene variations in a reduced manner, but its acceptability is dependent upon the scene; if the out-of-gamut colours cover a significant or important element of the scene and are just out of gamut, the result is acceptable. However, if the out-of-gamut colours occupy a small area, are highly saturated and unimportant in perceiving the scene, then they can cause a dramatic reduction of saturation, which is particularly unacceptable if only a very small unimportant area of the scene is driving the reduction in the saturation of the remainder of the scene.

A fourth approach, which is a modification of that described above, is to use a compression algorithm to successively reduce the saturation of those colours which lie close to and beyond the display gamut, just bringing the most saturated colours into the display gamut and thus preserving the saturation of the majority of colours in an average scene.

It is clear from the above that the strategy to be adopted for best results will depend to a large degree on the scene and the distribution of the out-of-gamut colours within it. This makes it difficult to adopt a particular strategy for a range of scenes and implies that for rapidly changing sequential scenes such as those appearing in television or cinematography, to achieve optimum results requires a choice of approach. Either each scene is analysed individually in post-production and an appropriate rendering intent is applied or a very sophisticated algorithm is used to monitor the out-of-gamut situations and apply an appropriate rendering intent automatically, which may be a complex mix of those strategies described above.

Such sophisticated solutions are unlikely to be practical within the display device at the present time, implying that gamut mapping for the cinema and DVDs will, for the time being, be undertaken prior to distribution. Where media are prepared for a range of different display populations, this implies different gamut mappings to accommodate the possibly different display gamuts of each population.

Ultimately, one approach (Stauder et al., 2007) would be to analyse the extent of the *scene* gamut on a shot-by-shot basis and signal this information via gamut identity (ID) metadata to a mixed population of displays. It would then be possible for a display to interpret the metadata and apply both the appropriate gamut transformation and the rendering intent. Metadata is data which are sent along with the RGB signal values and is used to describe parameters relating to the shooting of the scene.

12.4   A Colorimetrically Ideal Set of Camera Spectral Sensitivities

As we have seen in Section 12.2.2, since we can matrix the signals from one colour space to another, we no longer need the camera spectral sensitivities to match the CMFs of the display primaries. Nevertheless, it must be recognised that before we are in a position to matrix signals representing one set of primaries to that representing a different set, we first need to process the camera signals derived from the all positive characteristic native camera spectral sensitivities, to represent different signals associated with the CMFs of a recognised set of primaries.

Although an approximate match of the native spectral sensitivities to an established set of CMFs of a recognised set of primaries can be achieved by matrixing, the result is almost invariably a compromise; the shapes of the native spectral sensitivity characteristics are generally not conducive to enabling any set of matrix parameters providing a good match. Nevertheless, by recognising that by initially selecting primaries with associated CMFs which are easier to match with appropriate optical filtering, significantly better compromises can be achieved.

In the following, three alternative approaches to deriving an ideal set of camera spectral sensitivities based on primaries located outside of the spectrum locus are considered.

12.4.1   The CIE XYZ Primaries Approach

In order to avoid negative lobes, the primaries must be selected to be outside of the spectrum locus of the chromaticity diagram, as was apparent from the definition of the position of the CIE XYZ primaries in Chapter 4. In fact, the spectral sensitivities of the camera could be made to be the XYZ CMFs, and a suitable matrix could then be used to emulate the signals being derived from a camera matching the display primaries. Such an approach of using *imaginary* primaries effectively overcomes the impossibility of designing a camera with negative lobes.

The CIE XYZ primaries are illustrated on the u′,v′ chromaticity diagram in Figure 12.8. It will be noted that whilst the Y and Z primaries are located reasonably close to the spectrum

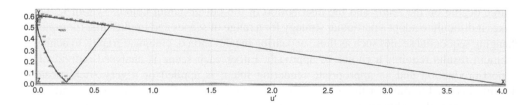

Figure 12.8   The CIE XYZ primaries located on the u′,v′ chromaticity diagram.

locus, the X primary is, in relative terms, very much further away, to the extent that the diagram must be extended by over five times in the *x*-direction in order for it to be accommodated. The reason the X primary is so much further away from the spectrum locus on the u′,v′ as opposed to the *x,y* chromaticity diagram is because of the stretch which was applied to this projection of the diagram to make the circles of constant JND in the red-to-blue area of the diagram more equally match those in the complementary area of the diagram.

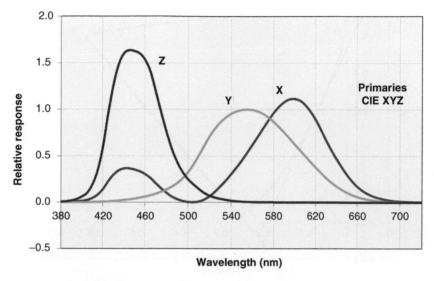

Figure 12.9 CIE primaries camera spectral sensitivities.

The matching camera spectral sensitivities for the XYZ primaries with a system white point of D65 are illustrated in Figure 12.9. The relative heights of these curves differ slightly from the classic X, Y, Z curves as a result of using D65 rather than equal-energy white as the system white.

However, as we shall see in Chapter 14, there are pros and cons in adopting such an approach for the camera. For the reasons outlined in Chapter 4, the XYZ primaries are situated some way from the spectrum locus and, as a result, produce 'red' and 'green' colour mixture curves which seriously overlap, which in turn will reduce the sensitivity of those cameras using three separate image sensors.

12.4.2 Primaries Derived from a 'Symmetrical' Approach to the u,v Chromaticity Diagram

The constraints for positioning the location of the XYZ primaries detailed in Chapter 4 are not relevant to a colour camera, and therefore, the imaginary primaries for a camera may be located close to the spectrum locus. As described in Chapter 14, such an approach ensures that less code values are 'wasted' on values with no real colour significance. One such set, which may be referred to as an 'Ideal 1' set of camera analyses characteristics, is based upon the primaries which are illustrated in Figure 12.10.

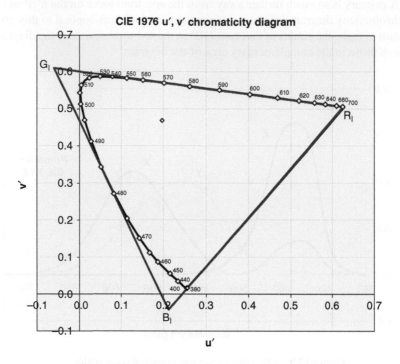

Figure 12.10 Gamut of 'Ideal 1' camera primaries.

The colour gamut is first constructed and the coordinates of the primaries are defined by the apexes of the resulting gamut triangle.

Two extended straight lines are constructed to overlay the two straight line sectors of the enclosed spectrum locus, which are the straight section between the red and the green areas of the spectrum locus and the line which joins the two ends of the spectrum locus.

The third line of the gamut is arbitrarily chosen to just graze the spectrum locus and provide roughly equal areas of the gamut outside of the spectrum locus adjacent to the green and blue primaries, respectively.

The resulting primary chromaticities are detailed in Table 12.5.

Table 12.5 'Ideal 1' camera chromaticity coordinates

'Ideal 1' camera	x	y	z	u'	v'
Red	0.7347	0.2653	0.0000	0.6234	0.5065
Green	−0.3258	1.3258	0.0000	−0.0666	0.6100
Blue	0.1413	−0.0104	0.8691	0.2180	−0.0360
White D65	0.3126	0.3290	0.3584	0.1978	0.4683

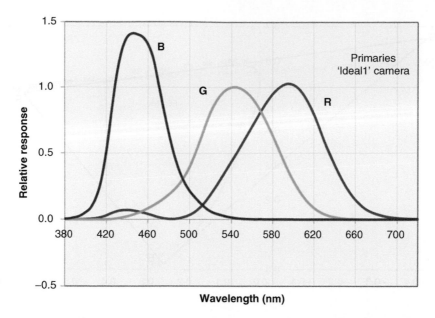

Figure 12.11 Camera spectral sensitivities of the 'Ideal 1' camera primaries.

The camera spectral sensitivities for the 'Ideal 1' camera primaries are illustrated in Figure 12.11.

You will note that there are no negative lobes to complicate matters and only one minor secondary positive lobe on the red characteristic. In addition, there is a greater separation of the red and green curves when compared with those derived from the XYZ primaries. Both of these factors will ease the optical filter design of the camera.

A further factor, addressed in Chapter 14, is that real colours utilise more of the available signal code values than when XYZ primaries are used.

Nevertheless, the overlapping red and green curves mean that light over a considerable portion of the spectrum between about 520 nm and 600 nm must be more closely shared between the red and the green sensors than is the case when spectral sensitivities are based upon primaries within the spectrum locus, leading to a drop in three-sensor camera sensitivity.

12.4.3 Primaries Derived from a 'Single External Area' Approach to the u,v Chromaticity Diagram

This approach is based upon a relatively minor amendment to the previous approach.

As illustrated in Figure 12.12, the colour gamut is first constructed and the coordinates of the primaries are defined by the apexes of the resulting gamut triangle.

As for the 'Ideal 1' camera, two extended straight lines are constructed to overlay the two straight line sectors of the enclosed spectrum locus, which is the straight section between the red and the green areas of the spectrum locus and the line which joins the two ends of the spectrum locus.

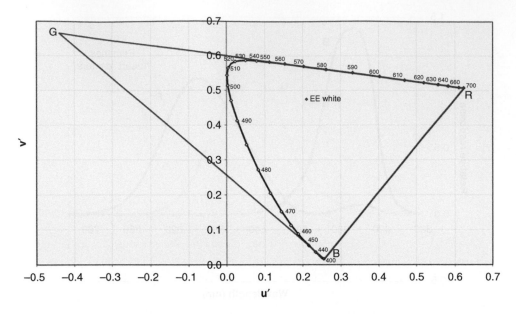

Figure 12.12   Gamut of 'Ideal 2' camera primaries.

One end of the third line of the gamut is located at the extreme of the blue end of the spectrum locus and meets the extended RG line in such a manner that the BG line is just tangential to the blue end of the spectrum locus. Thus, only one area external to the spectrum locus results, albeit it a considerably larger area than that which resulted for the 'Ideal 1' camera approach.

The resulting primary chromaticities are detailed in Table 12.6.

Table 12.6   'Ideal 2' camera chromaticity coordinates

'Ideal 2' camera	x	y	z	u'	v'
Red	0.7347	0.2653	0.0000	0.6234	0.5065
Green	−2.0578	3.0593	−0.0015	−0.4400	0.6659
Blue	0.1741	0.0050	0.8209	0.2568	0.0166
White D65	0.3126	0.3290	0.3584	0.1978	0.4683

The camera spectral sensitivities for the 'Ideal 2' camera primaries are illustrated in Figure 12.13.

As with the 'Ideal 1' solution, there are no negative lobes to deal with. However, a further factor which will aid the optical filter design of the camera is that there are no secondary positive lobes.

The red and green curves are closer together however, to the point where they share much of the spectrum and thus lead to a loss of camera sensitivity in those cameras where the light is shared between the image sensors.

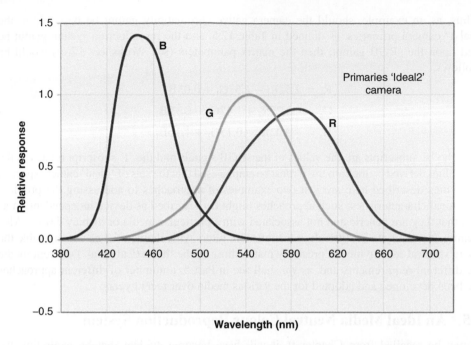

Figure 12.13 Camera spectral sensitivities of the 'Ideal 2' camera primaries.

A further factor, addressed in Chapter 14, is that real colours utilise more of the available signal code values than when XYZ primaries are used.

It is interesting to note that the general shape of the 'Ideal 2' spectral sensitivities illustrates a marked similarity to the colour characteristics of the eye, though the separation of the red and green characteristics is greater than it is for the eye.

12.4.4   General Considerations

It must be emphasised that native camera spectral sensitivities are usually proprietary and their characteristics are not published by the manufacturer. However, it can be hypothesised on the basis of the above that the camera native gamut primaries used as the basis for the design of the camera optical spectral sensitivities will probably be situated either very close to or just outside the spectrum locus in order to ensure that the resulting CMFs do not exhibit significant negative lobes. It may be that there would be some compromise between the 'Ideals' proposed here and that which would result from primaries located just within the spectrum locus, in order to produce for a particular camera model the best compromise between colour fidelity and sensitivity.

With the much improved sensitivity of modern sensors, one could envisage that for critical colour work, it would be a satisfactory compromise to produce a camera which met the above criteria by sacrificing a degree of sensitivity for the opportunity of near-perfect colour rendition for colours within the gamut of chosen system primaries. When the camera native gamut does not match the system gamut, then the manufacturer will embed within the camera a matrix which converts the native RGB signals to those that would appear as if they had been derived from a camera with the system-defined gamut.

Thus, as an example, should the camera native chromaticity gamut be based upon the 'Ideal 1' camera primaries, as defined in Table 12.5, and the reproduction system gamut be based upon the sRGB gamut, then the matrix parameters (see Worksheet 12(c)) would be as follows:

$$R_s = 2.52R_I - 1.57G_I + 0.05B_I$$
$$G_s = -0.27R_I + 1.42G_I - 0.15B_I$$
$$B_s = -0.02R_I - 0.15G_I + 1.16B_I$$

where the 's' subscripts are the values of the sRGB signals and the 'I' subscript relates to the RGB values derived by the camera. It must be emphasised that the sets of 'ideal' camera spectral sensitivities described here are just two examples of approaches to addressing the problem of the ideal characteristics; such approaches might be described as device independent, in as much that they are generic and not associated with a particular media or display device. Also in reality, it is likely to be difficult to manufacture filters with characteristics which enable the camera spectral sensitivities to precisely match those of the theoretical ideal. Different media have different requirements and, as we shall see in Part 5, a number of different approaches have been developed and adopted for the various media over recent years.

12.5   An Ideal Media Neutral Colour Reproduction System

As may be recalled from Chapter 9, it will have become evident that by exploiting the functionality of matrixing, it is possible with the assumptions outlined above to design a colour reproduction system which is capable of perfect colour rendition for those scene colours circumscribed by the range of Pointer surface colours.

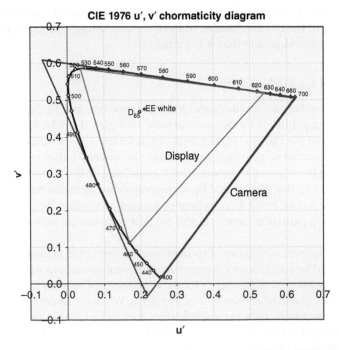

Figure 12.14 The ideal camera and display gamuts for perfect reproduction of the Pointer surface colours.

Thus, if signals from a camera with spectral sensitivities matching those of the 'Ideal 1' camera are matrixed to emulate a camera with spectral sensitivities matching the 'Ideal' display primaries derived in Chapter 9 and subsequently used to drive such a display, then perfect colour rendition for all surface colours will be achieved.

These ideal camera and display chromaticity gamuts are illustrated in Figure 12.14.

The matrix required is derived in Worksheet 12(c) and is as follows:

$$R_D = 1.38R_C - 0.39G_C + 0.01B_C$$

$$G_D = -0.10R_C + 1.21G_C - 0.11B_C$$

$$B_D = 0.00R_C - 0.05G_C + 1.05B_C$$

where the subscripts 'C' relate to the camera and 'D' to the display, respectively.

12.6   Using System Primaries or Device-Independent Encoding

As technology improves to the point where incorporating individual matrices in advanced displays can be achieved cost-effectively and display devices appear with extended gamuts but are also capable of receiving signals from historic colour systems, the requirement to consider the concept of using system primaries or device-independent encoding becomes a priority. Thus, one may envisage the situation where the camera designer is not limited by a specific set of display primaries in specifying the spectral sensitivities of the camera, which might be designed to complement the greatest possible display gamut, whilst the display designer may design the display with a set of primaries to suit either a cost or performance criterion. As long as there is a set of recognised standardised system primaries, which encompass all surface colours, between these elements of the system, good colour reproduction should result, with the superior displays providing enhanced results for the situation where the system gamut equals or exceeds the display gamut.

Frequently, the population of displays are remote from the source of the signals and thus require a common distribution system to transfer the signals from the source to the display. The gamut of the signals to be transferred is called the system gamut and it is likely that a display may be required to receive signals from different systems with different system gamuts.

Ultimately, one way of ensuring the recognition of which system primaries are in use for a particular system is to signal their identity in terms of their intended display chromaticity coordinates by encoding the information as a profile within the signals themselves. In this way, a display system will use the current profile to identify the system primaries and set up the matrix processing appropriately to match the system primaries to the display primaries.

In an ideal situation, the system gamut will be based upon a set of standard imaginary primaries to ensure that no avoidable colour distortion will occur as a result of any constraints in the transmission system or at the display matrix which matches the system gamut to the display gamut. (It may be remembered that once a system has been constrained by a limited gamut, it is not possible to recover the RGB values of out-of-gamut colours.)

In legacy systems however, which were standardised before it was economic to install a matching matrix in each display, the display chromaticity coordinates were first standardised on the then currently available primaries and the system gamut then made to be identical to the display gamut. Such a practical approach works well within the constraints of the system; however, if modern displays with wider gamuts than the system standard are introduced into the system, then as discussed in Section 12.2, a matching matrix will be required to avoid artificially enhanced saturation of the scenes transmitted.

13

Preserving Tonal Relationships – Tone Reproduction and Contrast Laws

13.1 Introduction

As noted in Chapter 9, the universal image display up to the turn of the last century was based upon the cathode ray tube (CRT), which has an electro-opto transfer function (EOTF) which follows a power law function, and it was necessary therefore to correct for this characteristic in the processing of the RGB signals which drive the display.

Traditionally, the power law exponent of display characteristics has been expressed by the Greek character γ, pronounced 'gamma'; thus historically, the signal flow element in the camera responsible for providing the complementary characteristic to the display is described as the gamma corrector. However, it has long been recognised that the gamma corrector has other benefits, and thus despite the demise of the CRT, the gamma corrector has been retained. In addition of course, unless a totally new system is introduced, the gamma corrector is necessary to service the legacy population of displays. A consequence of this decision is that when linear displays are in use in a legacy system, a de-gamma process is required in the display device in order to obtain linear signals for further processing before presenting them to the display.

With few exceptions, all reproduction systems are limited to providing a scaled-down linear relationship between the luminance of the scene and the luminance of the display, since only very recently have highly specialised displays become available which can begin to match the luminance of the original daylight or studio scene. Inevitably, this limitation means that a reproduced image of limited luminance is viewed under a range of different levels of environmental lighting, which, with the exception of a completely darkened room, compete with the luminance level of the image in exploiting the adaptation characteristic of the eye. As described in Chapter 10, this will reduce the effective contrast range of the display and, in turn, limit the ability of the eye to produce an accurate representation of the contrast range of the scene. The effect can be partially compensated for by adjusting the overall contrast law of the system away from a linear relationship, such that the overall gamma of the system is slightly above unity.

Colour Reproduction in Electronic Imaging Systems: Photography, Television, Cinematography, First Edition. Michael S Tooms.
© 2016 John Wiley & Sons, Ltd. Published 2016 by John Wiley & Sons, Ltd.
Companion Website: www.wiley.com/go/toomscolour

One of the most important criteria of a colour reproduction system is the ability for a range of tones from black to white in a scene to be perceived in the same relative relationship in the reproduced image.

In order to achieve this outcome, it might appear at first sight that the overall system contrast law between the scene contrast and the reproduced image contrast should be linear, that is, that the display should reproduce precisely the complete range of scene luminances; however, when considering the range of contrasts of various scenes, it becomes clear that such an objective is generally impractical because of the limited luminance of the display.

Nevertheless, an inspection of Figure 1.6, which illustrates the ability of the eye to adapt to a range of contrasts over a wide range of luminance levels, indicates that subject to certain constraints, it should be possible to render images of a scene at lower levels of luminance and contrast range than the original, which will be regarded as perceptually good reproductions of the original and this is indeed the case. These constraints are to varying degrees:

- a proportional relationship between the luminances of the scene and of the reproduced image;
- the adaptation of the eye to the average luminance of the display;
- the adoption of the recommendations described in Chapter 9 regarding the illumination of the surrounding viewing environment.

However, as will be seen later in this chapter, an improved rendition of the scene, in terms of perceptual acceptability, will often result if a small degree of non-linearity is introduced into the relationship between scene luminance and image luminance. The extent of this non-linearity is dependent upon the lighting of the viewing environment, the highlight luminance and contrast range of the display, and the angle of view the reproduced image subtends at the eye.

In addition to the above considerations, elements of the signal chain between the scene and the reproduced image may themselves fundamentally have non-linear characteristics and may in addition introduce artefacts, the perceptibility of which, because of the non-linear response of the eye, will be dependent upon where in the signal chain they are introduced.

As will be seen later in this chapter, a change in contrast law will change the colour of the reproduced image at the display, and therefore, management of the overall contrast law of a reproduction system is critical to the colour fidelity of the reproduced scene.

In order to determine the factors affecting the rendition of contrast range, we need to be aware of the parameters associated with the contrast ranges of:

- the scene;
- the screen as illuminated for viewing;
- the eye at the luminance levels of the reproduced image.

The manner in which the requirement to match these differing contrast ranges is achieved will be addressed in the remainder of this chapter.

13.2   Terms and Definitions

13.2.1   Non-Linear Transfer Characteristics and Contrast Laws

The literature on this topic uses a range of expressions to describe the parameters and relationships associated with the measurement and reproduction of contrast, with respect to both the opto-electronic image sensors in the camera and the electronic-optic display devices. Between these two devices, located at either end of the reproduction system, are electronic circuits which are designed to:

- introduce specific non-linearity into the system to correct for any non-linearity in system elements;
- ensure a controlled, and often adjustable, small degree of non-linearity, to compensate for the limitations of the viewing environment;
- exploit the non-linearity of the eye by introducing non-linearity into the system prior to the introduction of unavoidable artefacts and, subsequently, prior to the display, provide correction by the insertion of circuit elements with a complementary non-linearity characteristic.

Each of these three requirements for electronic correction will be reviewed independently later in this chapter, but before investigating further the various relationships between the input and the output of these elements in the reproduction system, it may be helpful to define the terms we intend to use.

13.2.2 Nomenclature

Where the physical units of the input and output of a system element are identical, we will use the expression 'contrast law' to describe this relationship. In mathematical terms, this function, when used in a media signal chain, is always described by a power law (or a variant of a power law) and, in the following example, the input (I) and output (O) units are voltages:

$$V_o = (V_I)^\gamma$$

where the exponent uses the Greek letter 'gamma' and will usually have a value which falls between 0.3 and 3.0.

The use of 'gamma' for the exponent has led to the word 'gamma' coming into common usage as a means of describing the various non-linear devices and circuits within the signal path of image reproduction systems. Thus, a CRT will be described as having a gamma of approximately 2.4, and a circuit to correct for this non-linearity will be described as a gamma corrector with an exponent equal to the reciprocal of 2.4, that is, approximately 0.42.

Where the input and output of a system relate to different physical units, the terms adopted by the international standards bodies will be used in the context of the chapter they appear in. The International Telecommunications Union (ITU), which sets the television standards, uses the term 'transfer functions', whilst the International Organisation for Standardisation (ISO) uses the term 'conversion functions'; for consistency in this chapter, we will use 'transfer functions'. Thus the optical image sensor in the camera will have a contrast law described by its *opto-electro transfer function* or OETF, and the contrast law of a display will be described by its *electro-opto transfer function* or EOTF. (International standards bodies use both 'electronic' and 'electro' in defining these transfer functions.) The simple phrase *transfer characteristic* remains in some conversations.

In mathematical terms, a camera image sensor may have an OETF described by the following function:

$$V = (E)^\gamma$$

where E is the level of illumination of the image sensor in lux and V is the output voltage.

A display device will have an EOTF described in the following manner:

$$L = (V)^{\gamma}$$

where V is the input voltage and L is the displayed luminance in nits[1] or candela/square metre (cd/m^2).

When the value of γ is unity, the transfer function is linear.

13.2.3   Characteristics of the Power Law Functions

Before reviewing the functions of the non-linear circuits and devices in the media chain, the characteristics of the functions themselves will be briefly reviewed.

As examples, in Figures 13.1 and 13.2, the characteristic curves of gamma functions are illustrated for gamma values of 2.5 and its reciprocal 0.4.

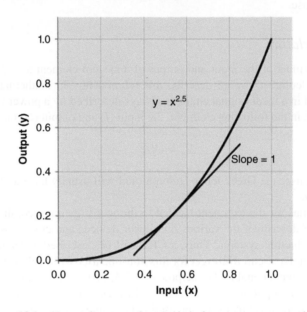

Figure 13.1   Shape of gamma characteristic for exponent greater than 1.

It will be noted that the slopes of the gamma curves vary continuously along their lengths, and since the gain of the characteristic is equal to the slope of the line which makes a tangent to the curve at that point, the gain varies accordingly. In the examples in Figures 13.1 and 13.2, the straight line which makes a tangent to the curves at 45 degrees indicates the point on the curves where the gain has a value of 1. For gamma values greater than 1, input values below this input level will be subject to ever-diminishing levels of gain until at close to zero-level input, the gain approaches zero. For input levels above this value, the gain increases until at

---

[1] See Appendix A. Using nits rather than cd/m^2 is more efficient, in the same way as we use amps or A to describe electric current rather than using coulomb per second or C/s.

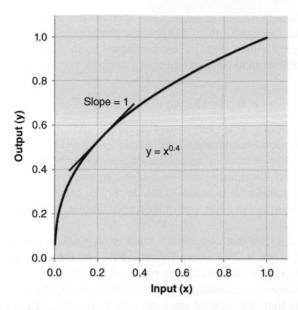

Figure 13.2 Shape of gamma characteristic for exponent less than 1.

an input level of 1, the gain reaches a maximum equal to the value of gamma, that is, 2.5 in this example.

For gamma values less than 1, as illustrated in Figure 13.2, the complementary situation exists; thus, for input levels below the input level where the gain has a value of 1, the gain increases rapidly such that at low levels of input level, the gain is very high, approaching infinite gain as the input approaches zero. The curve illustrated in Figure 13.2 was restricted to a lowest input level of 0.001, at which point the gain had increased to a value of 25. For input levels above the level at which the gain is equal to a value of 1, the gain diminishes until at an input value of 1, the gain as before equals the value of gamma, in this example 0.4.

Thus, for curves with an exponent greater than 1 at low levels of input corresponding to the darker elements of a scene, the gain at black level is very low and only increases to a value of 1.0 at an input level of 54.3% of white. Such a system element would cause severe tonal compression of the darker parts of a scene, often referred to colloquially as 'black crushing'.

When these characteristics are plotted on log–log graph paper, the result is a straight line, with the slope of the line being related to the gamma value. In certain circumstances, particularly when large contrast ranges are being considered, the log–log plot produces a more informative graph than the linear variant.

The data for the curves illustrated in Figures 13.1 and 13.2 are calculated in Worksheet 13(a).

13.3   Contrast Ranges

13.3.1   Contrast Range of the Scene

As we saw in Section 1.4, a critical factor is generally the very high level of illumination of outdoor scenes, which, together with the associated areas of shade, often results in a correspondingly high surface contrast range. Thus, surfaces in the scene on a sunny day may have luminances in the range from about 10 nits in the shadows to over 20,000 nits in the high lights, a contrast range of some 2,000:1. These luminance levels and static contrast ranges are generally not obtainable from the displays currently available for rendering media images.

13.3.2   Contrast Range of the Display

The contrast range of the display is likely to vary depending upon the form of pattern presented to the display, the amount of environmental lighting incident upon the screen and the percentage of the incident light reflected from the screen. In addition, some displays are technically poor, in that they are unable to sustain the same peak highlight luminance for patterns of variable screen area. In what follows, it is assumed the display is capable of providing the same highlight luminance irrespective of the area of the screen illuminated or the duration for which the pattern is displayed.

When an area of a screen that comprises a number of pixels is operating at peak white intensity, some of the light will be internally reflected from the structural front surface of the screen and fall back onto the surrounding pixels where a proportion of that light will be reflected forward through the transparent material to add to the light generated by the screen. The level of this flare light will depend upon the optical density of the transparent material and the reflectivity of both the pixels and the material surrounding the pixels. Generally, the level of flare will be at its highest immediately adjacent to a highlight area, falling off rapidly with distance.

Unless the screen is mounted in a totally darkened room, a percentage of the environmental lighting falling on the screen will be reflected forward, adding to any residual luminance from the active area of the screen representing black, such that the contrast ratio of the screen, measured by the ratio of peak white to black, is effectively diminished. The level of stray light reflected will be highly dependent upon the display: in a cinema, a large proportion of the light incident upon the screen will be reflected, whereas in a direct viewing screen environment, some light will be reflected from the front surface whilst the remainder will suffer the same attenuation from the optical density of the transparent medium and the reflectivity of the active surface as did that producing the flare described above.

Even in the situation of a darkened room with no environmental lighting, the display itself will generate sufficient light to reflect from surfaces in the environment, back on to the screen. A tinted screen and treatment of the pixel surrounding area to reduce reflections significantly reduce the reflection of ambient light. In a direct view display this is unlikely to seriously impinge upon the contrast ratio, but for a projector display where the screen is designed for maximum reflection, this stray light can seriously reduce the contrast ratio. Thus, in order to preserve the contrast ratio, surfaces in a position to reflect light towards the screen should be of very low reflectivity.

Backlit direct view screens based upon LCD technology are fundamentally limited in the contrast ratio they can provide since the level of attenuation provided by the polarising cells that control the light levels is limited, such that at maximum attenuation, some backlight is still passed through the cells, setting the minimum black level of the screen. In order to combat this limitation, 'dynamic' contrast control is sometimes introduced, which makes use of one or both of the following techniques. In the first approach, when an averagely dim background scene is detected, the level of the backlight is reduced, which reduces the level of black proportionately and thus provides a first degree of improvement.

An alternative means of reducing the black level, which requires a considerably more sophisticated approach, utilises a large number of LED backlights to mimic in a relatively coarse manner the spatial layout of the pixel elements, such that the light level corresponding to a localised dark area of the image may be reduced in a spatially dynamic manner. Depending upon the level of sophistication of the application of the technique, the improvements can be quite dramatic; however, it will be appreciated that controlling the level of the light from the LEDs should ideally be at frame rate if artefacts are to be avoided; also the number of LEDs cannot match the number of pixels in the screen, so the result will be limited in its effectiveness for image spatial resolutions greater than that represented by the spatial resolution of the LEDs.

Displays based upon plasma technology do not suffer from providing a luminance level limited to some value above black; nevertheless, their contrast range is somewhat limited by other factors. As the technology of display production improves, so the individual pixels themselves will become LEDs of one form or another, such as organic LEDs (OLEDs), enabling the light they emit to be controlled down to zero level.

Evolving from the above considerations, two different parameters have emerged for measuring screen contrast ratios.

13.3.2.1   Sequential Contrast or Inter-Frame Contrast

The sequential contrast should be measured in a totally dark room with no extraneous light allowed to fall upon the screen of the display. The sequential contrast describes the ratio of the highlight luminance of the screen when displaying a large area of peak white compared with the screen subsequently displaying a full screen representing black. Some current display devices are able to achieve sequential contrast ratios of several thousand to one.

13.3.2.2   Simultaneous Contrast or Intra-Frame Contrast

Simultaneous contrast is so called because the white and black are presented to the screen simultaneously, usually in the form of a chequerboard pattern. Thus, the black squares of the pattern will be illuminated to a degree by the flare light from the white squares, usually significantly reducing the contrast compared with the sequential contrast. However, simultaneous contrast provides a more realistic value of what the actual contrast is when viewing typical scenes. Typical simultaneous contrast ratios of current good-quality screens are in the order of several hundred to one.

The perception of the quality of the reproduced image is greatly affected by the effective contrast range of the display, which takes into account all the above factors. Generally speaking, the higher the sequential and simultaneous contrasts of the screen in its 'native' state, without the 'improvements' of dynamic contrast, the better will appear the reproduced image.

13.3.3   Perception of Contrast – Contrast Relationships in the Eye

In the introduction to this chapter, the various elements of the system which affect the contrast in the perceived image were briefly covered. However, perhaps the most important of all these elements is the eye itself; unless the contrast relationships of the eye are understood, it is not possible to determine how the contrast relationships in the remainder of the system may be managed to ensure that the tone relationships in the original scene are either preserved or mapped in a manner which produces the best compromise in the reproduced image.

13.3.3.1   Spatial Dynamic Contrast Ratio

Once the eye has adapted to an average luminance in an elemental area of either the scene or the reproduced image, there is a limit to how small a change in luminance it is able to perceive. Within this elemental area, the ratio of the highlight luminance to the smallest perceived change in luminance may be defined as the spatial static contrast ratio of the eye. However, the adaptation capabilities of the eye mean that if the focus of the eye is moved to another elemental area of the scene of a different average luminance, it will immediately and subconsciously adapt to the new range of luminances of that area but once more be limited to the same spatially static contrast ratio. However, as the eye focusses on each elemental area in the scene in turn, it would appear that the eye–brain complex remembers each perceived elemental image and builds up a complete image of the scene in the mind in such a manner that the eye appears to have a contrast ratio which exceeds the spatial static contrast ratio. This contrast ratio is defined here as the spatial dynamic contrast ratio and has a value which, depending upon the method of measurement, is in the order of thousands to one.

13.3.3.2   Spatial Static Contrast Ratio

Measuring the spatial static contrast range of the eye is relatively straightforward since we can simplify the problem of defining the elemental area of the scene the eye is focused upon by ensuring that the field of view of the test scene is very large compared with the elemental area. Such a test scene would comprise a large area of luminance, the intensity of which can be controlled and on which a variable intensity pattern is imposed. A series of measurements would then be taken over a large range of intensities of both the large area of luminance and the intensity of the luminance of the pattern which was just discernible to the eye. The ratio of the small change in luminance, ΔL to the luminance of the large area L, that is, $\Delta L/L$, would provide a measure of the sensitivity of the eye to a change in luminance over the range of luminances tested.

We have seen in Section 1.4 that according to Weber's law, the sensitivity of the eye to a change in luminance, that is, $\Delta L/L$, is a constant over the range of adaptation of the eye. The value of this constant has been known to be at a level of about 1%, depending upon the perceptibility of the pattern or artefact causing the change in intensity.

The $\Delta L/L$ relationship is illustrated in Figure 13.3. In earlier discussion on the Weber law relationship, it was assumed that the minimum perceptible change in $\Delta L/L$ was in the order of 1% or lower. Naturally, the threshold will depend upon the visibility of any pattern causing the change in luminance; nevertheless, if the response of the eye is truly logarithmic, then the threshold will be a straight line on a logarithmic plot.

In Figure 13.3, the straight lines encompass values of $\Delta L/L$ from 0.1% to 1%, and experience indicates that, depending upon the sensitivity of the pattern chosen, the corresponding Weber law characteristic will fall between these limits.

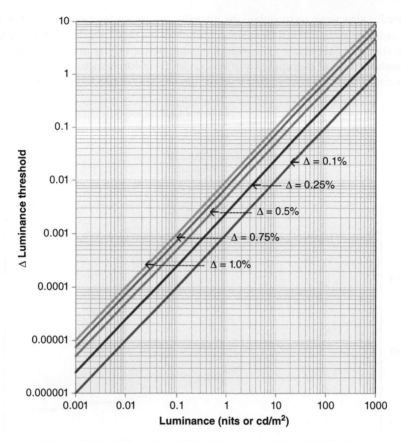

Figure 13.3 Weber's law $\Delta L/L$ – likely range of threshold limits.

13.3.4 Threshold of Perceptibility

Although the general form of this relationship has been known since the nineteenth century, the level of luminance where the Weber relationship ceases to hold to a constant has been somewhat indeterminate. However, since these levels are likely to be within the limit of the contrast range of the reproduced images of current reproduction systems, it is important to establish the manner of its departure from a constant in order that the values of critical system design parameters may be calculated to avoid the perception of artefacts which may be inherent in the system.

What is required therefore is a definitive description of the luminance levels where the $\Delta L/L$ relationship begins to depart from Weber's law. Much work has been carried out in this area, most latterly and comprehensively by Barten (1999), whose work on the human visual modulation threshold (HVMT) has been used by Maier[2] et al. within the SMPTE Standards Community to highlight its relevance to contrast law parameters and the perception of artefacts in media systems.

---

[2] T. Maier, SMPTE Engineering Guideline 432-1:2010, Digital Source Processing – Colour Processing for D-Cinema.

Barten used a sine wave to modulate the intensity of a plain background level of neutral luminance, ensuring sufficient displacement amplitude of the pattern and the minimum number of cycles of sine wave necessary to provide maximum perceptibility in order to establish the absolute threshold level of perceptibility at various levels of luminance.

The modulation m of the background level of luminance is defined as follows:

$$m = \frac{L_{high} - L_{low}}{L_{high} + L_{low}}$$
$$= \frac{\Delta L}{2 * L_{average}}$$

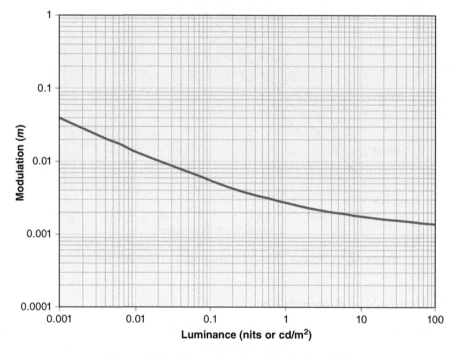

Figure 13.4 Human visual modulation threshold.

The plot of m against luminance is illustrated in Figure 13.4. Any artefact of modulation, such as noise or digital contouring, occurring below the level of the curve will not be perceived.

It will be noted that at higher levels of luminance, the curve is tending to become asymptotic to a constant value of modulation m, which is to be expected as $\Delta L/L$ is a constant at higher luminance levels.

One would anticipate that at the higher levels of luminance in the range illustrated in Figure 13.3, the results would follow Weber's straight line law, whilst deviating from the line at low levels of luminance.

In order to establish whether this is the case, we need to plot the Barten results on the same form of graph as the Weber relationship. To do this, we must first express $\Delta L/L$ in terms of modulation m.

From the formula in the text associated with Figure 13.4:

$$m = \frac{\Delta L}{2 * L} \quad \text{and} \quad \frac{\Delta L}{L} = 2m$$

The values of m and L from Figure 13.4 were used to form a table in Worksheet 13(a) from which the $\Delta L/L$ values were calculated and used as the basis of a log–log plot as illustrated in Figure 13.5.

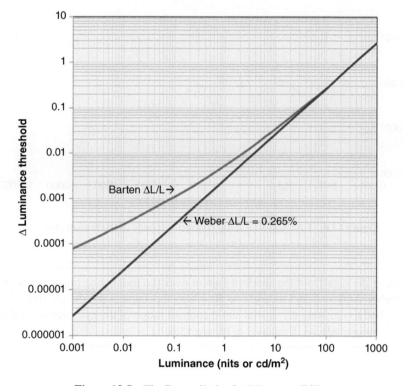

Figure 13.5   The Barten limit of $\Delta L/L$ perceptibility.

Though the data are limited to a maximum luminance of 100 nits, nevertheless, from approximately 50 nits, the plot is found to be asymptotic to a straight-line Weber law.

In the worksheet, a Weber $\Delta L/L$ plot was added to the graph and, by empirically, adjusting its $\Delta L/L$ value, it was found that the straight line of $\Delta L/L = 0.265\%$ is the line to which the Barten curve is asymptotic, as illustrated in Figure 13.5.

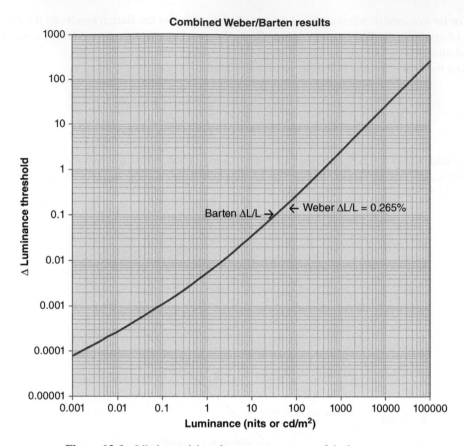

Figure 13.6 Likely spatial static contrast response of the human eye.

By combining the data from the work of Barten over the luminance range below 50 nits with the Weber $\Delta L/L = 0.265$ data above 50 nits in the same table, the combined results may be graphed to illustrate the spatial static contrast range of the eye over several orders of luminance magnitude, as illustrated in Figure 13.6.

Thus, it would appear the minimum perceptibility of change in luminance by the eye, using the Weber's law threshold for the most perceptible pattern, may be considered to be $\Delta L/L$ equal to about 0.25%, some four times smaller than the value of 1% previously considered; though it had been appreciated that the higher value did not necessarily correspond to the most perceptible of patterns.

At values less than about 50 nits, the sensitivity of the eye to changes in luminance decreases; this in logarithmic terms is beginning to approach the level where the photopic response of the eye is approaching the limit of its adaptation capabilities. From Figure 13.6, at a luminance level of 1 mnit, $\Delta L \approx 100$ μnit, a value some 40 times higher than the value of 2.5 μnit, the Weber value at the same luminance.

Further work by Maier and his colleagues indicated that for less critical interfering patterns, a threshold some 10 times higher than the HVMT established by Barten could be tolerated, implying that the delta luminance perceptibility associated with Weber's law would be about 2.5%. Thus, given that the range of delta luminance perceptibility in practice is between 0.25% and 2.5%, depending upon the perceptibility of the pattern causing the change, it seems reasonable to adopt the traditional value of 1% as reasonably representative of normal situations, preserving the value of 0.25% for very critical appraisals.

13.3.5   Displayed and Perceived Contrast Range

The contrast range of the eye is a constantly moving variable depending upon both the contrast range of the scene and where the eye is focused within the scene; notionally at any point in the scene, the eye appears to have a *spatially static* contrast range of a few hundred to one, but as it focuses on areas of widely different average luminances, the eye will accommodate immediately to the new situation, giving it effectively a *spatially dynamic* contrast range which is likely to be in the order of several thousand to one.

When viewing a reproduced image, the situation described above is influenced both by the restricted angle of view of the image compared with that of the scene and also the luminance of the surrounding environment when compared with the luminance of the image. For example, as a reproduced image subtends an increasingly small angle of view to the eyes, tending towards the elemental area discussed earlier, the whole of that image will eventually be perceived by the eye as one area of an average luminance and thus will be restricted to the notional 100:1 spatial static contrast range. Conversely, when viewing a reproduced image which fills a significant percentage of the field of view, the eye will focus on different objects of interest within the scene with possibly widely different average luminances, enabling the eye–brain combination to effectively exploit its larger spatially dynamic contrast range. *It would seem therefore that for critical viewing and appraisal of tone relationships, particularly in a picture-matching context, the field of view of the reproduced image should correspond to that of the environment in which it is intended the image will eventually be displayed to an audience.*

In the introduction to this chapter, it was assumed that if there was a linear relationship between the perceived luminance range of the scene and the luminance range of the display, then, to a first degree, a satisfactory reproduction of the scene would have been achieved. However, it would appear from an inspection of the delta luminance threshold in Figure 13.6 that below about 10 nits, the response of the eye to changes of luminance increasingly fails to meet the Weber's law threshold, indicating that at low levels of image display luminance, there is likely to be an increasing risk of failing to replicate the contrast range of the scene as perceived by the eye.

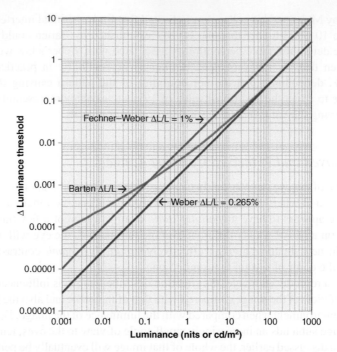

Figure 13.7 Determining the Barten limit perceived contrast range.

> As is illustrated in Figure 13.7, for a display with a typical highlight luminance of 100 nits, then with increasing range of contrast beyond about 10:1 there would be an increasingly perceived black crushing, such that as one approached a contrast range of 1000:1, ΔL changes of 1% or less would no longer be perceived. (This is where at a display luminance of about 0.1 nits, the Barten delta luminance threshold crosses Weber's law delta luminance 1% threshold.)

Thus it would appear that a display with a highlight luminance of 100 nits and describing a small field of view to the eye would be only marginally compromised in providing a satisfactory match to the contrast range of the eye but for displays with larger fields of view, where in the original scene the spatially dynamic contrast range of the eye could be operating, the perceived contrast range would continue to be compressed by the Barton delta luminance curve and could therefore be significantly compromised. In a cinema situation where the field of view was large enough for the spatially dynamic contrast range of the eye to be capable of reaching a value of perhaps 5,000:1, then, since the lower limit for perceiving a 1% delta luminance change is 0.1 nits, for the eyes to perceive a Weber's law delta luminance of 1% throughout the contrast range, the screen highlight luminance would need to be at least 500 nits and the display plus reflected ambient light would also require to have a contrast range of at least 5,000:1.

13.4 Gamma Correction

13.4.1 The Requirement for Gamma Correction

In Section 13.2, the three requirements for the use of non-linear circuits were outlined. In this section we will deal with the requirement to correct for unavoidable non-linearity in either/or both the image sensor or/and the display device. Specifically, it is worth highlighting that in the event that both of these devices are linear in operation, then the correction described in this section would not be required; however, it would still be necessary to address the other two requirements which were outlined and which are dealt with later in this chapter.

As an example, in some very early colour television systems, the image sensors, which were vidicon image sensor tubes, had an OETF characteristic which, over much of its operating range, approximated to a power law with an exponent of about 0.5, whilst the display device was a CRT with an exponent of about 2.4. Thus, since these two functions were approximately complementary, gamma correction was not necessary and colour cameras of this type in that era did not incorporate gamma correction circuits.

However, from the mid-1960s, the image sensors in cameras for all forms of media have been very close to linear in operation (i.e. the OETF has a gamma value equal to 1), but the CRT remained the only practical display device for a further 40 years and set the scene for gamma correctors, which became necessary to correct for their EOFT characteristic. The current standards (2013) for both television and electronic photography were set in the era of the CRT and thus reflect its continued use, albeit flat-panel displays, which have native EOTFs more nearly linear, have since become the norm. As we shall see later in this chapter, the continued use of gamma correctors, which emulate the original requirement of compensating for non-linearity of the CRT, serendipitously satisfy the third requirement for non-linear circuits identified in Section 13.2, where non-linearity is introduced to minimise the perception of artefacts in the RGB signals which drive the display.

Image displays have different EOTF characteristics and, in particular, historic displays, such as those based upon the CRTs, have a transfer characteristic which is non-linear. This non-linearity is a characteristic of the three electron guns which produce the controlled beams of electrons which strike the three sets of phosphors deposited on the screen. Under normal, non-overloaded conditions, the mechanism of producing photons by the beam electrons is a linear process as we saw in Section 6.7.

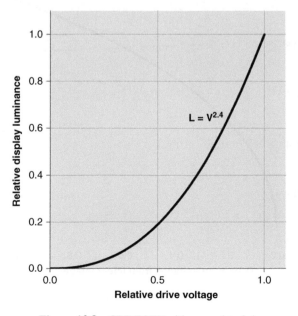

Figure 13.8 CRT EOTF with γ equal to 2.4.

The transfer characteristic of the electron guns follows a power law, and the equation relating light out L to drive voltage V in a CRT is given by $L = V^\gamma$ and has a characteristic illustrated in Figure 13.8. The value of gamma varies slightly depending upon the drive arrangements of the CRT but is now normally taken to be about 2.4. (Measuring the value of CRT gamma precisely is a difficult process and, over the years, several values have been obtained between 2.2 and 2.8.)

It can be seen from the graph that if the luminance signal were to be applied directly to the CRT without gamma correction, changes in the low level of the signal would produce very small changes in the displayed luminance, leading to 'black crushing'; conversely, similar levels of change in the higher levels of luminance would lead to exaggerated changes in the lighter areas of the image. The result would be the appearance of an overly contrasted image.

13.4.2   The Practicalities of Gamma Correction

13.4.2.1   The Required Correction Transfer Characteristic

Notionally, in order to correct for the gamma law of the CRT, it is necessary to apply the inverse correction to the RGB signals in the signal path prior to applying the signals to the CRT. The effect of the transfer characteristics of the gamma corrector and the CRT in series should be to produce a linear rendition of the levels of luminance in the original scene, albeit at a lower level of luminance.

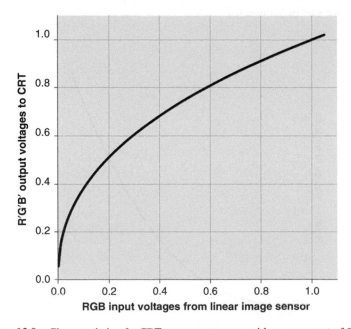

Figure 13.9   Characteristic of a CRT gamma corrector with an exponent of 0.417.

Thus, in order to ensure that the linear RGB signals from the image sensors in the camera are reproduced proportionately on the display, it is necessary to apply electronic gamma correction to the signals such that the transfer characteristic of the corrector is the inverse of the display device, as is illustrated in Figure 13.9.

Thus, if V_c represents a normalised signal from the linear image sensor in the camera, V_d represents the normalised signal from the gamma corrector which drives the display and ε is the exponent of the gamma corrector, then for an overall linear transfer characteristic

$$V_c^\varepsilon \times V_d^\gamma = 1; \text{ thus, } \varepsilon \times \gamma = 1 \text{ and } \varepsilon = \frac{1}{\gamma} = \frac{1}{2.4} = 0.417.$$

Gamma correction is usually described in terms of the exponent of the law of the corrector rather than using the inverse of γ; thus, $\varepsilon = 1/\gamma$ and, in the case of a theoretical corrector for the CRT, $\varepsilon = 0.417$.

Once the RGB signals have been gamma corrected, they are differentiated from the linear signals by the addition of a prime to the letters so that they appear as R', G' and B'.

Consideration of the curve in Figure 13.9 will indicate that there are problems in the application of the full theoretical gamma correction. As the level of the input signal is reduced, the slope and thus the gain of the characteristic increase rapidly; by differentiating the equation of the characteristic, we can calculate the gain G at any point on the curve:

$$V_d = (V_c)^\varepsilon; \text{ thus, } G = \frac{\delta V_d}{\delta V_c} = \varepsilon V_c^{\varepsilon-1} = \frac{\varepsilon}{V_c^{1-\varepsilon}}.$$

The plots of gain against camera signal voltage are illustrated in Figure 13.10, both for the full range of input voltage and, in order to illustrate the characteristic in more detail, for the critical low input signals in the range of 0–10%. At the 1% level, the gain is about 6, but below this figure, the gain increases rapidly to very high values. The gain diminishes slowly towards white, where the relative input voltage is equal to 100% and where the gain becomes equal to the exponent ε, in this case to 0.417.

The increasing high gain of the gamma correction circuit below an input level of about 2%, as illustrated in Figure 13.10, is likely to cause practical problems in its implementation. It is important therefore both to establish the criteria governing the perception of luminance modulation[3] at these levels and to discern between the wanted modulation and that which occurs as a result of technological limitations, in order to address how the characteristic of the gamma corrector can be modified to accommodate the practicality of the situation.

---

[3] In this context modulation is defined as small changes in luminance level at a particular average level of luminance.

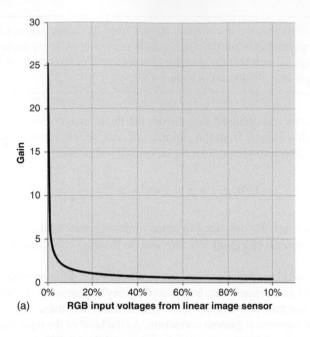

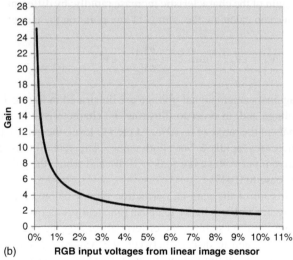

Figure 13.10   Gain of a gamma corrector for $\varepsilon = 0.417$ for different input ranges.

13.4.3   The Solution to Limiting the Gain of the Gamma Correction Function

Clearly it would not be acceptable to determine some low level of luminance where, in order to limit the gain of the characteristic, the gain of the gamma corrector changed abruptly. The solution is to select an acceptably low level of luminance and then ensure that the gain at this point becomes a constant which is applied between that point and zero luminance level. Effectively this requires two characteristics, a power law and a linear characteristic which are joined at the common gain point in order to ensure no discontinuity in the overall characteristic.

It is in this range therefore that colour reproduction systems select an input level where the transfer characteristic of the gamma corrector changes from a power law to a linear law. In general terms, the more critical are the viewing conditions, and thus at low levels of luminance, the more perceptible are the changes in luminance level, the lower is the level selected for the change in characteristic. Fortunately, reference to Figure 13.10 indicates this is precisely the range at which the gain of the characteristic of an unmodified gamma correction curve starts rapidly to increase; thus with care, a judicious compromise can be made between too high a gain and a satisfactorily perceived undistorted contrast range.

As an example of the approach used to determine the solution for the combined curves, we will use the CRT inverse exponent of $\varepsilon = 0.417$ and specify the break point between the two curves at the 1% input point. This curve is calculated and plotted in the Gamma worksheet 13(b), from which it is determined that the gain at the 1% input point is 6.11 and the difference in the output voltage between the linear and the power law curves at the 1% input point is 8.5%.

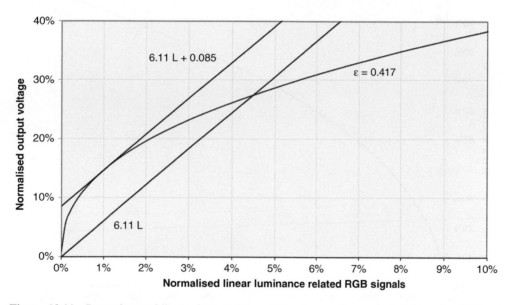

Figure 13.11 Power law and linear characteristics. Gain of correction curve is 6.11 at the 1% input point.

Figure 13.11 illustrates the gamma correction curve over the first 10% of the input luminance in order to emphasise the desired area of interest. The exponent of the curve is $\varepsilon = 0.417$ and has a slope or gain equal to 6.11 at the 1% input point. The linear straight line for a gain of 6.11 is also shown, as is the same linear curve displaced with the addition of a constant of value 8.5%, which makes it tangential to the gamma curve at the same slope.

For the two curves to cross at the point of the same slope on both curves, defined as the break point, the power law curve must be dropped by the application of a negative pedestal of 8.5%, which will cause its amplitude to also drop by the same amount and will therefore require an increase in gain from 1 to $m = 1/(1 - 0.085) = 1.093$ in order to make a 100% input provide a 100% output.

The mathematics required to establish a general solution for the various parameters which relate to the combined transfer characteristic are detailed in Appendix H. It transpires that if two of the parameters, ε the exponent of the power section of the characteristic and G_B the gain at the break point (and thus the gain of the linear section of the characteristic), are first defined, then the value of the remaining parameters, the luminance level L_b at the break point, the gain m of the power element of the characteristic and the pedestal p to be removed from the power law element of the characteristic, follow automatically. These two groups of parameters are termed the independent and dependent parameters, respectively, in the remainder of this chapter.

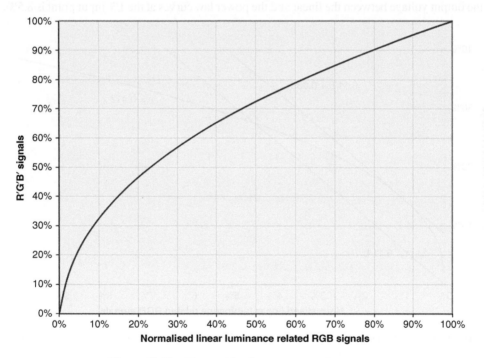

Figure 13.12   The combined gamma correction curve.

When the calculations in the above paragraph are undertaken, the resulting gamma correction curve is illustrated in Figure 13.12.

These adjustments are such that the gain of the combined curve at the break point relating to an input of 1.0% has increased from 6.11 to 6.70% and the value of m, the gain applied to the power section of the characteristic, is 1.093.

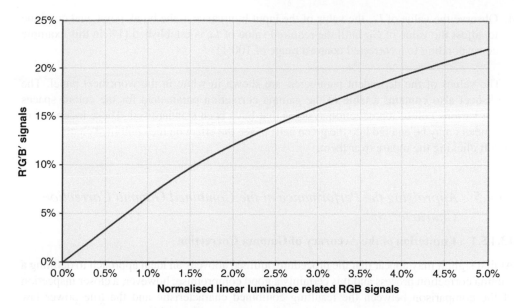

Figure 13.13   Combined gamma correction over first 5% of input.

In Figure 13.13, only the input range of 0–5% is illustrated in order to highlight the linear portion of the combined curve over the 0–1% input range.

As the gain of the linear portion of the curve is 6.7, the output corresponding to an input of 1% is 6.7%.

The primary element of the Gamma Worksheet 13(b) has also been configured such that by selecting the value of the two independent parameters for a colour reproduction system (highlighted in green in the worksheet), the two dependent parameters are calculated and presented – as is also the corresponding combined transfer characteristic curve.

13.4.4   Specifying the Gamma Correction Parameters for a Colour Reproduction System

The general approach to using the worksheet to obtain the required values for the parameters of a gamma correction stage of any specific colour reproduction system is as follows:

1. Determine the required contrast range of the reproduced image – in the above example, 100:1.
2. Enter the exponent ε of the gamma correction required.
3. Enter the estimated gain of the linear portion of the characteristic, G_B. (Usually between 3 and about 30, dependent upon the contrast range of the display or print, with higher values for larger contrast ranges.)

4. Observe the value of L_b, the value of the input luminance at the break point, and continue to adjust the value of G_B until the required value of L_b is established (1% in this example corresponding to a corrected contrast range of 100:1).

The values of the dependent parameters are shown in white in the worksheet panel. The worksheet also contains a table of the gamma correction parameters for the colour spaces of the various colour reproduction systems that have been standardised. These independent parameters may be entered into the green cells to see the effect on the combined characteristic by left clicking the mouse over them.

13.4.5   Appraising the Performance of the Combined Gamma Correction Characteristic

13.4.5.1   Limitation of the Accuracy of Gamma Correction

At first sight it may appear that the procedure outlined above is an ideal approach to defining a gamma correction process and it is certainly a good compromise; however, a closer inspection of the comparison between the resulting combined characteristic and the true power law characteristic in Figures 13.14 and 13.15 indicates quite critical mismatches.

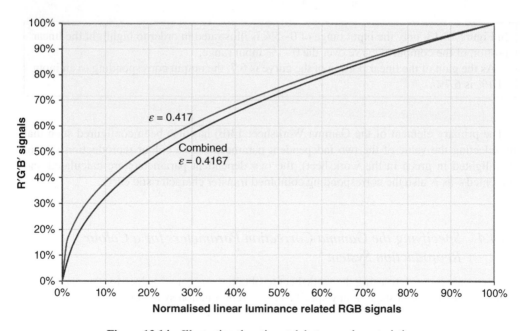

Figure 13.14   Illustrating the mismatch between characteristics.

In Figure 13.14, the combined characteristic for the example exponent of $\varepsilon = 0.417$ and a linear gain of 6.7 is compared with a true power law characteristic of exponent $\varepsilon = 0.417$.

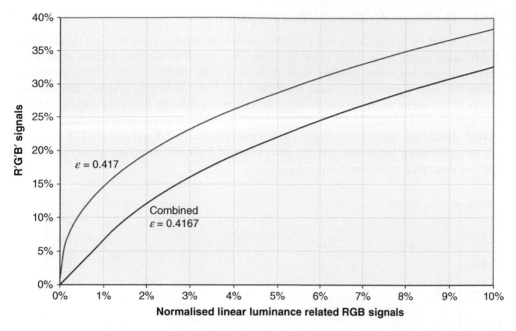

Figure 13.15   Illustrating the large errors at low levels of luminance.

Though the criteria of selecting the gain of the corrector to be equal to the gain of power law at a suitably low level of signal (equal to the same percentage of scene luminance) appeared a reasonable basis for a compromise characteristic, it is clear from Figure 13.15 that the actual output level at 1% input level is substantially in error at 6.7% rather than the correct figure of 14.7%, a difference of 8% and an error of over 100%.

Thus, although the ability to display changes in luminance at low levels of luminance has been preserved, the actual intensity of the low level of luminance is seriously compromised.

Worksheet 13(b), with its formulaic layout and its 'Dynamic' Figures A.1 and A.2, can be used to establish that a lower value of exponent ε and a higher level of linear gain will provide a better match to the required power law characteristic.

It does indicate however that the combined characteristic should not be described as having a gamma relating to the exponent initially chosen to establish the parameters of the combined characteristic; as the above example illustrates, the difference between this characteristic and a true power law of $\varepsilon = 0.417$ is substantial in the critical dark areas of the image, as is shown in Figure 13.15. In fact the nearest power law characteristic to match the combined characteristic here described has an exponent of about 0.5, as can be shown by entering the appropriate values in Cell G31 and inspecting Figures A.1 and A.2 in Worksheet 13(b). A match in this case is a compromise as the two curves cannot be made to overlay each other; thus, a 'best match' is a subjective best appraisal of the closeness of the two curves and the level of luminance at the crossover point.

13.4.5.2   The Observable Errors in Tone Reproduction

There are therefore limitations to this procedure which are more serious than at first contemplated because of the logarithmic response of the eye, which as we have seen in subjective terms emphasises luminance changes in the dark areas of an image.

Using the same parameters as used for Figure 13.15, the following graphs illustrate the situation as the signal from the linear image sensor passes through firstly the various stages of the reproduction system and then the eye–brain complex. It is assumed that the gamma-corrected signals are applied to a display with a true gamma of 2.4 and that the CIE-defined relationship between luminance and the perception of lightness based upon the cube root power law described in Section 4.6 is used.

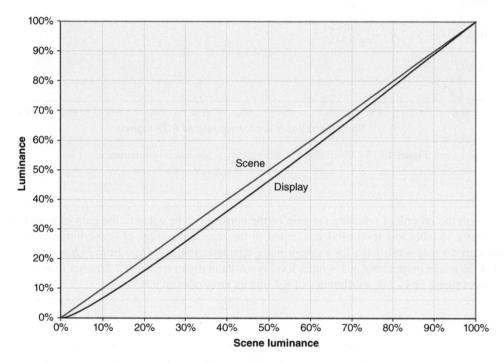

Figure 13.16   Comparing display and scene luminance.

Figure 13.16 illustrates the comparison of the original scene luminance and the displayed luminance after the application of gamma law correction and the CRT gamma, where $\gamma = 2.4$.

In Figure 13.17, the two graphs illustrate the difference between using a true correction curve and the combined curve on the perceived lightness of the display. The limited range graph clearly shows very large errors in the perception of lightness, differences of 10% at about the 20% brightness level and 5.7% at about the 46% brightness level, representing errors of about 50% and 12%, respectively. These are very large errors in perception and go some way to explaining why when using a CRT for display, images often appear 'black crushed' and

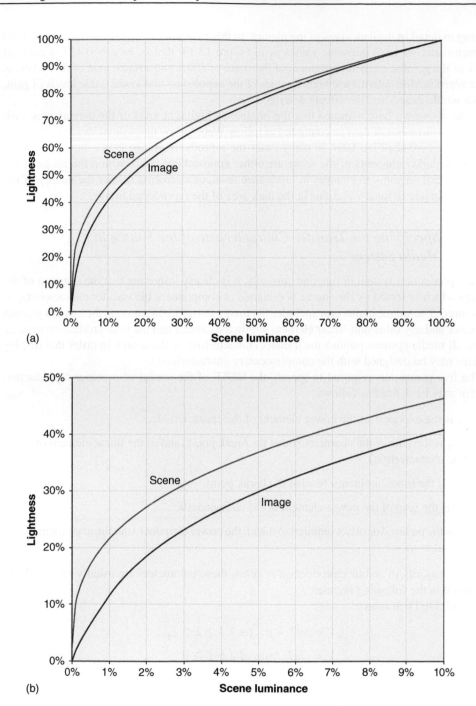

Figure 13.17 Comparing perceived lightness of the scene with the image.

lacking in detail in the dark areas of the picture. In this hypothetical example for the combined characteristic, we used the same values as in Figure 13.14, that is, an $\varepsilon = 0.417$, a contrast range of the gamma correction segment of the curve of 100:1 and a linear gain of 6.7, whereas some reproduction systems use higher values of the exponent ε and lower linear levels of gain, which would exacerbate the effects described here.

In the above we have assumed that the luminance highlight level of the display was such that the performance of the eye continued to follow Weber's law throughout its contrast range. However, as we shall see later, in many cases the reproduced image is at a luminance level where the darker elements of the scene are either approaching or are within the range of the eye's reduced response to changes in luminance level, exacerbating further the quality of the perceived image to tonal variations in the dark area of the reproduced image.

13.4.6   Specifying the Transfer Characteristic of the Source in a Media System

For a reproduction system to function correctly, it is clearly important that the segment of the system which is served by the source is designed to complement the transfer characteristic of the source. The source in this case includes the image sensor in the camera, the camera gamma corrector and any other non-linear element within the camera and the post-production system. Thus, all media systems publish the transfer characteristic of the source in order that display devices may be designed with the complementary characteristic.

The five parameters required to specify the OETF of the source of a colour reproduction system may be defined as follows:

ε is the exponent of the power element of the characteristic.

G_B is the gain of the characteristic at the break point (and of the linear element of the characteristic).

L_b is the input luminance level at the break point.

m is the gain of the power element of the characteristic.

p is the pedestal or offset required to match the power element to the linear element, equal to m − 1.

In the majority of colour reproduction systems, these parameters are usually specified and formatted in the following manner:

Overall OETF at source:

$$V = mL^{\varepsilon} - p \quad \text{for} \quad 1 \geq L \geq L_B$$

$$V = G_B L \quad \text{for} \quad L_B > L \geq 0$$

where:

L: Luminance of the image $0 \leq L \leq 1$

V: Corresponding electrical signal

The first line of the specification indicates that for a luminance signal equal to or above the L_B level and not greater than the 100% level, the power law equation should be used, and the second line indicates that for a luminance signal level below the L_B level and equal to or above the zero level, the linear gain equation should be used.

Thus, for the example of gamma correction used in this section, the formal overall OETF would be described as follows:

$$V = 1.093L^{0.4167} - 0.093 \quad \text{for} \quad 1 \geq L \geq 0.010$$

$$V = 6.70L \quad \text{for} \quad 0.010 > L \geq 0$$

13.5 Standard or Reference Displays

The use of the CRT as a display device was retained for several years for the critical assessment of picture quality in television control rooms after the introduction of quasi-linear displays as its characteristics were well known and standardised, whereas the then new linear displays incorporated electronic 'de-gammering' processors which were not standardised and led to the displayed image looking different on displays from different manufacturers, also a real problem for photographers.

One might question at this stage, now that nearly linear displays have virtually replaced the CRT in computer monitors and television sets, why gamma correction is retained as an accepted requirement in colour reproduction systems. There are two primary reasons for this: firstly, the standards for television and photographic systems were established during the era of the almost universal use of the CRT and for many years after the introduction of linear display devices the CRT remained in widespread use; thus there were huge legacy problems: at which point would the source EOTF be changed to a linear or near-linear characteristic? Without the introduction of a new improved system with incentives for the public to upgrade their displays, there is no time when this could be done without making the huge investment by the public in television and computer displays redundant. Furthermore, the play-out of archival material would require significant processing to make it compatible with the characteristics of the new system. Lastly, there is another important advantage of the use of a non-linear characteristic at source, as described in the next section. As a consequence, gamma correction as described in this chapter continues as a critical concept in colour reproduction systems which carry an on-going legacy of display devices.

Nevertheless, the introduction of 'non-standard' displays needed to be addressed, particularly for those interested in accurate colour reproduction. Of course gamma circuits were introduced into linear displays from the beginning to compensate for the camera gamma correction circuits, but there were no standards as to what parameter values to adopt for the EOTF. As indicated above, this lack of a standard delayed the use of the linear displays in critical environments, such as the vision control room of television production centres where pictures from different cameras are critically matched under a controlled illumination environment. However, as CRT displays became obsolescent, it became critical to formulate a standard for the EOTF of displays used for critical appraisal. (It would of course also be beneficial if such a standard were to be universally adopted by the manufacturers of displays for the television and computer industries.)

This situation has been addressed by the appropriate professional bodies associated with the various media reproduction systems and standardisation has been or is being approved for the EOTF in each case, as is described in the appropriate chapters of Part 5.

13.6   Masking Artefacts

In media systems the term 'artefacts' is used to describe visually perceived disturbances to the reproduced image which were not present in the scene. They are caused either by limitations in the technologies of implementation of the reproduction system or by the injection of interference from the external environment.

13.6.1   Source Noise

In a practical colour reproduction system the image sensors in the camera generate a signal proportional to the luminance of each element of the scene. However, these electronic devices also generate electronic 'noise', random signals at a low level which are related to the physics of the operation of the device and the temperature at which it is operated. In the early days of colour reproduction systems, this noise was at a level which could regularly cause impairment of the reproduced image, and though technological improvements have significantly improved the signal-to-noise ratio, under low light level conditions it can still be a factor in image impairment.

In Chapter 1 we noted that the logarithmic response of the eye leads to the eye perceiving equal *percentage* changes in luminance as equal changes in lightness, irrespective of where in the contrast range of the image the percentage change occurs. Thus, a 1% change at a luminance level of 5%, that is, a change of 0.05% of white, will be equally perceived as a 1% change at 90%, that is, a change of 0.9% of white, a difference in luminance change of 18 times. Thus, noise which appears at equal signal levels on either a dark grey or light grey signal will appear in this example 18 times more noticeable in the dark areas of the scene compared with the light areas.

It is apparent therefore that any processing of the signal, such as gamma correction, where high gain will occur in the dark areas of the scene and where the eye is considerably more sensitive to change in luminance, must be approached with caution. Of course in a perfect situation where the exponents of the display and the corrector are the inverse of each other, the perceived noise would be no different to that perceived on a linear display with no gamma corrector. However, the practicality of the situation is that near black level, the situation in not perfect; noise transients occur equally in both the positive and the negative direction, and if the noise is amplified at these low luminance levels, the peaks of the random noise will extend not only positively well into the lighter grey region but also negatively in the opposite direction beyond black. Thus, the positive transients of the noise will be displayed but the compensating negative transients will only partially be so, since the display device cannot produce negative light. The result is that a gamma corrector operating too closely to the ideal power curve may impair the displayed image. In electronic terms this form of passing only one polarity of the signal is termed 'rectification'.

In addition, irrespective of the above, there is an argument for reducing the gain of the system near black in order to reduce the visibility of the noise, albeit at an apparent cost to the accuracy of the greyscale reproduction.

13.6.2   Determining the Location of Gamma Correction in the System Path

As has been seen in Section 13.4, historically, the primary purpose of gamma correction is to correct for the non-linearity of the CRT; it follows therefore that logically the gamma corrector should be located in the display device, and in colour reproduction systems, where the display device is relatively expensive and designed to serve a large number of people simultaneously and thus the cost of the gamma corrector is relatively insignificant, this is where it is located.

However historically, the three electronic circuits required for gamma correction were relatively complex and expensive in terms of providing a very stable form of correction, which would be particularly necessary in a domestic environment. Thus, in a situation where the colour reproduction system was based upon serving possibly millions of relatively cheap display devices from a source of a few cameras, it made economic sense to place the gamma correctors in the camera. This approach provided an ideal solution in the days of monochrome television; however, although this procedure was cost-effective and did provide the advantage of assisting in the masking of transmission artefacts, such as signal path noise, with the introduction of colour, it did lead to a compromise in the overall system design, as we shall see in Chapter 14. This compromise could have been avoided had the more cost-effective current technology been available at the time, where to include the gamma correction circuitry in the large-scale integration of the display device electronics would not have significantly enhanced the cost of the device. In fact, processing to de-gamma the $R'G'B'$ signals is now built into virtually all current displays to provide the native quasi-linear screen with complementary near-linear RGB signals.

It was also realised that gamma correction located at source actually introduced a benefit to the performance of the overall system because of the serendipitous close match between the inverse law of the CRT and the logarithmic transfer characteristic response of the eye. In television terms, in the days of analogue processing and transmission, the signals were much more prone to the addition of noise and to distortion, particularly in the transmission path, and as was shown in Section 13.6.1, this noise is far more disturbing in the dark areas of the image. However, by locating the gamma corrector at the source, the transmission noise subsequently added will, when reaching the display device, be subjected to the inverse characteristic of the display or de-gamma circuitry, where the low level signals and the accompanying noise will be severely attenuated. The result is that if the gamma correction characteristic is broadly similar to the tonal response of the eye, the transmission noise will appear uniformly perceptible at all luminance levels, rather than emphasised in the darker tones of the reproduced image.

It would seem that the very real advantages of gamma correction in analogue systems appear to become less so in digital systems, where the effects of noise and distortion in the signal path are either not visible or cause a total loss of the signal; however, as will be seen below, that is not the complete picture.

13.6.3   Digital Contouring and Perceptible Uniform Coding

The digital coding of the RGB or $R'G'B'$ signals is a technological procedure which at first sight appears to be transparent to the colour reproduction process but in fact can impinge on the process by the introduction of artefacts, which under certain conditions can become perceptible. Also, since it has become apparent that often confusion has crept in regarding the use and terminology of 'gamma correction' and 'perceptible uniform coding', and since the latter relies on aspects of human vision for its use, some words to clarify the situation are in order.

Despite the relative immunity of digital systems from signal path distortions, electrical noise and interference, there is however the quantisation effect of the digital encoding system itself on the RGB signal levels to be considered.

The quantisation process is a technological topic and as such is beyond the scope of this book; however, in order to provide sufficient information for the comprehension of the remainder of this section, the process is briefly described; for those who wish to study the subject in more depth, it is well described by Poynton (2012). The analogue signal level for each pixel is sampled and measured and the value obtained is converted into a digital integer number. The digital numbers available are limited by the number of digital bits used to describe each number, usually a figure of 8, 10, 12 or more bits. Thus, in digital terms, 8 bits will provide a discrimination of 2^8, that is, 256 levels, so all the thousands of variations of analogue signal level, which fall between 0% and 100%, will be directed into the nearest of these 256 digital levels, a process referred to as quantisation. Thus, in a scene with gradually changing tones across the image, there will be an abrupt change in tone of 1/256 of luminance level at each digital sample. When the tone level is changing only slowly, the result will be that for many adjacent samples the bit level will be identical, although over the same area, the analogue signal is slowly changing in level. As a consequence, the image is displayed with contour lines representing the eventual change in digital level, which in certain circumstances may be perceptible.

In order to explore the critical parameters associated with quantisation level, the resulting contouring and the perceptibility of the contours, we will initially envisage a simple colour reproduction system where both the OETF and the EOTF are linear and therefore there is no requirement for gamma correction between the camera and the display device.

Being aware that the static contrast sensitivity of the eye is about 1%, then it might be reasonable to assume that in order to avoid the perception of quantisation contouring, we would need to select a quantisation system where the number of bits available for quantisation is in excess of 100. Seven bits provides 2^7, that is, 128 levels of quantisation, which is perhaps a little marginal, and eight bits provides 2^8, that is, 256 levels.

By plotting the change in levels of luminance in the display against a 1-bit change in quantisation level, the results can be compared with the Weber–Barten human vision threshold (HVT) derived in Section 13.3. These plots are calculated in Worksheet 13(c), where the full range of parameters on which the resulting plot depends is available for experiment.

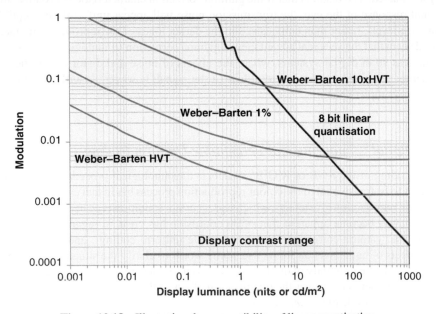

Figure 13.18   Illustrating the perceptibility of linear quantisation.

> The plots are illustrated in Figure 13.18. Three levels of sensitivity of the HVT are plotted: the threshold limit, the limit associated with 1% $\Delta L/L$ change over the normal adaptation range of the eye and, finally, a 10 times threshold limit to indicate the limit for contour artefacts, as discussed in Section 13.3.

The white luminance of the display is set to 100 nits and the contrast range of the display is assumed to be 5,000:1.

The contrast range of the display is based upon the parameters selected and is illustrated by the horizontal blue line at the bottom of the chart. Only in this range are the curves on the chart of interest to us.

The plot of the delta luminance changes caused by the 8-bit quantisation is illustrated by the line which is straight over the major portion of its length. With only 256 levels available, quantisation begins to fail at luminance levels below 1 nit, causing the kinks in the straight line.

It can be seen that this quantisation strategy is at best marginal. The quantisation contouring at luminance levels below about 150 nits is above the HVT level, albeit below the 10 times level, and at luminance levels below 3 nits, the contouring artefacts are well within the range of perceptibility.

It would seem from the above that there are two approaches to ensuring that contour artefacts are below the level of perceptibility; firstly, to digitally encode the RGB signals in a manner which exploits the contrast law of the eye and, secondly, to increase the number of quantisation levels.

We have seen earlier in this chapter that the response of the eye is logarithmic and approximates to a power law with an exponent of 1/3, which makes the perception of $\Delta L/L$ changes constant throughout the range of accommodation, as illustrated by the straight line section of the HVT curve between about 80 nits and 1,000 nits. Thus, by adopting a power law characteristic with a fractional exponent rather than a linear characteristic prior to the quantisation process, and a power law of a complementary characteristic following the digital–to-analogue decoding process, the contouring artefacts will be less perceptible.

Figure 13.19 illustrates that by using a power law in the quantisation process, the slopes of the contour perceptibility curves are significantly reduced from that of the linear case in Figure 13.18.

It can be seen from Figure 13.19 that the introduction of non-linear coding has significantly reduced the perceptibility of the contours, particularly in the darker areas of the image. All three exponent curves are close to the Weber–Barten 1% curve but well below the 10 times HVT curve. The three straight lines represent 1-bit change values for a 10-bit digital coding system using 95% of the available bits between peak white and black and assuming a display luminance of 100 nits with a 5,000:1 contrast range and with exponent values of 2.0, 2.4 and 2.8.

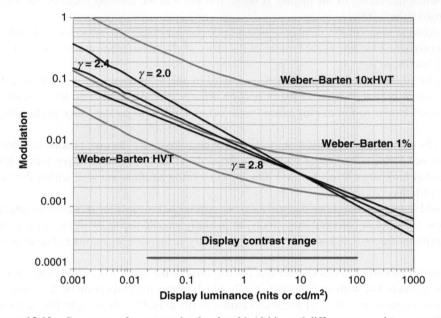

Figure 13.19   Contour artefact perception levels with 10 bits and different power law exponents.

In Figure 13.20, the exponent of the power law is kept constant at a value of 2.4 and the curves for quantisation levels corresponding to 8 bits, 10 bits and 12 bits, that is, 256, 1024 and 4096 quantisation levels, respectively, are illustrated.

As would be anticipated, the larger the number of quantisation levels, the smaller is the effect of a single-bit change, and therefore the less are higher-bit digital systems likely to cause perceptible contouring artefacts. It has to be borne in mind, of course, that data rates are directly proportional to the number of quantisation levels and therefore higher-bit rate systems are more demanding of storage and conveyance capacity (see Chapter 14).

It is clear from the above that the adoption of non-linear digital coding provides very significant advantages in the masking of the perceptibility of quantisation contouring. Because it is based upon the aim of matching the perceptibility limits of the eye, it is referred to as 'perceptibly uniform coding'. Generally, it is different from the gamma correction described in the previous section only in as much as the exponent of the power law selected is complementary in the coding and decoding process (with no limitations on the gain of the encoding law at low luminance levels). Its value is selected to ensure that on the modulation charts as used in Section 13.3, the chosen critical HVT function is not crossed by the quantising curve.

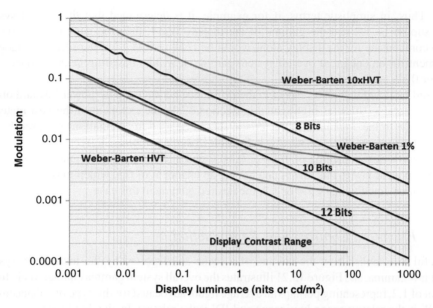

Figure 13.20 Contour artefact perception levels with power law exponent equal to 1/2.4 and different quantisation levels.

It is evident however that in reality, there is very little between the two procedures. In effect, the gamma correction introduced to correct for display non-linearity was serendipitously also very close to what is required for perceptibly uniform coding. Generally speaking, traditional media systems such as television and photography continue to use gamma correction as described in Section 13.4, whereas for the cinema, where the media signal path is split into a well-defined set of processes, gamma correction and perceptibly uniform coding are likely to be used independently of each other.

13.7   Matching the Contrast Law to the Viewing Environment

It was noted in Chapter 10 that in a range of circumstances, including:

- where the reproduced image has a lower highlight luminance than the original scene;
- where the viewing conditions associated with the viewing of the reproduced image are such that the image occupies only a small fraction of the field of view;
- where the surrounding illumination has a relatively high average level compared with the average level of the image,

then the perceived image is improved if the overall contrast law of the system has an exponent which is slightly greater than unity. Exponent values of 1.1–1.5 depending upon the circumstances listed above are proposed by a number of workers in this field, the most quoted of which are Bartleson and Breneman (1967), (Hunt 2004) and Liu and Fairchild (2007). Values of exponent between 1.1 and 1.3 tend to be adopted for colour reproduction.

Where it is determined that an overall system gamma should be greater than unity, the adjustment is usually included in the same processor which compensates for the display gamma. Thus, in simplistic terms, if for example, it is determined that the overall system gamma should be 1.2 and the gamma of the display is 2.4, then the gamma correction in order to compensate for the display alone would be 1/2.4 or 0.417, but in order to include the requirement for an enhanced overall system gamma, it will be based upon an exponent ε of 1.2/2.4 or 0.50.

It should be noted that with respect to the difference between a pure power law and one of combined characteristics (as highlighted in Section 13.4), the exponent of the best match of the power law to the combined characteristic should be used in calculations required for the overall system gamma. Thus, as we saw in Section 13.4, if a combined correction characteristic is based upon a design gamma of 1/2.4 or 0.4167 and a linear gain of 6.7, the nearest match of a true exponent law to this law is one having an exponent of 0.5, which makes the overall system gamma equal to 0.5×2.4 or 1.2, the desired gamma in this case.

13.7.1   Enhanced Contrast Law Reproduction

It is useful to establish how the colours of a scene are distorted by the increase in system contrast law gamma, and Figure 13.21 illustrates the overall system gamma characteristic for an exponent of 1.2, representing perhaps an average figure advocated for this type of enhancement, where 'S' is the relative scene luminance and 'D' is the relative display luminance.

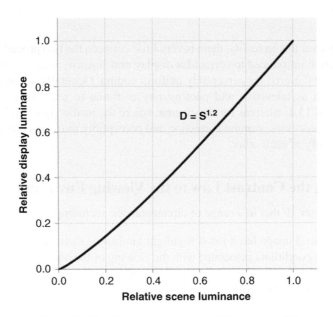

$$D = S^{1.2}$$

Figure 13.21   System contrast law with gamma = 1.2.

An inspection of Figure 13.21 indicates low luminance signals will receive less gain and signals above about 40% will receive higher gain than they would otherwise do with a linear system. In tonal terms, the lighter greys will be brighter and more delineated, whilst the darker greys will be less bright and less delineated or black crushed; that is, the image will appear to have more contrast.

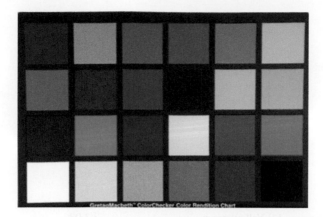

Figure 13.22   The ColorChecker Chart.

It will be appreciated that in applying this characteristic to the individual RGB signals, unless they are of equal levels, then their levels will be differentially affected, which will cause a change in the chromaticity of the colour. In Worksheet 13(e), the spectral reflective distributions (SRDs) of the coloured patches of the ColorChecker chart form the basis of the calculations to establish the values of the R, G and B signals from a camera with spectral sensitivities which match the ITU/sRGB primaries defined in Section 11.2, both in a linear system and in a system with an overall gamma of 1.2. The ColorChecker chart is illustrated in Figure 13.22 and the pairs of chromaticities are plotted onto a chromaticity diagram as illustrated in Figure 13.23.

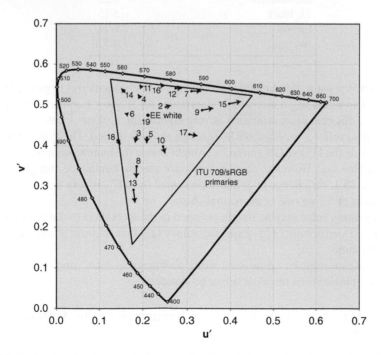

Figure 13.23   Indicating the change of chromaticity of the ColorChecker chart patches with a change in overall system gamma from unity to a value of 1.2.

Table 13.1 ΔE^{*}_{00} values

	ΔE^{*}_{00}
1. Dark skin	6.20
2. Light skin	4.06
3. Blue sky	6.58
4. Foliage	6.52
5. Blue flower	5.97
6. Bluish green	3.17
7. Orange	4.78
8. Purplish blue	6.08
9. Moderate red	5.12
10. Purple	5.63
11. Yellow green	3.03
12. Orange yellow	4.14
13. Blue	5.44
14. Green	5.51
15. Red	4.25
16. Yellow	2.64
17. Magenta	5.20
18. Cyan	6.22
19. White	0.00
20. Neutral8	1.80
21. Neutral6	4.04
22. Neutral5	6.72
23. Neutral3	6.22
24. Black	5.15
Largest	6.72
Average	4.770

The arrowheads point to the direction of the change in chromaticity; in general terms, the major change is in terms of an increase in saturation; for the low luminance colours on the red-to-blue axis, this change in saturation is very significant. Colours located on the yellow-to-orange axis also suffer a considerable hue change towards the red. The cyan patch, which is located just outside the ITU/sRGB gamut, is brought inside the gamut and moved to a slightly more saturated blue hue. However, equally important are the luminance changes, and as can be seen in Table 13.1, the change in colour represented by ΔE^{*}_{00} is very significant, reaching a maximum value of 6.7 on one of the neutral chips.

The overall gamma value may be simply changed in the worksheet to illustrate the effect of values greater and smaller than 1; as expected values less than 1 cause chromaticity changes of reduced saturation.

It is clear that care must be taken in adopting a strategy of increasing the overall system gamma if unacceptable shifts in colour are to be avoided.

13.8   Overall Opto-electro Transfer Characteristics in Actual Reproduction Systems

Each of the colour reproduction systems described in this book, television, photography and cinematography, have widely different means of both displaying the image and setting the viewing conditions, both of which are central to the choice of the parameters upon which the gamma corrector parameters (or opto-electro transfer characteristic) of the source is based. Thus, the specific gamma corrector parameter values relating to each of these systems will be described in the appropriate chapters in Part 5.

13.9 Producing a Greyscale Test Chart

13.9.1 An Exercise in Comprehending Perceived Contrast Range

The material in this chapter has covered a very wide range of related topics and it would not be surprising if the reader was beginning to wonder how it all related to the reality of the situation. This exercise brings together many of these topics in a manner which explores their interdependence and in doing so provides the opportunity to gain a greater depth of understanding of the subject.

Before beginning a session of viewing and adjusting pictures displayed on a screen, one of the tools often used to provide a quick check that in contrast terms all is reasonably well with the set-up of the display is to first switch to a digitally generated grey scale to ensure that one is able to perceive all the steps of the scale with no black or white crushing.

Designing such a grey scale and checking that it matches up in subjective terms to evenly displaying all the steps of the scale provides a valuable exercise in bringing together the topics of this chapter and comprehending how the instrumentation of the exercise and the perception of the resulting grey scale are interrelated.

13.9.2 The Structure of the Greyscale Chart

As a first step we will define our grey scale in such a manner that when viewed in a well-defined viewing environment with a properly set-up display, it will have 10 chips of equal perceptible difference, distributed in lightness order between perceptual black and white. Perceptual black is at a luminance level such that in the defined viewing environment, lower levels of luminance level would not indicate any perceptual difference in lightness. Including black, this will give us 10 values of lightness. These chips will be arranged in a broad strip horizontally across the screen with a background whose lightness will be set at mid-lightness level.

Figure 13.24 illustrates the approach and gives the projected perceived lightness level of each chip level within the grey scale. In order to provide a simple check that the full range of signal levels is displayed, the black and white chips are double width to incorporate two additional mini-chips at equal perceptible differences between the black chip and the adjacent chip and between the white chip and the adjacent chip, respectively. These mini-chips will therefore be in increments of 1/27th or 3.70% of white level lightness. An additional mini-chip in the black chip with a signal code value of zero assists in appraising the accuracy of the monitor set-up in a manner which will be described later.

13.9.3 The Basic Procedure

To produce such a digital grey scale signal, which is able to provide equal perceptible lightness steps, we need to work back from the wanted values of the lightness of the chips to the test signal digital code values required to achieve this. This process requires the following steps:

1. Calculate the luminance of each step using the relationship between luminance and lightness.

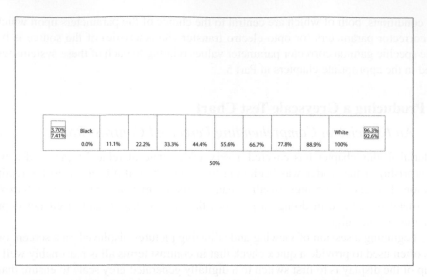

Figure 13.24 Greyscale chart outline indicating perceived lightness levels.

The CIE relationship between lightness (L) and luminance factor (Y) for reflective surfaces was defined in Section 4.6:

$$\text{for } (Y/Y_n) > 0.008856 \quad L^* = 116(Y/Y_n)^{1/3} - 16$$
$$\text{for } (Y/Y_n) < 0.008856 \quad L^* = 903.3(Y/Y_n)$$

Transposing:

$$\text{for } L^* > 0.08 \quad Y = Y_n((L^* + 16)/116)^3$$
$$\text{for } L^* < 0.08 \quad Y = Y_n(L^*/903.3)$$

However, although this relationship has been found to work well for lightness values between 10 and 100, below the value of 10, for luminous surfaces in particular, the relationship appears to be more accurately defined by using $Y = Y_n((L + 16)/116)^3$ for all values of lightness L. As there is as yet no formal definition of this relationship, it is this formula which will be used to determine the luminance of the greyscale steps.[4]

2. Assuming a monitor which faithfully follows a prescribed gamma law, calculate the level of drive signal to provide the level of luminances calculated in (1) using the relationship:
Voltage $= (\text{Luminance})^{1/\gamma}$

3. Calculate the signal integer code values from the voltages which define each lightness level using the number of code values associated with the bit depth selected in the application generating the grey scale.

---

[4] The CIE have instigated a technical committee, TC1-93 Calculation of Self-luminous Neutral Scale, with the following Terms of Reference: To recommend a formula or computational method for an achromatic, neutral or greyscale for self-luminous (i.e. non-reflective) surfaces. (This computation complements CIE Lightness, L^*, which serves a similar purpose for reflective surfaces.)

13.9.4 Conditions for Viewing the Grey Scale

In testing this concept, we had access to both a legacy CRT display and a professional photographic post-production suite with D65 environmental lighting and an Eizo 30″ CG301W LCD monitor. The monitor was calibrated to a peak white of 80 nits at a colour temperature of D65, a display gamma used by many professional photographers of 2.2 and a 'monitor black' of 0.12 nits. (Current display technology is generally incapable of producing a zero level luminance for a zero level input signal without using dynamic techniques to adjust the back light, which in turn introduces other problems; see Section 8.3. In consequence, these dynamically adjustable back lighted screens should not be used for critical picture adjustment.)

The inter-frame contrast range of the display was therefore 80/0.12, that is, about 660:1.

The environment in which the display was located had the window almost completely blacked out and the subdued environmental D65 lighting could be switched off, a useful approach to illustrating how the adaptation of the eye changes to enable the perception of darker tones.

13.9.5 Some Initial Considerations

Because of the complex interrelationship between all the factors involved in ensuring the specification of a grey scale produces the required subjective results, it is advisable to first review these parameters in the context of viewing the grey scale.

13.9.5.1 Perceived Contrast Range

It is clear that if all the chips of the grey scale are to be perceived with equal differentiation, their luminance values must fall within the perceived contrast range of the eye under a prescribed set of viewing conditions. When the screen fills only a relatively small fraction of the field of view, this contrast range is likely to be close to the spatial static contrast range defined earlier; however, as the screen field of view increases, then the effect of the spatial dynamic contrast range of the eye will become influential and the effective overall contrast range of the eye will increase.

In a critical viewing environment with low levels of environmental lighting, the luminance content of the greyscale image itself is most critical in influencing the perceived contrast range of the eye. Assuming that the grey scale produces an average luminance level which corresponds closely to the average screen lightness level of 50%, we know from the considerations in Section 13.3 of this chapter that the perceived contrast range of the eye under these conditions is likely to be in the range between 100:1 and 500:1.

At the limit of perceptibility of perceived contrast range is *perceptual black*, which may be defined as equating to a subjective black luminance level where a superimposed low level luminance pattern which extends above subjective black luminance level is perceived whilst the same level of pattern which extends below this level is not perceived.

13.9.5.2 Screen Contrast Range

Assuming for the moment that the screen is characterised by its defined gamma law over the whole of its characteristic and that therefore the luminance value of screen black is zero, then its contrast range will be dependent upon the code depth of the signal which drives it.

As an example, assuming a screen with a gamma of 2.2 and a test signal with an 8-bit code depth, the luminance and resulting contrast range for some low code values are given in Table 13.2.

Table 13.2 Screen relative luminance against code value assuming adherence to defined gamma law

Code value	Volt	Luminance	Contrast
0	0.000	0.00000	∞
1	0.004	0.00001	196965
5	0.020	0.00018	5710
10	0.039	0.00080	1243
15	0.059	0.00196	509
20	0.078	0.00370	270
25	0.098	0.00604	166
30	0.118	0.00902	111
255	1.000	1.00000	1

From Table 13.2, it is evident that the contrast range of the screen far exceeds that of the eye. A code value of 1 produces a contrast range of nearly 200,000:1 and there are 30 code values below the code value which provides a contrast range of 100:1.

Thus, if the lightness values defined in Figure 13.24 were to be matched linearly with code values, it is clear that the low luminance level chips would be well below the perceived contrast range of the eye and would therefore be either black crushed or not perceived at all. For code depths of 12 or 14 bits, the mismatch situation is further exacerbated in terms of the greater number of code values producing luminances below the level of perceptibility.

It is useful therefore to envisage the contrast range of the screen in similar terms as to that of a very deep pool in which only the top two metres at the surface are required in which to swim, and occasionally several more in which to dive, but below this level are greater depths rarely, if ever, explored.

13.9.6 Matching the Greyscale Lightness Values to the Signal Code Values

Remembering that the eye always adapts to the lighter elements of the image, one way to correctly envisage the approach to the objective of achieving a greyscale with equally differentiated perceived lightness level steps is to start at white level and determine the level of *perceptual black* from the assumed perceived contrast range. Since the monitor highlight luminance is 80 nits, then perceptual black will occur at some point between 80/500 and 80/100 nits, that is, between 0.16 and 0.80 nits. By definition, any luminance levels below 0.16 nits are beyond the perceived contrast range of the eye and will appear as blacks.

The only practical way of determining the actual value of perceptual black for a given set of viewing conditions is to construct a number of greyscale charts, with the luminance representing perceived black level set in the range between 0.16 and 0.80 nits, and to display them in turn to find the chart which matches the criteria. The actual perceived contrast range of the eye for these conditions will then have been determined.

However, in designing the range of charts, we have to take account of the reality of current displays which are generally incapable of producing a real black. Thus, we will define *monitor black* as the level of black displayed by the monitor when receiving *digital signal black,* which in turn is represented by a code value of zero. As long as monitor black is below the luminance level of perceptual black, its effects can, to a first degree, be compensated for in the design of the greyscale chart. In this case, monitor black has been measured to be 0.12 nits and is thus below the projected lowest level of perceptual black at 0.16 nits (but not much below, which is likely to cause some small distortion of one or possibly both of the levels of the black mini chips when viewing the higher contrast grey scales).

13.9.6.1   Calculating the Code Values for the Greyscale Charts

The general approach is first to establish the luminance level for perceptual black and then calculate the lightness this luminance level represents by using the defined relationship between luminance and lightness defined in Section 13.9.3. In the lightness scale, this value now represents perceptual black and thus the lightness values required to give even steps between this value and white can be calculated. It is these lightness values which then form the basis of the calculations which lead, in turn, to obtaining the corresponding luminance values, the voltages corresponding to the luminances and, finally, the code values corresponding to the voltages.

A grey scale based upon a contrast range of 500:1 will be used as an example for calculating the code values required. These calculations are illustrated in Table 13.3, which is derived from Worksheet 13(d), and are described in what follows.

Remembering that we have nine steps and that at black and white, we have two further steps at fractions of one third of the lightness values of the adjacent steps, then it will be seen that the steps correspond to a number of fractions of 9×3, that is, 1/27th of peak white. These fractions are listed in column 1 of Table 13.3. The lightness of these steps is calculated in column 2.

Now, step 0 needs to be made equal to the particular perceptual black represented by the chosen contrast range of the chart. Notionally this is calculated as 'Perceived black' at 0.16 nits in the top of the table; however, it must be remembered that with the exception of the CRT display, the screen black luminance level is also contributing to the perceived black luminance level. Thus, the luminance level of screen black at 0.12 nits should be subtracted from the luminance relating to perceived black level in order to establish the level of luminance to be provided by the chart code value. The table calculates the difference to be 0.04 nits and the lightness level corresponding to this luminance is calculated to be 7.94. (For the table to be used for the charts for the CRT display, the 'Screen black' level is set to zero.)

In column 3, the equal lightness levels between greyscale black at 12.60 and white at 100 are calculated. Column 4 calculates the screen luminance contribution from the greyscale values corrected by the contribution from screen black, and column 6 calculates the voltage required to produce the luminance. Finally, in column 7, the code values are calculated.

Table 13.4 is a check table which calculates the lightness values from the code values to monitor that the calculations are reasonably sound. The minor differences between the initial and the calculated lightness values relate to rounding errors in the digital coding calculations.

The luminance corresponding to perceptual black, as we have seen from above, is 0.16 nits, corresponding to a relative lightness of 12.6, but monitor black is already providing 0.12 nits, so we need only 0.04 nits from the chart value to add to monitor black in order to provide our perceptual black luminance level.

Table 13.3   Calculating greyscale code values

Perceived lightness to code value					
Assumed perceivable contrast ratio				Luminance (nits)	Calculated lightness
500:1			Perceived black	0.16	12.60
			Screen black	0.12	11.45
			Start black	0.04	7.94

Super white code values: 0			White code value 255		
	Lightness exponent 3.00	Screen white (nits) 80		Display-exponent 2.20	Bit depth 8

Step value in 1/27th	Normalised relative lightness	Calculated lightness	Corrected luminance (nits)	Contrast ratio	Normalised drive voltage	Code value
0	0.00	12.60	0.040	1997	3.16	8
1	3.70	15.84	0.198	404	6.54	17
2	7.41	19.07	0.435	184	9.35	24
3	11.11	22.31	0.768	104	12.11	31
6	22.22	32.02	2.507	32	20.73	53
9	33.33	41.73	5.695	14	30.11	77
12	44.44	51.44	10.772	7.4	40.22	103
15	55.56	61.16	18.177	4.4	51.02	130
18	66.67	70.87	28.352	2.8	62.45	159
21	77.78	80.58	41.734	1.9	74.45	190
24	88.89	90.29	58.763	1.4	86.97	222
25	92.59	93.53	65.326	1.2	91.26	233
26	96.30	96.76	72.360	1.1	95.61	244
27	100.00	100.0	79.880	1.0	100.00	255
13.5	50.00	56.30	14.16	5.64	45.54	116.00

Furthermore, the minimum black level luminance of the monitor will be added to the simple calculated value of luminance from lightness and, where it is comparable in value, it will affect the appearance of the darker steps.

In order to complete the flexibility of Table 13.3, the feature is also provided for the entry of reserved codes for super white, which are code levels above system white to accommodate for example minor overloads and transients. Photography does not use super white codes, but the television ITU-R BT 709 standard specifies 20 super white codes for the 8-bit code depth system, for example, thus making system peak white equal to a code value of 235 rather than 255.

The final column in Table 13.4 indicates that this grey scale should provide equal perceptible lightness steps.

To assist with appraising the display of the different contrast range charts, an additional digital black mini-chip was added in the extended perceptual black chip with a code value of zero.

The completed chart appears approximately as illustrated in Figure 13.25.

Table 13.4 Greyscale perceived lightness values

	Code value to perceived lightness				
		Screen black 0.12 nits			
	Screen white (nits)	Gamma		Lightness exponent	
	80	2.20		3.00	
Code value	Normalised screen drive	Screen luminance (nits)	Contrast ratio	Chart-based lightness	Normal relative lightness
8	3.14	0.159	503	12.58	0.00
17	6.67	0.326	245	15.98	3.89
24	9.41	0.561	143	19.14	7.51
31	12.16	0.894	89	22.36	11.19
53	20.78	2.640	30	32.08	22.30
77	30.20	5.852	13.7	41.82	33.45
103	40.39	10.991	7.3	51.60	44.64
130	50.98	18.264	4.4	61.12	55.52
159	62.35	28.377	2.8	70.79	66.58
190	74.51	41.933	1.9	80.63	77.84
222	87.06	59.008	1.4	90.35	88.96
233	91.37	65.619	1.2	93.61	92.69
244	95.69	72.615	1.1	96.82	96.37
255	100.0	80.000	1.0	100.00	100.00
116	45.49	14.24		56.25	49.96

Figure 13.25 Approximate appearance of the grey scale. (The black mini-chips are unlikely to be visible in print.)

13.9.7 Appraisal of the Charts

One of the main problems in determining the code values to be given to the chart is that the contrast range of the eye when viewing a screen will have a value somewhere between that for the spatial static and that for the spatial dynamic contrast range values, depending upon the average and highlight luminance of the screen, its angular field of view, and the luminance of the surrounding surfaces. Thus, to produce the required perceptually even range of lightnesses, ideally, the chart should be designed for a contrast range which matches a specific set of viewing conditions. However, generally speaking, we would wish to view the chart in reasonably critical viewing situations where the screen fills a considerable percentage of the field of view and the ambient and surround lighting are at a low level. In these circumstances, the effective contrast range of the eye, as we have seen in Section 13.9.5.1, is likely to fall between 100 to 1 and 500 to 1.

Thus, five charts were produced which embraced these contrast ranges and each was viewed critically in turn with very low levels of ambient light at a desk viewing distance such that the screen filled the central field of view.

The criterion for selecting the correct chart is that the digital black mini-chip should not be visible but the remaining two dark mini-chips should be at their most perceptible.

Both the CRT and the LCD displays were appraised with their appropriate charts. The 100:1 contrast range chart clearly showed the digital black mini-chip indicating the eye perceiving a greater than 100:1 contrast range. The 500:1 contrast range chart clearly showed a loss of perceptibility of the two dark steps, and the 400:1 contrast range chart met the above criterion.

Thus, using these criteria, the charts may be used to determine the usable contrast range for a particular combination of screen black level, average scene luminance, the angle the screen projects at the eye, and the level and distribution of the luminance of the surrounding surfaces. Once the appropriate chart has been established, it may be used on a day-to-day basis to quickly affirm that the contrast criteria of all monitors are satisfactory before undertaking critical adjustment, picture matching or appraisal work. For example, in a critical television viewing environment, where the field of view of the monitor is likely to be less than in the example used for this exercise, the greyscale chart contrast range is likely to be in the 300:1 to 400:1 range.

It should be appreciated that since the contrast range of the eye under the critical viewing conditions outlined is dependent upon the average luminance of the screen over relatively short time periods, that is, its level of adaptation, then broadly speaking, for scenes of higher than average luminance, the contrast range of the eye will be diminished and, for scenes of lower than average luminance, the range will be enhanced on the figures obtained above.

13.9.8 Implications for Bit Depth Requirements

Recognising that display screens will eventually become widely available, which are capable of producing a highlight luminance of possibly up to fifteen hundred or so nits and also a display black of zero luminance, in the light of the findings above it would be prudent to review the criterion for the required bit depth of digital systems for colour reproduction.

In an 8-bit depth system, notionally 255 bits are available; however, assuming a contrast range of 350:1, from Worksheet 13(d), some 18 code values are below black and are therefore not perceived; if in addition, the system specifies 20 code values for super white excursions of the signal, then only 217 code values are available to represent lightness changes. This may be considered a marginal number of code values to portray all scenes without introducing contouring artefacts.

If however a 10-bit system is selected, 1,023 code values are available and the perceptual black code value is 71; if a further 83 code values are reserved for super white excursions, this would leave 869 code values to portray the range of luminance levels, a significantly improved situation.

14

Storage and Conveyance of Colour Signals – Encoding Colour Signals

14.1 Introduction

What one may ask have storage and conveyance to do with colour in reproduction? We have seen that connecting the RGB signals from the camera to the display can produce excellent results, so from the point of view of the reproduction of the colours in the scene, why should storing and conveying these signals require addressing? Well, it depends. As it turns out, storing and conveying RGB signals is not only very inefficient but in the days of analogue systems particularly, could also lead to a loss of colour balance due to the variation in gain in the different circuits carrying the RGB signals from the camera to the display.

Furthermore, at the time of the introduction of colour to television, there was a large population of black and white systems in use in photography, television and cinematography, which would need to accommodate the new colour signals. It was crucial therefore that some means be found to make the colour signals compatible with the black and white systems then in existence.

In order to address these various issues it became necessary to determine how the three RGB signals could be processed to overcome these shortcomings. Since the driving imperative at the time was the introduction of colour television in the United States, it was the working parties of the National Television Systems Committee (NTSC), which, in exploiting the characteristics of the eye and a number of emerging electronic techniques, evolved an efficient and sophisticated approach to encoding the RGB signals. Encoding in this context describes a number of processes which may be used to both improve the integrity of the signals during conveyance and storage and reduce the amount of information needed to define a colour image in an electronic form, including matrixing, filtering of spatial detail the eye does not perceive and multiplexing.

The colorimetric aspects of matrixing and filtering are addressed in the following sections. However, multiplexing is the process of combining the processed RGB signals into a single stream for storage and transport, and since it should not fundamentally affect the colour rendition of the final image, it falls into the category of the supporting technology (Poynton,

---

Colour Reproduction in Electronic Imaging Systems: Photography, Television, Cinematography, First Edition. Michael S Tooms.
© 2016 John Wiley & Sons, Ltd. Published 2016 by John Wiley & Sons, Ltd.
Companion Website: www.wiley.com/go/toomscolour

2012), which is outside the scope of this book. (Nevertheless for completeness, multiplexing in early colour television systems is briefly alluded to in Section 17.2.6.)

However, depending upon the strategy adopted for encoding the RGB signals, some of the signal parameters may be compromised. To assist in being aware of how these compromises affect the quality of the reproduced image it is helpful to understand the mechanisms of encoding and their limitations.

It is clear from the above that in addressing the requirements for the storage and conveyance of colour signals, three areas need to be addressed:

- Retention of colour balance
- Compatibility with monochrome systems
- Efficiency improvement in terms of data storage capacity and conveyance data rate

14.2   The Imperatives for Encoding RGB Colour Signals

14.2.1   Retaining Colour Balance

We have seen in Chapter 10 that in ensuring a colour camera produces the correct signals, it is necessary, either manually or automatically, to ensure that the camera is white balanced; that is the levels of the RGB signals are made equal when the camera is imaging a white in the scene. Any imbalance will show as a colour cast on the reproduced image; this may be acceptable, if not desirable, when viewing a photographic print in isolation but in television terms, where, as the viewer is presented with images from different cameras in sequence, any minor imbalance is immediately noticeable and disturbing. It is essential therefore that the white balance is preserved when the signals are processed, stored and transferred from one element in the signal chain to another. This was particularly difficult to achieve in the days of analogue systems where to stabilise the gain of the innumerable items of equipment in the three signal paths to the desired level was all but impossible.

Thus, it is essential that the encoding of the RGB signals into the three new signals produces a result that ensures that any differences in the gain of the three channels carrying the new signals does not change the colour balance of the reproduced image.

14.2.2   Ensuring Compatibility with Monochrome Systems

If an image from a monochrome camera is to produce an image which is perceived to represent in all respects with the exception of chrominance, the original scene, then the camera spectral sensitivity should follow the photopic response of the eye. In this manner, assuming an overall linear relationship in the reproduction system, surfaces in the reproduced image will be displayed with the same brightness relationship as those in the original scene.

In order to derive a suitable monochrome signal, it is therefore essential that one of the new encoded signals has a characteristic that emulates the signal derived from a monochrome camera. Thus, when a composite colour signal is made available to a monochrome system, all that is required is for the monochrome system to use the signal emulating a monochrome signal and dispense with the two remaining signals. The term 'composite colour signal' describes a signal which contains the three new encoded signals in a single multiplexed format.

14.2.3 Improving the Efficiency of the Colour Signals

Before investigating how the efficiency of colour signals may be improved, we first need to have an understanding of just which parameter(s) we are looking to improve. Fundamentally we are looking at the amount of data required to describe a colour signal and we saw in Section 8.4 that this translates to the number of pixels required to ensure that the resulting image is capable of satisfying the recognition acuity of the eye. The number is dependent upon the size of the image and the viewing distance, and since reproduction systems serve a range of requirements in this context, then the number of pixels required for the various reproduction systems which serve television, photography and cinematography also varies.

In practice reproduction systems were, and to an extent still are, defined around the level of technology available at the time of the introduction of the system and how this limited the achievable viewing angle the display subtended at the eye. As an example, the 405-line system, one of the earliest television systems to be standardised in the 1930s, defined the number of luminance changes per picture height at about 375, which corresponded in a current system terms to 375 pixels per picture height. This was an adequate number for displaying the image on a cathode ray tube (CRT) of maximum diagonal dimension of about 20 inches at a minimum viewing distance of about 2.8 metres. However, as the technology has advanced to make available display devices of 50 inches and more, so new systems have been standardised to match the number of pixels to the increased area of display.

When standardising a new reproduction system for a particular environment, the approach is to define both a critical viewing distance associated with the maximum screen size and the projected advances in technology likely to be available at the time of implementation. The number of pixels is then calculated to satisfy the recognition acuity of the eye by a factor which accommodates slightly larger screens and/or shorter viewing distances.

In Worksheet 8, the formula developed for relating the acuity of the eye, the viewing distance, the screen size and the minimum number of pixels required, is used to generate graphs which express viewing distance and screen size for current standards associated with various system pixel numbers, as illustrated in Figures 14.1 and 14.2. Following usual practice, the screen size is given in inches.

In recent times it has become the practice to express television and cinematography system standards in shorthand terms as the number of thousands of pixels per picture width and denote this by using the letter 'K'. In fact the K used actually represents 960 pixels for historic reasons; thus the current world standard high-definition television system (HDTV) is referred to as a 2K system, with 1,920 pixels per picture width. Newly proposed systems are usually expressed in terms of multiples of the 2K system, as shown in the graph in Figure 14.2.

As an example of using the graphs, if we assume that when viewing a largish print with an aspect ratio of 3:2, which had been shot and printed with a camera and printer capable of the equivalent of 7.1 megapixels, at a viewing distance of about 0.7 metre, then from Figure 14.1, the maximum print diagonal which could be used before the eye was able to detect a loss of resolution would be about 31 inches.

It is evident from the above that generating a single picture requires a considerable amount of data; each composite pixel comprises effectively a red, green and blue pixel and thus there is an R, G and B value for each composite pixel.

Table 14.1 illustrates the total bits of data which would be required to describe an image derived from the 7.1-megapixel camera prior to encoding.

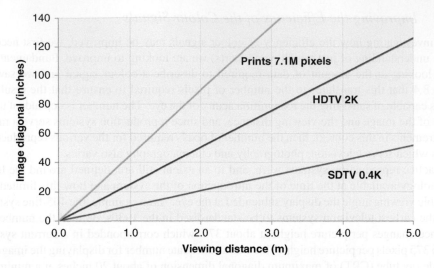

Figure 14.1   Screen size for group viewing.

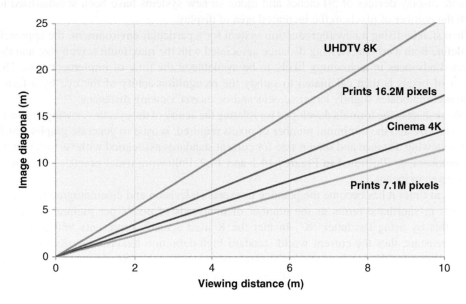

Figure 14.2   Screen size for public viewing.

Table 14.1   Establishing the total amount of data for each picture or television frame

Vertical pixels	Horizontal pixels	Total composite pixels	Total RGB pixels	Bit no. after quantisation at 8-bit depth
2,176	3,264	7,102,464	21,307,392	170,459,136

Prior to the advent of digital technology, considerable effort was brought to bear to reduce these data quantities, and the processes developed then continue to be used today. Once the signals have been quantised, then compression technology (Watkinson, 1999) is used to very significantly reduce the quantity of data required. In television and cinema terms, the frames are conveyed at a minimum rate of 25 or 30 times per second in current systems, with higher rates being considered for projected systems, producing very high data rates.

Large quantities of data are expensive to store and conveyance capacity for high data rates is limited; there is thus a strong incentive to evolve the means to reduce the data required to describe an electronically generated image.

14.3 System Compatibility and Retention of Colour Balance

As noted in the Introduction to this chapter, it was essential that any colour reproduction system should be compatible with monochrome photographic and television systems already in widespread use. Furthermore, the colour balance of the system should not change with differential change in the gain of the three paths carrying the colour signals.

14.3.1 The Luminance Signal

The essence of a monochrome system is obtaining an electrical signal derived from a camera whose optical analysis system has a spectral sensitivity characteristic which broadly follows the luminosity function of the eye as illustrated by the V_λ curve in Figure 14.3.

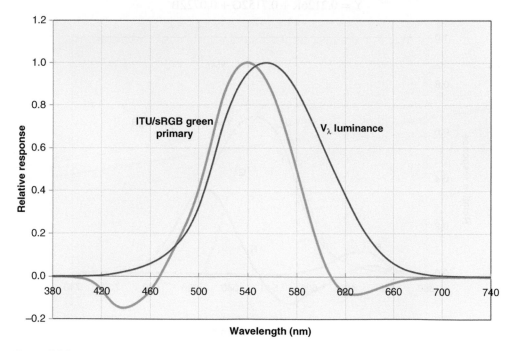

Figure 14.3 The photopic response of the eye and a typical ideal green camera spectral sensitivity characteristic.

Thus, what is required from a colour reproduction system is a means of deriving a *luminance* signal from the three R, G and B signals which emulates the luminance response in Figure 14.3. Intuitively, consideration of the colour camera spectral sensitivities and the luminosity function of the eye indicates that as the green characteristic is closest to the luminosity function, this would form the basis of the luminance signal, with the addition of diminishing contributions from the red and blue characteristics respectively. However, we need to find a method of calculating precisely the values of the contributions or weighting factors of the RGB signals derived from the three spectral sensitivities to match the luminance response.

It may be recalled from Chapter 4 that the 'Y' characteristic of the CIE XYZ colour-measuring system was made to follow the luminosity function of the eye in order that the value of Y always measured the luminance of a colour. Thus, by using the inverse of the matrix procedures outlined in Appendix F to derive the XYZ characteristics from the camera RGB characteristics, we can use the resulting factors of Y in the matrix to establish the RGB weighting factors. In Worksheet 14, Matrices 1 and 6 illustrate the procedure for obtaining this inverse matrix and, following the selection of the appropriate primaries, we can use the values derived there, to establish the values of the RGB weighting factors required.

From Worksheet 14, after selecting the ITU/sRGB primaries, which are used in both tele-vision and photography, the Y row of Matrix 6 provides the following values for the RGB luminance weighting factors:

$$L_R = 0.2126, L_G = 0.7152 \text{ and } L_B = 0.0722.$$

Thus, the luminance signal, designated by the symbol Y, can be derived as follows:

$$Y = 0.2126R + 0.7152G + 0.0722B$$

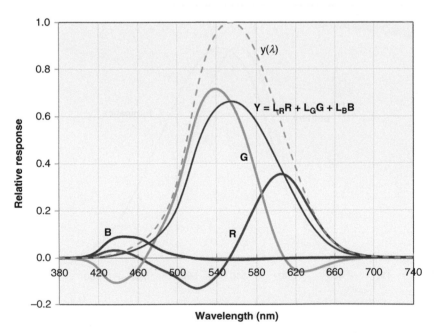

Figure 14.4 Illustration shows that summing the weighted RGB curves produces a true luminance response.

In Worksheet 14, the luminance weighting factors are applied to the RGB camera spectral sensitivities to provide the weighted curves in Figure 14.4.

These curves are then added in Worksheet 14, Table 1, to show that they sum to produce the Y curve. The CIE $y(\lambda)$ curve is also shown to confirm that the Y curve is of identical shape.

This signal would be used by a monochrome display to produce results which would compare identically to that derived from a monochrome camera whose optical response matched the V_λ curve. Systems using different primaries would need to use different appropriate weighting factors for the RGB signals comprising the luminance signal. These weighting factors can be determined from Matrix 6 using the selection buttons to select the appropriate primaries or copying the chromaticity coordinates from the Primaries Worksheet to the relevant cell range in Worksheet 14.

14.3.1.1 Gamma Correction

As explained in Section 13.4, the linear RGB signals are gamma corrected at source and the luminance signal is no exception to this requirement, so, as in a monochrome system, gamma correction would also be applied.

Thus the gamma-corrected luminance signal is:

$$Y^{1/\gamma} = (0.2126R + 0.7152G + 0.0722B)^{1/\gamma}$$

Now, for the sake of consistency, we would normally designate the gamma-corrected luminance signal with a prime in the same manner as used to designate gamma-corrected RGB signals; however, Y' has historically been used to describe the 'luma' signal, which is defined in terms of the addition of the luminance-weighted $R'G'B'$ signals; thus:

$$Y' = 0.2126R' + 0.7152G' + 0.0722B'$$

It is perhaps not surprising that the use of both of these signals in various colour reproduction systems has led to much confusion, so a few words of clarification would not go amiss.

Traditionally, probably for simplicity and overlooking the problems which can occur in encoding and decoding by its use, the Y' signal has been almost universally used in reproduction systems and was loosely and incorrectly referred to as the luminance signal. (We shall investigate the problems referred to above in Section 14.6.) Poynton (2012) has been at pains to attempt to clarify the situation by introducing the term 'luma' to describe the Y' signal, a recommendation we have adopted throughout this book.

The use of the luma signal can lead to compromises in the quality of the reproduced image, particularly with regard to the lack of detail in saturated areas of the image. This compromise has been recognised from the beginnings of electronic colour reproduction but it is only recently that newly defined systems offer the option of using the luminance $Y^{1/\gamma}$ signal in preference to the luma Y' signal. In these systems, $Y^{1/\gamma}$ is sometimes designated as Y_C' and the subscript 'C' is used as an abbreviation of 'constant luminance' to differentiate it from Y', which is not a constant luminance signal as we shall see subsequently.

14.3.2   The Complementary Colour Difference Signals

Having derived a luminance signal for compatibility purposes, in order to avoid duplication of data it will also be used as one of the principal signals for the colour reproduction system. But what are the requirements for the two signals to complement the luminance signal? We have seen that in order to describe a colour, three values are required; these may be values of red, green and blue; values of luminance, hue and saturation; or values of luminance and chromaticity, where as we have seen, specifying chromaticity requires two values.

Since we have already derived a luminance signal, then two further signals are required, which ideally would describe the chromaticity of the scene. The principal criterion is that any changes in the relative amplitudes of the three signals as they pass through the signals chain do not change the colour balance of the reproduced image.

14.3.2.1   The Linear Case

We will first derive some basic properties of a linear-based system before addressing a system using gamma-corrected signals.

As the luminance signal is comprised primarily of the green signal, then the two other signals are generated by subtracting the luminance signal from the red and blue signals respectively. Thus, using the ITU/sRGB primaries to two decimal places:

$$R - Y = 1R - 0.21R - 0.72G - 0.07B = +0.79R - 0.72G - 0.07B$$

$$\text{and } B - Y = 1B - 0.21R - 0.72G - 0.07B = -0.21R - 0.72G + 0.93B$$

Since the luminance signal is subtracted from the colour signals, these new signals are referred to as *colour difference* signals.

Thus, the three new signals are Y, R – Y, and B – Y and we shall explore their properties in order to show that they meet the requirements listed in the Introduction to this chapter.

Recovering the RGB Signals

Firstly, we need to show how we can extract the original RGB signals from the luminance and colour difference signals.

The R and B signals are recovered by direct addition of the luminance signal to each of the colour difference signals respectively. To recover the G signal:

$$Y = 0.21R + 0.72G + 0.07B \tag{14.1}$$

and also

$$Y = 0.21Y + 0.72Y + 0.07Y \tag{14.2}$$

Subtracting (14.2) from (14.1)

$$0 = 0.21(R - Y) + 0.72(G - Y) + 0.07(B - Y)$$

Thus

$$G - Y = -\frac{0.21}{0.72}(R - Y) - \frac{0.07}{0.72}(B - Y)$$

and

$$G = Y - 0.30(R - Y) - 0.10(B - Y) \tag{14.3}$$

Thus

$$G = 1.40Y - 0.30R - 0.10B \tag{14.4}$$

Colour Difference Signal Amplitudes
When the camera is scanning a white in the scene:

$$R = G = B = 1 \text{ and } Y = 1$$

Thus

$$R - Y = 0 \text{ and } B - Y = 0$$

Furthermore, if $R = G = B = 0.5$, then $Y = 0.5$ and again the colour difference signals are zero.

In the general case, whenever the camera is scanning a white or neutral grey in the scene, the colour difference signals will be zero. Furthermore, it can be shown that the amplitude of the colour difference signals increases with the saturation of the colour or conversely diminishes as the saturation approaches zero. This property was very important in the analogue days of colour reproduction, since for much of the time an average scene has low levels of saturation and thus the corresponding low levels of the colour difference signals were less likely to cause mutual interference in systems of encoding where they shared frequency bands with the luminance signal.

It can also be shown by a few examples that the colour difference signals increase in level with increasing luminance; thus both luminance and saturation cause a rise in the levels of the colour difference signals. It may be recalled that in Section 2.4, this property of colour is defined as chroma and in consequence when the colour difference signals are taken together, they are usually referred to as *chrominance* signals. A reproduction system comprising signals in a luminance and chrominance format is often described as being a YC format system and, with the exception of the initial electronic system for cinematography, all current and proposed photographic and television systems utilise this format in one way or another.

Colour difference signals are often plotted with the B–Y signal on the *x*-axis and the R–Y signal on the *y*-axis as illustrated in Figure 14.5, for the additive and subtractive primaries, based on the figures in Table 2 of Worksheet 14. The resulting vector length of the colour is proportional to both the saturation and the luminance of the signal, as can be seen from the inner shape, which shows the same colours at 50% amplitude but the same saturation. (If these were chromaticity values the vectors for each colour would be the same length for both sets of levels.)

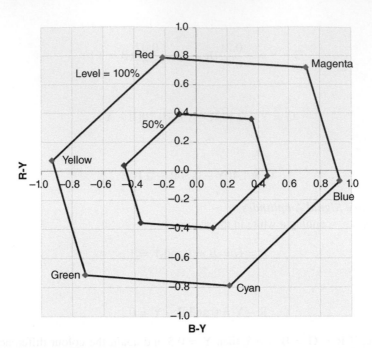

Figure 14.5   Chrominance values for each set of primaries at both 50% and 100% levels.

Changes in Channel Gain

Any differential changes in gain between the three channels carrying the YC signals will not change the colour balance of the image since on a grey-scale the colour difference signals are zero. This property satisfies our criterion that the colour balance is not modified by changes in gain between channels. Thus the critical neutral colours in the scene will not change; clearly however, a differential change in gain between the two colour difference signals will change the chromaticity of the non-neutral colours in the scene.

14.4   A Simple Constant Luminance Encoding System

In Section 14.5, we will be looking in a little more detail at a characteristic of the eye, first alluded to in Section 1.4, in which the eye is much more sensitive to changes in luminance than it is to changes in chrominance. This inspired early workers responsible for defining colour reproduction systems to engineer a system where, when it became necessary to trade off unwanted interference or noise in the signals of a system, then more of the interference or noise was directed into the chrominance channels. This strategy will operate only at maximum efficiency if none of the luminance information is carried in the chrominance channels.

The term 'constant luminance' is used to describe a system whereby all the luminance information is *constantly* carried in the luminance channel; the converse of this statement is that there are no circumstances when the chrominance channels carry any luminance information.

A consequence of this property is that the addition of any signal to the colour difference signals, such as noise or interference, will change only the chromaticity and not the luminance of the image, making it far less visible than it would otherwise be.

As an example of how this operates, let us assume we commence with a colour C represented by:

$$C = 0.6R + 0.4G + 0.5B$$

Thus

$$Y = 0.21R + 0.72G + 0.07B$$
$$= 0.21 \times 0.6 + 0.72 \times 0.4 + 0.07 \times 0.5$$
$$= 0.126 + 0.288 + 0.035$$
$$= 0.449$$

and

$$R - Y = 0.6 - 0.449 = 0.151$$
$$B - Y = 0.5 - 0.449 = 0.051$$

During the passage between the camera and the display, let us assume that a noise signal of level 0.1 is added to the colour difference signals.

Then the new value of the signals will be:

$$Y = 0.449 \qquad R - Y = 0.251 \qquad B - Y = 0.151$$

Therefore

$$R = (R - Y) + Y = 0.700$$
$$B = (B - Y) + Y = 0.600$$

And, using equation (14.3) derived in Section 14.3:

$$G = Y - 0.30(R - Y) - 0.10(B - Y)$$
$$= 0.449 - 0.30 \times 0.251 - 0.10 \times 0.151$$
$$= 0.3594$$

Thus the new luminance will be:

$$Y = 0.21R + 0.72G + 0.07B$$
$$= 0.21 \times 0.70 + 0.72 \times 0.3594 + 0.07 \times 0.600$$
$$= 448 \text{ (an error of 0.001 due to rounding to two decimal places above)}$$

Ignoring the rounding error, this is the same luminance level we started with, confirming that as long as the luminance signal carries the full luminance information, the constant luminance system is immune to error signals in the colour difference signals causing errors in the display of the luminance information.

14.5   Exploiting the Spatial Characteristics of the Eye

In Section 1.4, where the characteristics of the eye relevant to colour reproduction were listed, reference was made to the spatial response of the eye and in Section 8.4 the acuity of the eye was defined and the relationship between acuity and the number of pixels required in order that the acuity of the eye is not compromised was established.

This work was related to the maximum acuity of the eye, which early experiment indicated was between changes in the luminance of objects in the scene. It was found that if the luminance is kept close to a constant but the chromaticities of the objects are changed the level of acuity diminishes. If two sets of saturated complementary colours from opposite sides of the chromaticity diagram are placed adjacent, it is possible to evaluate in qualitative terms how the colour acuity of the eye is affected.

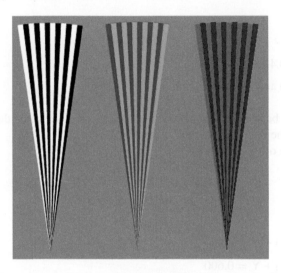

Figure 14.6   Resolution wedges indicate the different acuities of the eye to luminance and chrominance.

Figure 14.6 Illustrates three equal frequency wedges: for luminance, for reddish orange and bluish cyan, and for yellow-green and purple respectively. The darkest colour of each pair was adjusted for maximum colourfulness and the lightness of the complementary colour in each wedge was subjectively adjusted for equal lightness.

If this image is viewed at an appropriate distance, it is found that at the distance where the luminance bars appear to merge halfway down the wedge, the colours merge considerably higher up the wedges.

Detailed measurements indicate that in broad terms, the orange to cyan colours merge at half the resolution of the luminance and the yellow-green to purple colours merge at about a third of the resolution of the luminance.

Using the same rationale as used in Section 14.2 to establish the amount of data required to be included in the signal to match the luminance acuity of the eye, it will be clear that only half the amount or less will be required for the chrominance information. Thus, the colour difference signals may be filtered or subsampled to a fraction of the information capacity of the luminance signal without impairing the reproduced image. In pixel terms, the colour difference values of only alternate pixels, in both the horizontal and vertical directions, need to be included in the composite signal.

Different reproduction systems, depending upon the level of performance required and the limitations in their channel capacity, select different fractions of the luminance signal data rate for the chrominance signals in both the horizontal and vertical directions. Since the fractions chosen in digital systems are always factors of 1, 2 or 4, compared with luminance, then in the digital domain the luminance signal sampling rate is always considered to be at a base rate of 4 and the colour difference signals at a base rate of 4, 2 or 1.

A nomenclature has evolved to describe the variants of subsampling used for the colour difference signals, which is written in the form of 4:4:4. The first number indicates the luminance sampling rate and the next two numbers indicate that the colour difference signals are sampled at the same rate as the luminance in both the horizontal and vertical directions; thus, 4:4:4 describes the signals as they are following the matrixing of the RGB or $R'G'B'$ signals to the Y, R–Y and B–Y signals.

The form 4:2:2 indicates that the colour difference signals are sampled at half the rate of the luminance signals in the horizontal direction only; thus the composite data rate is reduced by a third.

The form 4.2.0 indicates the sampling at half the luminance rate in both the horizontal and vertical directions, thus the composite data rate is reduced by a half.

Both the above formats will generally produce images with no perceptual impairment.

Finally, the form 4.1.1 indicates that the colour difference signals are sampled at a quarter of the luminance rate in the horizontal direction only. Such a low chrominance sampling rate is marginal in terms of not affecting the quality of the perceived image and so is not generally used in quality reproduction.

In the early analogue systems where the multiplexing parameters were more stringent, in order to minimise interference or crosstalk between channels, different fractions were sometimes applied to different versions of the colour difference signals. These variants are briefly reviewed in Chapter 17.

14.6 A Practical Constant Luminance System

As we saw in Chapter 13, for a number of reasons the signals from the camera are gamma corrected and as a consequence the simple system described in Section 14.4 becomes somewhat more complicated to implement.

The gamma-corrected versions of the components of the YC signal defined earlier are:

$$Y^{1/\gamma}, R' - Y^{1/\gamma}, B' - Y^{1/\gamma}$$

In what follows the various configurations of the variants of the YC system will be illustrated in a schematic form.

14.6.1 A Constant Luminance Camera

In a camera designed for constant luminance operation, the pertinent processes required between the derivation of the linear RGB signals from the image sensors and the YC output of the camera are illustrated in Figure 14.7.

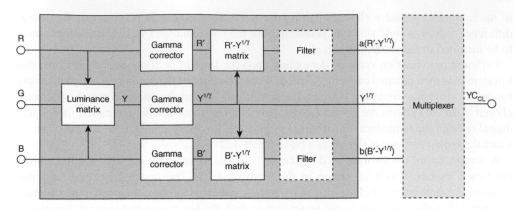

Figure 14.7   Constant luminance camera encoder.

It will be noted that the luminance signal is formed from the linear RGB signals before gamma correction, ensuring that it truly represents the luminance of the scene and emulates the signal from a well-designed monochrome camera.

The colour difference signals are formed by simple subtractive matrices which usually also contain scaling factors 'a' and 'b' for the $R'-Y^{1/\gamma}$ and $B'-Y^{1/\gamma}$ signals respectively. These scaling factors vary according to the particular colour system in use and are designed to reduce the amplitude of the colour difference signals, which in peak-to-peak terms would otherwise exceed the maximum value of the luminance signal, so ensuring they do not exceed the signal level capacity of the multiplexer.[1] Since the constant luminance colour difference signals are not symmetrical around zero level, as are the non-constant luminance colour difference versions, then where it is important in encoding to retain extent of polarity symmetry, the weighting factors for the positive and negative excursions may differ to make the signals symmetrical. This procedure is explained in detail in Section 20.4.2.4.

Depending upon whether subsampling is used, the colour difference matrices may be followed by filters which reduce the information content in the manner described in the previous section to produce signals appropriate to the sampling standards of the particular colour reproduction system.

The multiplexer does not change the colour content of the signals in any way and the technical description of its operation is therefore beyond the scope of this book. The output of the multiplexer is a single signal in YC format, sometimes with a subscript to indicate whether it is a constant luminance or non-constant luminance signal.

14.6.2   A Constant Luminance Display

In Figure 14.8, the YC_{CL} signal is fed to the de-multiplexer, not described here, which outputs the $Y^{1/\gamma}$, $a(R' - Y^{1/\gamma})$ and $b(B' - Y^{1/\gamma})$ signals. The luminance signal is added to the scaling corrected colour difference signals in the red and blue matrix respectively to recover the

---

[1] The multiplexer is a device that utilises the structure of the luminance and colour-difference signals in a manner that enables all three signals to be combined and subsequently separated with the minimum of interference between the signals.

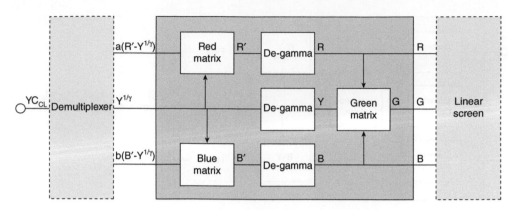

Figure 14.8   Constant luminance display decoder.

gamma-corrected R' and B' signals, which, together with the $Y^{1/\gamma}$ signal, are fed to the gamma circuits with a characteristic of $V = V^\gamma$ and thus produce the linear Y, R and B signals, respectively. These linear signals are then used by the green matrix to recover the G signal.

The three linear signals are then fed to the linear screen. It should be remembered that following the demise of the CRT, most displays operate in a linear fashion and, though often equipped with inverse gamma correctors, do so only in order to complement the gamma correctors inserted at source to emulate the legacy CRT displays.

This section has described the configuration of a constant luminance system designed with the linear screen in mind; in these circumstances, the gamma and de-gamma elements are more likely to take on the role of 'perceptible coding' elements where the encoding and decoding characteristics are fully complementary, as described in Section 13.6.

Other approaches to a constant luminance system are designed to use legacy equipment which significantly complicates the resulting configuration. In the case of a legacy camera, linear RGB signals are not available to the encoder and legacy display devices have built-in gamma circuits to emulate displays based upon the CRT. Such an approach therefore requires a plethora of gamma and gamma corrector circuits to provide the correct signals before and after multiplexing respectively.

One of the principal advantages of the constant luminance system is that since all the high resolution luminance information is carried in the $Y^{1/\gamma}$ signal, at maximum sample rate, no high resolution luminance information is lost. This is in contrast to the non-constant luminance system described in the next section, where, as some of the high resolution luminance information in saturated colours is carried in the chrominance signals, it is removed by the chrominance filters in non 4:4:4 systems.

14.7   A Non-Constant Luminance System

The essence of a non-constant luminance system is the use of the luma signal, Y', comprising the addition of the gamma-corrected R'G'B' signals, rather than the true luminance signal $Y^{1/\gamma}$ derived from the addition of the linear RGB signals.

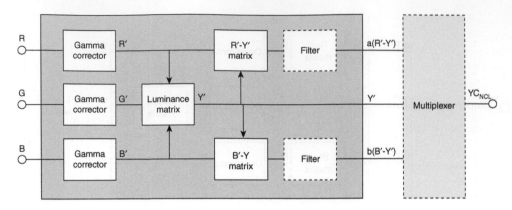

Figure 14.9   A non-constant luminance camera encoder.

14.7.1   A Non-Constant Luminance Camera Encoder

The non-constant luminance encoder is very similar in the type and number of processes it contains to the constant luminance encoder, as a comparison of Figure 14.9 with Figure 14.7 illustrates. The only difference is that the matrix for the Y signal is positioned following rather than preceding the gamma correctors.

The output from the multiplexer is designated the YC_{NCL} to differentiate it from the YC_{CL} signal from the constant luminance system.

14.7.2   A Non-Constant Luminance Display Decoder

A schematic diagram of a non-constant luminance display decoder is illustrated in Figure 14.10 and a comparison with the constant luminance decoder shown in Figure 14.8 illustrates why the non-constant luminance approach has been the *de facto* method adopted by virtually all colour reproduction systems to date. Because the CRT was the only practical way of displaying electronically generated images up to the turn of the century, and its nonlinear characteristics

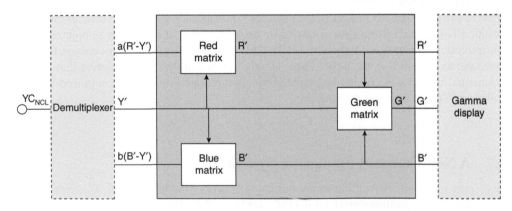

Figure 14.10   A non-constant luminance display decoder.

negated the requirement for gamma circuits in the display, it meant that these expensive circuits of the time were unnecessary. In consequence, the decoders for television receivers and computer displays were significantly simpler and cheaper than the alternative for constant luminance signals.

Such a system has served the photographic and television industries well. As can be seen from Figures 14.9 and 14.10, the signals delivered to the gamma display are apparently not compromised and good-quality pictures generally result. However, under certain conditions artefacts are introduced and these are detailed in the next section.

14.8 The Ramifications of the Failure of Constant Luminance

As we have seen constant luminance fails because the Y' signal carries all the luminance information only on neutral colours; for other colours, increasing levels of saturation lead to an increasing percentage of the luminance information being carried by the colour difference signals; conversely, the Y' signal diminishes in level with increasing saturation. Table 14.2, generated in Worksheet 14, gives the values of Y' and $Y^{1/\gamma}$ for primary and complementary colours at levels of 50% and 100% and the ΔL column illustrates the level of the inconstancy.

14.8.1 Loss of Compatibility with Monochrome Systems

It may be recalled that one of the important criteria for a colour reproduction system in the early days was compatibility with the large number of monochrome systems already in use. However, it can be seen from Table 14.2 that the use of the Y' signal for luminance, though accurate for neutral colours, will result in increasing error as the level of the saturation of the colours in the scene increases.

Thus, for colours of increasing saturation, monochrome displays will render the scene with increasingly diminished levels of luminance compared with the correct value. Since the viewer

Table 14.2 The loss of constant luminance as a result of using Y'

Colour	R	G	B	Y	$Y^{1/\gamma}$	Y'	ΔL
White	1.000	1.000	1.000	1.000	1.000	1.000	0.000
Red	1.000	0.000	0.000	0.213	0.498	0.213	0.286
Magenta	1.000	0.000	1.000	0.285	0.568	0.285	0.283
Blue	0.000	0.000	1.000	0.072	0.306	0.072	0.234
Cyan	0.000	1.000	1.000	0.787	0.898	0.787	0.111
Green	0.000	1.000	0.000	0.715	0.860	0.715	0.145
Yellow	1.000	1.000	0.000	0.928	0.967	0.928	0.039
Grey	0.500	0.500	0.500	0.500	0.732	0.732	0.000
Red	0.500	0.000	0.000	0.106	0.365	0.156	0.209
Magenta	0.500	0.000	0.500	0.142	0.416	0.209	0.207
Blue	0.000	0.000	0.500	0.036	0.224	0.053	0.171
Cyan	0.000	0.500	0.500	0.394	0.657	0.576	0.081
Green	0.000	0.500	0.000	0.358	0.630	0.524	0.106
Yellow	0.500	0.500	0.000	0.464	0.708	0.679	0.029

Figure 14.11   The photograph illustrates the loss of detail in saturated colours when the chrominance signal is filtered in a non-constant luminance system.

was usually unaware of the original colour for most of the time, the distortion was often not perceived, particularly since those things we do recognise, such as the human face, do not generally contain saturated colours and were therefore reproduced satisfactorily. However, there are occasions when the general rule does not apply, examples being the use of saturated red lipstick, which would cause the lips to appear very dark, and product packaging, which often contained highly saturated colours with which the viewer would be familiar.

14.8.2   Loss of Detail in Colours of High Saturation

One of the main advantages of YC systems is the ability to exploit the acuity characteristics of the eye by ensuring that the full detail of the scene is carried by the luminance signal. However, as the non-constant luminance encoder appearing in Figure 14.9 illustrates, the colour difference signals, which carry increasingly larger percentages of the luminance information on saturated colours, are often filtered, which removes all the finer detail present. This includes the fine detail of the luminance information carried in these signals on saturated colours.

When the colour difference signals are filtered, the result of the loss of this detail information on saturated colours in the reproduced image is very obvious and is particularly noticeable on flowers, where most of the image is clearly in focus but the saturated flowers appear out of focus, despite being in the same image plane as the remainder of the scene. Figure 14.11 attempts to illustrate this effect.

The opportunities for introducing the constant luminance approach in television systems are discussed in Section 20.2.4.

15

Specifying a Colour Reproduction System

15.1 Introduction

In Part 4, we have reviewed the procedures and processes required for the implementation of practical colour reproduction systems and have indicated the parameters which need to be defined in order that users of the systems can ensure good quality images are displayed. However these parameters, which are widely dispersed throughout the text of Part 4, will benefit by being brought together in order to provide a coherent specification of the colour reproduction system.

15.2 Deriving the Specifications

The approach to deriving a system specification is dependent upon whether an open or a closed system is to be specified, or in media terminology, a scene-referred or an output-device-referred system is to be specified. Many current systems are output-device-referred, that is, the signals derived at source are specified to serve a display population which had a fixed common set of display characteristics, whereas scene-referred systems are based on source signals with characteristics which are device independent (see Section 12.6) and are designed to serve a population of displays which may have or are likely to have different display characteristics. Such systems would have been impractical prior to the digital era. Some systems are variations between these two extremes, whereby sources of different characteristics serve a display population with a range of characteristics. Examples of each of these system types appear in Part 5.

15.2.1 Specifying an Output-Device-Referred System

Although formal specifications are usually arranged in an order where the parameters associated with image capture and source processing only are presented, it should be remembered

Colour Reproduction in Electronic Imaging Systems: Photography, Television, Cinematography, First Edition. Michael S Tooms.
© 2016 John Wiley & Sons, Ltd. Published 2016 by John Wiley & Sons, Ltd.
Companion Website: www.wiley.com/go/toomscolour

that in terms of colour reproduction, it is the environment associated with the viewing of the rendered image that determines the value of the parameters associated with image capture and source processing. These independent and dependent parameters are illustrated in the first and second columns of Table 15.1, respectively.

Table 15.1   Illustrating the independent and dependent parameters of a colour reproduction specification

Viewing environment parameters	Camera/Source parameters
Display primaries chromaticity coordinates Display white point	Camera spectral sensitivities[1]
Display gamma exponent Display highlight luminance Display surround surfaces luminance Display contrast range in environment	*System transfer characteristics*[2] Notional gamma correction exponent Gain of linear element of characteristics
Decoding format – constant luminance?[3]	Encoding format – constant luminance?

Notes:

1. It is common in formal specifications to quote the display primaries chromaticities in the camera or system specification rather than provide the colour-matching functions which define the camera spectral sensitivities. It then becomes the responsibility of the camera manufacturer to calculate the camera spectral sensitivities.
2. In contrast to Note 1, the values of the elements of the system transfer characteristic specification are usually provided from calculations based upon the values of the parameters in the 'Viewing environment' column. In this example, it is assumed that perceptibly uniform coding is not used; if it were, then the value of the parameters would need to be calculated independent of any gamma correction required.
3. The decision as to whether or not to use constant luminance decoding is usually dependent upon the cost premium of the decoder in the monitor or receiver and therefore becomes an independent viewing environment parameter.

Each media colour reproduction system has its own form of specification format, but nevertheless, the majority of the parameters required to specify the various systems are identical, albeit their values are different.

In addition to the parameters listed in Table 15.1, other parameters define the picture spatial characteristics, the digital representation and the digital coding characteristics.

In Section 15.3 is a generic collection of the parameters used in these specifications, laid out for simplicity in the tabular format used by current television system specifications. Only those parameters which pertain directly or indirectly to the reproduction of colour are included, and the values taken as a whole are not intended to relate to a specific television system; they are representative only.

Specifications relating to particular media systems are addressed in the appropriate chapters of Part 5.

15.2.2   *Specifying a Scene-Referred System*

Ideally there are two criteria for source signals intended to service a scene-referred system:

- The characteristics of the source signals should not limit in any way the ability of any display to provide an optimum rendition of the original scene within the capabilities of its operating characteristics.
- The means of storing and delivering the signals in such a manner as to avoid any perceptible artefacts appearing in the display of the optimum rendered image of the original scene.

With regard to source signal limitations, the implication is that the colour space to which the source signals are encoded encompasses all colours within a contrast range at least as great as that of the spatial dynamic contrast range of the eye (see Section 13.3) when viewing any display in its environment. However, such a system, unless constrained, would not necessarily provide the optimum rendition on displays of limited contrast ratio. Thus, the compromises necessary, which are also dependent upon the viewing environment, are described in the appropriate chapters of Part 5.

In order to avoid the introduction of artefacts during storage and delivery whilst meeting the first criterion, it is implied that a perceptibly uniform coding system be adopted (see Section 13.6).

15.3   A Representative Closed Colour Reproduction System Specification

15.3.1   Camera/Source Parameters

The table below lists the parameter values:

Table 15.2   Camera/source parameters

Item	Parameter	System values	
1.1	Opto-electronic transfer characteristic before non-linear pre-correction	Assumed linear	
1.2	Transfer characteristic at source	$V = 1.099L^{0.45} - 0.099$ for $1 \geq L \geq 0.018$ $V = 4.500L$ for $0.018 > L \geq 0$ Where: L: Luminance of the image $0 \leq L \leq 1$ V: Corresponding electrical signal	
1.3	Chromaticity coordinates (CIE, 1931) Primary	x	y
	Red (R)	0.630	0.340
	Green (G)	0.310	0.595
	Blue (B)	0.155	0.070
1.4	Assumed chromaticity for equal primary signals (reference white)	D65	
		x	y
	$E_R = E_G = E_B$	0.3127	0.3290

15.3.2   Picture Spatial Characteristics

The table overleaf lists the parameter values:

Table 15.3   Picture spatial characteristics

Item	Parameter	System values
2.1	Aspect ratio	16:9
2.2	Samples per active line	1920
2.3	Sampling lattice	Orthogonal
2.4	Active lines per picture or pixels per picture height	1080
2.5	Pixel aspect ratio	1:1 (square pixels)

15.3.3   Signal Coding Format

The table below lists the parameter values:

Table 15.4   Signal coding format

Item	Parameter	System values	
		Constant luminance	Non-constant luminance
3.1	Signal format		
3.2	Derivation of Y'_C and Y' $(Y'_C = Y^{1/\gamma})$	$Y'_C = (0.2627R + 0.6780G + 0.0593B)'$	$Y = 0.2627R' + 0.6780G' + 0.0593B'$
3.3	Derivation of colour difference signals	$C'_{BC} = (B' - Y'_C)/1.9404,$ for $-0.9702 \le B' - Y'_C \le 0$ $C'_{BC} = (B' - Y'_C)/1.5816,$ for $0 < B' - Y'_C \le 0.7908$	$C'_B = (B' - Y'_C)/1.8814$
		$C'_{RC} = (R' - Y'_C)/1.7184,$ for $-0.8592 \le R' - Y'_C \le 0$ $C'_{RC} = (R' - Y'_C)/0.9936,$ for $0 < R' - Y'_C \le 0.4968$	$C'_R = (R' - Y'_C)/1.4746$

15.3.4   Digital Representation

The table below lists the parameter values:

Table 15.5   Digital representation

Item	Parameter	System values		
4.1	Coded signal	R', G', B' or Y', C'_B, C'_R or Y'_C, C'_{BC}, C'_{RC}		
4.2	Sampling lattice R', G', B', Y', Y'_C	Orthogonal, line and picture repetitive co-sited		
4.3	Sampling lattice C'_B, C'_R or C'_{BC}, C'_{RC}	Orthogonal, line and picture repetitive co-sited with each other The first (top-left) sample is co-sited with the first Y samples.		
		4:4:4 system	4:2:2 system	4:2:0 system
		Each has the same number of horizontal samples as the $Y'(Y'_C)$ component.	Horizontally subsampled by a factor of 2 with respect to the $Y'(Y'_C)$ component.	Horizontally and vertically subsampled by a factor of 2 with respect to the $Y'(Y'_C)$ component.
4.4	Coding format	10 or 12 bits per component		

15.3.5   The Viewing Environment Specifications

The viewing environment parameters differ considerably with each media type; thus, the specifications of these parameters are covered separately in the appropriate chapters of Part 5.

Part 5

The Practicalities of Colour Reproduction – Television, Photography and Cinematography

Introduction

All the material appearing in Parts 1–4 of this book has, with only one or two exceptions where the differences have helped with the understanding of the concepts being described, dealt generically with the topic of colour reproduction, that is, the material presented has been of equal relevance to all forms of colour reproduction by electronic methods, whether for television, photography or cinematography.

Part 5 dealing with the practicality of colour reproduction is divided into three parts, A, B and C for television, photography and cinematography, respectively, each with its own dedicated chapters. There are two reasons why this is necessary, the most fundamental of which relates to the viewing conditions under which the reproduced images are viewed for

Colour Reproduction in Electronic Imaging Systems: Photography, Television, Cinematography, First Edition. Michael S Tooms.
© 2016 John Wiley & Sons, Ltd. Published 2016 by John Wiley & Sons, Ltd.
Companion Website: www.wiley.com/go/toomscolour

the three different media systems and the correspondingly different criteria which are selected at source to compensate for these differences, as indicated in Section 9.3, and Chapter 13. The second reason is that although there is now a much broader understanding between those responsible for setting the technical specifications and evolving the standards for what were three completely different media, it has to be remembered that these media developed separately over several decades with often little interaction between the experts in each of the areas of specialisation. In consequence, inevitably specifications were proposed and standards adopted which appeared best suited to the particular media at that time, without there always being much interaction between the experts of each specialist media group.

In more recent years this situation regarding the specialisation of experts within the various media groups has been largely overcome with much sharing and understanding of what is being proposed for the evolution of future industry specifications. Nevertheless, current standards do reflect these different legacy specifications, both with regard to their limitations and also to the different approaches adopted to ameliorate them in the specifications they support.

For those who consider they already have a relatively good grasp of the fundamentals of colour reproduction but wish to learn or refresh their memory of material relevant to the media in which they operate, whether it be television, photography or cinematography, then the appropriate chapters in this Part are written in a largely self-contained manner with reference to the fundamentals in earlier chapters only being alluded to when the complexity or detail of the material is likely to warrant such an approach. There is little more annoying to a reader than interruptions to the continuity of explanation through continual reference to earlier material; thus although sometimes it may seem that a topic repeats that which appears earlier in the book, it will be presented in a more summarised form with only occasional references to the more fundamental approaches of the earlier material.

In contrast to those seeking information only about their particular media, are those who wish for an overview of colour reproduction in all three media types and for this purpose, it is important to provide a continuity of description which follows the thread of development as it progressed through those three media types. Historically, significant electronic development was required before suitable equipment would become available to support all three media types; the costs of the early electronic colour cameras were equivalent to some 40 times the average annual salary of the time, with another even higher amount required to purchase recording equipment. Only when the resulting pictures could be shared amongst a very wide population could such costs be justified, so it was inevitable that television led the way in the development of electronic colour reproduction.

Colour television was introduced to public service broadcasting in the 1950s, based at that time on analogue technology; it was a further 30 years before advances in chip design led to the practicability of introducing digital technology in a limited fashion into the demanding television domain; another 10 years before digital cameras became the order of the day and yet another ten years before the complete signal path was digitised with the introduction of digital transmissions.

The spur of the availability of miniaturised solid state integrated circuits for digital processing of video signals and solid state opto-electronic image sensors led to the production of the first practical digital photographic stills cameras for use by the general public in the early 1990s, some 40 years after the introduction of colour television.

Finally the availability of electronic projectors capable of producing displays bright enough for viewing in public cinemas led the film industry to adopt standards for electronic cinematography in the first decade of the twenty-first century followed by the operation of digital cinemas shortly thereafter.

The stability and accuracy of digital processing led to the introduction of circuitry to carry out the processing of video signals in an ever more accurate manner which in turn led to the systems and specifications of colour reproduction systems being upgraded as the technology developed.

From the foregoing it is apparent that if the historical thread is to be preserved as a basis for describing the continuing development of colour reproduction then it must be initiated by television and followed by photography and cinematography in that order.

Part 5A

Colour Reproduction in Television

Introduction

In the context of this book, television production is defined in terms of the live editing of scenes using historic television multi-camera techniques, primarily for broadcast.

This part commences with an outline description of the signal flow, highlighting those elements which are in a position to influence the colour of the viewer's display. The following chapter briefly describes the introduction of colour into television; which is relevant since the concepts evolved at that time for capturing the colour information from the scene and processing it into a form suitable for storage, distribution and display have since been adopted by all current media systems, television, electronic photography and cinematography.

The established colour rendering index (CRI) used for evaluating the performance of electrical discharge and LED lamps used for scene illumination has long been found to be a poor indication of its suitability for this purpose. The newly introduced television lighting consistency index (TLCI) which overcomes these limitations is described in detail and supported by a worksheet which enables the TLCI to be calculated from the spectral power distribution of the lamp.

The current high definition television system, in terms of its colorimetric performance, is reviewed at length, which provides the basis for appraising the potential of the ultra-high definition system. UHDTV is currently under development and holds the promise of significant improvements in the rendition of the displayed image, as the limitations in legacy systems are addressed with the introduction of a wider colour gamut and contrast range and a constant luminance approach to the delivery of the signals to the display. In the final chapter the approach to colour management in television is described in the context of achieving satisfactory results when the final image is displayed in widely different viewing environments.

Colour Reproduction in Electronic Imaging Systems: Photography, Television, Cinematography, First Edition. Michael S Tooms.
© 2016 John Wiley & Sons, Ltd. Published 2016 by John Wiley & Sons, Ltd.
Companion Website: www.wiley.com/go/toomscolour

16

The Television System and the Image Capture Operation

16.1   The Television System Workflow

In order to identify those elements within the system which influence the manner in which images of the scene are reproduced, Figure 16.1 illustrates the workflow of the overall television system with those elements of the workflow which can influence the portrayal of the pictures displayed in the home of the viewer, highlighted in blue. This is a typical operation; the detail of the arrangement will vary from operation to operation. Since those elements of the workflow with a grey background do not influence the grading of the picture, they are not described further but are included here to assist in understanding the context of the overall system operation.

In Figure 16.1, for simplicity of display, the shooting operation is illustrated separately from the television centre, which of course it is for an outside broadcast (OB) operation, but otherwise, all the blocks of the shooting operation are replicated in each studio within the television centre.

For simplicity and clarity, only a three-camera operation is illustrated, whereas in reality, the number of cameras used, depending upon the complexity of the production, would likely be in excess of this number.

Generally speaking, cameras designed for the type of television operation described above are configured as a number of separate packages, which together comprise a *television camera channel*. The head end of the camera contains only the lens, the colour analysis optics if appropriate, the image sensors, the analogue to digital converters[1] and a viewfinder; the raw red, green and blue signals are sent down the cable to the camera control unit (CCU), which contains all the signal-processing elements. The signal for the viewfinder is returned after processing from the CCU to the camera head.

The CCU is usually located in the room designated as Vision Control or Picture Control or sometimes historically as 'Racks' and is where the raw signals from the camera head

---

[1] Digital processing of television signals within the camera channel did not occur until the 1980s; see Chapter 17 for the phasing of the introduction of digital processing.

Colour Reproduction in Electronic Imaging Systems: Photography, Television, Cinematography, First Edition. Michael S Tooms.
© 2016 John Wiley & Sons, Ltd. Published 2016 by John Wiley & Sons, Ltd.
Companion Website: www.wiley.com/go/toomscolour

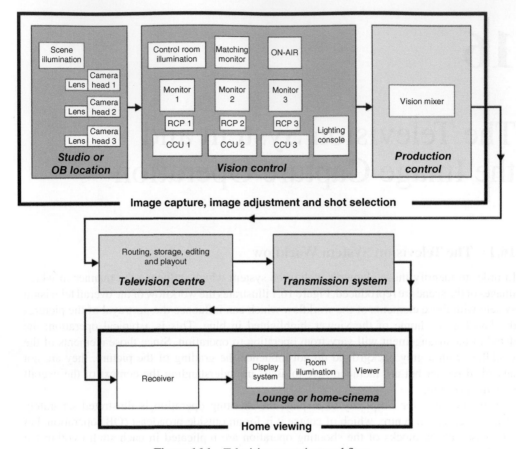

Figure 16.1 Television operation workflow.

end are processed and output to both the associated picture monitor, on which the picture is adjusted, and the vision mixer in the production control room. The vision mixer operator then selects which camera shot will be selected for the programme. The controls associated with the adjustment of exposure and the processing circuits parameters of the camera channel are located on a remote control panel (RCP) connected to the CCU.

Vision Control is at the heart of the picture adjustment and matching operation and provides two operational positions, for the lighting director and the vision controller respectively, ideally adjacently located in order to facilitate communication between these two individuals who are jointly responsible for the lighting, adjustment and matching of the pictures.

The lighting director manages the positioning of the *key*, *fill*, *backlight*, *set dressing* and *effects* luminaires, and operates the lighting console, which enables the intensity of each of the luminaires in the studio to be adjusted for the best contrast composition of the scene whilst matching the exposure and illuminant colour balance requirements of the cameras.

In an ideal environment, the vision control position comprises a row of RCPs on the desk facing a picture monitor stack which includes a monitor for each camera and a monitor for picture matching, whose input is automatically switched between the 'on-air' camera and the picture from the camera whose RCP is currently under adjustment, usually the next camera to be selected to air. The RCP will normally have controls for colour balance, exposure and black level or 'lift', the operation of which will be further described in Section 21.6.

The positioning and level of the environmental lighting in Vision Control is critically arranged and set to match recommended ambient light levels and has a colour temperature to match the white point of the adopted television system.

The home viewing section comprises the final elements of the workflow and the blue background indicates those elements which influence how the final displayed picture will be perceived.

The operational procedures which are undertaken to ensure well-matched high grade pictures are delivered to the viewer are described in the appropriate sections of Chapter 21.

16.2   The Television System Signal Path

The full signal path of the television system may be derived from the workflow diagram illustrated in Figure 16.1; however, we are only interested in those elements of the system which influence the perception of the displayed image and these are detailed in Figure 16.2.

The configuration of the elements in Figure 16.2 has hardly changed since the introduction of colour television and serves as a template for all the systems described in the following chapters. Since the introduction of digital television, the appropriate Analogue to Digital (A–D) and Digital to Analogue (D–A) converters have been added at the beginning and end

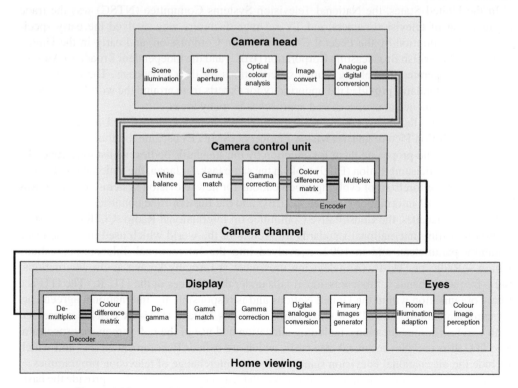

Figure 16.2   Television signal path elements influencing image perception.

of the signal chain, respectively, as the image sensors and image generators are fundamentally analogue in operation.

The description of the functionality of each of these elements has been described in detail in Part 4 and will be addressed further in the context of current practice in Chapters 19 and 21.

The specific arrangement of the elements in the display, particularly with regard to the gamma corrector, is dependent upon whether a constant luminance system or a non-constant luminance system is in use, as described in Sections 14.6 and 14.7.

16.3 The Television Standards Organisations

From the viewers' standpoint, television is by its nature a multi-source single-destination system; that is, a single television set in the viewer's home is capable of receiving signals from different cameras in a production, different studios within a television centre and different broadcasters. Thus, if the viewed pictures are to be perceived as realistic and matching from source to source, it is essential that the signal path elements described in Section 16.2 operate to the same specifications; inevitably therefore, from the beginning, standards were established by broadcast authorities and imposed upon the broadcasters.

Initially these standards were set by national bodies, but as the technology progressed to the point where the interchange of programmes could take place at national level, new bodies were incorporated and professional organisations took up the challenge to evolve specifications for adoption as national standards.

In the United States, the National Television Systems Committee (NTSC) was the trade organisation of television camera and TV set manufacturers who evolved the early specifications for adoption by the Federal Communication Commission, and early in the United Kingdom, the British Broadcasting Corporation (BBC) and the Independent Broadcast Authority (IBA) cooperated in evolving mutual specifications for their broadcasts. The requirement for common standards provided the opportunity for experts from around the world to exchange views and commence to evolve shared approaches in many areas.

Eventually this led to fewer but more international orientated bodies; in the United States, the Society of Motion Picture and Television Engineers (SMPTE) became the focus for evolving specifications and proposing them for adoption by the standards bodies, whilst in Europe, the European Broadcasting Union (EBU) took on the same task. In Japan, the national broadcaster NHK was at the forefront of evolving specifications for new television systems, and there was an interchange of information between the EBU and the Eastern Bloc countries.

For many decades, the Consultative Committee on International Radio (CCIR) was instrumental in setting international standards for sectors of the world which used the same basic scanning parameters, and in 1992, it evolved into the International Telecommunications Union – Radio or ITU-R. (As television uses what is formally known as the international radio frequency bands for transmission, it falls under the auspices of the ITU-R.) The ITU is a treaty organisation within the United Nations and is responsible for international agreements on communications. The ITU Radio Communications Bureau (ITU-R/CCIR) is concerned with wireless communications, including allocation and use of the radio frequency spectrum. The ITU also provides technical standards, which are called 'Recommendations' and which include the international television standards for the interchange of television programmes.

The ITU undertakes studies, the results of which form reports, which in turn provide the basis for the formal recommendations. In broadcast television, these reports and recommendations follow a naming sequence, ITU-R BT.601 for example, where BT stands for 'broadcasting service (television)' and the number is a sequence number. In order to avoid lengthy repetition, once these document titles have been introduced, subsequent reference to them in the text is abbreviated to, for example, Rec 601.

Many of the world's television experts are now also members of the SMPTE, which helps to ensure that the specifications developed by them, the EBU and NHK share as many common values for the critical parameters as is practical and thus smooth the path for these specifications to be adopted as ITU recommendations.

The technical standards for television broadcast have evolved over the decades as the progress in technological development has provided the opportunity to greatly enhance the quality of the pictures generated, transmitted and displayed. These enhancements embrace all the factors contributing to picture quality but the most perceptible improvement with each enhancement has been in the increased spatial resolution of the reproduced image. For this reason, generations of new technical specifications for broadcast are often referred to by the resolution they provide; in the early days by the number of scan lines, and latterly, by the number of pixels which comprise an image. In the following chapters, the specifications for the resolution parameters are not discussed in great detail despite the recognition of their prime importance since they are not parameters which affect the fidelity of colour in reproduction.

Though not essential to the understanding of current and future specifications discussed in the latter chapters of this part on television, the next chapter on the history of colour in television does provide a sound basis for that understanding and deserves at least to be scanned if not read in depth.

17

A Brief History of Colour in Television

17.1 The Beginnings

Work on experimental television systems commenced in a serious manner at the beginning of the 1920s in several countries throughout the world. The EMI Company in the United Kingdom was amongst the first to develop an electronic image sensor which made the development of a television camera a practical proposition and which led the British Broadcasting Corporation (BBC) to adopt the EMI system in order to commence the world's first public television service in November 1936. By the end of the 1930s several other countries had also commenced public service television broadcasting, and in 1941, the Federal Communications Commission (FCC) in the United States authorised the adoption of the specification proposed by the National Television Systems Committee (NTSC) for the commencement of television broadcasting from July 1941. The NTSC was a committee established by the radio industry trade association to derive and recommend for adoption by the FCC a technical specification for broadcast television.

The Second World War brought to an end public television broadcasting in Europe and with it the development work on colour television by the television equipment manufacturers. However, in the United States, work on experimental colour television systems continued apace and by end of the 1940s, pressure was mounting on the FCC to approve a technical specification as a standard for a public colour television service. After a false start, when the FCC authorised the commencement of a service using the CBS colour system, which was incompatible with the some 10 million black and white receivers then in use in the United States, the NTSC was reactivated in January 1950 to derive and agree a specification for a compatible system to be recommended for adoption by the FCC.

It is difficult to exaggerate the extent and importance of the work undertaken by the NTSC; at that time amongst the various disparate colour television systems which had been developed by the principal manufacturers of television equipment in the United States, there was no common thread and no system which met all the criteria to enable it to be recommended for adoption. In consequence, under the auspices of the Committee, extensive development work and system tests were undertaken by some 300 engineers and colour scientists drawn

Colour Reproduction in Electronic Imaging Systems: Photography, Television, Cinematography, First Edition. Michael S Tooms.
© 2016 John Wiley & Sons, Ltd. Published 2016 by John Wiley & Sons, Ltd.
Companion Website: www.wiley.com/go/toomscolour

from a broad spectrum of those organisations with a contribution to make; these experts were organised into 10 panels and 55 subpanels (Fink, 1955). Following an interim report in 1951, a reorganisation of the committee to take account of the conclusions of that report and considerable further work, the NTSC agreed the final form of the compatible colour television signal specification in July 1953. The FCC approved this specification, which became the standard for public service colour television broadcasting in December 1953 and which led to the first national broadcast utilising the NTSC system on 1 January 1954. However, it was not until the mid-1960s that there was a sufficient uptake of colour television sets by the public to support all programmes being produced and transmitted in colour during prime viewing time.

Since the 1950s, the performance of colour television systems has come a long way and much has been written on the differences and improvements of these latter systems when compared with the NTSC system; nevertheless, an objective analysis of all these new systems will show that at a fundamental level, the essential elements of the NTSC system remain in use. In colour terms, as opposed to spatial and temporal resolution, it is only in the manner in which the colour difference or chrominance signals as derived by the NTSC are conveyed through the multiplex[1] that have continued to improve with each new system introduction and which have led to systems with different names but which are fundamentally variants of the NTSC system.

In Europe, by the mid-to-late 1950s, experimental colour transmissions were taking place during the close down times of the regular monochrome services. The BBC broadcast colour test films using a 405 line variant of the NTSC system during this period and discussions commenced between the principal experimenters, notably those in the United Kingdom, France and West Germany, who were working on various solutions, to overcome the problems of the NTSC system of that era. In the early 1960s under the auspices of the European Broadcasting Union (EBU), work commenced on the selection of a common system for Europe which would avoid these problems.[2] During this period, the *Sequentiel Couleur à Memoire* (SECAM) system, which used a variant of NTSC whereby the multiplex used frequency modulation for the colour difference signals, was developed in France, and later in West Germany yet another variant, even closer to the NTSC system, was developed based upon alternating the phase of the chrominance subcarrier on a line-by-line basis, which became known as the 'Phase Alternation Line' (PAL) system. After an initial lack of agreement, most of Europe adopted the PAL system, which overcame the phase sensitivity issues of the NTSC system, and in the United Kingdom, the BBC commenced the first European colour television service using PAL in July 1967, quickly followed by Independent Television (ITV). In the same period other European countries commenced broadcasting in PAL, with the exception of France, Luxembourg and the USSR, who commenced services using the SECAM system.

The remaining countries of the world selected one of these three systems as their standard, depending to a large extent on which monochrome line standard had been previously adopted and to a degree on their political affiliations.

These three systems then served the world until the introduction of the high definition television system (HDTV) in the 1990s in Japan and in the remainder of the world in the

---

[1] The multiplex in this context is the circuit component which, in a variety of different ways for different systems, combines the same three luminance and colour difference components of any compatible system into a single signal for storage and distribution.

[2] The basis of these problems are briefly addressed in Section 17.2.6.3.

2000s; see Chapter 19. However, within the television broadcast centres, digital component systems started to replace these traditional systems from the late 1980s onwards.

17.2   The NTSC, PAL and SECAM Colour Television Systems

One of the principal requirements of the NTSC system was that it should be compatible with the large population of black and white television sets already established in the field. In consequence as noted above, a number of the NTSC-defined processes as applied to the camera RGB signals were, because of their fundamental nature to the solution of establishing any compatible system, also adopted by all subsequent systems. Thus, in dealing with the fundamentals of colour reproduction in Part 4, and in particular the encoding of television camera signals in Chapter 14, we have by default already described many of the essential elements which were originally derived by the NTSC, and in consequence, it may be appropriate for the reader to review that chapter before progressing further. Nevertheless, in order to avoid repetition, the approaches to encoding described in detail in Chapter 14 are only summarised below using the values for the universal parameters specified by the NTSC or the EBU as appropriate.

These systems are fully described elsewhere (Carnt & Townsend, 1961; Carnt & Townsend, 1969; Wentworth, 1955); our interest is limited primarily to reviewing those features of the systems which influence the colour characteristics of the reproduced image. However, Section 17.2.6.3 will attempt to explain the multiplex associated problems of the NTSC system in the early days of its use.

17.2.1   The System Primaries and White Point

Television display devices have historically been based on primaries derived from the excitation of phosphors deposited on the faceplate of cathode ray tubes. Thus, since the introduction of colour television, the primaries have been based upon the colorimetry of the somewhat limited range of phosphors then available.

The original choice of primaries by the NTSC was made on the basis of the chromaticity of silicate phosphors, which gave the widest chromaticity gamut of the limited range of phosphors then available. Though apparently based upon sound colorimetric grounds, in fact the green primary chosen, though an excellent choice from the point of view of the large colour gamut obtained, was associated with an inefficient phosphor, which enabled only relatively dim pictures to be reproduced. Receiver manufacturers soon ignored the standard and produced display devices with considerably more efficient green and red phosphors but with chromaticities which produced a relatively limited colour gamut well removed from the specification. Thus, since the RGB signals which drove the display were derived to match the NTSC primaries, the colours were inevitably reproduced with less accuracy than they would otherwise have been. Nevertheless, the colour gamut volume within the colour space was dramatically improved, and it was generally accepted that this improvement was an acceptable compromise for the loss of chromaticity fidelity.

The introduction of colour television into Europe followed much later in 1967 and gave the EBU the opportunity to opt for an improved compromise in the selection of the display chromaticities between luminous efficiency and chromaticity gamut. In the knowledge of more than a decade of phosphor development, a set of phosphors based upon sulphides for the green and blue primaries and a rare earth for the red primary were specified. Although the chromaticity of these phosphors was a compromise, there was recognition of the need to restrain the impetus for ever brighter displays at the cost of seriously compromising the display chromaticity gamut.

The failure of the original NTSC primaries chromaticities specification was recognised and superseded by the SMPTE 'C' RP145 (Recommended Practice) primaries chromaticity specification during the 1980s, which generally reflected the chromaticities of the phosphors then being used by the receiver industry. The SMPTE and the EBU primaries have very similar chromaticities. In fact they are so close, it is a pity that the SMPTE did not adopt the EBU primaries; should they have done so, they would have effectively achieved a world standard for television system primaries chromaticities, which would have eased the standards conversion requirements when programmes were interchanged between these different areas.

One of the principal design criteria for the colour systems of this era was compatibility with monochrome television, the signal format standards of which did not accommodate negative excursions of the signal, thus preventing the transmission of data relating to those saturated colours beyond the gamut of the chosen phosphors. This situation will be explored further in Chapter 20.

The primary chromaticities chosen by the NTSC, the EBU and the SMPTE, together with the chromaticities of their adopted system white points, are given in Table 17.1 and their chromaticity gamuts are illustrated in Figure 17.1.

At the time the NTSC specification was being agreed, the recommendation for the specification of daylight was Illuminant C, but by the time the EBU and the SMPTE had defined the new primaries, the CIE had introduced new daylight 'D' illuminants based upon the specifications described in Section 7.3, in consequence the new D65 illuminant was selected as the system white for these new system specifications.

Table 17.1 Historic television system primaries

	Historic system primaries and white points			
	x	y	u'	v'
NTSC				
Red	0.67	0.33	0.477	0.528
Green	0.21	0.71	0.076	0.576
Blue	0.14	0.78	0.152	0.196
Illuminant C	0.3101	0.3162	0.2009	0.4610
EBU Tech 3213				
Red	0.640	0.330	0.451	0.523
Green	0.290	0.600	0.121	0.561
Blue	0.150	0.060	0.175	0.158
D65	0.3127	0.3290	0.1978	0.4683
SMPTE RP 145				
Red	0.630	0.340	0.433	0.526
Green	0.310	0.595	0.130	0.562
Blue	0.155	0.070	0.176	0.178
D65	0.3127	0.3290	0.1978	0.4683

As is illustrated in Figure 17.1, there is very little difference between the chromaticities of Illuminant C and D65; however, the D65 specification has a higher ultraviolet content, which enables it to be used to more accurately simulate daylight when illuminating fluorescent surfaces.

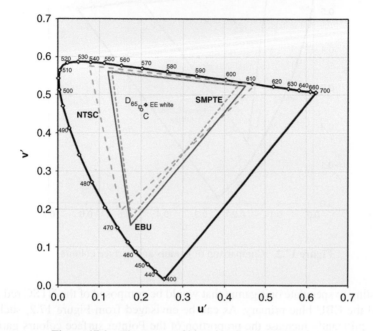

Figure 17.1   The gamut of historic television primaries.

In Figure 17.1, it can be seen that the SMPTE gamut is very slightly smaller than the EBU gamut; however, both are close enough to be considered the same for practical purposes.

A comparison with the NTSC gamut indicates a much improved blue primary, a red primary slightly inferior (SMPTE is worse) and a much inferior green primary, which significantly reduces the size of the chromaticity gamut that may be reproduced. This is the price which had to be paid for brighter pictures.

It is instructive to compare the historic chromaticity gamuts with the surface colours gamut of Pointer as shown in Figure 17.2 (see also Figure 9.4). Clearly, although the television colour system is capable of good-quality reproduction of the common range of colours in a scene, its performance on saturated colours such as costumes and flowers can be disappointing; some of these colours will appear relatively desaturated in the display.

Compared with the NTSC primaries, large areas of saturated green and cyan chromaticities are not reproducible by the EBU primaries, and similarly, compared with the EBU primaries, the NTSC primaries are unable to reproduce a broad band of saturated blue-to-magenta hues.

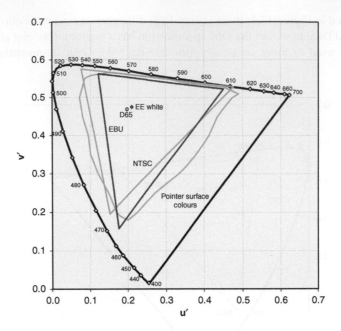

Figure 17.2   Comparison of gamuts and surface colours.

It is interesting to speculate on a gamut that would be composed of the NTSC red and green primaries and the EBU blue primary. As can be envisaged from Figure 17.2, such a gamut would very significantly increase the proportion of the Pointer surface colours gamut which could be reproduced and thus begin to approach an ideal set of primaries. Unfortunately, after some 60 years of research and development, no phosphors have been found with an acceptable level of efficiency which also match these chromaticities.

17.2.2   Derivation of the Ideal Camera Spectral Sensitivities

The procedure for establishing the relationship between the RGB primaries of a system and the camera spectral sensitivities in terms of the XYZ colour matching functions (CMFs) was described in Section 9.3, and together with the relationships derived in Appendix F, may be used to establish the television system spectral sensitivities for any triple set of primaries as first described in Chapter 9.

By entering the values given in Table 17.1 into the formulae derived in Appendix F (which is embedded in Worksheet 17[3]), we can calculate the coefficients of the XYZ CMFs for the NTSC, the EBU and the SMPTE RP145 sets of primary chromaticities. As we have seen in Section 9.3, these coefficients lead directly to providing the camera spectral sensitivities for any set of primary chromaticities, and in Worksheet 17, these coefficients and the corresponding camera optical spectral sensitivities are derived and illustrated as follows.

---

[3] Worksheet 17 also has a number of macro-driven 'keys' which when selected automatically enter the appropriate primary chromaticities into the formulae and produce the corresponding chromaticity charts and camera spectral sensitivities.

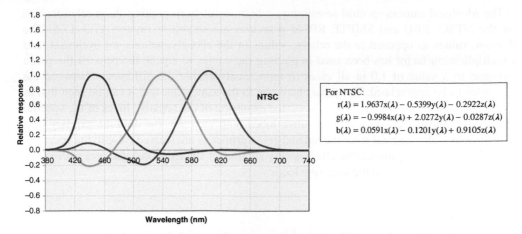

Figure 17.3 NTSC 'idealised' camera spectral sensitivities.

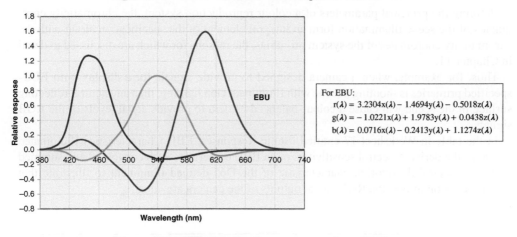

Figure 17.4 EBU 'idealised' camera spectral sensitivities.

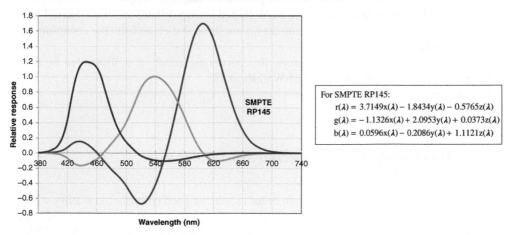

Figure 17.5 SMPTE 'idealised' RP145 camera spectral sensitivities.

The idealised camera spectral sensitivities which result from plotting these relationships for the NTSC, EBU and SMPTE RP145 primaries are shown in Figures 17.3–17.5. The absolute values as opposed to the relative values in the above equations are irrelevant, and a multiplication factor has been used in plotting the curves to equate the peak of the green response to a value of 1.0 in all cases to more easily enable the differences between the responses to be appreciated. The three charts are to the same scale and it can be seen that as the size of the chromaticity gamut reduces, the extension of the negative lobes of the spectral sensitivities and the corresponding positive lobes both increase; for the blue primary, this effect is relatively marginal since the red and green primaries are located close to the spectrum locus, but for the red primary, the effect is very significant since the blue and green primaries are located well away from the spectrum locus.

17.2.3   Matching Scene Illumination to the Spectral Sensitivities

In defining the principal parameters of a colour reproduction system, the chromaticity coordinates of the scene illumination form an integral element of the specification along with the chromaticity coordinates of the system primaries, the reasons for which are discussed in detail in Chapter 11.

Thus, for example, when a camera designed to provide signals for a display with EBU-specified primaries is shooting a scene with Illuminant D65 lighting, then any neutral reflecting surface in the scene will cause a colour-balanced camera to provide equal levels of the RGB signals.

In the table in Worksheet 17 entitled 'Static Responses, EBU Characteristics Illuminated by D65', the derived spectral sensitivities of an EBU camera are each convolved and summed with the spectral distribution characteristic of the D65-defined illuminant to illustrate that under these conditions, the RGB signal outputs of the camera are equal.

Figure 17.6 illustrates in graphical terms the process described in the previous paragraph.

The broad line curves labelled EBU + EEW (Equal Energy White) are those derived for the EBU primaries, whilst the narrow line curves are those that result from the convolution by the D65 SPD.

The areas enclosed by the three narrow line curves are equal, as is illustrated by the equal totals of the columns in Table 4 of Worksheet 17.

17.2.4   Matching the Camera Spectral Sensitivities to the Display Primaries

As we have seen in the previous section, in order to achieve good colorimetric fidelity within the gamut of real colours described by the colour primaries triangle on the chromaticity diagram, the camera should exhibit spectral sensitivity characteristics which match the CMFs of the primaries.

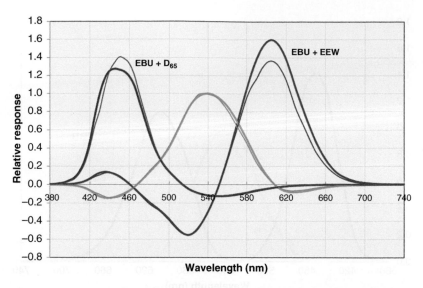

Figure 17.6   Showing the weighting of the EBU characteristics with the SPD of Illuminant D65.

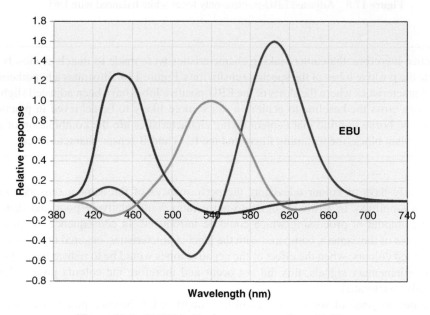

Figure 17.7   EBU idealised camera spectral sensitivities.

There are practical difficulties to achieving such a match. Using the EBU primaries as an example, an inspection of the curves in Figure 17.7 shows that over certain portions of the spectrum, all three characteristics at different wavelengths require the camera sensors to provide a negative output. No sensors exist or are foreseen which will provide such characteristics.

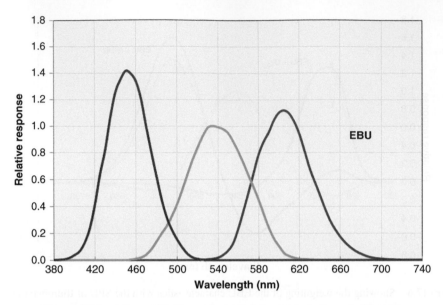

Figure 17.8   Adjusted EBU-positive-only lobes white balanced with D65.

In practice therefore, the camera taking characteristics were made to match as closely as possible the positive lobes of the spectral sensitivities. Figure 17.8 illustrates a hypothetical set of characteristics where the skirts of the EBU-positive lobes have been adjusted slightly where they cross the baseline to achieve a shape more likely to be achieved in practice. It should be borne in mind that camera taking characteristics are the combination of the dichroic prism block, the trimming filters and the R, G and B sensor characteristics.

In the early days of colour television, the performances of the camera sensors and the subsequent amplifiers in terms of signal-to-noise ratio were relatively poor and therefore limited the amount of processing which could be introduced. In consequence the practical characteristics of the camera, compared with the ideal characteristics, ensured that on narrow-band saturated colours, when the effect of the negative lobes would be to reduce the amplitude of the complementary signals, this did not occur and therefore the colours produced were significantly desaturated.

As the performance of sensors and amplifiers improved, it became practical to consider techniques which could begin to compensate for the lack of the negative lobes in the responses of the camera. An inspection of the camera responses in Figure 17.8 indicates that the ideal negative lobes illustrated in Figure 17.7 may be crudely matched by adding or subtracting appropriate levels of the complementary signals to the required primary signal. For example, in emulating the idealised red response, subtracting a small amount of the green signal from the red signal would make an approximate match to the negative red lobe in Figure 17.7. Similarly, adding a small proportion of the blue signal to the red signal would begin to simulate the minor red positive lobe centred on 440 nm. Thus, by electrically adding or subtracting proportions of the complementary signals from each primary, a better match can be made to the idealised spectral sensitivities. Effectively, this is an empirically derived approximated version of the matrixing techniques described in Section 12.2 and is a technique which is adopted in all modern colour cameras.

In Worksheet 17, the positive lobes of the EBU characteristics are transformed by such a matrix to derive an approximate match to the idealised characteristics. The matrix was adjusted empirically to provide the match whilst ensuring the coefficients of the primaries in each line summed to a value of 1.0 in order that the matrix does not change the colour balance when the scene is illuminated by D65. The resulting matrix coefficients are shown in Table 17.2, and the comparison of the resulting characteristics with the original white-balanced idealised characteristics are illustrated in Figure 17.9.

Table 17.2 Matrix for correcting practical camera spectral sensitivity characteristics to match system CMFs

	R_{in}	G_{in}	B_{in}
R_{out}	1.4700	−0.5600	0.0900
G_{out}	−0.0900	1.1880	−0.0980
B_{out}	−0.0200	−0.1200	1.1400

Thus, the overall response of the camera, resulting from the prism, trimming filters, sensors and matrix, approximates reasonably closely to the ideal response, and the system colour reproduction fidelity for colours not too close to the periphery of the primaries triangle on the chromaticity diagram is acceptable to most viewers, especially since the viewers generally do not have access to the original colour in the scene for comparison. For this reason, where

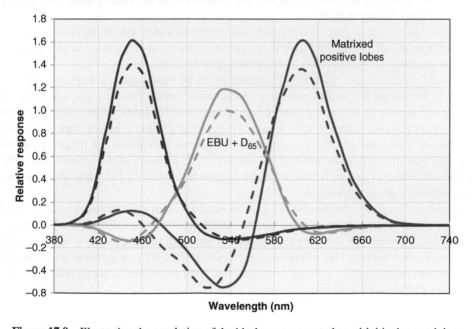

Figure 17.9 Illustrating the emulation of the ideal camera spectral sensitivities by matrixing.

compromises were made in shaping the characteristics of the camera spectral sensitivities or the matrix parameters, then the criteria was to make the best match to flesh colours, since it is these colours with which the viewer is most familiar and often has a local reference.

17.2.5   Lighting for Colour Television

Over the period in which the systems described in this chapter dominated, tungsten luminaires were the standard illuminant for television studios. At the commencement of the period, these luminaires were based upon lamps using only a relatively simple tungsten filament, but from the end of the 1950s, tungsten halogen lamps with their increased efficiency and higher operating colour temperature of 3,400 K were increasingly used.

Both of these variants of tungsten illumination have CRIs of 100 and therefore present no problems in terms of rendering the colours of the scene accurately, subject to the camera having a suitable colour temperature correction filter in place; see Section 11.3.

In practical terms care must be taken to ensure that when dimming a luminaire to provide a balanced level of illumination to an element of the scene, the point is not reached where the lower colour temperature produced does not change the colour balance of the camera output.

17.2.6   Gamma Correction

The primary television audience is the public, and during the period being considered, all domestic television displays were based upon the CRT in one of its forms. As seen in Section 13.4, the electro-opto transfer function (EOTF) of the CRT is a power law with a gamma value which, at the beginning of the period, was somewhat difficult to determine in a precise manner. Furthermore, it was recognised that the gamma varied depending upon the circuitry arrangements for driving the CRT.

Nevertheless, the use of gamma correction was universal, and it was appreciated that it was necessary to establish a figure for the display gamma in order that the correction characteristic applied at source could be specified in the system specification.

The NTSC specified the correction characteristic to be based upon a display gamma of 2.2; that is in notional terms, the correction characteristic would follow a law which had an exponent the inverse of this value, that is approximately 0.45. In the United Kingdom, the System I specification updated in 1971 specified the display gamma to be 2.8, with a tolerance of +0.3 or −0.3, and various other countries adopted gamma values between these two values. It was later established that the value of 2.8 was not a true measure of the actual value of the CRT gamma, which is generally now acknowledged to be close to 2.35[4] when driven from a sufficiently low cathode impedance.

However, the carefully measured CRT gamma established in the laboratory conditions of a fully darkened room with even more careful adjustment of the critical 'lift', 'brightness' or 'black level' control, as it is variously called, to establish the true black level, bore little resemblance to the reality of the effective characteristics of a CRT in a domestic environment, where the viewer has control of black level, and that level in subjective terms is impaired by the reflection of ambient light from the screen.

---

[4] Private discussion with Alan Roberts relating to his work at BBC Research.

Furthermore, before the introduction of digital processing in cameras in the late 1980s, the gamma correction was carried out by analogue circuit techniques, which varied significantly in their ability to accurately track the specified law, particularly at low luminance levels where a compromise had to be reached between the required gain of the corrector and the limited signal-to-noise ratio of the camera signals.

In order to establish a satisfactory compromise between these various limitations, the vision control operator is situated in a critical lighting environment aimed at representing a reasonably critical domestic viewing environment and provided with carefully set-up monitors (see Chapter 21) and the black level control of each camera to allow him or her to provide well-set-up pictures which are matched on a camera-by-camera basis to the studio or outside broadcast (OB) output.

17.2.7   A Brief Description of Historic Encoding Techniques

17.2.7.1   The Luminance Signals

In Section 14.7 the advantages and disadvantages of constant and non-constant luminance systems were outlined. The advantages of the non-constant luminance system, in terms of achieving simplicity and therefore relatively low cost of implementation in receivers of the time, ensured that non-constant luminance systems were adopted universally.

As described in detail in Section 14.3, the luminance signal is derived from the addition of the red, green and blue signals in appropriate proportions, relating to the contribution each camera spectral sensitivity, respectively, makes to the luminance response of the eye. These contributions will in turn be dependent upon the chromaticity of the primaries which are used to derive the camera spectral sensitivities.

In Worksheet 14, Matrix 6 provides the required coefficients in the formula for the luminance or Y signal, and by selecting the NTSC 'button', the coefficients for the three NTSC primaries are found to be, to three significant figures:

$$Y = 0.299R + 0.587G + 0.114B$$

However, as the non-constant luminance systems were adopted, it is the luma signal which is derived using the gamma-corrected versions of the RGB signals:

$$Y' = 0.299R' + 0.587G' + 0.114B'$$

This is the formula for luma used in the NTSC specification and it continued to be adopted later, not only by the SMPTE in the adoption of the new primaries described earlier but also by the PAL system, despite the formula no longer representing the luminance coefficients of the new primaries. This is not surprising; the composition of the luma signal does not affect the displayed colour since the matrixing process on the colour difference signals in the receiver ensures the luma signal cancels out of the equations. The luma signal will not provide an accurate luminance signal for monochrome displays, but the effect of the inaccuracy will be marginal at worst. Most importantly there is no requirement to change the receiver design to match the new coefficients of the contributions to the luma signal.

The coefficients of the RGB signals required to achieve an accurate representation of luminance may be found for any set of primaries by selecting their chromaticity coordinates for entering into the Matrix 6 formula in Worksheet 14.

17.2.7.2   The Colour Difference Signals

In Section 14.2 the basis for deriving the three signals used in all non-constant luminance colour systems was set out as the luma signal, Y', and the two colour difference signals, $R' - Y'$ and $B' - Y'$. In this respect, the different colour systems differ only in the manner in which these colour difference signals are prepared for multiplexing.

In both the NTSC and the PAL systems, each colour difference signal is balance modulated onto a subcarrier of the same frequency but differing in phase by 90 degrees; the two modulated carriers are then summed, a technique known as quadrature modulation. The resulting combination is referred to as the chrominance signal. By using a reference subcarrier of precisely the same frequency in the receiver, the two original signals may be demodulated with no mutual interference between them. The subcarrier frequency is arranged to be at the upper end of the video spectrum and be an odd multiple of the line scanning frequency in order to both minimise interference with the luma signal and reduce visibility on monochrome receivers.

The U' and V' Colour Difference Signals

The composite video signal is composed of the luma and chrominance video signals and the synchronisation signals. The frequency of the subcarrier is chosen in such a manner that at the fine spectral level, the components of the chrominance and luma signals interleave in order to minimise mutual interference.

In order to avoid this combined signal extending beyond the signal level capacity of following equipment and particularly the transmitter, the colour difference signals are attenuated prior to modulation, which limits the excursions of the composite signal in such a manner that the positive excursion beyond the peak white value of 1.0 is limited to 1.33 and the negative excursion below the black level is limited to -0.33.

The attenuation factors calculated to meet the above criteria are as follows:

$$U' = \frac{(B' - Y')}{2.03} \text{ and } V' = \frac{(R' - Y')}{1.14}$$

The letters U' and V' are used as shorthand to describe the specified attenuated versions of the colour difference signals.

Thus, in terms of the R', G', B' signals:

$$U' = (1B' - 0.299R' - 0.587G' - 0.114B')/2.03 = -0.103R' - 0.289G' + 0.436B'$$
$$\text{and } V' = (1R' - 0.299R' - 0.587G' - 0.114B')/1.14 = +0.615R' - 0.515G' - 0.100B'$$

The criteria for the attenuation factors are based upon the ability to accommodate the signal level excursions associated with the peak level colour bar signal. This signal, which has become ubiquitous, is an electronically generated video signal representing the display of eight vertical stripes, comprising all combinations of the $R'G'B'$ colour signals at levels of either 100% or 0%, that is, in luminance order: white, yellow, cyan, green, magenta, red, blue and black.

As a consequence of the quadrature modulation of the colour difference signals, the resulting subcarrier vector will vary in amplitude and phase in accordance with the colour represented by the U',V' signals and will fall to zero when a neutral is scanned by the camera. This property of the chrominance signal can therefore by portrayed by a vector diagram as illustrated in Figure 17.10.

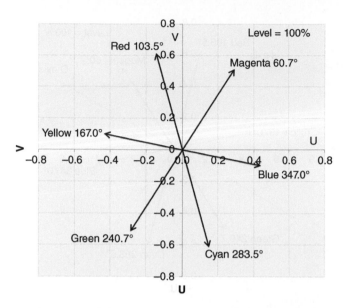

Figure 17.10   The chrominance U', V' vector diagram illustrating the peak primary and complementary colours for 100% colour bar signals.

Figure 17.10 illustrates how the vector phase represents the hue of the colour and the amplitude its saturation.

The positive U' axis is defined as the zero-degrees axis, and thus, the vectors for the colour bar waveform described above appear with the amplitudes and phases shown in the diagram, as based upon the calculations and chart derived in Worksheet 14.

The NTSC I' and Q' Signals

Because of the limitation in the bandwidth available in the 525 line systems, the chrominance signal is limited to a frequency which is critical in terms of being close to the limit of the corresponding colour acuity of the eye, as discussed in Section 14.5. In consequence, the NTSC specification applies a further layer of sophistication to the manner in which the U and V signals are processed.

It is beyond the scope of this book to explain the reasons why it is possible under certain limitations to allow one of the colour difference signals to have a larger bandwidth and therefore improved exploitation of the colour acuity of the eye than that enjoyed by the other signal, but it is a feature of balanced subcarrier modulation systems which enables this to be so.

Unfortunately, neither set of colours represented by the U' and V' signals align with the axes of colours of minimum acuity of the eye as identified in Section 14.5, that is, the yellowish green to purple axis. This axis of minimum colour acuity is called the Q axis, and the axis at 90 degrees to the Q axis is referred to as the I axis, along which the reddish orange to blueish cyan colours lay.

These colour axes are overlaid on the U',V' vector diagram as illustrated in Figure 17.11.

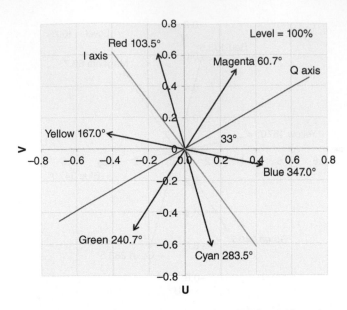

Figure 17.11   Illustrating the position of the I′ and Q′ axes.

The Q axis of minimum acuity is the critical axis for establishing which colours should be encoded with the lower bandwidth signals; the I axis colours are less critically selected, though as Figure 14.6 illustrates, this axis will be close to the axis of maximum colour acuity.

From the diagram it can be seen that the Q axis is at an angle of 33 degrees to the U axis. By using this angle and simple trigonometry, it is apparent that the colour vectors can be resolved into values of U′ and V′ which are aligned to the I and Q axis.

The projection of a colour vector onto the Q′ axis is a combination of the components from the U′ and V′ axes, thus:

$$Q' = V' \sin 33° + U' \cos 33° \text{ and similarly } I' = V' \cos 33° - U' \sin 33°$$

Since $\sin 33° = 0.545$ and $\cos 33° = 0.839$, then substituting for U′ and Y′ in these formulae enables us to define the three signals which form the composite signal of the NTSC specification in terms of the original gamma-corrected R′, G,′ B′ signals:

$$Y' = 0.299R' + 0.587G' + 0.114B'$$
$$I' = 0.596R' - 0.275G' - 0.322B'$$
$$Q' = 0.211R' - 0.523G' + 0.312B'$$

The bandwidth available for the I′, Q′ signals was dependent upon the characteristics of the radio frequency channel allocations, which differed on a country-by-country basis. In the

United States, where at that time, the nominal luma bandwidth was 3.0 MHz, the bandwidths allocated to the I', Q' signals were 1.0 MHz and 0.34 MHz, respectively.

In 625 line countries, where generally the PAL system was adopted, the channel allocations were usually more generous in terms of the bandwidth allocated to each channel and it was therefore unnecessary to adopt the I', Q' approach in formulating the chrominance signal; thus, the U,V signals were transmitted with the same bandwidth.

It must be emphasised that this section of the chapter has been a very cursory overview of the encoding systems; for more in-depth and broader descriptions of the systems, the reader is directed to the books already cited in Section 17.2.

17.2.7.3 Early Experience Using the NTSC System

On the basis of the principle that the technological aspects of reproducing colour would only be discussed in this book if that technology impinged upon the reproduced image, the reader would be correct in questioning why we have gone into such detail of the NTSC system in the previous sections of this chapter. If the practical implementations of the technology were working correctly, then indeed there should be no influence on the colour of the displayed image as a result of encoding and decoding the R', G,' B' signals. Unfortunately, in the early days of the operation of the NTSC colour system, this was not the case and the system was beset with problems.

Figures 17.10 and 17.11 highlight how the hue of the signal is dependent upon the phase of the subcarrier vector. so it is implicit that the phase of the reference subcarrier used to demodulate the chrominance signals is aligned correctly with the I', Q' axes respectively, during the demodulation process. If this alignment is incorrect, the signals will be demodulated along other axes, which will cause the de-matrixed R', G,' B' signals to bear increasingly little resemblance to the original signals as the reference error angle increases.

Naturally, the system designers were aware of the requirement for accurate phasing and a 'colour burst' of a short period of reference phase subcarrier is specified to be transmitted adjacent in time to the line synchronising signal to enable the receiver to lock the phase of its reference subcarrier to this colour burst signal.

However, the combination of processing circuits with poor phase stability and downstream equipment such as video tape recorders, which re-inserted the colour burst after processing the signal off tape, meant that colour TV sets had to be furnished with a phase control to enable the viewer to subjectively vary the phase to produce a displayed image of acceptable hues. The problem was exacerbated when the source of the signal changed, which often required a readjustment of the phase control.

These changes of hue earned the system the epithet, based upon a different interpretation of the letters of the NTSC, often referred to as 'Never Twice the Same Colour', the result of the sophistication of the system being several years in advance of the technology required to implement it in a fully stable manner.

It was during this early period that the European broadcasters were planning to introduce colour systems and led directly to them introducing a phase reversal of the U' signal on alternate lines in time sequence as transmitted, which, when two adjacent lines were added together, cancelled out any phase errors, that is, the PAL system.

Eventually, the technology developed to the point where it was sufficiently stable to enable NTSC signals to be displayed continuously with the correct hue, but by that time, the PAL system had been introduced into a large number of the countries of the world.

It was unfortunate that an initial weakness in the technology supporting the multiplex operation of the NTSC system led to it initially having a poor reputation, which detracted from what was otherwise a very sophisticated specification evolved from a superbly organised development project with complex and diverse criteria to satisfy.

17.3   The Introduction of Digital Television

17.3.1   The Evolution of Digital Specifications

By the mid-1970s digital technology had developed sufficiently for digital processing to be introduced into some of the television equipment within the production workflow, and by the end of this decade, in Europe, Japan and the United States, attention was being focused on the requirements for specifying the digital coding format of composite television signals for use in the production centre.

The importance of relating the digital sampling frequency to the line and frame structure of the system and also to the subcarrier frequency was recognised. For the NTSC system, this did not represent a problem, and the SMPTE had draft recommendations for a digital encoding specification for NTSC composite signals ready by the end of the decade. However, the PAL subcarrier is related to the line frequency in a more complex manner, which prevented the derivation of a digital sampling frequency geometrically aligned to the scanning format of the image, which in turn opened up the opportunity out of necessity for the EBU to take a more ambitious step and consider a digital component system as an alternative to a digital composite system.

By this era the two principal organisations for evolving television specifications, the EBU and the SMPTE, were aware of the advantages of developing specifications which shared as far as possible common parameters, since not only would this keep the cost of broadcast centre equipment down but would also ease the standards conversion requirements on programme interchange between the two areas. The SMPTE therefore held back their draft composite system specification in order to allow time for the two organisations to work closely together to see whether it was possible to develop a common digital component specification that would satisfy the requirements of both the 525 and the 625 line systems.

During 1980 the two organisations each undertook extensive programmes of work, interspersed with meetings between them, to derive a specification which in digital sampling terms shared a common specification. By early 1981 proposals had evolved for adopting a common set of parameters, and during the Annual SMPTE Television Conference in San Francisco in February 1981, the SMPTE organised a joint set of tests, using three proposed sets of parameters, attended by interested parties (including the author) from around the globe. The work undertaken during this period[5] may be favourably compared in extent to the original work of the NTSC in terms of identifying and agreeing the parameters for an international specification for interfacing equipment directly in the digital domain (Tooms, 1981). During 1981 the Japan Broadcasting Corporation (NHK) had completed their tests of the proposed specification and concurred with the proposals, which led to the EBU and the SMPTE submitting their versions of the common specification to the ITU. This led to the specification being adopted as an

---

[5] An excellent summary of this work is contained in the EBU/SMPTE document 'Rec. 601 – the origins of the 4:2:2 DTV standard' (Baron & Wood, 2005): http://tech.ebu.ch/docs/techreview/trev_304-rec601_wood.pdf

international standard in 1982, under the title: 'Recommendation ITU-R BT.601',[6] commonly referred to as Rec 601.

A review of the digital technical characteristics of Rec 601 is beyond the scope of this book, but it is worth emphasising that the basic digital sample rates adopted have formed the basis of every television standard that has evolved since. Rec 601 stipulates 720 luminance samples and 360 samples for each of the colour difference signals per line for 4:3 aspect ratio formats and uses the shorthand 4:2:2 to describe the ratios of luminance to colour difference sampling rates. It was agreed that square pixels were desirable, which led to the number of lines per frame for a 4:3 aspect ratio system being 540. (Later, when 16:9 aspect ratio systems were introduced, it was assumed that only the picture width changed and therefore 540 lines were retained; however, to retain square pixels and effectively the same horizontal resolution in terms of pixels per angular field of view, it would have been necessary to change the number of luminance samples per line to 540 × 16/9, i.e. 960. Thus, all future 16:9 aspect ratio systems are built upon simple multiples of this 'standard definition' picture frame size of 960 × 540 pixels, albeit such a format was never adopted in practice as far as the author is aware. Since 960 pixels per picture width approximated to 1,000, it has also become referred to as a '1K' system and future systems are described in terms of multiples of this 1K system.)

As with the introduction of colour, it took a number of years for the digital standard to permeate throughout the television centre. Individual pieces of equipment such as cameras and video tape recorders would generally remain connected via the analogue infrastructure of the installation, a situation which came to be described as digital islands in an analogue sea. The first all-digital camera was introduced in the early 90s, albeit a 'camcorder' for location work; studio cameras retained an analogue camera head for a longer period, though the CCU processing was also converted to digital at about this time. As the technology developed, so the digital islands became larger, first encompassing a complete studio operation, and finally in 1993,[7] a complete studio production centre, including the studios, the edit suites, the play-out operation and most importantly the digital switching infrastructure, which connected these various facilities.

17.3.2   The Digital Colour Parameters

As reference to Figure 16.1 will indicate, of all the facilities referred to above, it is only in the camera that the colour of the reproduced image is affected, so we need to establish the values of those particular parameters in this new digital specification which are relevant in this context.

Since Rec 601 is limited to studio centre operations, it was necessary at the end of the studio centre signal path to transcode the digital component signal into a classic NTSC or PAL signal to service the transmission systems and the then current millions of analogue home viewing systems. In consequence, the legacy sets of display primaries then in existence were adopted for the Rec 601 standard, that is, the EBU and SMPTE primaries described in Section 17.2. The system white reference was not changed, using D65.

---

[6] The standard has been updated a number of times and the latest (2011) version can be found at http://www.itu.int/dms_pubrec/itu-r/rec/bt/R-REC-BT.601-7-201103-I!!PDF-E.pdf
[7] The Wharf Cable Television Centre in Hong Kong commenced service in October 1993.

Being a component system, the colour difference signals were not time or frequency shared with the luma signal in the encoder, so it was necessary to define the signal level excursions of the three signals prior to component quantisation and it was agreed that the peak-to-peak signal levels of the three signals should be identical.

The luma signal is given by:

$Y' = rR' + gG' + bB'$, where the luminance coefficients: $r + g + b = 1$.

For reasons of compatibility with legacy equipment, it was decided to retain the coefficients used previously for the analogue systems, despite these coefficients no longer representing the luminance contribution of the system primaries, thus:

$Y' = 0.299R' + 0.587G' + 0.114B'$, identical to the PAL and NTSC systems and using this notation, the maximum level of Y' is 1.00, for R, G and B equal to 1.00.

To bring the colour difference signal amplitudes in line with that of the luma signal, the scaling factor x and y in the following formulae require to be solved for the largest excursions of these signals:

$$C_B = \frac{(B' - Y')}{x} = 1 \text{ and } C_R = \frac{(R' - Y')}{y} = 1$$

In solving for x, we saw in Figure 14.5 that the largest excursions of B – Y occurred for the 100% saturated colours blue and yellow, respectively.

$$\text{Thus, } x = (B' - Y')_{blue} - (B' - Y')_{yellow}. \tag{17.1}$$

Noting that $B' - Y' = B' - (rR' + gG' + bB') = B'(1 - b) - (rR' + gG)$, and $r + g = 1 - b$ and that for blue: $R' = 0, G' = 0, B' = 1$, and for yellow: $R' = 1, G' = 1, B' = 0$.

Substituting in (17.1) above:

$$x = (1 - b) - (0 - (r + g))$$
$$= 2 - 2b$$

A similar approach to solving for y leads to:

$$y = 2 - 2r$$

These equations hold irrespective of the values of the luminance coefficients. The attenuation factors x, y calculated to meet the above criteria, where $b = 0.114$ and $r = 0.299$, are therefore as follows:

$$C_B = \frac{(B' - Y')}{1.772} \text{ and } C_R = \frac{(R' - Y')}{1.402}$$

The letters C_B and C_R are the notation used to describe the specified attenuated versions of the digital colour difference signals. Care needs to be taken in their use however as the luminance coefficients will be dependent on the chromaticities of the system primaries and thus the scaling factors will change accordingly.

Thus, in terms of the R', G,' B' signals:

$$C_B = (1B - 0.299R' - 0.587G' - 0.114B')/1.772 = -0.169R' - 0.331G' + 0.500B'$$
$$\text{and } C_R = (1R' - 0.299R' - 0.587G' - 0.114B')/1.402 = +0.500R' - 0.419G' - 0.081B'$$

The composite format of these three signals uses the notation YC_BC_R.

It will be noted that the format of the signals, apart from the colour difference scaling factors, is identical to the original NTSC specification.

17.3.2.1 Mapping the YC_BC_R Signals onto the Digital Bit Stream

The Rec 601 specification offers either 8- or 10-bit quantisation levels, and as the television signal may from time to time exceed the notional RGB range of values of $0 - 1.00$, the mapping of the YC_BC_R signals onto the bit stream allows for foot and head room to accommodate these occasional out-of-tolerance signal levels. Thus, the 8-bit stream has bit levels between 0 and 255, and the Y' signal is arranged to use a quantisation range of 220 levels between 16 (black) and 235 (white), and the C_B, C_R signals a quantisation range of 225 levels between the level of 16 and 240, where black sits at a level of 128. For the corresponding 10-bit stream with bit levels between 0 and 1,023, the Y' and C_B, C_R signals use the quantisation ranges of 877 and 897, respectively, leading to Y' residing between quantisation levels of 64 and 941, and C_B, C_R signals between 64 and 961.

17.3.3 The Source OETF

During the early years of the ITU evolving specifications for television systems, a somewhat anomalous practice was adopted of treating the complete camera transfer characteristic as the signal source; thus, the characteristics of the image sensor and the gamma corrector were combined under the ITU nomenclature as the system Opto-Electronic Transfer Function (OETF). Since the actual OETF of current image sensors is linear, then the OETF quoted in ITU specifications prior to Rec 2020 is actually the transfer function of the gamma corrector.

The source OETF or, more correctly, the gamma correction transfer function is defined in Rec 601 using the approach described in Section 13.4:

$$V = 1.099L^{0.45} - 0.099 \text{ for } 1 \geq L \geq 0.018$$
$$V = 4.500L \text{ for } 0.018 > L \geq 0$$

Worksheet 13(b) is used to produce the shape of the characteristic illustrated in Figure 17.12. The power law curve follows the characteristic with an exponent of 0.45 over the input range greater than 1.8% and below this figure has a linear gain component with the gain set at 4.5.

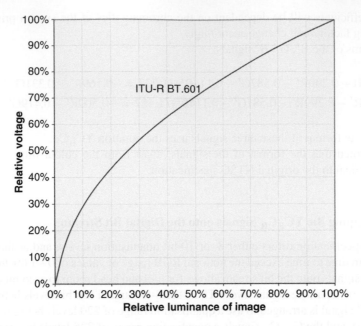

Figure 17.12   The Source OETF characteristic of Rec 601.

17.3.4   Standards Conversion

Although the digital sampling frequencies are a world standard, there are still significant differences between the signals used in the 525 line and 625 line areas of the world, particularly the use of different display primaries. Thus, if programme interchange between these areas is required, then the programme must be standards converted, which includes the requirement to process the RGB signals in such a manner as to make them appear as if they were derived from a camera with spectral sensitivities related to the displays of the population to be serviced, rather than the displays of the original population.

The technique for achieving this emulation of a different camera spectral sensitivity was evolved for a different set of circumstances in Section 12.2, and we can use Worksheet 12(a) with the appropriate primaries chromaticities to calculate the matrix coefficients necessary to transfer one set of RGB signals derived from the original set of primaries to a second set of RGB signals which emulate signals derived from an alternate set of primaries.

Table 17.3   Matrix for converting between EBU and SMPTE RGB signals

	R_{EBU}	G_{EBU}	B_{EBU}
R_{SMPTE}	1.1124	−0.1025	−0.0100
G_{SMPTE}	−0.0205	1.0370	−0.0165
B_{SMPTE}	0.0017	0.0161	0.9822

Table 17.3 gives the coefficients which need to be applied to the RGB signals of an EBU-derived programme in order that the image will be displayed with the correct colours on a display with SMPTE RP145 primaries.

Table 17.4   Matrix for converting between SMPTE and EBU signals

	R_{SMPTE}	G_{SMPTE}	B_{SMPTE}
R_{EBU}	0.9005	0.0888	0.0106
G_{EBU}	0.0178	0.9658	0.0164
B_{EBU}	−0.0019	−0.0160	1.0178

Similarly, Table 17.4 provides the coefficients for converting from SMPTE primaries to EBU primaries.

17.4   The Rise of High Definition Television

Since the 1920s, following the experimental Baird system, any significant improvement in the system which displayed images of significantly enhanced resolution was described as a 'high definition' television system.

In order to avoid ambiguity we will adopt the ITU definition of high definition television provided in Report ITU-R BT.801:

'A high-definition system is a system designed to allow viewing at about three times the picture height, such that the system is virtually, or nearly, transparent to the quality of portrayal that would have been perceived in the original scene or performance by a discerning viewer with normal visual acuity'.

On the basis of this definition, it was the Japanese who were instrumental in introducing HDTV to the public with their MUSE system in 1979. The congestion in the very high frequency (VHF) and ultra high frequency (UHF) bands in most advanced countries of the world prevented the broadcast of high definition (HD) signals because of the much higher bandwidth required in the analogue domain and so acted as a damper on development. Nevertheless, experimental HD cameras were developed both in the United States and in Europe during the 1980s.

The advent of advanced digital compression systems in the 1990s and the corresponding reduction in the bandwidth requirements was the spur that led to work being undertaken to agree a world standard for the fundamental characteristics of an HD system. In the early 1990s the ITU-R BT.709 standard was agreed, comprising a 16:9 aspect ratio and a square pixel format and broadly based upon a '2K' system, that is, an image comprising 1920 horizontal pixels by 1080 vertical pixels.

HDTV public service transmissions based upon this standard commenced in the United States in 1996 but were delayed in Europe until 2004, due primarily to the requirement to totally reorganise the radio frequency spectrum to avoid interference in an environment where the coverage of major transmitters overlapped due to the close proximity of European cities. During the 2010s HDTV services have become widespread and are rapidly becoming the current world standard, albeit standard definition television (SDTV) is still broadcast in many countries to service the legacy population of standard definition receivers. The colour aspects of this system will be described in more detail in Chapter 19.

18

Lighting for Colour Television in the 2010s

18.1 Background

In Section 7.2 the fundamental approach to deriving an index to classify the suitability of an illuminant for illuminating a scene for colour reproduction was dealt with in some detail and the internationally agreed (CIE) procedure for measuring the Colour Rendering Index (CRI) of a number of different sources was outlined. It was also noted in that section that for a number of decades, the limitations of the CRI in terms of predicting a satisfactory level of colour reproduction had been recognised and a number of informal alternative approaches had been adopted in attempts to establish results more consistent with experience.

By the late 2000s, with the introduction of more efficient light-emitting diodes (LEDs) taking over from tungsten-based sources, often with poor results despite relatively good CRIs, the situation had become critical. In television particularly, attention was turned to revisiting approaches originally proposed by workers in BBC Research in the United Kingdom in the 1970/1980s based upon using the characteristics of a camera rather than the eye to ascertain the index of suitability of sources of illumination.

A small team of independent broadcast specialists (Roberts et al., 2011) presented a paper at the International Broadcasting Convention (IBC) in 2011 outlining a new lighting index and was invited by the EBU to form a working party to establish an EBU recommendation based upon this index. Since the basic work had already been undertaken, by December 2012, the EBU were able to formalise the procedure in Recommendation R137 – The EBU Television Lighting Consistency Index – 2012 (TLCI). In recognition of this work the SMPTE has now (2013) established a committee to consider adopting the basis of the EBU proposals within an SMPTE recommendation.

The author is grateful for the support of the team and for their permission and that of the EBU to draw upon their work in briefly describing the Recommendation and giving examples of its use in calculating the TLCI of a number of current luminaires.

In order to avoid repetition in what follows, it is assumed the reader is familiar with the contents of the introductory paragraphs to Section 7.2.

Colour Reproduction in Electronic Imaging Systems: Photography, Television, Cinematography, First Edition. Michael S Tooms.
© 2016 John Wiley & Sons, Ltd. Published 2016 by John Wiley & Sons, Ltd.
Companion Website: www.wiley.com/go/toomscolour

18.2   The EBU Television Lighting Consistency Index – 2012

The EBU recommends in Recommendation R137 that the Television Lighting Consistency Index – 2012 (TLCI) be used for evaluating the suitability of luminaires for lighting scenes for television. The Recommendation is strongly supported by three technical papers:

- Tech 3353 – Development of a 'Standard' Television Camera Model Implemented in the TLCI-2012
- Tech 3354 – Comparison of CIE Colour Metrics for Use in the TLCI-2012
- Tech 3355 – Method for the Assessment of the Colorimetric Properties of Luminaires

Since these highly informative and detailed technical papers can be easily accessed on the Web,[1] this chapter will provide only a synopsis of the Recommendation. The reader with a specific interest in this field is strongly recommended to review these excellent papers.

As fully described in Section 7.2, the general approach to deriving an illuminant index is to measure the reproduced colours under both a reference illuminant and the test illuminant for a range of colour samples and express the difference in the reproduced colours in a subjectively meaningful manner, which then provides the basis for the index.

The use of the TLCI is described in EBU TECH 3355, and in methodology terms, differs from the CRI only in the following manner:

- The colour samples of the ColorChecker colour rendition chart form the test colour samples.
- The measurement system for deriving the XYZ values of the colours, rather than being based upon the characteristics of the eye as hitherto, is based upon the characteristics of a specified television reproduction system, which includes the camera, the signal processing chain and the display device.
- The colour measurements are based upon the CIELAB method and the algorithm used for establishing the visual magnitude of colour differences is the CIEDE2000 metric.

The significance of these three differences will be described in the following three sections of this chapter.

18.3   The ColorChecker Chart

The colour test samples used for the TLCI are those of the first three rows of the ColorChecker chart, first described in Section 7.2.2.

The chart is illustrated in Figure 18.1, where the top two rows are intended to be illustrative of common colours found in the scenes around us. The first two colours are representative of skin colour and the remainder have representative samples of foliage, the sky, etc. The third line is representative of highly saturated additive and subtractive primaries, making the chart a very useful representation of the gamut of colours likely to be met in practice.

The spectral reflectances of the additive and subtractive primaries are illustrated in Figure 18.2.

---

[1] http://tech.ebu.ch/docs/tech/tech3353.pdf, http://tech.ebu.ch/docs/tech/tech3354.pdf, http://tech.ebu.ch/docs/tech/tech3355.pdf

Figure 18.1 The ColorChecker Chart.

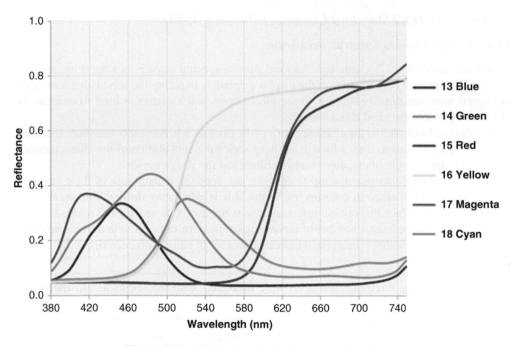

Figure 18.2 Chart spectral reflectances of primaries.

18.4 The TLCI Standard Television Reproduction System Model

Whereas the reproduction model of the CIE CRI index is based upon the characteristics of the eye–brain complex, the TLCI model is based upon a closely defined television reproduction system set of characteristics representing a typical system, as illustrated in Figure 18.3.

The characteristics of each of the processes appearing in Figure 18.3 are defined in the following paragraphs.

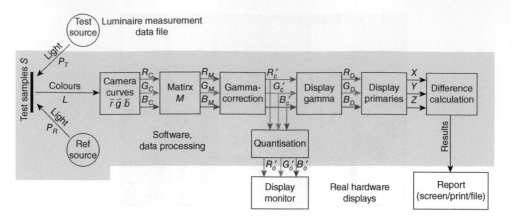

Figure 18.3 The standard television reproduction system model. (After Figure 5 of EBU Tech 3353.)

18.4.1   The TLCI Standard Camera (EBU Tech 3353)

18.4.1.1   The Camera Spectral Sensitivities

As we saw in Section 17.6, although the ideal camera spectral sensitivities are those derived from the chromaticity coordinates of the display primaries, in reality the actual characteristics are based upon the positive lobes of these ideal curves and a matrix is used to emulate as closely as possible the ideal characteristics.

In order to closely define the characteristics of the camera, it is necessary therefore to define its actual spectral sensitivities, which as we have seen are a convolution of the characteristics of the lens, the dichroic filters, any trimming filters and the image sensor.

Unfortunately camera manufacturers consider the spectral sensitivities of their cameras to be a trade secret, since in colorimetric terms, this is what may differentiate them from their competitors. Since this critical information was not available to the EBU team, the only alternative was to measure a number of representative modern cameras to determine whether it was realistic to consider specifying a representative set of camera characteristics.[2]

Nine cameras from three manufacturers were measured, with the results as illustrated in Figure 18.4.

In contrast to the results from similar tests carried out in the 1970/1980s, these curves show a remarkable similarity across the range of cameras – perhaps an indication that over the years, there has been a steady move towards the ideal compromise for a particular sensor characteristic.

It was considered that the small disparity between the different cameras justified the adoption of a standard set of characteristics based upon the average of the measured results being representative of colour television cameras currently in use.

---

[2] This work was undertaken by Per Böhler of NRK, the Norwegian public service broadcaster, and is described in EBU Tech 3353.

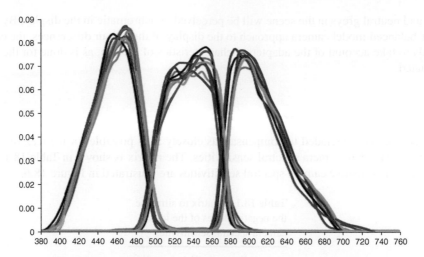

Figure 18.4 Camera spectral sensitivities measured by NRK. (After Figure 9, EBU Tech 3353.)

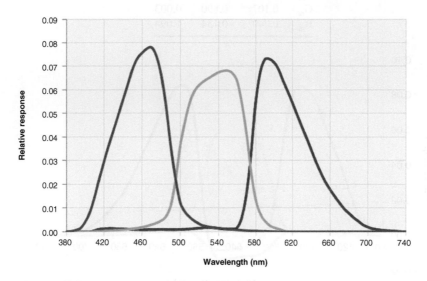

Figure 18.5 The TLCI standard camera spectral sensitivities. (Bohler et al., 2013.)

The set of standard camera characteristics adopted by the EBU in Recommendation 137 are listed in the 'Camera' Worksheet and illustrated in Figure 18.5.

18.4.1.2 Camera Adaptation or Colour Balance

The camera has an important advantage over the eye–brain complex in dealing with illuminants of different colour temperature; whereas the eye has to adapt to the new illuminant colour temperature, the camera effectively adapts by undertaking a colour balance which ensures that white and neutral greys in the scene will be perceived as achromatic in the display. By using a colour-balanced model camera approach to the display of the colour differences, the complex formula to take account of the adaptation characteristics of the eye, as is done for the CRI, is eliminated.

18.4.1.3   The Camera Matrix

A camera matrix is included to compensate as closely as is possible for the lack of negative lobes in the standard camera spectral sensitivities. The matrix is shown in Table 18.1 and the corresponding effective camera spectral sensitivities are illustrated in Figure 18.6.

Table 18.1   Matrix to simulate the negative lobes of the ideal camera spectral sensitivities

	R_{in}	G_{in}	B_{in}
R_m	1.182	−0.209	0.027
G_m	0.107	0.890	0.003
B_m	0.040	−0.134	1.094

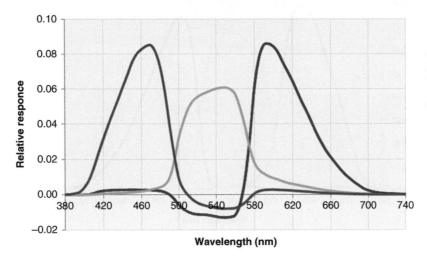

Figure 18.6   Effective camera spectral sensitivities with matrix in place.

18.4.1.4   Gamma Correction or Opto-electronic Transfer Characteristic

The gamma correction characteristic used in the standard camera is that defined for use in ITU-R BT.709, the current (2013) world high definition television standard, as described in Chapter 19.

$$R' = 1.099R^{0.45} - 0.099 \text{ for values of R } > 0.018, \text{ otherwise } R' = 4.5R$$

and similarly for G' and B'.

18.4.1.5 Saturation Adjustment

Though not illustrated in Figure 18.3, the TLCI Standard Model also includes a saturation adjustment which is set to a value of 90%, apparently to make sure that there is enough space around the chromaticity diagram for colour errors.

18.4.1.6 Display Gamma or Electro-opto Transfer Characteristic

The display gamma is assumed to be 2.4, that is, the relationship between input voltage and displayed light is given by:

$$L_R = R'^{2.4}$$

and similarly for L_G and L_B.

18.4.1.7 Display XYZ Values

The display is assumed to have primaries with chromaticities as defined by Rec 709, as shown in Table 18.2.

Table 18.2 Rec 709 primaries and white chromaticities

	x	y
Red	0.7007	0.2993
Green	0.1142	0.8262
Blue	0.1355	0.0399
White D65	0.3127	0.3290

A conversion matrix is required in order to obtain the XYZ values of the light emitted by these display primaries from the R'G'B' drive values, as illustrated in Table 18.3 (see Chapter 12 and Worksheet 18).

Table 18.3 Conversion matrix for obtaining XYZ values from the RGB light values

	L_R	L_G	L_B
X	0.4124	0.3576	0.1805
Y	0.2126	0.7152	0.0722
Z	0.0193	0.1192	0.9505

18.5 Selecting a Colour Metric for the TLCI (EBU Tech 3354)

As discussed in Section 7.2, the original CIE CRI used the CIE1964 colour metric based upon the obsolete u,v chromaticity diagram; however, since that time, the CIE has been active in evolving a number of colour metrics which continue to improve the relationship between the results perceived by the eye and those produced by the metric. These range from the u',v' diagram (1976), through the $L^*a^*b^*$ metric and a number of ever more complex metrics, to the current CIEDE2000 metric.

The work that was undertaken to determine which of these metrics was best suited for adoption for measuring the colour differences for the TLCI is fully described in the Tech 3354 document. The tests were fully comprehensive and clearly showed that the current CIEDE2000 metric produced ΔE^*_{00} values which were most closely related to the perceived colour differences and therefore led to the decision to adopt this metric for the TLCI. (It is also interesting and worthy of mention that the $L^*u'^*v'^*$ metric performed surprisingly well, whilst the $L^*a^*b^*$ metric did not. Since all the modern metrics are based upon the $L^*a^*b^*$ metric, with their ever more complex formula, in an endeavour to match the results to those perceived, it seems possible that if the $L^*u'^*v'^*$ metric had formed the basis of these later metrics, less complex formulae for deriving the CIEDE2000 ΔE^*_{00} values may have resulted.)

18.6   Measuring the TLCI of Luminaires (EBU Tech 3355)

The calculations required to derive the TLCI from the SPD of a test illuminant are extensive and relatively complex, as is shown in detail in the Tech 3355 document. Worksheet 18, which is described in Appendix J, Guide to the Colour Reproduction Workbook, also provides a basis for calculating the TLCI. However, the EBU provides a Windows application[3] which not only undertakes the calculation to establish the TLCI but also provides additional useful information to a colourist in the form of a comprehensive display, an example of which is illustrated in Figure 18.7.

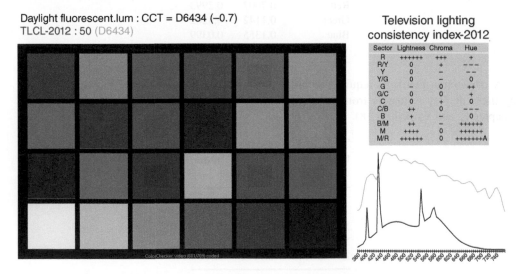

Figure 18.7   Example of TLCI-2012 output. (From Tech 3355 Figure 4.)

The use of the application and the resulting display is fully described in EBU Tech 3355 and as can be seen from Figure 18.7 it provides the CCT of the illuminant, the TLCI itself, an

---

[3] https://tech.ebu.ch/tlci-2012

indication of the perceived difference in colour using the inset squares of the ColorChecker chart, a table for use by the colourist and the SPD of the test illuminant against the matching CCT 'D' reference illuminant. The 'daylight fluorescent' example luminaire illustrated in Figure 18.7 was chosen as the reference to adjust the TLCI formula such that a figure of 50 was obtained. As described in Tech 3355, this illuminant produces a TLCI ΔE^*_{00} value of 3.16; errors greater than this, corresponding to lower TLCIs, are difficult to correct in post-production colour correction equipment.

Worksheets 7(a) and 18 were used to calculate and compare the CRI and the TLCI figures of a number of different types of luminaires, with results as shown in Table 18.4.

Table 18.4   CCT and index figures for a range of different luminaires

Luminaire	CCT	CRI	TLCI	Luminaire	CCT	CRI	TLCI
Tungsten	2857	99.6	100.0	**EE White**	5459	95.3	99.4
Daylight	6500	100	100.0	**Xenon**	6044	93.9	99.2
FL1	6430	75.8	49.8	**HMI 1**	6002	87.9	65.6
FL4	2942	51.4	19.5	**HMI 2**	5630	88.2	75.6
FL7	6597	91.0	93.6	**LED 1**	6536	96.4	97.4
FL 3.15	6508	98.5	99.8	**LED 2**	6686	62.7	36.1

The SPDs of these lighting sources are tabled in the 'Illuminants' worksheet and copied into Worksheets 7(a) and 18 to produce the figures in the table. The 'FL' luminaires are defined CIE SPD fluorescent lamps; FL1 is described as 'daylight' and is the source used by the EBU to produce the TLCI figure of 50. (There are minor differences in the figures produced by the EBU TLCI application and Worksheet 18, leading to a figure in the latter case of 49.8.)

It is interesting to note that in these examples, CRI values above 90 always lead to higher TLCI values, indicating that CRI values in this range may be relied upon. However, CRI values below 90 always correspond to TLCI values which are worse, often considerably worse, and therefore should not be relied upon. The Xenon lamp measured had an excellent TLCI of 99.1, but both the HMI luminaires were in the middle of the just acceptable range despite appearing to have good CRI values; both the LED luminaires are in use for television lighting, but as can be seen, LED 2, which in CRI terms is in the acceptable range, is clearly rated poorly in terms of its TLCI at 36.1.

19

Colour in Television in the 2010s – The High Definition Colour Television System

19.1 The High Definition System Specification

From Section 17.4, the ITU definition of 'high definition' television (HDTV) is defined as follows:

'A high-definition system is a system designed to allow viewing at about three times the picture height, such that the system is virtually, or nearly, transparent to the quality of portrayal that would have been perceived in the original scene or performance by a discerning viewer with normal visual acuity'.

The world technical standard for HD is embodied in Recommendation ITU-R BT.709-5[1] and entitled 'Parameter values for the HDTV standards for production and international programme exchange'. The specification was first established in the early 1990s and this fifth current revision is dated April 2002 and is often referred to as Rec 709; it effectively supersedes the Rec 601 standard, which defined digital standard-definition television (SDTV).

The specification is split into Parts 1 and 2; the former provides compatibility with legacy systems, whilst the latter describes a common image format and is effectively the new world standard. Thus, whilst Part 2 continues to have options to service the different frame rate systems supporting the different television broadcasting operations around the world, in terms of its essential elements such as the picture format of 1920 pixels wide by 1080 pixels high (2K) and the colour-dependent parameters, for the first time, there is a world standard which supports its title.

HDTV services were commenced in the United States in the late 1990s and gradually introduced to most of the world during the 2000s, such that by the 2010s, it had become virtually fully operational around the world, albeit in many countries SDTV continued to provide a service to legacy receivers.

---

[1] http://www.itu.int/dms_pubrec/itu-r/rec/bt/R-REC-BT.709-5-200204-I!!PDF-E.pdf.

Colour Reproduction in Electronic Imaging Systems: Photography, Television, Cinematography, First Edition. Michael S Tooms.
© 2016 John Wiley & Sons, Ltd. Published 2016 by John Wiley & Sons, Ltd.
Companion Website: www.wiley.com/go/toomscolour

The tables relevant to the colour reproduction aspects of the specification are extracted and laid out in the section below. The original table numbers are retained at the beginning of each section in order to retain consistency with the specification but are supplemented by chapter table numbers in order to provide reference to the text which follows the tables.

Table 19.1   1 Opto-electronic conversion

Item	Parameter	System values									
		60/P	30/P	30/PsF	60/I	50/P	25/P	25/PsF	50/I	24/P	24/PsF
1.1	Opto-electronic transfer characteristics before non-linear pre-correction	Assumed linear									
1.2	Overall opto-electronic transfer characteristics at source	$V = 1.099L^{0.45} - 0.099$ for $1 \geq L \geq 0.018$ $V = 4.500L$ for $0.018 > L \geq 0$ where: L: Luminance of the image $0 \leq L \leq 1$ V: Corresponding electrical signal									
1.3	Chromaticity coordinates (CIE, 1931)	x					y				
	Primary – Red (R) – Green (G) – Blue (B)	0.640 0.300 0.150					0.330 0.600 0.060				
1.4	Assumed chromaticity for equal primary signals (reference white)	D65									
	$E_R = E_G = E_B$	x 0.3127					y 0.3290				

Table 19.2   3 Signal format

Item	Parameter	System values									
		60/P	30/P	30/PsF	60/I	50/P	25/P	25/PsF	50/I	24/P	24/PsF
3.1	Conceptual non-linear pre-correction of primary signals	$\gamma = 0.45$ (see item 1.2)									
3.2	Derivation of luminance signal E'_Y	$E'_Y = 0.2126E'_R + 0.7152E'_G + 0.0722E'_B$									
3.3	Derivation of color difference signal (analogue coding)	$E'_{CB} = \dfrac{E'_B - E'_Y}{1.8556} = \dfrac{-0.2126\,E'_R - 0.7152\,E'_G + 0.9278\,E'_B}{1.8556}$ $E'_{CR} = \dfrac{E'_R - E'_Y}{1.5748} = \dfrac{0.7874\,E'_R - 0.7152\,E'_G - 0.0722\,E'_B}{1.5748}$									

19.1.1 The Colour-Dependent Parameter Values of Part 2 of the ITU BT.709 Specification

19.1.2 Observations on the ITU BT.709 Recommendation Parameters

To place the values of the Rec 709 parameters into the context of the previous material, in this book, they will be reviewed in turn. Since all previous new television system specifications had to take account of the legacy population of domestic TV receivers, in order to avoid compatibility problems, it had not been possible to reconsider the values of the basic parameters. However, the change to HDTV was fundamental; in order to receive the HD signals, new TV sets had to be purchased with different characteristics, thus providing the opportunity to review those parameter values which may have become obsolescent.

19.1.2.1 The System Primaries and White Point

The system primaries are effectively based upon a new set of display primaries whose chromaticities are listed in Table 19.3, which unfortunately are very little different to those discussed earlier for SDTV; effectively, a compromise between the SMPTE and EBU primaries defined in Rec 601 – in hindsight, an opportunity missed to potentially extend the display gamut using the techniques described in Chapters 12 and 20. Some manufacturers of display panels are already using more highly saturated primaries than those of the HD standard and these panels are therefore potentially capable of a wider gamut, which the HD system is unable to exploit.

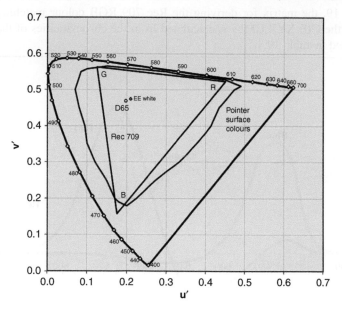

Figure 19.1 Comparison of Rec 709 primaries and Pointer surface colours gamuts.

It is instructive to compare the current Rec 709 standard gamut with the surface colours gamut of Pointer as shown in Figure 19.1. Clearly although the television colour system is capable of good-quality reproduction of the usual range of colours in a scene, its performance on saturated colours such as costumes and flowers is disappointing; these colours will appear relatively desaturated in the display.

Table 19.3 Rec 709 primaries chromaticities

	x	y	u′	v′
Red	0.640	0.330	0.451	0.523
Green	0.300	0.600	0.125	0.562
Blue	0.150	0.060	0.175	0.158
D65	0.3127	0.3290	0.1978	0.4683

The primaries chromaticity coordinates and the system white point of D65 are listed in Table 19.3.

Table 19.4 Matrix for deriving the Rec 709 RGB CMFs from the CIE XYZ CMFs

	X	Y	Z
R =	3.4177	−1.6212	−0.5258
G =	−1.0221	1.9783	0.0438
B =	0.0587	−0.2151	1.1146

In Worksheet 19, the matrix for deriving the Rec 709 RGB colour matching functions (CMFs) from the CIE XYZ CMFs is calculated from the chromaticities of the primaries and is illustrated in Table 19.4.

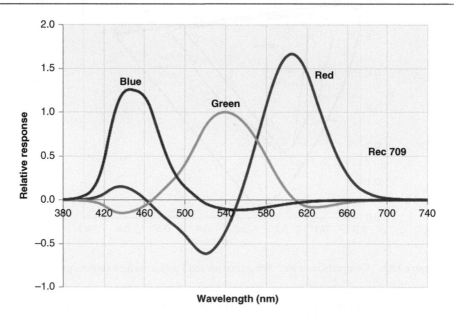

Figure 19.2 Idealised camera spectral sensitivities for Rec 709 primaries.

The CMFs, which become the idealised camera spectral sensitivities, are illustrated in Figure 19.2.

19.1.2.2 The Luma Signal

The coefficients for the RGB contributions to the luminance signal are calculated in Worksheet 19, where the matrix derived for Table 19.4 is inverted to provide the XYZ values in terms of the RGB values. Since the Y CMF represents the luminance response of the eye, the coefficients of Y are those that are required to generate the luminance signal, as is shown by Matrix 6 in the worksheet.

The luma signal is defined somewhat ambiguously in Table 19.2 of the specification as the luminance signal and is derived as:

$$E'_Y = 0.2126E'_R + 0.7152\,E'_G + 0.0722E'_B$$

It is interesting to note that this is the first time in a new specification, since the introduction of the original NTSC specification, that the coefficients of the luma signal have properly represented the luminance contribution of the display primaries; previously, as the display primaries have been changed in updated specifications, the original NTSC coefficients had been retained in order for the signal to remain compatible with legacy receivers and monitors. Since the balance of these coefficients only affects the appearance of the display on black and white receivers, which generally are no longer in use, the lack of accuracy in this respect was not important.

19.1.2.3 Gamma Correction

The specification is not entirely unambiguous in this area; Table 19.2 refers to the 'Conceptual non-linear pre-correction of primary signals: $\gamma = 0.45'$ without defining what γ represents, though to the knowledgeable it may be implied from Table 19.1 where the definition of the 'Overall opto-electronic transfer characteristics at source' is given as:

$$V = 1.099L^{0.45} - 0.099 \text{ for } 1 \geq L \geq 0.018$$
$$V = 4.500L \text{ for } 0.018 > L \geq 0$$

where:

L: Luminance of the image $0 \leq L \leq 1$

V: Corresponding electrical signal

The above specification is identical to that which appears in the preceding Rec 601 specification.

This is the format for defining gamma correction (or what the specification refers to as the 'opto-digital transfer function') described in Section 13.4.6 and indicates that for signals below a level of 1 and equal to or above a level of 1.8%, the transfer characteristic is a power law with an exponent of 0.45, and for signals above 0 but below a level of 1.8%, it is based upon a straight line law with a gain of 4.5. As Figures 19.3 and 19.4 from Worksheet 13(b)

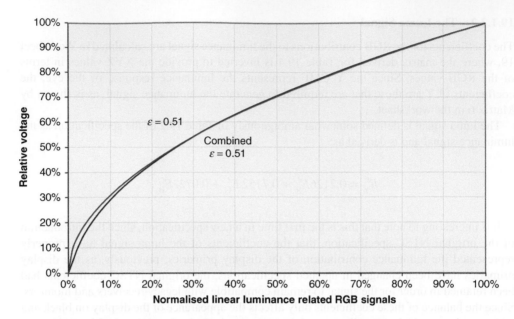

Figure 19.3   Comparison of Rec 709 and true power law with an exponent of 0.51.

illustrate, the resulting characteristic is a close match to a power law curve based upon an exponent of about 0.51 for values of signal level above 20%, but departs increasingly from a match for values below 20%.

It is not categorically stated what the transfer characteristic of the display is, which this law is intended to complement; however, by implication, the display would have characteristics

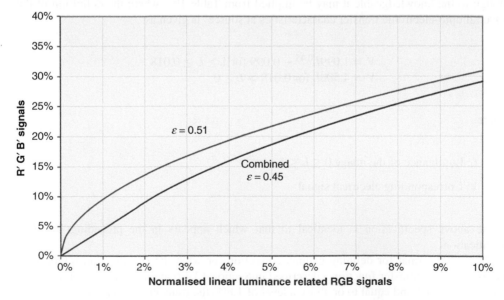

Figure 19.4   as Figure 19.3 but over input signal range from 0% to 10%.

as described in Rec 1886,[2] which gives a power law transfer characteristic with an exponent or gamma of 2.4. Thus, assuming that the viewing conditions partially mask the perception of signals below a level 20%, the overall system gamma would appear to be 2.4 × 0.51 or 1.224; however, below 20%, the effective system gamma changes with signal level, becoming a system point gamma which varies from about 1.3 at a signal level of 20% to 1.6 at a signal level of 1%.

Mismatches of this level in the visible range cannot be ignored and will produce significant errors in the rendered image, as will be illustrated in Section 19.2.

19.1.2.4   Colour Difference Signals

In Table 19.2, the colour difference signals are defined traditionally but with new scaling factors. Using the formula for generating the scaling factors x and y for the blue and red colour difference signals, respectively, developed in Section 17.3 to ensure both the luma and colour difference signals have the same amplitude of 1.0 on maximum chroma signals:

$x = 2 - 2b$ and $y = 2 - 2r$ where b and r are the luminance coefficients of the blue and red primaries, respectively, then as $b = 0.0722$ and $r = 0.2126$:

$$C_B = \frac{B' - Y'}{1.8556}$$

$$C_R = \frac{R' - Y'}{1.5748}$$

It is interesting to note that the encoding format for colour television signals originally evolved for the NTSC system is still in use with current colour television systems.

19.1.2.5   Signal Levels

Table 4 of the specification is not reproduced here since all but one of the parameter values it defines are not relevant to the reproduction of colour; however, one parameter, 4.1 Nominal level, deals with signal level, which is described in terms of having values only between the value representing black and that representing white. At first, this might seem reasonable; however, if one considers a camera with a matrix included to emulate the negative excursions of the camera spectral sensitivities, then it becomes apparent that for colours outside of the system primaries gamut, negative signals and signals above the peak white value will be produced, that is, they will extend below the level defined for black and above the level defined for white in one or other of the RGB signals. Thus, since the system is unable to accommodate negative signals, they will be clipped to black level, which is not a problem for a display with the system primaries but does mean that extended gamut displays will not be able to accurately portray colours which they would otherwise have been able to do so following an appropriate correcting display matrix operation.

This issue is explored in the next section and revisited in Chapter 20.

---

[2] Recommendation ITU-R BT.1886 'Reference electro-opto transfer function for flat panel displays used in HDTV studio production'.

19.2 Evaluating the Performance of the HDTV System

We have considered the ramifications of the specification of the HDTV system at some length but this approach falls short of giving an indication of the performance of the system in terms of accurately reproducing the colours in a scene.

In order to evaluate the system, we need to be in a position to measure the colour of samples in the scene and compare the results with those effectively taken from the display. By establishing a representative model of the system, with the parameters of each element in the signal path between the scene and the displayed image fully specified, it is possible to build a mathematical model incorporating the values of the parameters to be varied within the limits representing the practicability of actual situations.

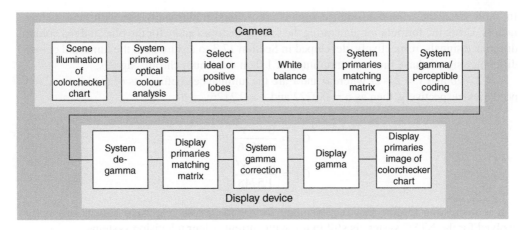

Figure 19.5 Model of the colour signal path of a Rec 709 system.

The signal path of the generic mathematical model is illustrated in Figure 19.5, where variability of the levels of the parameter values for each signal path element ensures it may be used to reflect the parameter values of any foreseeable colour television system. As the model is fully flexible, a few words of description for each of the elements will avoid ambiguity.

In the camera:

- the Scene Illumination spectral characteristic may be selected from the wide range representing any likely situation;
- the System Primaries Optical Colour Analysis shapes the camera spectral sensitivities to provide the best match to those of the specified system primaries;
- the Select Ideal or Positive Lobes element enables the selection of the ideal system primaries spectral sensitivities; the positive lobes only of these characteristics or a set of curves representing the Television Lighting Consistency Index (TLCI) camera characteristics;
- White Balance adjusts the levels of the RGB signals to be equal on the white in the scene;
- System Primaries Matching Matrix provides a number of options to select the best match of the practical camera spectral sensitivities to the system primaries spectral sensitivities;
- System Gamma/Perceptible Coding enables the selection of a specified gamma law/ perceptible coding characteristic.

In the display device:

- System De-gamma applies the inverse law of the System Gamma element to provide linear RGB signals to the Display Primaries Matching Matrix element, which, should it be necessary, modifies the signals from those representing the system primaries to those matching the display primaries;
- the System Gamma Correction element notionally has a characteristic which complements the Display Gamma characteristic, that is, a simple power law; however, in the emulation of the Rec 709 specification, it would have a System Gamma characteristic;
- finally, the Displayed Image ColorChecker Chart is a matrix element based upon the display primaries chromaticities which converts the light generated by the display from RGB values to XYZ values.

Effectively, we already have available most of the elements of this mathematical model; in Part 4, each element of the system was described and illustrated with examples from the worksheets dedicated to that element, so the model appearing in Worksheet 19 is composed of sections, each drawn from earlier worksheets. Each section mathematically represents an element in the model, with the output of successive elements being the input of the next element in the signal path. The worksheet model commences with the selection of the spectral power distribution (SPD) of the illuminant of the ColorChecker chart and terminates with the XYZ values of the display light from each sample of the displayed image of the ColorChecker chart. The worksheet includes both a u′,v′ chart with vectors illustrating the chromaticity errors and a CIEDE2000 metric calculator to provide ΔE^*_{00} values for each sample. Further guidance on the use of the worksheet is contained in in the section entitled 'Using the Colour Reproduction Worksheet' at the end of the book.

The ColorChecker chart illustrated in Figure 19.6 is used to represent a wide gamut of surface colours in the scene, as already described in Sections 7.2 and 18.3.

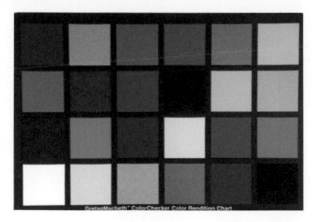

Figure 19.6   The ColorChecker chart.

In order to provide a reference colour against which the displayed colour is measured, in Worksheet 19, the XYZ values of each sample in the chart are calculated under the system white illuminant by integrating the convolution of the spectral reflectivity of each sample with the illuminant SPD and the CIE XYZ colour mixture curves.

The worksheet provides options at each elemental stage to vary the value of the associated parameters, including values which effectively neutralise the effect of the stage. For example, by selecting the values in Table 19.5, which reflect the ideal values of the parameters, the veracity of the mathematical model may be checked, since under these conditions, the colour differences for each sample should be zero. However, there is an exception on the cyan sample, whose chromaticity lies outside of the gamut of the Rec 709 primaries, as is illustrated for sample number 18 in Figure 19.7.

The ColorChecker samples are numbered from 1 to 18 starting at the top left of the chart.

The last two rows of Table 19.5 indicate the sample with the worst value of ΔE^*_{00} and the value of ΔE^*_{00} for Sample 2, the light flesh colour, which is regarded as the most critical sample colour.

The model is useful for indicating the extent of chromaticity and colour difference errors for different values of the parameters in the signal chain in addition to providing values for an ideal *practical* range of parameter values, as illustrated in the last of the following examples.

Table 19.5   Ideal parameter values to produce no model errors

Parameter	Parameter value
Illumination	D65
Camera spectral sensitivities	Rec 709 primaries
Ideal or Positive lobes	Ideal
System primaries matrix	Unity
System gamma	Unity
System de-gamma	Unity
Display Matrix	Rec 709 primaries
System gamma	Unity
Display gamma	Unity
Worst ΔE^*_{00}	2.2
Light skin (2) ΔE^*_{00}	0.1

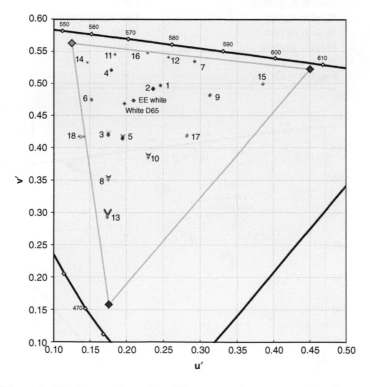

Figure 19.7   ColorChecker samples colour differences – element parameter values set to ideal.

19.2.1   Non-Standard Illuminants

19.2.1.1   Illuminant A

A camera capturing the image under Illuminant A without first applying an optical correction filter but after colour balancing to white will exhibit significant colour errors, as illustrated by the values in Table 19.6 and Figure 19.8. Applying an appropriate colour correction filter will restore the performance to give low ΔE^*_{00} errors.

Table 19.6   Scene illumination set to SA, other parameter values set to ideal

Parameter	Parameter value
Illumination	Illuminant A
Camera spectral sensitivities	Rec 709 primaries
Ideal or Positive lobes	Ideal
System primaries matrix	Unity
System gamma	Unity
System de-gamma	Unity
Display matrix	Rec 709 primaries
System gamma	Unity
Display gamma	Unity
Worst ΔE^*_{00}	12.2
Light skin (2) ΔE^*_{00}	3.9

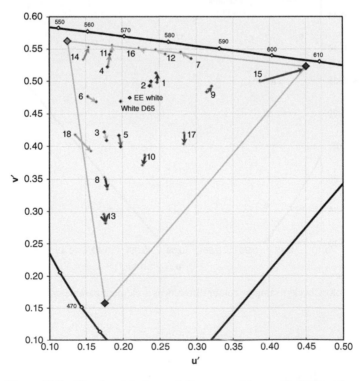

Figure 19.8   Illuminant A chromaticity errors with no optical correction.

19.2.1.2   Illuminant FL1 – Fluorescent Daylight – TLCI = 50

This example, illustrated in Figure 19.9, is useful for indicating the level of errors associated with what is often regarded as a just satisfactory illuminant.

Table 19.7   Scene illumination set to FL1, other parameter values set to ideal

Parameter	Parameter value
Illumination	Illuminant FL1
Camera spectral sensitivities	Rec 709 primaries
Ideal or Positive lobes	Ideal
System primaries matrix	Unity
System gamma	Unity
System de-gamma	Unity
Display matrix	Rec 709 primaries
System gamma	Unity
Display gamma	Unity
Worst ΔE^{*}_{00}	6.2
Light skin (2) ΔE^{*}_{00}	4.2

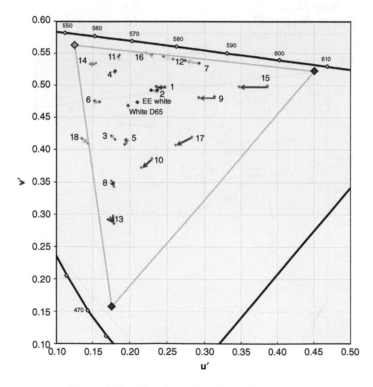

Figure 19.9   Illuminant FL1 chromaticity errors.

19.2.2 Other Parameter Variables

19.2.2.1 TLCI Practical Camera Spectral Sensitivities

These are the spectral sensitivities typical of current television cameras as described in Chapter 18 but with no correcting analysis matrix and thus all samples are significantly desaturated, as illustrated in Figure 19.10.

Table 19.8 Camera spectral sensitivities set to TLCI, other parameter values set to ideal

Parameter	Parameter value
Illumination	D65
Camera spectral sensitivities	Rec 709 primaries
Ideal or Positive lobes	TLCI practical
System primaries matrix	Unity
System gamma	Unity
System de-gamma	Unity
Display matrix	Rec 709 primaries
System gamma	Unity
Display gamma	Unity
Worst ΔE^*_{00}	8.4
Light skin (2) ΔE^*_{00}	3.7

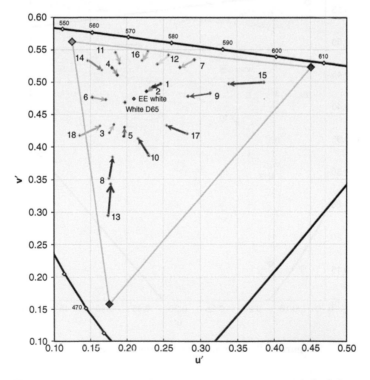

Figure 19.10 'TLCI' practical analyses curves with no negative lobes.

19.2.2.2 TLCI Positive Lobes with 'Best Adjustable' Matrix

With the adjustable matrix in place and optimally adjusted, the undersaturation is virtually eliminated on light colours and much improved on others, as illustrated in Figure 19.11.

Table 19.9   Camera spectral sensitivities set to TLCI, system primaries matrix set to best adjustable, other parameter values set to ideal

Parameter	Parameter value
Illumination	D65
Camera spectral sensitivities	Rec 709 primaries
Ideal or Positive lobes	TLCI practical
System primaries matrix	Best adjustable
System gamma	Unity
System de-gamma	Unity
Display matrix	Rec 709 primaries
System gamma	Unity
Display gamma	Unity
Worst ΔE^*_{00}	4.8
Light skin (2) ΔE^*_{00}	3.1

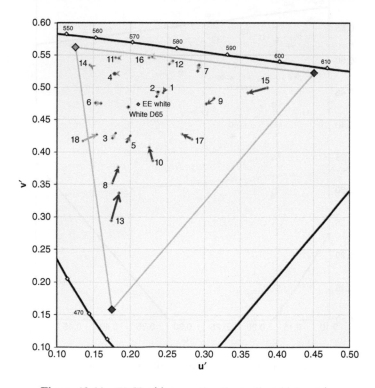

Figure 19.11   TLCI with correcting 'best adjustable' matrix.

Table 19.10   Parameter values set to reflect the best practical situation for a Rec 709 system

Parameter	Parameter value
Illumination	D65
Camera spectral sensitivities	Rec 709 primaries
Ideal or Positive lobes	TLCI practical
System primaries matrix	Best adjustable
System gamma	Rec 709 gamma
System de-gamma	Inverse Rec 709
Display matrix	Rec 709 (Unity)
System gamma	Rec 709 gamma
Display gamma	2.4
Worst ΔE^*_{00}	9.1
Light skin (2) ΔE^*_{00}	5.9

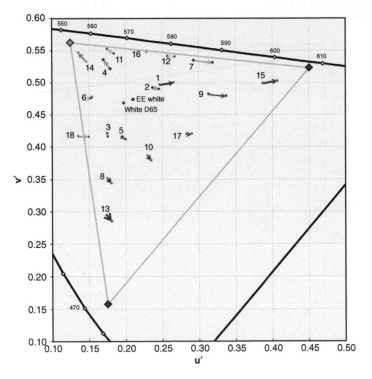

Figure 19.12   Rec 709 colour difference errors.

19.2.2.3   System Gamma and Display Gamma set to the Rec 709 Specification

We have now reached the point where the parameter values for the camera and display device reflect in a realistic manner the conditions defined in Rec 709. As described in Section 13.4, there is a significant mismatch between the Rec 709 gamma correction characteristic, or EOTF characteristic, and a straightforward exponential characteristic as required by Rec 1886. In this case, the situation is exacerbated as the inverse of the closest exponential match to the Rec 709 characteristic has a notional exponent of 1.96, whilst the display exponent is 2.4, leading to a system gamma of about 1.2 and a consequent significant increase in saturation, as illustrated in Figure 19.12. As discussed in Section 13.7, it may be considered desirable for a number of reasons, associated with the average television viewing conditions, to adopt an overall system gamma of a little greater than unity for domestic viewing.

Nevertheless, this approach does produce relative high values of colour difference, a worst ΔE^*_{00} value of 9.1 when compared with the value of 5.9 achieved when this unmatched combination of gamma correction and display gamma are not in the signal path.

19.3   Appraisal of the Rec 709 Recommendation

It is undeniable that when viewing a well-set-up television display, most viewers would regard the quality of pictures reproduced by a professional camera capturing well-lighted scenes in a Rec 709 system as excellent. Nevertheless, it must be remembered that generally speaking, the television viewer is unable to compare the viewed reproduction with the original scene and is thus not as critical as he or she might otherwise be.

At the time the Rec 709 specification was being evolved, the emphasis was on the 'high definition' aspect of the specification, that is, both the greater number of pixels and of finally achieving, in picture format terms, international agreement, and in this respect, the specification was a success. However, it was somewhat disappointing that the colour-related parameter values of the preceding Rec 601 specification were adopted virtually without significant amendment, thus missing the opportunity to improve the quality of the rendered image.

20

Colour in Television in the 2020s

20.1 The Potential for Improved Colour Reproduction

20.1.1 Introduction

In Part 4, where the fundamentals of colour reproduction were developed, ideal system design criteria were explored before investigating the parameters and their values for practical systems. In reviewing the requirements for colour in television in the 2020s, it is incumbent upon us to determine whether the supporting system technologies have advanced sufficiently to embrace these ideal approaches.

In the last chapter, the performance of the current HD system was analysed in some depth and a number of areas where the performance was limited emerged. These limitations when compared with those which can be ideally achieved may be listed as follows:

- Limited chromaticity gamut
- Limited portrayal of contrast range
- Distorted portrayal of contrast range and therefore also of saturation
- Failure of constant luminance

The first two items in this list may be considered together as 'limited colour gamut', that is, a limitation in accurately portraying the colours of the colour space described in Section 4.6.

A further related factor to consider is how television is currently viewed. Some parameter values, particularly those associated with contrast range, were selected when traditionally domestic viewing took place on what is now regarded as a small screen in rooms with lighting suitable for day-to-day living rather than on considerably larger screens with often more subdued lighting better suited to an immersive experience.

In legacy systems, in order to keep the cost of the domestic receiver as low as possible, the systems were configured accordingly, with essential processing being undertaken at the camera once rather than millions of times in each receiver. In addition, the advantages of adopting a systems gamut, in terms of the flexibility of accommodating improvements in the chromaticity gamut of display devices, were again sacrificed for cost-effectiveness. Once such

Colour Reproduction in Electronic Imaging Systems: Photography, Television, Cinematography, First Edition. Michael S Tooms.
© 2016 John Wiley & Sons, Ltd. Published 2016 by John Wiley & Sons, Ltd.
Companion Website: www.wiley.com/go/toomscolour

an inflexible system is in widespread use, it is not possible to change the parameters without making large numbers of legacy receivers incompatible.

When a completely new system is being considered, with parameter values which are so incompatible with the current system that a new population of receivers are required, the opportunity presents itself to review *all* the current parameters, and where technology and cost have provided the opportunity to move towards more ideal solutions, to adopt them. Such a new system is the Ultra High Definition Television System (UHDTV) currently under development and which is likely to be introduced in one form or another in the 2020s. The current specifications for this system are reviewed in the final section of this chapter, but first in order to put the proposals for the UHDTV system into context, we will:

- evolve the colour reproduction specification of a practical ideal colour television system using current technology, initially without consideration for the legacy receiver population and subsequently accommodating those requirements without compromising the original solution;
- review the proposals which have been put forward in the recent past to overcome some of the limitations of previous legacy systems and to ameliorate those of the current HDTV system.

20.2   Colour Specification of a Practical Ideal Colour Television System

20.2.1   The Configuration of an Ideal System

By initially putting to one side the legacy approaches which have evolved through two generations of system development, that is, the SDTV and HDTV systems, and exploiting the improvements in technology which have occurred during that period, it is possible to configure a system which not only has a superior colour reproduction performance but is also comparatively simple to implement at the fundamental level.

The key factors which are instrumental in enabling this approach are:

- Much improved image sensor sensitivity
- A universal population of linear display devices in the foreseeable future
- The ability to incorporate processing circuits within the display device with a minimal cost premium
- Display devices with ever-improving colour gamuts, that is, in terms of both chromaticity and contrast range.

As a result of these advances, a different philosophical approach may be adopted in the ideal system configuration, with the critical differences being:

- the adoption of an optimum set of *system* primaries rather than those based upon a particular set of *display* primaries;
- the removal of the ambiguity between the requirements for gamma correction and perceptible uniform coding by adopting a symmetrical pair of codecs dedicated to perceptible uniform coding;
- the elimination of display gamma correction;
- the provision of a very simple constant luminance system, which falls into place as a consequence of the above approaches.

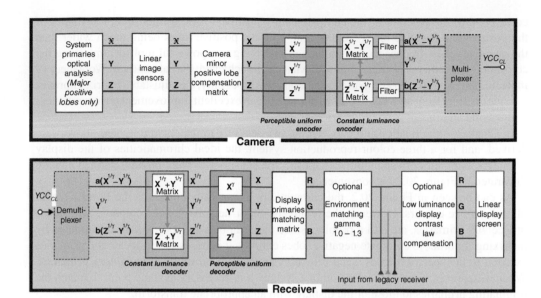

Figure 20.1 Configuration of an ideal colour television system.

The configuration which results from this approach is detailed in Figure 20.1, which illustrates the relevant elements of the overall signal path of the system from the light-splitting elements of the camera to the displayed image in the receiver.

The shaded processors have complementary characteristics and therefore effectively have no influence on the colour rendition of the system.

It will be noted that despite a number of significantly different approaches to the system design, the original luminance and colour difference configuration developed for the NTSC system for storage and distribution is retained.

The Optional Environmental Matching Gamma element has an adjustable gamma between 1.0 and 1.3 to match the overall system gamma, which is linear up to this point in the signal chain, to the viewing environment as described in Section 13.7. Thus, for a relatively small screen in a bright environment, the value of gamma might be adjusted to 1.2, whilst in a home theatre situation it would likely approach a value of unity.

The Optional Low Luminance Display Contrast Law Compensation element is a processor which only applies to those particular displays with limited highlight luminance, which may benefit from some adjustment of the contrast law in the darker areas of the picture, as described in the next section, where some of the critical changes are described in more detail.

20.2.2   The System Chromaticity Gamut

The practical chromaticity gamut of a system is primarily dependent upon two factors:

- The gamut defined by the chromaticities of the display primaries
- The effective camera spectral sensitivities

As we saw in the previous chapter, although compensation for the lack of negative lobes in the ideal spectral sensitivities of real primaries considerably reduces errors, it is by no means perfect. Unfortunately, the locations of the negative lobes along the wavelength axis do not correspond to the position of the complementary positive lobes; furthermore, the shapes of the positive and negative lobes are different, with the positive lobes being much wider than the negative lobes. In consequence, although matrixing is a powerful tool for improving the colorimetry of cameras, there are limits to what can be achieved; the calculator in Worksheet 19 indicates that with Rec 709 primaries, major positive lobes only, the adjustable matrix improves the colour errors from an ΔE^*_{00} of 8.12 to one of 3.02, an excellent improvement but nevertheless still leaving significant perceptible errors.

In Chapter 9, the relationship between the two sets of parameters highlighted above were derived, and for a three-colour reproduction system, the ideal chromaticities of the display primaries and their corresponding camera spectral sensitivities were broadly defined and are illustrated in Figure 20.2 for convenience.

As can be seen in Figure 20.2, the chromaticity gamut fully embraces the Pointer gamut of real surface colours. However, the relatively minor negative lobes compared with those of Rec 709 primaries, are still a problem since they can be only approximately emulated by matrixing. In order to eliminate negative lobes entirely, it is necessary to locate the primaries external to the spectrum locus; such primaries are often referred to as *imaginary* primaries. When different primaries are used within a system, it is necessary at some point to match the originating primaries to those of the display with an appropriate transform.

In Section 12.2, the technique for manipulating the colour spaces of chromaticity gamuts was outlined, and in Section 12.5, an ideal system was defined using a combination of imaginary *system* primaries and display primaries which embraced the Pointer surface colours.

One set of these imaginary system primaries sets is illustrated in Figure 20.3(a) together with the ideal display primaries for comparison. In Figure 20.3(b) is the resulting idealised set of system spectral sensitivities with no negative lobes. Other external sets of primaries may be envisaged as illustrated in Chapter 12, including a set which eliminates the minor red-positive lobe altogether and also the XYZ primaries set, which in this context may be seen as just another set of imaginary RGB primaries. However, the XYZ primaries have the added advantage that, since the Y spectral sensitivity is identical to the luminous efficiency function of the eye, it may be used directly in the downstream coding as the luminance signal without the need for it to be obtained by deriving appropriate portions of the RGB signals.

There are a number of other advantages of such a system; these system spectral sensitivities may be adopted directly in the design of the camera spectral sensitivities where, with no negative lobes to take into account, a much closer emulation of the ideal characteristics will be achieved, possibly so close that a correcting matrix will not be necessary. Furthermore, the signal at the receiver carries the correct information for *all* colours in a scene, and with the aid of a simple display matching matrix, as described in Chapter 12, the signal may be processed to *accurately* display *all* the colours in the scene which fall within the display chromaticity gamut, irrespective of which particular display chromaticities these are. It is only at the output of the display matrix that negative signals will occur for those scene colours which are outside the display gamut and these will be clipped by the display.

As we saw in the previous chapter, the mismatch between the matrix-corrected practical spectral sensitivities of the camera and the ideal characteristics, produced significant colour errors. By eliminating this source of error, the system is able to serve a mixed population of receivers with different display primaries chromaticities without compromise, always reproducing colours within the various display gamuts with no chromaticity errors.

Which particular set of imaginary primaries is selected for the system gamut depends on the selection criteria; the system primaries illustrated in Figure 20.3 are relatively very efficient in

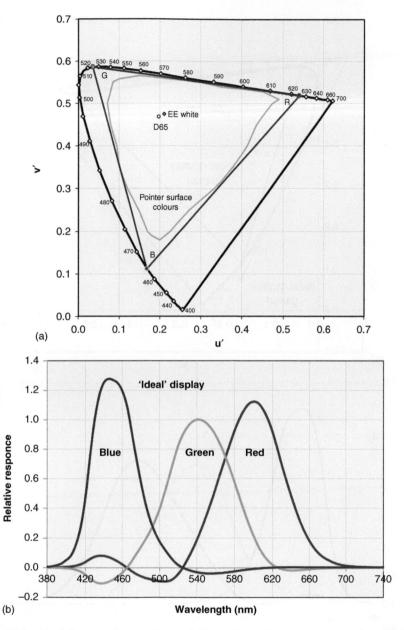

Figure 20.2 Ideal display primaries chromaticities and matching camera spectral sensitivities.

code utilisation, whilst the primaries set referred to as 'Ideal Camera 2' in Chapter 12 produces no secondary positive lobe in the red characteristic to be emulated but is less efficient both in the utilisation of code values and in the sharing of available light. The CIE XYZ primaries set are even less efficient in the use of code values,[1] as may be reasoned from an inspection

---

[1] Later work showed that coding efficiency is not a significant factor.

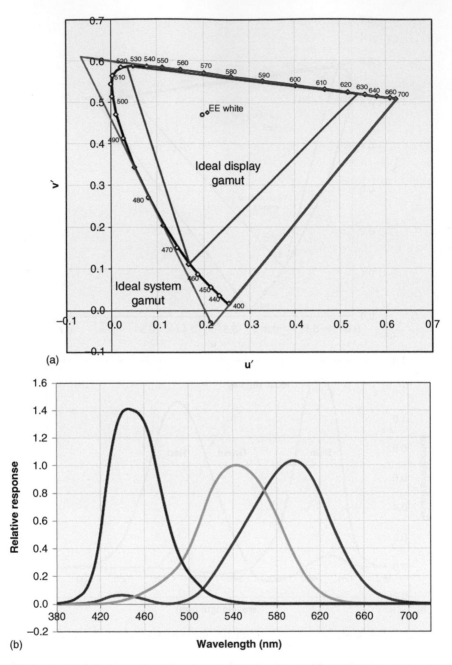

(a)

(b)

Figure 20.3   An ideal display and system primaries design with matching system spectral sensitivities.

of Figure 12.8, but produce directly, without the requirement for matrixing, a true luminance signal. Such an approach ensures optimum resolution with less dependence on the accuracy of alignment of the image sensors in separate sensor cameras; also since it is a true luminance signal rather than a luma signal, it may be used directly in a constant luminance system without further processing.

Some of the improved sensitivity of the image sensors may be utilised by sharing the light in the upper band of the visible spectrum evenly between the red (X) and green (Y) sensors, which will enable a more accurate match of the camera responses to the positive lobes of the ideal spectral sensitivities than can be achieved by the compromise which occurs when splitting the spectrum in the manner associated with traditional three sensor cameras.

In what follows, using the mathematical model incorporated into Worksheet 19 relating to calculating colour difference errors of the ColorChecker chart at the rendered image, it is assumed that a good physical match of the camera responses to the ideal principal positive lobes has been achieved.

The 'Ideal Camera 1' with matrix and the 'Ideal Camera 2' with no secondary positive lobes described in Chapter 12 produce very acceptable ΔE_{00}^* values of 0.24 and 0.26, respectively. However, when using the XYZ primaries, even the relatively extended secondary positive lobe of the X colour matching function can be very well simulated with the simple matrix detailed in Table 20.1:

Table 20.1 Matrix giving XYZ values from major positive lobes of the XYZ CMFs

	$\mathbf{\mathfrak{X}}$	Y	Z
R = X	0.830	0.000	0.170
G = Y	0.000	1.000	0.000
B = Z	0.000	0.000	1.000

In Figure 20.4, it is assumed that the native camera spectral sensitivities are the major positive lobes of the XYZ CMFs. The ideal camera spectral sensitivities are shown with dotted lines and the practical responses after matrixing are shown with continuous lines. (The actual

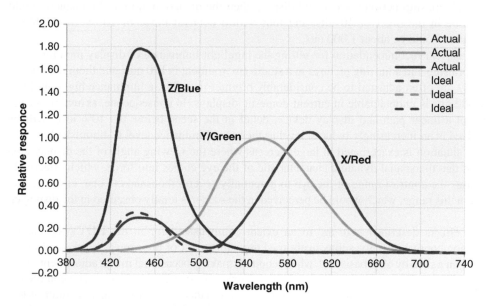

Figure 20.4 XYZ camera spectral sensitivities.

and ideal responses are identical for the green and blue responses.) The old English $\mathfrak{X}$ character appearing in the matrix and in the configuration of Figure 20.1 indicates that the response does not yet relate to a true CIE X response. Using this matrix produces a worst ΔE^*_{00} value of 0.91 on the ColorChecker samples, a barely noticeable difference to that from the ideal responses illustrated in Chapter 12 and a three to one improvement on the best achieved by the HDTV system. Since these responses have the added advantage of producing a Y signal without matrixing, they will be adopted for the ideal system as the best compromise of the three imaginary primaries sets considered.

20.2.3   Contrast Range

20.2.3.1   Limited Portrayal of Contrast Range

The understanding of the treatment of contrast range in Chapter 13 is pertinent to the rationale which follows in this section.

In Section 13.3, it was shown that in a practical television-viewing environment, the perceived contrast ratio of the displayed reproduction is dependent upon:

- the display contrast ranges, both sequential and simultaneous;
- the contrast range of the eye, both spatial static and spatial dynamic;
- the luminance range over which the eye adapts to conform to these contrast ranges.

It was noted that the spatial static contrast ratio of the eye varied from about 40:1 to about 400:1 depending upon the sensitivity of the eye to the pattern of the low level changes in luminance being perceived, with a value of 100:1 being the accepted average over the range of luminances to which the eye is capable of adaptation. This range starts to be compromised at a luminance level below that of about 10 nits, where the sensitivity of the eye to changes begins to reduce. Thus if the luminances ranges in the scene are to be satisfactorily reproduced in a linear relationship between scene and display, then the display minimum luminance should ideally be in the order of 10 nits, and thus with a contrast range of 100:1, the highlight luminance would be about 1,000 nits.

The current recommendation for setting the highlight luminance of display monitors used for appraising and adjusting pictures in a studio environment is 100 nits, and though domestic displays are often adjusted to be considerably brighter than this, a luminance highlight level of 1,000 nits is unobtainable in current domestic displays. In consequence, assuming for the moment a linear path and display device, detail in the scene below the 10% level will be perceived at an increasingly reduced level as the scene luminance level is diminished.

The situation is exacerbated in larger screens where the viewing angle of the display area is such that the spatial dynamic characteristic of the eye comes into force, which effectively increases the contrast range of the eye to potentially several thousands to one, ensuring that detail in this range, which would be perceived in the scene, is totally obscured in the displayed image.

The effects described here are to an extent compensated for in the HDTV system by the vision control operator subjectively tracking adjustment of the camera black level or lift control on a shot-by-shot basis and, when required, making occasional minor adjustment of the camera gamma correction characteristic. In this example system where there is a completely linear relationship between the scene and display luminance, it is for the manufacturer of low luminance highlight displays to determine whether to incorporate some low luminance level compensation, as indicated by the penultimate element in the system configuration diagram of Figure 20.1. As the highlight luminance level of displays increases (see High Dynamic Range later in this section), there will be an ever diminishing requirement for this form of compensation.

20.2.3.2 Distorted Portrayal of Contrast Range

The gamma correction regime in HDTV systems uses parameter values which fail to adequately protect the contrast range of the scene, even assuming that the luminance range of the display was such as to overcome the problem described above. The Rec 709 parameter values of 0.45 for the exponent and 4.5 for the linear gain element produce a source characteristic which when combined with the power law display characteristic, which has an exponent of 2.4, will provide, for a scene luminance of 1%, a display luminance of only 0.06%, roughly a factor of 17 times in error. The errors for lower levels of scene luminance are progressively worse. However, the non-linear response of the eye partially compensates for this error, giving an error in terms of lightness at this luminance level of about 2.5 to 1. These figures are taken from the table in Worksheet 13(b), where they are borne out by the increase in ΔE_{00}^* value from 3.0 for a linear system to 9.4 with the gamma correction regime in place, as illustrated in Section 19.2.

Interestingly, other media use gamma correction parameters which produce lower errors in the darker tone ranges of the scene. The sRGB and Adobe RGB gamma law parameters used in photography both use the same exponent value of 0.4167, whilst the linear gain element has values of 12.92 and in excess of 30 for the sRGB and Adobe RGB parameters, respectively. These parameter values produce display luminance errors at 1% scene luminance of 2.5 times and 2.0 times, respectively, and perceived lightness of 1.36 times and 1.27 times, respectively, a considerable improvement over the error figures produced by the Rec 709 gamma correction routine.

Thus, in the event that the same gamma correction regime were to be retained, the adoption of lower values of correction exponent and higher levels of linear gain would ensure very much lower levels of distortion at the limits of the spatial static contrast ratio of the eye.

20.2.3.3 Avoiding the Perils of Legacy Gamma Correction Regimes

The use of gamma correction is, as has been noted often previously, directly related to providing correction at source for the transfer characteristic of the legacy CRT. At the time the HDTV specification was agreed, the CRT was still very much in evidence, albeit it was clear at that time that flat-screen displays of notionally linear transfer characteristics would soon replace the CRT. However, at the current time the linear display is the *de facto* standard, and as we have seen, the use of a simulated gamma element prior to the display to emulate a CRT characteristic causes significant distortion of the displayed colours.

Nevertheless, as described in Chapter 13, there is a requirement for a set of complementary circuits with appropriate power law characteristics between the source and the destination to mask the effects of quantisation. The current situation is unsatisfactory, but has continued to be used because in legacy terms, the traditional gamma correction also acted serendipitously as a perceptible uniform coding scheme. The situation has been exacerbated by the difficulty of accurately emulating the CRT characteristic, which with this approach was essential for linear devices replacing traditional monitors as the reference monitor for picture adjustment and matching.

In order to avoid the legacy of the ambiguous gamma correction regime used to date, the time has come to define a Perceptible Uniform Coding (PUC) regime separate from gamma correction and address any gamma correction requirements independently.

Thus, apart from the PUC complementary characteristic elements in the camera and display device, the ideal system is linear throughout, avoiding the legacy problems and simplifying the system.

20.2.3.4   High Dynamic Range

In recognition that, particularly for reproduced images of small fields of view, a linear relationship between the contrast range in the scene and the contrast range of the image does not result in a satisfying image, work in recent years, particularly in the field of photography, has explored means of portraying scenes with contrast ranges which exceed the static dynamic contrast range of the eye.

Taking an original scene of high contrast, where the spatial dynamic range characteristics of the eye is able to operate unhindered and an image of limited field of view, where the small size of the image dictates against the spatial dynamic contrast range of the eye being active, then in essence, the problem is to emulate the perceived contrast of the original scene in the limited contrast range of the image.

There are two solutions to this problem. The first, which is the solution adopted in photography, is to capture the high dynamic scene contrast in a number of shots which cover a range of exposures and to select different bands of tones from each exposure which best capture the required contrast detail and combine them into a single picture (Reinhard et al., 2006). Historically, such an approach was beyond the capabilities of the real-time portrayal of television; however, with much improved sensor sensitivity and the option of exposure times which are a fraction of the standard television frame time, engineers are now experimenting with these techniques in television systems.

Rather than addressing the problems of a limited field of view and a limited image luminance, the second solution is to avoid them by adopting a trend to brighter and larger screens.

First, it is necessary to overcome the compression due to the luminance range of the display overlapping the range where the $\Delta L/L$ contrast range of the eye is limited by the human visual modulation threshold, that is, the static dynamic range of the eye is not fully exercised. This can only be achieved by increasing the highlight luminance of the display, as noted earlier. The second approach is to increase the field of view of the display, either by reducing the viewing distance or increasing the size of the display; above a minimum critical viewing angle, the spatial dynamic contrast range characteristic of the eye will begin to come into play, becoming more effective as the field of view approaches that of the eye itself. Thirdly, the eye should be able to accommodate to the average screen luminance, and to ensure this occurs, the average ambient luminance of surfaces, particularly those close to the display, should be below that of the display.

The current trend towards larger and brighter television screens indicates that by the 2020s we shall be at least part way to satisfying these requirements.

20.2.4   Accommodating a Constant Luminance System

The poor compatibility of non-constant luminance systems with monochrome systems has been overcome as the vast majority of the colour system signals are viewed on a colour display where the luminance values are properly portrayed.

The loss of fine detail in those elements of scenes containing highly saturated colours, where much of the luminance detail is carried by the limited bandwidth colour difference signals, is however perhaps one of the most serious impairments when critically viewing an image containing saturated colour detail, such as flowers and costumes, and therefore it seems reasonable that whenever the opportunity arises to remedy the situation, it should be taken.

These opportunities occur only very rarely. Clearly it is not practical to introduce such a change into an established system since chaos would reign as the modified signals were displayed on legacy equipment. Furthermore, whilst the display technology which led to the introduction of non-constant luminance systems remained, such as the CRT, there was no incentive to make a change. A further factor was the cost of the gamma correction circuits in the display device, but clearly since the advent of integrated circuits the cost of such elements can now be factored into the processing chips and thus is no longer a challenge.

In consequence, with the proposals for a completely new system, together with a new population of receivers, there would no longer be any valid reason for not adopting a constant luminance system, particularly since with the other strategies adopted above such a system occurs naturally and simply.

20.2.5   Matching the System Primaries to the Display Primaries

In Figure 20.1, the 'Display Primaries Matching Matrix' matches the system primaries to the display primaries.

In Section 9.2, a set of ideal display primaries were defined, ideal in as much that they fully encompassed the gamut of Pointer surface colours as illustrated in Figure 9.5, and cited earlier. The chromaticity coordinates of these primaries are listed in Table 20.2, and the matrix coefficients required to match the XYZ signals to these display primaries are listed in Table 20.3.

Table 20.2   Display primaries chromaticity coordinates

	x	y	u	v
Red	0.7007	0.2993	0.5400	0.5190
Green	0.1142	0.8262	0.0360	0.5861
Blue	0.1355	0.0399	0.1690	0.1119
White D65	0.3127	0.3290	0.1978	0.4683

Table 20.3   Matrix coefficients for matching system primaries to display primaries

	R_{CAM}	G_{CAM}	B_{CAM}
R_D	1.4590	−0.1946	−0.2645
G_D	−0.6572	1.6187	0.0384
B_D	0.0299	−0.0736	1.0437

The display matrix coefficients for a wide range of system primaries and display primaries are calculated automatically in the 'Display' panel of Worksheet 19 on selection of the appropriate set of display primaries and may be used to demonstrate the freedom available to manufacturers to supply displays with a range of primaries to suit the application.

20.2.6   Accommodating Legacy Services

Although a new 'ideal' service of the type described in this section will not be available to the population of legacy receivers currently in use, it is essential that the new receivers are able to display the legacy services, since in the early days of the new service, programme content produced in accordance with the new specification is likely to be in short supply.

The necessary elements of the legacy receiver illustrated in Figure 19.5 can, in colour specification terms, be made to fit comfortably into the ideal receiver illustrated in Figure 20.1 by the addition of the elements illustrated in Figure 20.5. (There would also be a requirement to up-sample the legacy signals to match the pixel geometry of the new service.) The connection between the two receivers is illustrated in both diagrams.

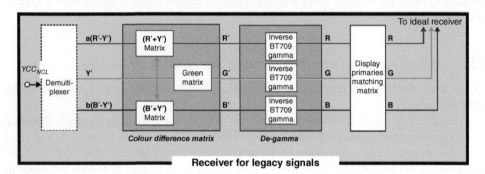

Figure 20.5   Configuration of a receiver for integrating legacy signals into the ideal receiver of Figure 20.1.

In comparing the configuration of this legacy receiver with current receivers, as represented in Figure 19.5, it will be noted that there is considerable simplification since, following the de-gamma circuits to provide linear signals for the Display Matching Matrix, there is no requirement to simulate a CRT display by inserting a power law characteristic element and thus no requirement to provide a system gamma-correcting component to proceed it.

This arrangement is potentially capable of improved colour reproduction since the mismatch of gamma law elements inherent in the current system, which was described in Chapter 19, has been removed. In consequence, the ΔE^* errors will be reduced from 8.7 to 3.0 on legacy material. However, there could be residual problems since the critical subjective camera shot adjustment carried out at the time the programme was made would have been undertaken with the mismatched gamma regime in place. Thus, it maybe that the operational black level or lift control (see Section 20.3) would have been adjusted to compensate for the mismatch of the gamma regime. If it is found that this is the case, then those two elements which were eliminated will have to be restored, that is, the System Gamma and Display Gamma elements illustrated in Figure 19.5.

Assuming that the display primaries have the chromaticity coordinates listed in Table 20.2, then from Worksheet 19, the coefficients of the Display Primaries Matching Matrix to convert the legacy Rec 709 primaries to the ideal primaries will be as shown in Table 20.4.

Table 20.4 Matrix coefficients for
converting from Rec 709 primaries to
'ideal' display primaries

	R_{CAM}	G_{CAM}	B_{CAM}
R_D	0.5870	0.3808	0.0321
G_D	0.0597	0.9146	0.0256
B_D	0.0158	0.0729	0.9113

20.3 Acknowledging the Requirement to Expand the Colour Gamut

Much of the material appearing in the last section on the 'Ideal' system specification has been accepted knowledge from the very early days of colour television, and since the technology to support such an approach has been available since the late 1980s, there have been proposals put forward for its adoption in one form or another by a number of those groups responsible for developing new system specifications.

This section first describes the Eureka proposal and then goes on to describe two subsequent international specifications for complementing the Rec 709 specification which enables the limitations caused by the failure to adopt this proposal by the body which formulated Rec 709 to be ameliorated.

20.3.1 The Eureka Proposal

In the late 1980s the television manufacturing companies and broadcasters from around the world were preoccupied with experimental work on HDTV and the major international groups which represented them were evolving specification proposals for submitting to the ITU as the basis for drawing together a specification for HDTV which would become a world standard.

One such group was the Eureka HDTV EU95 organisation, which included all the major manufactures and broadcasters of the nations of Europe, who had in the preceding years organised a number of projects to investigate the requirements for HDTV and had evolved a far-reaching set of proposals for the parameters and their values to be considered for adoption by the ITU in the emerging world standard specification.

This group had recognised that in the timing of these proposals, unusually, a number of conditions were right such that in the new specification there was the opportunity to not only set the 'high definition' parameters of HDTV but also to review the limitations of the colour specifications of the then current systems. These conditions were:

- It was clear that linear display devices would in the next decade supersede the CRT as the principle domestic display device.
- Gamma and matrix correction circuits could now be absorbed into the general electronics of the receiver with little or no cost premium.
- A new system, which would be fundamentally incompatible with the current system and which would require new source equipment such as cameras, recorders, infrastructure items and new displays for the public, was to be specified and thus the drag of compatibility with legacy systems could be largely put aside.

Thus, the Eureka proposals to the ITU which were submitted in 1988 included not only parameters for the high definition elements of the specification but also advocated new colorimetric standards for a wider chromaticity gamut and the adoption of a constant luminance system.

The chromaticities of the Eureka primaries are listed in Table 20.5.

Table 20.5   Chromaticity coordinates of the Eureka primaries

Eureka primaries	x	y	z	u'	v'
Red	0.6915	0.3083	0.0002	0.5203	0.5219
Green	0.0000	1.0000	0.0000	0.0000	0.6000
Blue	0.1440	0.0296	0.8264	0.1878	0.0869
White D65	0.3127	0.3290	0.3583	0.1978	0.4683

The colour gamut and system camera spectral sensitivities which originate from these primaries are derived in Worksheet 20 and illustrated in Figures 20.6 and 20.7, respectively.

The red and blue primaries are located on the spectrum locus nearer the spectrum ends than the SMPTE and EBU primaries, which were then in place, and the green primary is located just outside the extremity of the locus, making it an imaginary primary. The critical characteristic of this proposal however is that the new chromaticity gamut encompasses all of the Pointer surface colours, providing the potential for the system to accurately render all these colours on displays with appropriate primaries. Also most importantly, the concept of *system* primaries rather than primaries based upon a specific set of *display* primaries had been embraced for the first time, ensuring that the system would be fundamentally capable of supporting all displays irrespective of the chromaticity of their primaries, by the incorporation of a suitable matching matrix located in the display device.

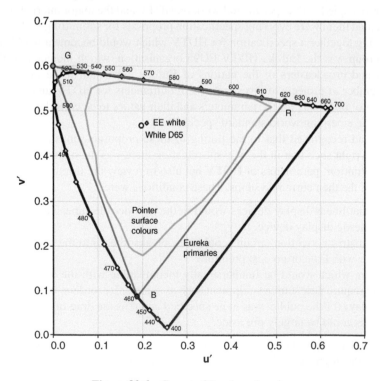

Figure 20.6   Gamut of Eureka primaries.

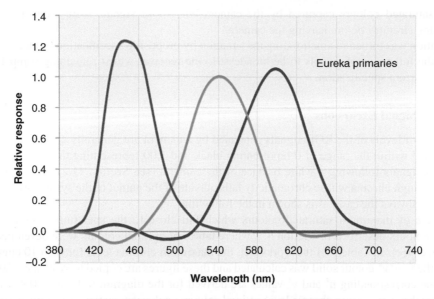

Figure 20.7 Eureka primaries spectral sensitivities.

The small secondary lobes of the camera spectral sensitivities is evidence of the wide gamut of the primaries, and although camera matrixing will be required to emulate these lobes, their small size means that errors resulting from any mismatch will be less perceptible than those derived from traditional characteristics. Nevertheless, no surface colour would be capable of generating signals which exceeded the 0% to 100% limits of the transmission system which, as we shall see, was in contrast to the then current system.

Since the green primary is located at the same position as the CIE Y primary, it might be assumed that the green spectral sensitivity characteristic would match the $\bar{y}$ colour matching function, that is, the luminance efficiency response of the eye, which would have been very convenient for coding the signals for transmission and particularly for a constant luminance system, as we have seen. However, it should be remembered from Section 4.4.4 that the shape of a particular CMF is dependent upon the chromaticities of the other two primaries in the trio rather than its own chromaticity.

The advantages of constant luminance systems had been advocated since the 1950s and four image sensor cameras, with one sensor having a spectral sensitivity matching the luminous efficiency response of the eye and thus producing a linear luminance signal directly, had been in common use from the 1960s. It was an obvious step therefore to include a constant luminance system as part of the Eureka proposals.

Unfortunately, there appears to have been insufficient awareness of the advantages of these approaches at the international ITU level, where the emphasis was placed on coming to agreement on the parameters for the high definition aspects of the specification, and for reasons which have eluded the researches of the author, the proposal was not adopted.

20.3.2 Accommodating Signal Excursions Which Fall Outside of the Rec 709 Specification

As a result of adopting camera primaries of limited chromaticity gamut, it eventually became clear to a broader consensus that the Rec 709 specification was limiting the ability to display highly saturated colours captured by the camera but whose signals were clipped by the processing circuitry before leaving the camera.

Attention was turned to ameliorating the situation by proposing specifications which would enable the full range of signals to be broadcast to the receiver whilst remaining compatible with the current specification.

20.3.2.1   Signal Excursions

The range of levels of the RGB signals generated by a camera are generally considered to be constrained within the range of 0 representing black and 1.00 representing the peak level of the RGB signals following a white balance of the camera. (see Section 11.2). However, for colours of high chroma whose chromaticity falls outside of the gamut of the system primaries, this is not always the case, as is shown in the following.

A range of maximum saturated colours which fall close to the spectrum locus are the *optimal* colours described in Section 4.7, where tables of optimal colours were developed in Worksheet 4(e). In Table 3 of the worksheet, the maximum chroma level for each 10 degrees around the $L^*u^*v^*$ colour solid was calculated and these figures are copied to Worksheet 20(b), where the corresponding u′ and v′ values are derived for the diagram in Figure 20.8, which illustrates these maximum chroma level optimal colours on the chromaticity diagram, together with the colour gamuts of the Eureka and Rec 709 systems.

The effect of a restricted chromaticity gamut on the levels of the RGB signals prior to processing is clearly shown in Figure 20.9, which illustrates the signal levels against the

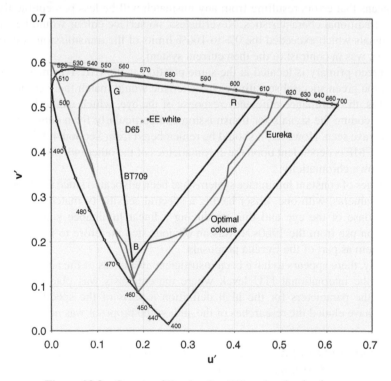

Figure 20.8   Gamuts of Eureka, Rec 709 and optimal colours.

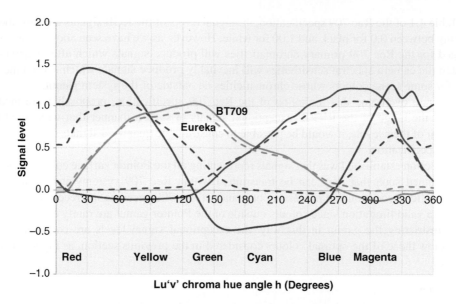

Figure 20.9   RGB signal levels for optimal colours before clipping.

chroma phase of the $L^*u^*v^*$ colour solid for maximum chroma optimal colours at every 10 degrees around the colour circle. The signal levels from the Eureka camera very nearly constrained within the 0–1.00 range, whilst those from the Rec 709 camera extend both below zero and above peak white by some 50%. The calculations to support these diagrams are contained in Worksheet 20(b).

Since the Rec 709 system is incapable of accommodating these *exceptional* signals, they are clipped at the 0.0 and 1.00 levels before encoding and are thus not available at the receiver. This is unfortunate as the signals contain all the information necessary to drive any extended gamut display, and with suitable matrixing, the full range of colours within the display gamut would be accurately rendered.

In the late 1990s it became more generally clear that the Rec 709 specification was limiting the capability of extended gamut displays to exploit those colours captured by the camera which were within an extended display gamut but which were being clipped before transmission. Consideration was therefore given to establishing a method of transmitting the exceptional level signals as a means of ameliorating the limited system gamut of Rec 709. The first specification to address these issues was ITU-R BT.1361 in 1998, which was followed by IEC 61966-2-4 in 2006 and finally ITU-R BT.2250 in 2012.

20.3.2.2   The ITU-R BT.1361 Specification

In the foreword to ITU-R BT.1361[2] 'Worldwide unified colorimetry and related characteristics of future television and imaging systems', the reasoning for considering the requirement for extending the limited gamut of the Rec 709 system is listed in some detail, though it is arguable, as its title suggests, whether it is appropriate for future systems.

---

[2] http://www.itu.int/dms_pubrec/itu-r/rec/bt/R-REC-BT.1361-0-199802-I!!PDF-E.pdf

In Table 4.1 of the Rec 709 specification, the signal levels of the RGB signals are described as falling between 0.0 for black and 1.00 for white. However, as we have seen above, a camera designed for the Rec 709 primary chromaticities will produce signals which after matrixing to match the camera spectral sensitivities will inevitably produce signals which exceed these limits for some scene colours whose chromaticities lie outside of the system gamut.

In order to overcome this limitation of the Rec 709 specification, proposals were made to restrict the range of the R'G'B' exceptional signals in such a manner that with suitable adjustment of their levels, it would be possible to constrain the quantised signals to within the Rec 709 specification.

The criterion established was that signals representing all the Pointer surface colours should be constrained such that they can be transmitted within Rec 709 systems and would not adversely affect reception and processing on standard Rec 709 specification receivers.

This is a valid limitation since colours outside of the Pointer gamut are rarely experienced, and by restricting the gamut in this way, the exceptional signal levels are constrained to levels below those of the optimal colours considered in the previous section, as Figure 20.10 illustrates.

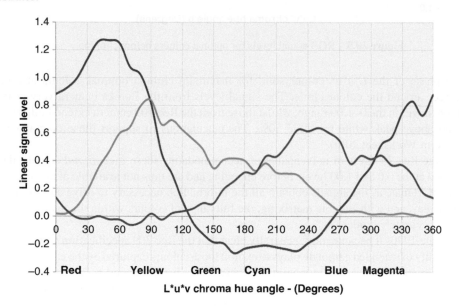

Figure 20.10   Rec 709 RGB signal levels for maximum chroma Pointer surface colours.

The maximum chroma value of the Pointer colour for each 10 degrees around the 360 hue angle was copied into Worksheet 20(b) from the 'Surfaces' worksheet and the corresponding RGB values calculated for the Rec 709 primaries. It can be seen that there is a considerable reduction in the exceptional signal levels; nevertheless, the red signal in particular exceeds the white value by 27% and extends negatively by about 25%.

The gamma correction regime of Rec 709 is inappropriate for these exceptional level signals since its characteristic is unable to respond to signals outside of the specified level range; in consequence, it was necessary to introduce a new gamma correction law. The approach was to use the same basic parameter values used for Rec 709 but to extend the range beyond those limits. The increasingly lower gain of the gamma law characteristic for the higher level signals assists the problem of limiting the exceptional level signals above 100%, but at the other end of the characteristic, where low level signals are negative in value, the high gain would extend the negative signals further. The solution to the problem was to introduce an attenuation of 4:1 to the negative signals within the correction process and provide a compensating gain in the inverse characteristic applied at the receiver.

The relevant tables from the Rec 1361 specification are reproduced as follows.

Rec 1361 TABLE 1 Colorimetric parameters and related characteristics

Parameter		Values		
1	Primary colours		Chromaticity coordinates (CIE, 1931)	
			x	y
		Red	0.640	0.330
		Green	0.300	0.600
		Blue	0.150	0.060
2	Reference white		Chromaticity coordinates (CIE, 1931)	
	(equal primary signal)	D65	x	y
			0.3127	0.3290
3	Opto-electronic transfer characteristics[1]	$E' = 1.099\,L^{0.45} - 0.099$ for $0.018 \leq L < 1.33$ $E' = 4.50\,L$ for $-0.0045 \leq L < 0.018$ $E' = -\{1.099(-4\,L)^{0.45} - 0.099\}/4$ for $-0.25 \leq L < -0.0045$ where L is a voltage normalized by the reference white level and proportional to the implicit light intensity that would be detected with a reference camera colour channel; E' is the resulting non-linear primary signal.		

[1]The non-linear pre-correction of the signal region below $L = 0$ and above $L = 1$ is applied only for systems using an extended colour gamut. Systems using a conventional colour gamut apply correction in the region between $L = 0$ and $L = 1$. A detailed explanation of the extended colour gamut system is given in Annex 1.

Rec 1361 TABLE 2 Analogue encoding equations

	Equations
Parameter	Conventional and extended colour gamut systems
4 Luminance and colour-difference equations	$E'_Y = 0.2126\,E'_R + 0.7152\,E'_G + 0.0722\,E'_B$ $E'_{CB} = \dfrac{E'_B - E'_Y}{1.8556}$ $= \dfrac{-0.2126\,E'_R - 0.7152\,E'_G + 0.9278\,E'_B}{1.8556}$ $E'_{CR} = \dfrac{E'_R - E'_Y}{1.5748}$ $= \dfrac{0.7874\,E'_R - 0.7152\,E'_G - 0.0722\,E'_B}{1.5748}$

Table 3 of the specification deals with the quantization levels of the signal and is not reproduced here.

It will be noted that apart from the OETF, the remaining parameters in Tables 1 and 2 of the specification have the same parameters and values as appeared in Rec 709.

The gamma correction transfer characteristic, or the misnamed Opto-Electronic Transfer Function (OETF), is specified in Section 3 of Table 1 of the specification, and it can be seen that the limits of the characteristic have been extended, encompassing values from −0.25 to 1.33. These figures are based upon the calculations carried out at the time for establishing the level limits of the Pointer surface colours. The corresponding figures derived in Worksheet 20(b) for the R signal are −0.26 and 1.27, respectively, indicating that the negative excursion of this exceptional signal just exceeds the Rec 1361 limit. Unfortunately, the method of calculating these figures from the lightness and chroma figures provided in Pointer's paper is not given in the specification, so it has not been possible to track down where this small exceeding of the negative limit on just one of the Pointer colours occurs.

The graph of the OETF is charted in Worksheet 20(b) and illustrated in Figure 20.11.

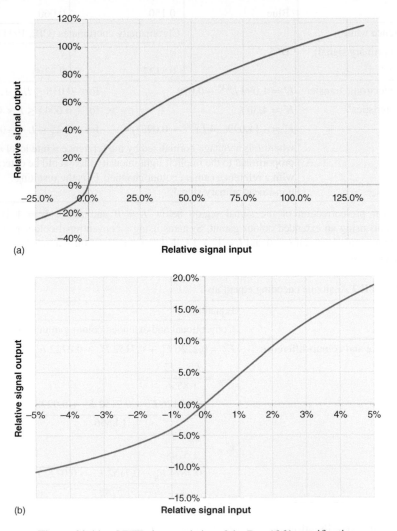

Figure 20.11   OETF characteristics of the Rec 1361 specification.

Figure 20(b) illustrates the smooth straight line section of the characteristic for low levels of positive and negative signals. When the linear signals from the camera matrix are applied to the gamma corrector, the resulting limits of the output levels for the Pointer surface colours are illustrated in Figure 20.12.

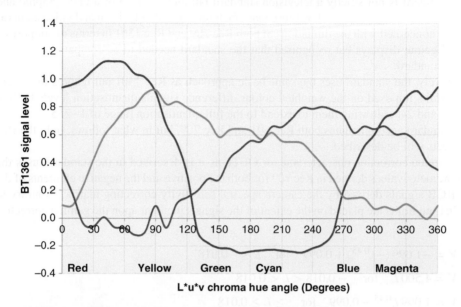

Figure 20.12 Level limits of the Rec 1361 gamma-corrected signals of the Pointer surface colours.

The formula for the OETF in the specification does not specify the output levels for input levels which exceed the limits for the exceptional levels; this appears to be an oversight since there are sources of light which do transcend the levels of the Pointer surface colours. In consequence, the formula based upon this characteristic in Worksheet 20(b) is modified to clip the output at the specified levels of −0.25 and 1.1505. The effect of this minor clipping of the red signal is not noticeable in Figure 20.12.

As explained in Rec 1361, the coding of the exceptional signals into luma and colour difference signals does not cause these encoded signals to exceed the normal dynamic ranges of 0–100% and +50% to −50%, respectively. However, for the digital signals, it is necessary to use different scaling factors in order to constrain the signals within the quantization signal level limits of Rec 709, at code levels of 16 and 235 for the luma signal and 16 and 240 for the weighted colour difference signals. In Worksheet 20(b), the signals derived from the Pointer colours are applied to the quantization process and the weighted colour difference signals are shown to be constrained to within the code level range 29–224.

The result of using these different scaling factors to squeeze the larger range of signals into the Rec 709 format is to limit the range of codes available for signals in the 0–100% range and therefore increase the risk that contouring of the image will be perceived on critical images using 8-bit quantization.

As far as the author is aware, the recommendations contained in Rec 1361 have never been adopted for use in the broadcasting of television signals.

20.3.2.3   The IEC 61966-2-4 Standard

The IEC 61966-2-4 Standard is entitled 'Multimedia systems and equipment – Colour measurement and management – Part 2-x: Colour management – Extended-gamut YCC colour space for video applications – xvYCC'.

This standard is not strictly a television standard but an extension of a photographic and videography standard developed by those manufacturers of both television and video cameras who were concerned with the limitations of both Rec 709 and Rec 1361 in terms of supporting extended gamut displays but recognised that the standard needed to be compatible with the Rec 709 standard.

Effectively, the standard uses the same basic approach as Rec 1361 but ignores the safety margin limits imposed on the weighted colour difference signals quantisation levels by Rec 709 at 16 and 240, allowing them to extend to the full quantisation range of 1–255.

The standard accommodates both Rec 601 and Rec 709, but in what follows, only the Rec 709 aspects will be described.

The gamma correction characteristic, or OETF as it is described in the standard, uses the same parameter values defined in Rec 709 for both the positive and the negative elements of the linear RGB signals derived by the camera spectral sensitivity correcting matrix. Contrary to Rec 1361, no limits are placed on the extent of the signals incident upon the gamma corrector:

$$V = -1.099\,(-L^{0.45}) + 0.099 \quad \text{for} \quad L \le -0.018$$

$$V = 4.500\,L \quad \text{for} \quad -0.018 < L < 0.018$$

$$V = 1.099\,L^{0.45} - 0.099 \quad \text{for} \quad \ge L \ge 0.018$$

where:

L: Level of the R, G and B components of the image

V: Corresponding RGB electrical signals

These parameters lead to the characteristic illustrated in Figure 20.13 and the corresponding R′G′B′ signals for the Pointer colours in Figure 20.14.

It will be noted that the positive excursions of the R′G′B′ signals are identical to those illustrated for Rec 1361 in Figure 20.12; however, because no attenuation factor is used for the negative signals as it was in Rec 1361, these signals extend considerably further into the negative domain.

As illustrated in Worksheet 20(b), when these signals are encoded and quantised using the same parameters used in Rec 709, the negative-going weighted colour difference signals extend over the full range of quantisation levels of the digital system. Thus, in contrast to the Rec 1361 specification, the exceptional signals of the Pointer colours have been accommodated without compromising the range of quantisation level used by those signals which fall within the 0–100% signal range.

Since this standard was introduced, a number of extended chromaticity gamut display systems have become available which it is claimed are compatible with the IEC 61966 standard.

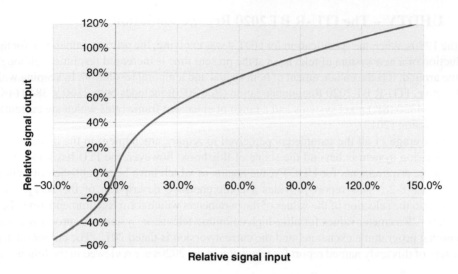

Figure 20.13   IEC 61966 gamma characteristic.

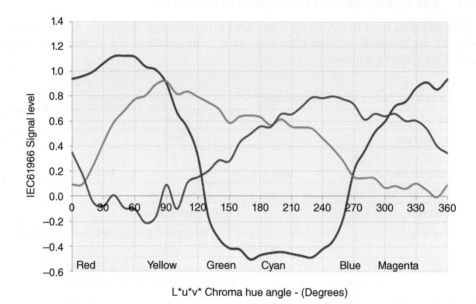

Figure 20.14   IEC 61966 R′G′B′ Pointer colours signal levels.

20.3.2.4   Report ITU-R BT.2250

In 2012 the ITU formally adopted the specifications of the IEC 61966 standard into Report ITU-R BT.2250 – 'Delivery of wide colour gamut image content through SDTV and HDTV delivery systems'.

The report lays out in terse formulaic terms the matrix equations and transfer characteristics required for each element of the signal chain from the camera to the display.

20.4   UHDTV – The ITU-R BT.2020 Recommendation

As in the 1980s, when the specification for HDTV was evolving, the primary imperative for the introduction of a new system of television at the present time is increased resolution, although this time around, it is the enhancement of both spatial and temporal resolution. In keeping with this objective, ITU-R BT.2020 Recommendation (Rec 2020) includes both 2160 × 3840 (4K) and 7680 × 4320 (8K) pixel systems and a range of enhanced frame rates which are currently under discussion (2013).

A full coverage of all the parameters perceived to require amendment in the definition of a new television system is beyond the scope of this book; however, the ITU has undertaken a study of the requirements for UHDTV, the results of which have been published in Report ITU-R BT.2246-2.[3] This report provides a comprehensive description of the background reasoning into the selection of the values of the parameters which form Recommendation ITU-R BT.2020 – 'Parameter values for ultra-high definition television systems for production and international programme exchange' and the current version is dated 2012. The colour related parameters of this aptly named report for television in the 2020s are reviewed in the following.

20.4.1   The Colour-Related Parameters of Rec 2020

The relevant tables from Rec 2020 are reproduced here. Although the picture spatial characteristics are not specifically colour related, it would be an oversight not to refer to the most critical of the system parameters.

Rec 2020 TABLE 1 Picture spatial characteristics

Parameter	Values	
Picture aspect ratio	16:9	
Pixel count Horizontal × vertical	7,680 × 4,320	3,840 × 2,160
Sampling lattice	Orthogonal	
Pixel aspect ratio	1:1 (square pixels)	
Pixel addressing	Pixel ordering in each row is from left to right, and rows are ordered from top to bottom.	

Both 3,840 × 2,160 and 7,680 × 4,320 systems of UHDTV will find their main applications for the delivery of television programming to the home, where they will provide viewers with an increased sense of 'being there' and increased sense of realness by using displays with a screen diagonal of the order of 1.5 m or more and for large screen presentations in theatres, halls and other venues such as sports venues or theme parks.

Presentation on tablet displays with extremely high resolution will also be attractive for viewers.
The 7,680 × 4,320 system will provide a more enhanced visual experience than the 3,840 × 2,160 system for a wider range of viewing environments.

An increase in the efficiency of video source coding and/or in the capacity of transmission channels, compared with those currently in use, will likely be needed to deliver such programmes by terrestrial or satellite broadcasting to the home. Research is under way to achieve this goal. The delivery of such programming will initially be possible by cable or fibre.

---

[3] http://www.itu.int/dms_pub/itu-r/opb/rep/R-REP-BT.2246-2-2012-PDF-E.pdf

Rec 2020 TABLE 3 System colorimetry

Parameter	Values			
Opto-electronic transfer characteristics before non-linear pre-correction	Assumed linear[1]			
Primary colours and reference white[2]	Chromaticity coordinates (CIE, 1931)		x	y
	Red primary (R)		0.708	0.292
	Green primary (G)		0.170	0.797
	Blue primary (B)		0.131	0.046
	Reference white (D65)		0.3127	0.3290

[1]Picture information can be linearly indicated by the tristimulus values of RGB in the range of 0–1.

[2]The colorimetric values of the picture information can be determined based on the reference RGB primaries and the reference white.

Rec 2020 TABLE 4 Signal format

Parameter	Values	
Signal format	$R'G'B'$[4]	
	Constant luminance $Y'_C C'_{BC} C'_{RC}$[5]	Non-constant luminance $Y'C'_B C'_R$[6]
Non-linear transfer function	$$E' = \begin{cases} 4.5E, & 0 \leq E < \beta \\ \alpha E^{0.45} - (\alpha - 1), & \beta \leq E \leq 1 \end{cases}$$ where E is voltage normalized by the reference white level and proportional to the implicit light intensity that would be detected with a reference camera colour channel R, G, B; E' is the resulting non-linear signal. $\alpha = 1.099$ and $\beta = 0.018$ for 10-bit system $\alpha = 1.0993$ and $\beta = 0.0181$ for 12-bit system	
Derivation of Y'_C and Y'	$Y'_C = (0.2627R + 0.6780G + 0.0593B)'$	$Y' = 0.2627R' + 0.6780G' + 0.0593B'$
Derivation of colour difference signals	$$C'_{BC} = \begin{cases} \dfrac{B' - Y'_C}{1.9404}, & -0.9702 \leq B' - Y'_C \leq 0 \\ \dfrac{B' - Y'_C}{1.5816}, & 0 < B' - Y'_C \leq 0.7908 \end{cases}$$ $$C'_{RC} = \begin{cases} \dfrac{R' - Y'_C}{1.7184}, & -0.8592 \leq R' - Y'_C \leq 0 \\ \dfrac{R' - Y'_C}{0.9936}, & 0 < R' - Y'_C \leq 0.4968 \end{cases}$$	$$C'_B = \frac{B' - Y'}{1.8814}$$ $$C'_R = \frac{R' - Y'}{1.4746}$$

[4]$R'G'B'$ may be used for programme exchange when the best-quality programme production is of primary importance.

[5]Constant luminance $Y'_C C'_{BC} C'_{RC}$ may be used when the most accurate retention of luminance information is of primary importance or where there is an expectation of improved coding efficiency for delivery (see Report ITU-R BT.2246).

[6]Conventional non-constant luminance $Y'C'_B C'_R$ may be used when use of the same operational practices as those in SDTV and HDTV environments is of primary importance through a broadcasting chain (see Report ITU-R BT.2246).

Rec 2020 TABLE 5 Digital representation

Parameters	Values		
Coded signal	R', G', B' or Y', C'_B, C'_R or Y'_C, C'_{BC}, C'_{RC}		
Sampling lattice $- R'$, G', B', Y', Y'_C	Orthogonal, line and picture repetitive co-sited		
Sampling lattice $- C'_B$, C'_R or C'_{BC}, C'_{RC}	Orthogonal, line and picture repetitive co-sited with each other. The first (top-left) sample is co-sited with the first Y' samples.		
	4:4:4 system	4:2:2 system	4:2:0 system
	Each has the same number of horizontal samples as the Y' (Y'_C) component.	Horizontally subsampled by a factor of two with respect to the Y' (Y'_C) component.	Horizontally and vertically subsampled by a factor of two with respect to the Y' (Y'_C) component.
Coding format	10 or 12 bits per component		

20.4.2   Observations on the Parameters of the ITU-R BT.2020 Recommendation

The following observations on the specifications in Rec 2020 will be in terms of a comparison between the values of the same parameters in both Rec 709 and in the ideal system described in Section 20.2.

20.4.2.1   The System Primaries and White Point

The requirement to provide suitable signals for extended chromaticity gamut display devices has been recognised by the definition of a new set of wide gamut system primaries, as illustrated in Table 20.6 and Figure 20.15.

Table 20.6   Rec 2020 system primaries chromaticities

	x	y	u′	v′
Red	0.7080	0.2920	0.5566	0.5165
Green	0.1700	0.7970	0.0556	0.5868
Blue	0.1310	0.0460	0.1593	0.1258
D65	0.3127	0.3290	0.1978	0.4683

The gamut of the Rec 2020 system primaries are illustrated by the full green line in Figure 20.15, together with the Rec 709 and 'Ideal' Display gamut derived in Section 9.2 for comparison. This is a very much improved gamut which embraces all but a few of the Pointer colours, the latter of which are however captured by the 'Ideal' gamut.

Report BT2246-1 describes at length the complex reasoning for selecting what are effectively display primaries, whilst appearing to miss the point that this reasoning would have been negated by defining an appropriate set of imaginary system primaries, which would have embraced the chromaticities of all colours.

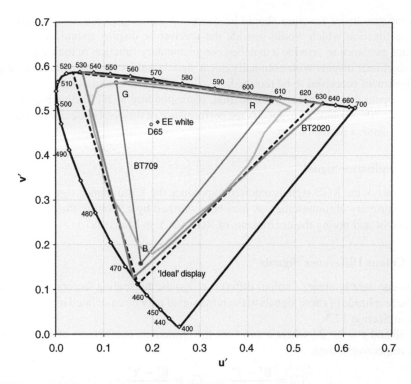

Figure 20.15 Comparison of the television primaries and the Pointer surface colours chromaticity gamuts.

It is inarguable that the Rec 2020 chromaticity gamut will embrace the majority of colours in the scene; nevertheless, the move to impose what is effectively a set of ideal *display* primary chromaticities as system primaries when with no compromise, imaginary primaries as described in Section 20.2 could have been specified is unfortunate. The adoption of imaginary primaries would have embraced *all* scene colours into the camera signals without the use of exceptional signals and would provide the freedom to manufacturers of displays to select the primaries best suited to their markets, including in the future, the possibility of four primary display gamuts to extend the reproduced gamut even further.

20.4.2.2 Non-linear Transfer Function

It is encouraging that the ambiguous practice of describing the non-linear transfer function as the OETF or gamma correction function has been dropped; nevertheless, the function itself retains the legacy parameter values associated with the HDTV and SDTV systems, with the potential for introducing the distortion described in Section 19.1 unless guidance is provided for display manufacturers.

In legacy terms, should the display manufacturers regard this non-linear function as a gamma correction function, they may conclude that for their linear display devices, a gamma circuit should be included in the signal path which emulates the CRT as described in BT2129, that is, a continuous power law characteristic, which would cause the distortion referred to above, whereas a gamma circuit with the inverse of the Rec 709 characteristic would avoid this distortion.

Ideally, the non-linear function should be described unambiguously as a perceptible uniform coding function which would provide the receiver or display manufacturers with the unambiguous guidance to provide a matched complementary function in their equipment and thus drive the linear display with notionally linear signals, subject to any enhanced subjective gamma adjustment perceived to be required.

It would appear that a new recommendation, corresponding to BT2129, which is complementary to Rec 709, is required to specify the performance of the display device to complement the Rec 2020 camera performance.

20.4.2.3 Luminance Signal

The coefficients of the RGB signals required to produce the Y luminance signal correspond to the specified primary chromaticities, as may be confirmed by activating the Rec 2020 button in Worksheet 14 and noting the coefficients of RGB for Y in Matrix 6.

20.4.2.4 Colour Difference Signals

For the non-constant luminance colour difference signals, the scaling factors x and y required to match the amplitude of these signals to the luma signal may be calculated using the formulae developed in Section 17.3:

$x = 2 - 2b$ and $y = 2 - 2r$, where b and r are the luminance coefficients of the blue and red primaries, respectively, then:

$$C_B = \frac{B' - Y'}{1.8814} \qquad C_R = \frac{R' - Y'}{1.4746}$$

It will be noted that for the constant luminance system, different scaling factors are required for the positive and negative elements of the signal. As noted in Section 14.6, this is because the constant luminance system produces polarity non-symmetrical signals around zero level, as illustrated for a colour bar waveform signal in Figure 20.16. The colour sequence of the waveform from left to right is white, yellow, cyan, green, magenta, red, blue and black.

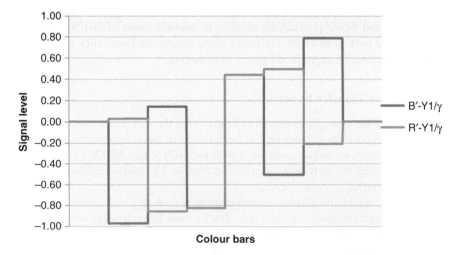

Figure 20.16 Constant luminance colour difference signal levels for a colour bar waveform.

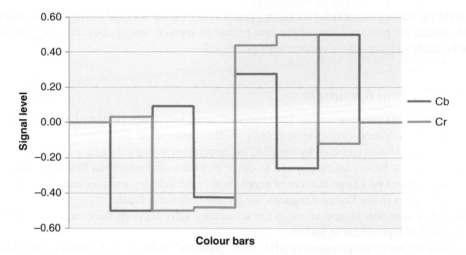

Figure 20.17 Weighted constant luminance colour difference signal levels for a colour bar waveform.

In Worksheet 20(c) the asymmetrical weighting factors contained in Table 4 of the specification have been applied to the constant luminance colour difference signals, and as shown in Figure 20.17, this results in the peak signal levels in both the positive and the negative directions being equal to a level of 0.5, ensuring that the full coding range of the digital system is occupied by both polarities of the colour difference signals.

20.4.2.5   Signal Levels

The Recommendations make no mention of accommodating exceptional signals of the type described in Section 20.3. Whilst it is acknowledged that few colours are likely to exceed the Pointer gamut, there are sources of colour that will do so and in so doing are likely to generate exceptional signal levels which will be clipped within the camera processing, thus preventing any wider gamut display devices from displaying the correct colour.

If an appropriate set of imaginary system primaries had been adopted, then of course there would never be a situation where exceptional signals would be generated.

20.4.3   *Potential Colour Performance of UHDTV*

By incorporating the parameter values of the UHDTV recommendations into the colour reproduction model of Worksheet 19, the potential performance of the system in terms of the ΔE_{00}^* colour difference values may be estimated.

On the assumption that the spectral sensitivities of the camera were the same as assumed for the Rec 709 evaluation of performance, that is, the TLCI curves, and that the display manufacturer assumed that the non-linear transfer function in the camera was for perceptible uniform coding purposes and therefore provides a fully complementary circuit in the display, then the value of ΔE^* would be calculated at about 5.1, a considerable improvement on the Rec 709 figure of 9.1.

Should the camera manufacturers be in a position to trim the spectral sensitivities to more closely match the positive lobes of the new primaries spectral sensitivities, the ΔE_{00}^* figure could be made to increasingly approach a value of 2.5.

20.4.4   Informal Appraisals

The BT2246 report cited earlier has been updated[4] to include a description of an 8K version of a Super High Vision (SHV) system built by the Japanese in accordance with the Rec 2020 recommendations and used by the BBC in cooperation with the Japanese manufacturers and the Olympics Broadcasting Service to cover elements of the London Olympic Games, which were viewed by a large number of people within the industry from around the world in screening theatres in the United Kingdom, the United States and Japan.

The author, who was present at one of the screenings, fully supports the conclusions of the report, which is reproduced as follows:

'We have advanced development of SHV with "presence" as its strongest feature, and these events have once again shown the extremely strong sense of presence delivered by SHV video and audio, and the unprecedented levels of emotion (that) can be imparted on viewers, giving them a sense they are actually at the Olympic venue. We also showed SHV can operate much like ordinary broadcasting, by producing and transmitting programs continuously, every day during the Olympic Games using live coverage and recorded and edited content. A completely different style was also used in production of the content, without using voice-overs (announcing or comments), and using mainly wide camera angles and long (slow) cut ratios. These were met with many comments of surprise, admiration, and of the new possibilities presented for broadcasting businesses.'

This appears to be truly a television system for the 2020s.

---

[4] Report ITU-R BT.2246-2 (11/2012).

21

Colour Management in Television

21.1 Introduction

There are two aspects to ensuring the good rendition of colour pictures; the first has been covered in the earlier chapters of this part of the book on television where the fundamentals and the specifications of the technology relating to colour have been described. Ideally, this should have led us to systems which could be switched on and all that would be required is to operate the camera in order to capture accurate reproductions of the scenes. However, although much progress has been made over the years in terms of stability of operation and automatic adjustments to compensate for the variables of the shooting environment, nevertheless, for critical work there remains the requirement for operational expertise during the shooting operation. It is the exercising of this expertise, to ensure the pictures rendered to the viewer are as accurate a representation of the scene as it is possible to achieve within the limitations of the operating environment, that goes under the description of colour management.

The immediacy of the television media operation, where pictures are captured in sequence from a number of live cameras, means that the extent of the colour management operation is constrained compared with other media such as photography or cinematography, where the opportunity exists in post-production to undertake more sophisticated colour management. Nevertheless, unless care is taken in the colour management process, the pictures produced are likely to suffer in terms of the quality of the picture that could have been achieved.

So, assuming that the camera is operating to its design specification, what are these variables which can affect the quality of the rendered picture in the home? They are:

- The characteristics of the scene illumination
- The setup of the camera for a particular scene or range of scenes
- The environmental illumination in the Vision Control room
- The performance and setup of the picture matching monitors
- The camera operational adjustments of exposure, black level and system gamma

Many of these items have been separately addressed previously in generic terms in Chapter 10 and Section 14.2; here, they will be addressed together specifically in terms of colour management in television. Once the decision is made as to where and under what conditions

Colour Reproduction in Electronic Imaging Systems: Photography, Television, Cinematography, First Edition. Michael S Tooms.
© 2016 John Wiley & Sons, Ltd. Published 2016 by John Wiley & Sons, Ltd.
Companion Website: www.wiley.com/go/toomscolour

the scene will be shot, the manner of dealing with the scene illumination can be dealt with. All the other items listed are dependent upon the Vision Control Room (VCR) operation that is the initial setup of the camera, the visual environment in which adjustments take place and the scene-by-scene operational adjustment of the camera controls.

For reasons which will become evident, each scene composed by the camera operator requires individual adjustment by the vision control operator. These are highly critical subjective adjustments which ensure, first, that each picture is pleasing in terms of gradation of tone reproduction and, just as importantly, that successive shots selected by the vision mixer are seen to visually match. Since the state of the level of adaptation of the eye can greatly influence the perception of the operator undertaking this task, it is essential that all factors within the environment on which the adaptation is dependent are strictly controlled. Thus, the elements contributing to attaining 'good' pictures fall into three categories: first, the establishment of the vision control room environment, which once achieved may be regarded as a static contribution; second, the line-up of the camera and display monitor parameters to agreed specifications; and finally, the shot-to-shot adjustment of each picture; these latter two when taken together may be regarded as dynamic contributions.

21.2 Scene Illumination

The camera is designed to capture scenes illuminated by lighting conforming to Illuminant D65, that is, by an illuminant with the spectral distribution described in Section 7.3, a simulation of a particular phase of daylight.

Daylight however varies significantly in spectral distribution gradually across the range of lighting phases defined by the CIE as the contributions of light from the sun, the blue sky and the clouds of varying density change. Generally speaking, a camera colour balanced for a specific daylight situation will provide satisfactory results across a range of different daylight conditions; however, in the extreme, for example for a camera colour balanced for full sunlight with a moderate amount of cloud, which pans into a fully shaded area where there is very little light reflected into the shaded area from surrounding surfaces, then the blue illumination only from a totally clear sky will cause significant perceived changes in colour balance. This is a difficult situation to control and, if practical, is best dealt with by a quick trial off camera, possibly relying on the auto colour balance feature of the camera.

Dealing with artificial illumination is dependent upon both the spectral distribution of the source lighting and the changes in the intensity of the lighting within a scene; the more even is the distribution of energy across the spectrum, the more easily can the camera be adjusted to compensate for the differences between the artificial illumination and D65. In Section 11.3, the effect of using the individual colour gain controls to correct for tungsten lighting by undertaking a colour balance was illustrated, indicating significant changes in the relative brightness of saturated reds and blues in the scene. To properly compensate for the well-understood and frequently experienced spectral distribution of tungsten (and tungsten halogen) lighting, a filter with the appropriate correcting characteristics to D65 should be employed. The use of an appropriate colour-correcting filter will produce results that are no different to that produced by the D65 phase of daylight.

The spectral distribution of other artificial lighting sources and its effects on the rendition of the image has been addressed in Chapters 7 and 18, respectively. Xenon and to a lesser extent HMI luminaires will produce acceptable results. The performance of fluorescent and LED sources is entirely dependent upon the quality of the luminaires, those with a value of TLCI below 50 are unlikely to produce satisfactory results, whilst the nearer the TLCI value is to 100, the more accurate will be the rendered image.

In a situation where relatively poor lighting is in use, for example a stadium with legacy lighting, it may be possible by the judicious use of a lens filter with characteristics which broadly compensate for the average distribution of the energy spectrum of the illumination to achieve improved results.

21.3   The Vision Control Operation

The name for the vision control operation varies from organisation to organisation, sometimes called 'picture control' and or 'racks', a legacy term relating to the time when the equipment supporting the cameras required racks or cabinets to accommodate them.

It is assumed in the following that the vision control operation follows that described in Section 16.1, where the picture derived from each camera is assigned to a dedicated picture monitor on which adjustments are carried out, together with a master picture monitor. The latter by default displays the picture selected by the vision mixer operator for transmission or recording but on touching an individual camera's exposure/black level control paddle will cause the picture from that camera to be switched to the master monitor.

This approach has two advantages: it enables the operator to rapidly switch between the 'on air' picture and the picture under adjustment without changing his or her line of focus, thus making any mismatch more critically perceived, and by undertaking the final match on the same monitor, it eliminates any residual difference in setup there may be between the monitors, an important advantage in the days when the setup of a monitor was less stable than it has become.

A waveform monitor (see Section 11.2) is also provided to both support the initial setup of the camera and to give an indication to the vision controller of the range of contrast explored by a particular scene.

21.4   The Vision Control Room Environment

The vision control environment is the total environment in which the vision controller makes critical shot-by-shot operational adjustments to the cameras in order to obtain the most satisfactory perceived pictures. This environment comprises all the lighting within the room, the reflection characteristics of the room surfaces within the field of view, the disposition of the monitors on which the pictures to be adjusted appear and the illumination falling upon the monitor screens.

Providing a suitable environment is more problematic than at first might appear because of the complex interaction of all the factors which influence the perception of the vision controller when adjusting the camera operational controls to produce the most satisfactory rendition of the scene. Accepting for the moment that minor adjustments to the operational controls of the camera can make dramatic differences to the perception of the image, it is critically important that the viewer is not aware of significant differences in the general appearance of the image on a shot-by-shot, programme-by-programme or broadcaster-by-broadcaster basis, that is, the pictures from all sources should in general terms match; there are occasions of course when creative requirements override these general rules.

Since the vision controller's perception of the displayed image is dependent upon:

- the environmental lighting;
- the reflection characteristics of the surfaces in the vision control room;
- the adjustment of the monitors on which the image is displayed

it becomes clear that if the above picture matching criteria are to be met, then all of these items need to be standardised and critically controlled.

21.4.1   Control Room Illumination

In Section 13.3, the adaptation characteristics of the eye were addressed in some detail, and it may be recalled that the eye accommodates and adapts to the average conditions of both the luminance level and the colour temperature of the field of view respectively; thus, in order to standardise the adaptation of the eyes, all illumination in the vision control room should match the system white, Illuminant D65.

21.4.2   Room Surfaces Reflection Characteristics

In the vision control environment, the field of view is composed of the surfaces surrounding the picture monitors and the picture monitor displays themselves. If the level of the surrounding surface luminance is comparable to or greater than the average luminance of the picture monitor displays, then depending upon the ratio of the area of the combined screens surfaces to the area of the surround brightness within the field of view, the eye will accommodate to the surround brightness and the darker tone detail in the rendered images will be lost. In contrast, if the room is darkened to the point where there is virtually no surround lighting, the accommodation of the eye to the average luminance of the screen will enable the vision controller to see increasing detail in the rendered image, which will be lost to the average viewer, who will be viewing in a significant level of surround lighting.

Clearly a compromise must be reached on the level of surround luminance and the approach is to select environmental lighting conditions which are regarded as slightly more critical than those of the average home viewer.

The surround surface colour will also cause a chromatic adaptation effect, to the extent that if the average chromaticity of the surround surfaces is significantly different from the system white and at a comparable or higher luminance level than the luminance of the screens, then the rendered image will appear to have an error in colour balance in the complementary direction to the surround average chromaticity, which in turn is usually dependent upon the chromaticity of the surround lighting. The solution in the control room is to ensure that the colour of the environmental illumination matches Illuminant D65 and that the surfaces surrounding the monitors are neutral in colour.

Before being in a position to specify the luminance of the surfaces surrounding the monitors, it is necessary to consider the respective fields of view of the monitor stack and the surrounding surfaces in the context of the adaptation of the eye to the relative dimensions of these quantities.

21.4.2.1 Viewing Distance

The critical viewing of a rendered image is clearly dependent upon the size of the picture and distance from which it is viewed, and in television, it is traditional to measure the distance in terms of the number of picture heights, which for a particular ratio, effectively defines the viewing angle the picture describes at the eye. In appraising pictures under maximally critical resolution conditions, the ratio of viewing distance to picture height is dependent upon the resolution of the image; too distant and detail in the scene will not be perceived, too close and the image will appear to loose definition. In Worksheet 8, the critical distance where the resolving power of the eye matches the resolution of the image for a particular picture size is calculated.

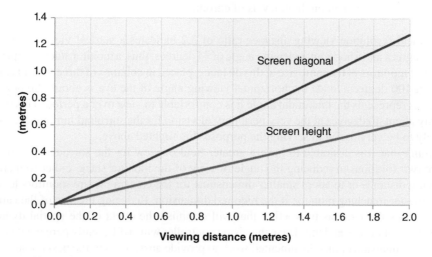

Figure 21.1 Screen dimension against viewing distance for critical viewing.

For the HDTV system, where the resolution is defined by 1920×1080 pixels, the results are illustrated in Figure 21.1, where the constant ratio between viewing distance and screen height is shown to be 3.2. Thus, for a screen height of 330 mm, the optimum viewing distance is just over 1 m; any closer and no further detail would be discernable, any greater distance and detail will begin to be lost. In the figure, the screen diagonal is also shown as a more familiar measure of screen dimension.

This is not to say that this is the ideal viewing distance for picture matching, but it does give the minimum distance. In a vision control room where often the number of cameras to be matched is between four and six, then if one includes the master matching monitor, a monitor stack of up to seven monitors could be required. Thus, the viewing distance in these circumstances is likely to be somewhat greater than the critical viewing distance in order to ensure the monitor stack is well encompassed in the field of view of the vision controller.

The point of these considerations is to arrive at a position where a broad figure can be estimated for the percentage of the field of view which is occupied by the surrounding surfaces of the room and thus consider the desirable luminance of these surfaces in the context of the accommodation of the eye and establishing an environment a little more critical than that of the average home viewer.

Arguments can be made that the luminance of the surrounding area to the picture matching monitors in a vision control room serves two roles:

- It contributes to the overall luminance of the scene and therefore should aim to set the accommodation level of the eye to correspond broadly to the accommodation level of the eyes of the home viewer.
- It provides the opportunity to tie the chromatic adaptation of the eyes of the vision controller to the chromaticity of the system white, that is Illuminant D65, thus helping to retain the correct chromatic adaptation in the extreme situations when the picture monitors may be displaying scenes containing high levels of chroma.

The critical resolution viewing distance ratio of 3.2 indicates a vertical viewing angle of about 18 degrees and a horizontal viewing angle of 34 degrees; thus, a monitor stack comprising a number of monitors of the same size at this distance, placed in columns of three, would extend over about 100 degrees. Since the horizontal viewing angle of the eye is about 180 degrees, 100 degrees represents the maximum angle it is comfortable to view over a period. This would leave only about 40 degrees at the very periphery of vision for the surround luminance, which is unlikely to be very useful in serving the purposes highlighted above.

Assuming that it is desirable to have a greater field of view for the surround luminance, there are two solutions to reducing the angle of view of the monitor stack, either to increase the viewing distance or to adopt smaller dimensions for the camera picture monitors leaving only the master matching monitor at the required dimension. Reducing the size of the camera monitors raises the issue as to whether that will diminish the effect of the spatial dynamic contrast range of the eye and thus leave the vision controller with a differently perceived image. The author is unaware of any documented work to provide answers as to the best compromise for the angle of view required for the surround luminance; however, it is reasonably evident that the critical viewing distance could be relaxed to provide a smaller field of view of the monitor stack without detracting from the ability to make satisfactory picture matching decisions.

Essentially the problem of these apparently conflicting requirements is due to the dichotomy between the high definition of the HDTV system and the home viewing practice, which fails to fully exploit the resolution available in a large percentage of homes, where to do so would imply a step towards a home theatre experience (e.g. a 50-inch diagonal screen would imply a viewing distance of only 2 m.), which in turn would likely lead to a more critical viewing environment, where dependence upon the luminance of surrounding surfaces becomes less critical as the display fills a larger percentage of the field of view.

On the basis it is assumed that in most homes the surround luminance is a strong influence on the accommodation status of the eye, it may be assumed that a significant portion of the field of view in the vision control room should be given over to the surround luminance. ISO 12608 - 1996 'Cinematography – Room and surround conditions for evaluating television display from telecine reproduction' recommends a surround area of eight times the screen area for a single monitor and five times the screens area for a picture monitor stack of two; presumably for larger monitor stacks, the area of the surround would reduce proportionately down to some minimum.

The Recommendation ITU-R BT.500-13 – 'Methodology for the subjective assessment of the quality of television pictures' provides parameter values for both the room and display device characteristics for both laboratory picture quality assessment and subjective appraisal of pictures in a home environment.

In Rec 500, the parameter values for laboratory picture quality assessment come closer to the values used in a vision control room but fall short by a large factor in emulating an environment which is 'slightly more critical than those of the average home viewer'. For example, the highlight luminance of the home display is given as 200 nits, the illumination incident upon it is given as 200 lux and the screen reflectance, which can vary considerably between screens of different manufacture, in a best-case scenario is given as 6%. The light reflected from the screen will therefore have a luminance of $200 \times 0.06/\pi$ or 3.8 nits, thus reducing the image maximum contrast ratio to about 50:1, well below the figure aimed for in the vision control room.

With regard to preferred viewing distance, BT500 recommends a range of viewing distance to picture height ratios commencing at 9 for a display of height of 180 mm and terminating at a ratio of 3–4 for a display height greater than 1.53 m. No reasoning is given as to why the ratio should change for monitors of different picture height, though comment is made that there is very little difference in appraisal between SDTV and HDTV, which is not surprising since these ratios are generally significantly greater than the critical resolution ratios discussed earlier in this section. The implication however is that a ratio of 6 or 7:1 would be acceptable in a vision control room environment. ISO 12608 recommends a viewing distance of 4–6 times picture height.

21.4.2.2   Monitor Surround Luminance Level

ISO 12608 recommends a level of 10% of screen highlight luminance for the surround. In order to achieve this level whilst ensuring as little light as possible falls upon the screen, the lamps providing the surface luminance are usually mounted behind the monitor stack. It is worth noting that a luminance of 15% translates to a lightness value of about 50%.

Rec 500 recommends that the luminance of the surfaces adjacent to the displays should be 15% of the screen highlight luminance, which as we shall see is likely to be in the order of 100 nits, considerably lower than for the home screen.

21.4.2.3   Illumination of Monitor Screens

Traditionally the ambient lighting in the vision control room which falls upon the monitor screens is arranged to be at a very low level in order to ensure the contrast ratio of the display is kept as high as possible. Thus, the room surfaces facing the screen should have a very low reflectance. EBU Tech 3320[1] – 'User requirements for video monitors in television production' indicates that the screen-reflected light from the room surroundings is likely to be in the range 0.05–0.01 nits, leading to an inactive monitor contrast ratio of between 2,000 and 10,000:1. In addition to the above parameters, ISO 12608 also recommends that desk and control console surfaces should be of a matte finish without dominant colours and have a level of illumination between 30 and 40 lux.

---

[1] ttps://tech.ebu.ch/Jahia/engineName/search/site/tech/publications?search=3320&x=0&y=0

21.5 The Line-up Operation

Historically, the relative instability of camera and monitor equipment meant that before the commencement of transmission or recording, a significant line-up procedure of cameras and monitors was required to ensure that the equipment met specification, in order that the following shot-to-shot adjustment required during shooting could take place with a high degree of confidence in the settings selected by the operator. The advances in the adoption of solid-state image sensors and in the stability of the electronics have greatly reduced the number of operations required to achieve a satisfactory line-up; nevertheless, the sensitivity of the eyes to very small changes in dark tone luminance makes it desirable to check the line-up before a transmission or recording takes place.

21.5.1 Camera Line-up

Prior to the beginning of a shoot, all cameras involved are lined up on a greyscale, or greyscales, illuminated by the lighting in which the scenes will be shot. A colour balance will be undertaken by the vision control operator on each camera, usually by adjusting the exposure of the cameras to make the maximum level of the green signal equal to 100%. The red and blue gain controls are then adjusted to make the level of the red and blue signals equal to the same 100% level.

The greyscale provides the opportunity to check that the characteristics of the RGB signal chains remain linear by ensuring that when the three signals are overlaid on the waveform monitor, there is no evidence that there is a departure by any of the three signals from the same level on each step.

21.5.2 Standard Displays for Picture Appraisal and Adjustment

Whilst the requirements of the camera have been described in some detail, the same attention has not yet been given to the picture monitor; thus, before describing the line-up of monitors, it is necessary to address the requirements of picture monitors, particularly those picture monitors to be used for critical picture evaluation.

21.5.2.1 The Requirements of a Vision Control Monitor

In contrast to the little information available for establishing the layout and design of the environment for undertaking the picture matching task, there is a plethora of information on the specification of the monitor used to appraise the rendered image. Monitors for undertaking critical picture appraisal and picture matching tasks are often referred to as grade 1 monitors or master monitors.

In understanding why the specification of a master monitor may appear somewhat convoluted, it is helpful to appreciate that specifying the performance of the monitor is beset by two fundamental problems:

- The characteristics and performance of legacy monitors
- The limitations in the technology currently available for the display device in master monitors

As discussed in Section 13.4, until about the turn of the century, the *de facto* display for both television sets and monitors was the CRT, which has an electro-opto transfer function (EOTF), which is a power law with an exponent or gamma of about 2.4. To cost-effectively compensate for this characteristic, a gamma correction circuit was added to the RGB signal paths in all television cameras. The HDTV system was introduced during the period prior to the general adoption of linear flat-screen displays, so the same arrangement of providing gamma correction circuits in the camera was continued. In consequence, as flat-screen displays were introduced, it was necessary for the monitor manufacturers to incorporate gamma circuits in each of the RGB signal paths in order to emulate the characteristics of a CRT.

However, it was found that the same signal displayed on both a CRT monitor and a flat-panel monitor did not match for a number of reasons, the prime one being that the LCD light control valve technology is based upon varying the amount of back light passing through a pair of polarisation filters by changing the angle of polarisation of one of the filters relative to the other. The angle of polarisation determines the amount of light passing through the filter pair, but at the maximum attenuation condition, a small amount of the back light continues to pass through the filter, thus limiting the contrast ratio. As a consequence of this limitation, CRTs have continued to be used for master monitors right up to the present day; however, with the advent of organic light-emitting diode (OLED) displays, which enable a true black to be displayed, it is evident that flat-screen displays will become available for picture matching in the future.

The precise requirements of a master monitor are comprehensively described in EBU - Tech 3320 already cited, and the accompanying document EBU - Tech 3325 'Methods for the measurement of the performance of studio monitors' describes the measurement methods to determine that user requirements have been met. These documents are a little more explicit than the corresponding ITU document, Report ITU-R BT.2129 – 'User requirements for a Flat panel display (FPD) as a master monitor in an HDTV programme production environment'.

The colour-related requirements of the master monitor described in Rep 2129 and Tech 3320 (in brackets) are summarised here without the tolerance values given in the Recommendation:

1. Luminance range: 100–250 nits (70 to at least 100 nits)
2. Black level: full screen black level signal 0.01 nits (0.05 nits). It must be possible to adjust black level with a picture line-up generator (PLUGE) test signal, (see Section 21.5.2.2) including sub-black according to the procedure outlined in ITU-R Rec. BT.814.)
3. Sequential contrast ratio: not specified (full screen 1% patch: above 2,000 to 1)
4. Simultaneous contrast ratio: 350:1 (with EBU box pattern: above 200 to 1)
5. Gamma characteristics: still under discussion (It is recommended that a nominal value of 2.35 is used.)
6. Tone reproduction: Greyscale tracking between colour channels shall be within ellipses defined by: + or –0.0010 $\Delta u'$, + or –0.0015 $\Delta v'$ from 1 to 100 nits, and deviation from grey should not be visible for luminances below 1 nit (0.5 Δu^*v^* for luminance from 1 to 100 nits and deviation from grey should not be visible for luminances below 1 nit.).
7. Colour gamut: The FPD should display images with colour gamut specified in Rec 709. (Colour primaries and reference white to the Rec 709 recommendation. All colours displayed within the system colour gamut must provide a metameric match to those displayed on an ideal CRT monitor.)

The user requirements outlined in the specification were laid down in order to provide guidance to monitor manufacturers, and a few of these have managed to develop flat-screen displays aimed at meeting the requirements of master monitors.

21.5.2.2   Monitor Line-up

Prior to shooting a scene, the setup of the picture appraisal monitors should be checked in terms of the values of highlight luminance, black level and chromaticity of display of system white on all steps of the greyscale.

Many master monitors are now supplied with an application which together with a specified light meter will enable the operator to specify the highlight luminance and chromaticity, the latter often in terms of the correlated colour temperature; some applications also enable the gamma to be specified. The light meter is suspended against the face plate of the monitor and the application then runs through a number of electronically generated test signals in sequence, measuring the level and chromaticity for each exposure. The results are used by the application to automatically change the settings of the circuits to bring them in line with the requested parameter values.

Traditionally, in the operational use of CRT monitors, the gamma law of the device made the accurate setting of black level difficult to achieve, that is the adjustment of the black level on a black level signal to be at just black without clipping detail in the dark areas of the picture. Furthermore, the instability of the electronic circuits meant that the setting would often need constant re-setting, a time-consuming task even for experienced staff.

In order to address this problem, the technical staff at the BBC developed the picture line up generator or PLUGE, an electronically generated test pattern which greatly eased the problem of accurately adjusting the black level of picture monitors. This approach to the line-up of picture monitors has been widely adopted around the world, and in Recommendation ITU-R BT.814-2 – 'Specifications and alignment procedures for setting of brightness and contrast of displays',[2] digital versions of the pattern were effectively standardised for both SDTV and HDTV displays. Figure 21.2 and Table 21.1, together with the following description of its use, are copied from Rec 814.

PLUGE for HDTV systems, as copied from Rec 814.

A PLUGE signal for HDTV displays is shown in Figure 21.2. *The peak white patch is used to set the peak luminance by means of the contrast control.*

Two types of signal can be used to set the brightness of the black level of the display by means of the brightness control.

The signal on the left-hand side of the picture consists of narrow horizontal stripes (a width of 10 scanning lines). The stripes extend from approximately 2% above the black level of the waveform to approximately 2% below the black level. The signal on the right-hand side of the picture consists of two coarse stripes (a width of 138 lines), one stripe is approximately 2% above black level, the other is approximately 2% below black level. This signal is suitable for setting display values for both CRT- and FPD-type displays.

The black level of the display is adjusted by the display brightness control such that the negative horizontal stripes disappear, whilst the positive horizontal stripes remain visible.

---

[2] http://www.itu.int/dms_pubrec/itu-r/rec/bt/R-REC-BT.814-1-199407-S!!PDF-E.pdf

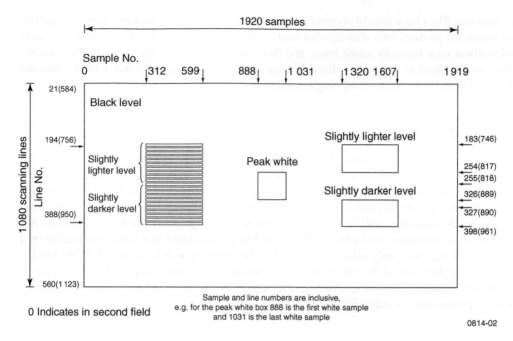

Figure 21.2   The PLUGE chart from Rec 814.

Table 21.1   Quantisation levels associated with the PLUGE pattern
(copied from Rec 814)

Parameter values, Figures 2 and 3	8-Bit digital value	10-Bit digit value
Peak white	235	940
Black level	16	64
Slightly lighter level	20	80
Slightly darker level	12	48

Rec 814 recommends that the highlight luminance be set to 70 nits, and there are other recommendations specifying figures up to 125 nits. The SMPTE is also currently defining parameters associated with the picture matching operation. It is apparent that there is a general trend towards accepting 100 nits as the current appropriate highlight luminance level.

It may be recalled however that in the final paragraphs of Section 13.3, it was noted that though a screen highlight luminance of 100 nits would provide a satisfactorily critical rendition of the scene on a screen of limited viewing angle, this level of luminance for screens showing a larger viewing angle would require to have a highlight luminance significantly higher than this value to prevent black crushing in the contrast range of the eye where the Fechner–Weber law begins to fail. Thus, as the technology becomes available to provide these levels of highlight luminance, it is likely in the future that the recommended level will increase accordingly.

21.6   Capturing the Scene

Once the camera and picture monitors have been adjusted as described in the foregoing, the operation of shooting and recording or transmitting the programme or programme sequence can commence. The vision control operator has three principal controls which are used dynamically to match the pictures on a shot-by-shot basis. Exposure and black level are likely to require adjustment on a scene-by-scene basis, and therefore to ergonomically optimise the controls, they are arranged in a 'paddle wheel' configuration comprising a joystick with a rotational knob mounted at its top for easy single-handed operation.

21.6.1 Exposure

The exposure of the camera to the light of the scene is controlled by moving the joystick of the paddle wheel linearly towards or away from the operator on the remote control panel, which varies the aperture of the camera iris. The amount of adjustment required will be dependent upon the production; for outdoor scenes with lighting changes dependent upon variable cloud cover, much adjustment will be required; for studio shooting, where the lighting director has mounted the luminaires and adjusted their levels to provide even highlight illumination on a scene-by-scene basis, only minor adjustment of the exposure will be required. Subject to the creative requirements of the production, the exposure is normally adjusted for best face or flesh tone lightness, subject to ensuring any required white detail in the scene is not lost by over-exposure and resultant signal clipping.

21.6.2 Black Level

The black level of the camera signal is adjusted by rotating the knob which resides at the top of the joystick. In attempting to obtain the best compromise in interpreting the contrast range of the scene into the limited contrast range the system is capable of rendering, it is likely that this is the adjustment most often used. Albeit the range of adjustment is usually minor, the effects of these small adjustments are perceptually very significant.

In Section 13.4, under the section entitled 'Appraising the Performance of the Combined Gamma Correction Characteristic', it was shown that the gamma correction characteristic of the camera is a poor match to the display gamma in the region close to the contrast range limit of the eye for scene luminances in the 0.5–20% range; at the 1% scene luminance level, the display is producing a lightness sensation in the eye of about half the value of the 21.5% that would be produced by the scene, as is illustrated in Figure 13.17.

It is this mismatch of contrast laws which is the primary reason that adjustment of black level is required so frequently. Many scenes have surface luminance values below the value of 20% of the peak white in the scene, and thus, without the compensation provided by the black level control, would appear black crushed. Nevertheless, the operation of the black level control is a compromise since by 'lifting the blacks', it will cause the gamma characteristics of the camera and display to cross over at just one point; at all other levels, there will be a mismatch, albeit a considerably smaller mismatch than the one compensated for.

21.6.3 Gamma

In the normal course of events, the gamma control is unlikely to be much used. However, sometimes, the difficulty of obtaining a satisfactory representation of the contrast range of the scene with the black level control or achieving a creative effect, justifies varying slightly the gamma of the overall signal chain. In practical terms, with current cameras, this is usually achieved by a relatively minor adjustment to the camera gamma circuits which provide the system gamma specified by Rec 709. In an ideal situation of the type described in Section 20.2, where gamma correction has been replaced by an inverse matching pair of perceptibly uniform codecs, a dedicated adjustable gamma circuit with a range of control of gamma between 1.0 and 1.3 could be considered. It may be recalled that in Section 13.7, it was noted that under home viewing conditions, it is considered by some that an overall system gamma slightly greater than unity is preferred.

21.7   Displaying the Image

The scene critically adjusted and captured in the vision control environment is eventually displayed in the home under a wide variety of environmental conditions embracing the setup of the television receiver and the room lighting, both of which are strongly influential in determining the quality of the perceived rendered image.

Discerning viewers will locate the display in an area of the room where the light falling on the screen is a minimum subject to the requirements of comfortable day-to-day living. More critical viewers may arrange different room lighting when viewing a programme in a committed manner. Nevertheless, even after these arrangements are in place, the quality of the displayed image often leaves much to be desired due to the poor setup of the receiver.

Historically, in visiting the television departments of the local stores, one would be beset by a large number of screens, often all displaying the same picture but with widely different appearances in terms of colour balance, saturation and black level. The situation has improved significantly in only recent years; the white point or colour balance of most receivers is now usually found to be close to 6500 K and the saturation variation from set to set is often noticeable but not objectionably so. However, all too frequently, the black level setting varies significantly, and all too often in the direction where the shadow detail is black crushed. It would also appear that manufacturers adjust their receivers before despatch to show a very bright picture to advantage in the often highly illuminated viewing area of the store and, in consequence, are incorrectly adjusted for home use.

Adjusting the receiver in the home environment can be problematic. The viewer is frequently offered a range of emulated picture styles to select from and often seems to select the option producing the most oversaturated pictures. The contrast and brightness controls seem to bear little relationship with the contrast and black level control of a studio monitor, presumably to prevent the viewer from selecting a totally unacceptable combination of adjustments; nevertheless, the apparent interaction of these controls make it difficult for the discerning viewer to obtain a satisfactory setup.

Ideally, in view of the sophistication now built into modern receivers, it would be relatively simple and cost-effective for the manufacturer to provide the option for the viewer to select in turn an electronically generated greyscale or the blue element of colour bars for display. With simple instructions, the viewer would then be enabled to adjust the receiver contrast, black level and saturation levels to a default ideal setup condition.

Part 5B

Colour Reproduction in Photography

Introduction

In keeping with the title of this book, Part 5B is constrained to describing *electronic* photography, or as it has become known, digital photography, which started to become a reality in the 1990s when digital techniques were adopted by the television industry. The quality of the rendered image soon competed with film-based photography and within a decade it became the dominant photographic technology.

However, it was not just the quality of the rendered image which was responsible for its widespread adoption. The flexibility it provided to the professional and amateur alike to manipulate just about every aspect of the captured image on the desktop computer, which was also coming into more widespread use at this time, made it increasingly popular and extended the photographic medium to a much broader base of users.

Since its introduction digital photography has blossomed to include not only the dedicated stills camera so familiar in the film period but also cameras integrated into mobile phones and tablet computers; the content of which, though often also viewed on desktop computer monitors and television screens, rarely reaches the stage of becoming a print. At one end of the user spectrum anybody using these devices is strictly a photographer but in the context of this book we will restrict the use of the term to those professionals and amateurs who follow through the operation to produce a fine print.

The success of digital photography was dependent upon the convergence of three established technologies: television, computers and cost-effective digital photographic printers. In principle, digital photography owes much to the technology of television but it was not until

Colour Reproduction in Electronic Imaging Systems: Photography, Television, Cinematography, First Edition. Michael S Tooms.
© 2016 John Wiley & Sons, Ltd. Published 2016 by John Wiley & Sons, Ltd.
Companion Website: www.wiley.com/go/toomscolour

that technology had developed sufficiently to bring about solid-state opto-electronic image sensors, miniaturised *digital* integrated electronic circuits and solid-state memory that it could be adapted to replace film in stills photography. At that time personal computers were being increasingly adopted by a broader range of users and in consequence software companies saw the opportunities and were developing applications to complement the capture of the scene by the camera. These applications enabled the contents of digital photographic files to be opened by the computer and adjusted on the CRT-based computer monitor before being processed to a form suitable for driving desktop colour printers, the third cornerstone of this new reproduction medium. One of the most comprehensive and widely used of the applications for manipulating the image is Adobe® Photoshop® and in consequence it is this application which will be used as the representative of all such applications in the various descriptions of the system which are provided in the following chapters.

Whereas the automated adjustment circuitry within digital cameras ensures that by and large the images rendered on the mobile phone or tablet computer are very acceptable, when viewed more critically on a desktop display it is often perceived there is room for improvement, especially in terms of colour balance and tone rendition. When it comes to appraising the rendered print, the results are often disappointing; this is particularly true for the enthusiastic amateur who is new to the operation.

The addition of a computer photographic processing application and a colour printer to what might otherwise be considered an extension of the relatively simple television workflow complicates the situation considerably. Furthermore with the increased flexibility available in the adjustment of a wide range of processing parameters comes the complementary situation of far more opportunities for maladjustment. In consequence, managing all the variables to ensure good colour rendition becomes an essential major element in the workflow under the title of 'colour management' and explains the emphasis given to this topic in the final chapters of this Part on photography.

The practicality of producing a fine printed image is central to the act of being a photographer and as such the chapters on colour management cover all the steps of adjustment required from the shooting of the scene, through each stage of Photoshop and the correct setting up of the interface between the computer and the printer in order to achieve, where appropriate, a print which is perceived to match the original scene. Where correct adjustment alone does not lead to a successful conclusion the process of producing profiles to more accurately match the stages of the workflow are also described.

The chapters on colour management are preceded by chapters dedicated to the photographic work and signal flow and to the application of colorimetry to the photographic operation.

22

An Overview of the Photographic System and Its Workflow

22.1 Introduction

In this chapter the basic elements of digital photography are reviewed in order to set the scene for a more detailed examination of the part colour plays in the rendition of the image captured by the camera both on the computer display and in the printed photograph.

The fundamental work on colour undertaken in Parts 1–4 of the book provides an underlying basis for understanding the chapters which follow; however, as the printer is not required for television or cinematography, it was not included in those chapters. Furthermore, since it uses principals of colour reproduction not considered since Chapter 2, the next chapter is dedicated to the fundamentals of the printing process.

22.2 An Overview of the Workflow

A simplified view of two of the four principal elements of the photographic workflow is illustrated in Figure 22.1, based upon Adobe® applications in the computer. The aim of this section is to provide an overview of the photographic workflow before later sections describe the elements of the signal flow in more detail. The computer monitor and the desktop printer are shown as composite items here, but are also described in more detail in later sections.

22.2.1 The Scene and the Camera

In Figure 22.1, the scene and the camera are shown at the top of the diagram, where the scene is illustrated by a flower and grey scale, and the light from the scene captured by the camera is technically represented by the spectral power distribution (SPD) of the scene illumination and the spectral reflectance of surfaces within the scene.

Although for simplicity the illustration implies three independent sensor devices, the majority of stills cameras utilise a single solid-state opto-electronic image sensor based upon the Bayer mosaic arrangement described in Section 7.2. However, there are alternative approaches

Colour Reproduction in Electronic Imaging Systems: Photography, Television, Cinematography, First Edition. Michael S Tooms.
© 2016 John Wiley & Sons, Ltd. Published 2016 by John Wiley & Sons, Ltd.
Companion Website: www.wiley.com/go/toomscolour

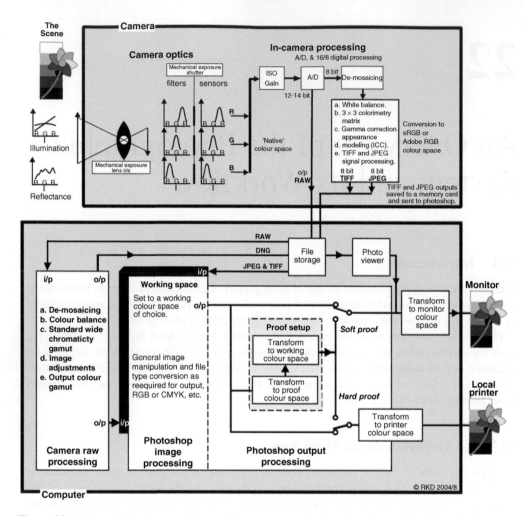

Figure 22.1   Simplified photographic workflow – an amended version of an original drawing by Ray Knight.

to the Bayer mosaic as exemplified by the Foveon[1] sensor, which has three separate image sensor layers arranged one above the other; the blue light is absorbed by the upper first layer, the green light by the intermediate layer and the red light by the lower layer.

The three red, green and blue spectral sensitivities of the camera are related to the CMFs of the display primaries, but since they are not generally at this stage a direct match, they are described as the 'native' characteristics of the camera and their spectral responses are regarded as proprietary by the camera manufacturer and thus their characteristics are not usually published. The aim of the designer is to make these responses as close a match as possible to a set of camera matching functions, in order that downstream the signals may be matrixed to a specific set of CMFs with the minimum of gamut clipping (see Chapter 12).

---

[1] http://en.wikipedia.org/wiki/Foveon_X3_sensor

The image sensor will have an opto-electro conversion function (OECF), which is normally linear and will output its signals to variable gain amplifiers which are adjusted for minimum gain, consistent with providing a standard level signal after the desired adjustment to the iris and shutter speed exposure settings. The gain adjustment is usually calibrated in terms of the ISO rating, which in turn is related to the ASA film sensitivity or speed rating familiar to traditional photographers.

The remainder of the camera processing is dependent upon the sophistication of the camera and its manufacturer. For simple consumer cameras, the signals are Analogue to Digital (A-D) converted using an 8-bit digital coding system, de-mosaicked and passed through a range of processors in tandem as listed in the diagram and including:

- colour balance on white or the average chromaticity of the scene;
- the colour matrix, which corrects the native spectral sensitivities to those which relate to the chromaticities of the standard sRGB or Adobe RGB primaries;
- gamma correction to compensate for the display actual or emulated CRT characteristic;
- proprietary appearance modelling;
- appropriate compression to reduce the file size for either JPEG or TIFF output files.

The output file is then stored on a memory card ready for transferring to a computer for further processing. The sRGB and Adobe RGB chromaticities of the display primaries and the JPEG and TIFF file formats are described Chapters 24 and 25, respectively.

More sophisticated cameras provide the option of storing a *raw* file, that is a file which has been digitised at a larger number of bits than used for the consumer-related output but has not otherwise been processed, separately on the memory card to enable the signal to be subsequently processed in accordance with the requirements of the photographer rather than that of the standard processing defined by the camera manufacturer.

22.2.2 *The Computer and the Adobe® Photoshop® Application*

The computer provides an environment in which the Photoshop application operates; it stores the processed and raw files from the camera in an appropriate folder and enables them to be previewed on the monitor using the simple photographic file viewer usually incorporated in the operating system of the computer, as illustrated by the Photo Viewer application in the top right of the computer element of the diagram in Figure 22.1.

Between the viewer application and the monitor is a processor for matching the characteristics of the camera to those of the monitor in terms of their respective colour spaces, a technique described in Chapter 12. This requirement to match different colour spaces occurs frequently throughout the photographic workflow, and the processor responsible for undertaking this activity is primed by two sets of values, one of which is contained in a source *profile*, which is incorporated within the photographic file and describes the colour space pertaining to the RGB values in the file, and the other in a display profile, which is usually loaded onto the computer when the monitor is first installed. *Profiles of this type are critical to the successful colour management of the photographic process, and their use in the workflow is described in some detail in Chapter 27.* In the default situation associated with previewing the images, often the computer monitors have display chromaticities matching the default chromaticities of the RGB signals within the processed file and thus the monitor profile is not required but if present causes the processor to act neutrally. In the event that no monitor profile was loaded at the time the monitor was installed, the operating system assumes the monitor has the standard sRGB primary chromaticities and processes the signals accordingly.

The Photoshop application will only accept specified photographic file formats, such as amongst many others, the Photoshop standard PSD files or the standard JPEG and TIFF files, which are described in Chapter 25. The raw files from the camera are usually generated to a manufacturer's proprietary specification and thus require to be converted to one of the standard formats before loading into Photoshop. Most manufacturers provide a camera raw processing application which may be loaded onto the computer for both this and preliminary adjustment purposes; however, Adobe also supply a generic raw file processing plug-in application, Camera Raw, which recognises a wide range of manufacturers' raw file formats and converts their files to a format which Photoshop recognises. It is assumed in what follows that the Adobe Camera Raw file processor is in use.

The Camera Raw processor carries out three principal tasks: firstly, it converts the native chromaticity gamut linear signals from the camera to a standard Photoshop wide chromaticity gamut format; secondly, it enables the operator to carry out a range of image adjustments on these linear signals; and finally, it processes the signals to the output colour space selected by the operator. It follows that for Adobe Camera Raw to be able to accommodate raw files from different camera manufacturers, Adobe must be confidentially informed by the camera manufacturer of the proprietary spectral responses of their native spectral sensitivities in order to enable Adobe Camera Raw to undertake the matrixing of the raw RGB signals to match that of the standard Adobe Camera RAW chromaticity gamut. Adobe uses this information to construct two profiles for each type of manufacturers' camera, one relating to a scene illumination of daylight at D65 and the other to a tungsten illumination of SA; the application then determines from an inspection of the raw file data which profile to apply.

Attempting to load a raw file directly into Photoshop will trigger the loading of the raw file processor which enables the operator to manually adjust the parameters which are otherwise set automatically in the camera processor, such as 'Exposure', 'White Balance', 'Contrast' and 'Blacks'. Once satisfied with the adjustments, the option is provided to store the file either in an Adobe standard 'Digital Negative' (DNG) raw file format for later Camera Raw adjustment or the output file colour space is selected and the file is opened directly into Photoshop.

The range of processes available in Photoshop is extremely extensive and continues to increase with every new version; our interest however is limited to the manner in which Photoshop manages the colour spaces at its input and in the interfaces to the computer monitor and the printer at its output. These interfaces are complex and the options associated with them are diversely distributed amongst the menu system, making the correct choice of option for a particular phase in the workflow far from a simple task. It is for this reason that Chapters 28 and 29 on colour management in the workflow are dedicated to providing a detailed description of the setting of these parameters, and thus in what follows, only an overview is given in order to provide an understanding of the system workflow.

Most image adjustments take place in the *working space* of Photoshop, a colour space which may be selected by the user as the default colour space in which he or she intends to operate. This colour space or gamut may or may not match the colour space of the file to be loaded from a computer storage folder or direct from the Camera Raw processor. Photoshop detects whether there is a match, and when there is not, offers the user a number of options to ensure a match is achieved.

In Figure 22.1, with the 'Soft Proof' switch in the upper position, the computer operating system detects which working colour space profile is in use within Photoshop and, together with the monitor profile, sets the parameters of the monitor conversion transform appropriately in order to match the image data to the monitor colour gamut.

The printer controls the amount of cyan, magenta, yellow and black (CMYK) dyes or pigments laid down to produce the rendered image; a printer driver is therefore necessary to convert the RGB signals to appropriate CMYK signals to drive the printer. All printers include such a driver in order that simple systems which do not use a colour processing system such as Photoshop can take an RGB file and use it to produce a satisfactory colour image. However, the Photoshop operator has the option of using either the Photoshop printer driver or the printer manufacturer's driver, but not both, as is discussed further in Chapter 29. For the sake of simplicity, only one printer driver is illustrated in Figure 22.1.

It should be appreciated that once the operator of Photoshop has adjusted the image on the monitor for the desired result, he or she will ideally also wish to see on the monitor screen a rendered image which is representative of how that image will appear in its final form, whether on a direct viewing or projection screen or in print. This can be a complicated process as these devices may well have displays with different colour characteristics to the computer monitor; however, if the monitor display has a wide colour gamut which encompasses the colour gamuts of the final viewing media, then it is possible to simulate the final appearance with a 'soft proof' on the computer monitor. The means of achieving this is shown in the 'Proof Setup' area of Figure 22.1; the signals from the working space are processed to the colour space of the final viewing media in order to impose any constraints caused by a smaller colour gamut, and then re-processed back to the working colour space. Photoshop allows the operator to select from a large number of destinations a final viewing media colour space which suits the situation. In the diagram, the proofing switches are shown in the default positions, allowing the adjusted image to be fed directly to the monitor and printer, respectively, for optimum results. In order to view on the monitor how an emulation of the image will be perceived on the selected final viewing space, the soft proof switch is moved to the lower position.

In a professional system where the final media is likely to be an external press printer, not only will the operator wish to view the soft proof on the computer monitor but may also be called upon to produce a 'hard' proof. Subject to the local desktop printer's colour gamut encompassing the gamut of the external printer, a good representation of the final rendered image can be captured on the local printer by selecting the characteristics of the external printer as the final media colour space and selecting the 'Hard Proof' position of the lower switch before requesting a print.

22.3   The Requirement for Technical Standards in Photography

Little consideration of the range of elements in Figure 22.1 is required to appreciate that if these elements, manufactured more often than not by different manufacturers, are to operate together satisfactorily in a complex workflow environment, whilst providing flexibility in terms of the colour spaces adopted at different stages in the workflow, then there must be an agreed means of describing not only the characteristics of the signal but also the manner in which the signal is processed when passing between environments operating with different colour spaces.

Consideration of the other parameters which need to be specified in order to achieve a comprehensive specification of a digital system of photography leads to a list which includes the following:

- Colour space occupied by the image
- Signal encoding format
- Digital format
- Compression format
- The means of managing any change of colour space required at the interfaces between elements of the workflow
- File format

In the 1980s, when it became clear that digital–based photography would become a reality, the technologies of other related industries were reviewed to see what could be adapted from them in specifying a system of digital photography. Prominent amongst these were the desktop publishing, graphics and television industries.

In television, the CRT which formed the display device for a market of millions of television receivers would clearly also be adopted as the display device for the computer monitor, and thus in practical terms, the chromaticity coordinates of its primaries and the system white point adopted would be a powerful contender in deciding upon the specification for a photographic colour space.

Also, the television industry had adopted, after much fundamental consideration, a luminance and colour difference signal format for conveying and storing RGB signals, as described in detail in Chapter 14 and briefly reviewed in Chapter 25. Since in fundamental terms there were no significant differences between the rendition of television and photographic images, the television signal format would be a strong contender for adoption by the photographic industry.

The desktop publishing and graphic industries had evolved a number of file formats for accommodating data derived from colour images representing what may be described as pictorial images, that is, images in which the data describing elements in the scene change on a gradual basis from pixel to pixel as opposed to those that change abruptly on a pixel-by-pixel basis, such as those representing diagrams and text. One of the leading contenders amongst these was the Tagged Image File Format or TIFF file.

Thus, in very general terms, three of the six parameters listed above could be specified without extensive research by adopting and adapting techniques used in associated industries. However, that would still leave the problem of specifying the means of managing the change of colour space between workflow elements and the means of digitising and compressing the signal. In television terms, there had been no requirement to signal the identity of the colour space, since for a particular national system, a single colour space was fully defined and no alternates were accommodated; there was thus no prevailing interchange format which could be adopted. Also, in compression terms, the approach adopted for television exploits redundancy in both the spatial and the temporal structure of the image and so was not suitable for adoption to the still photographic image structure.

In order to address these shortcomings, two committees were formed from interested industry and manufacturer groupings: the International Colour Consortium (ICC) was formed to address the requirement of specifying both the data which described the colour space and the means of converting it when required to a different colour space between elements of the workflow, and the Joint Photographic Experts Group (JPEG) was formed to address the compression requirements.

However, in reality the situation, though broadly as described in the above paragraphs, was unfortunately not as clear cut as indicated for a number of reasons, including: historical legacy considerations; the requirement to provide flexibility in the selection of: a colour space, a signal format and a compression format; and for reasons relating to the interdependence of the defined parameters across some of the six to be specified.

Thus, for the six parameters which are described in the following chapters, less emphasis will be given to the sections on compression and file formats, since where there are parameters associated with these areas that do directly affect the colour rendition of the final image, they will be covered in the explanations of the other sections.

As many of these emerging standards of the embryonic digital photographic industry became stable and accepted by a wider community, they were adopted as international standards by such bodies as the International Electrotechnical Commission (IEC) and the International Organisation for Standardisation (ISO).

23

The Printing Process

23.1 Introduction

In Part 3, the basic elements of the colour reproduction process which are common to all three media types were briefly described; however, as the printing process was only of relevance to the photographic media, it was omitted from those descriptions. Thus, before proceeding further with descriptions of the elements of the photographic workflow, we need to address the characteristics of the printing process, with descriptions of both the concept of printing and the colour processes behind the production of a photographic print.

The printing of images is a complex process, which has been continually refined over more than a hundred years, though it is only since the early 1990s that there has been a requirement to produce prints from digital photographic files and, even more recently, the requirement to make this facility available in a cost-effective manner to the broad photographic community. To meet these requirements, new printer types have been developed which interface directly into the photographic workflow, albeit based upon the same fundamental principles as those in the printing industry. Printer technology serves a diverse range of industries and in consequence has many different and complex forms, so much so that in this chapter only inkjet printers will be described.

23.2 Conceptual Considerations in Photographic Printer Design

23.2.1 Evolving Printer Concepts

As we have seen in Section 8.2, the camera-generated digital signals are produced by scanning the electric charge pattern of the image produced by the image sensor and, in turn, the displayed optical image is again produced by a digital scanning mechanism controlling the level of light required at each display pixel. To adapt this concept to a printer is not straightforward for two reasons: firstly, the strength of the ink in terms of the amount of light it absorbs cannot in practical terms be controlled, certainly not to the fine level of gradation possible with the voltage that drives a display; thus, effectively an ink spot is either on or off; there is no means of adjusting the amount of light it absorbs and reflects. Furthermore, the alternative

Colour Reproduction in Electronic Imaging Systems: Photography, Television, Cinematography, First Edition. Michael S Tooms.
© 2016 John Wiley & Sons, Ltd. Published 2016 by John Wiley & Sons, Ltd.
Companion Website: www.wiley.com/go/toomscolour

approach of controlling light absorption in the fine amounts relating to a contrast range of over 100:1, by varying the amount of ink deposited on the paper and thus the area of the spot on a pixel-by-pixel basis, is impractical. In addition, the scanning process of the ink-depositing mechanism is entirely mechanical rather than electronic, as it is for the camera and display, thus prohibiting the rapid repeat of scan lines which is possible in electronic scanning, making the generation of a print by individual scans of a single print head a very time-consuming exercise and therefore practically unacceptable.

Furthermore, there is a conceptual difference between producing a colour image by generating light from three primary colours and producing an image on a paper surface from the deposition of inks or pigments which reflect the spectral elements of the white light incident upon them after absorption. In the latter case, the colour perceived relates to the addition of the spectral colours which are reflected from the inks after the white light has been selectively absorbed. For example, yellow ink may be characterised by the blue light which it absorbs or subtracts from the incident white light, leaving a band of spectral colours from red through to green, which, being located along the straight line section of the chromaticity diagram, will result in the colour yellow being perceived.

This characteristic of inks, together with the perception of which colours are perceived as mixtures of these by the eye, was explored in some detail in Chapter 2, leading to the concept of subtractive primaries based upon cyan, yellow and magenta. However, before investigating further how these subtractive primaries might be exploited to produce a photographic image, we need to understand the fundamental principles involved in producing such an image, and in doing so it is helpful to first consider the production of a half-tone monochrome image produced with only black ink on white paper.

23.2.2   Controlling the Reflectance of Each Pixel

The first question which arises is: how are we to produce a range of greys between black and white to enable us to represent the fine gradations found in a well-lit scene using only black ink? The answer is to lay down onto the paper varying numbers of tiny black dots so close together that they are beyond the resolution of the eye to differentiate them. Traditionally, the size of the dots is varied to vary the reflectance, but in inkjet printers, by changing the ratio of the area of the black dots to the area of the surrounding white paper in an area defined as a cell, where the cell dimension is smaller than that which can be resolved by the eye, a range of grey luminances will be perceived. The larger the number of dots which can be accommodated in a cell, the greater will be the number of steps which may be obtained between black and white. This approach, described as half-tone printing, was first initiated by Talbot[1] in the nineteenth century and has been continually refined since with each new generation of printer.

In 1936 Alexander Murray of Eastman Kodak was working on characterising half-tone printing in mathematical terms only to find that his colleague E. R. Davies at the Franklin Institute had already developed a formula for predicting the density of dot coverage in simplistic terms. Whilst density is a useful concept in film photography, its inverse logarithmic relationship, 'reflectance', is more useful and intuitive in digital photography and thus the reflectance version of the formula is developed in the following.

---

[1] William Henry Fox Talbot (11 February 1800 to 17 September 1877) http://en.wikipedia.org/wiki/Halftone#History

In a unit area of half-tone print, if the reflectance of the paper is given as R_P and the reflectance of the ink is given as R_i and the area of the ink is given by a factor a of the unit area, then the remaining area will be $1 - a$. Thus, the total reflectance of the half-tone will be given by R_{HT}, where:

$$R_{HT} = (1 - a).R_P + a.R_i \qquad (23.1)$$

This formula has subsequently become known as the Murray–Davies formula (Murray, 1936) and enables the reflectance of a half-tone to be broadly described in terms of the relative dot area and the reflectance of the paper and ink.

The term 'broadly' in the above paragraph is used advisedly since in practice it does not provide an accurate measure of the actual reflectance obtained, the level of inaccuracy being to some extent dependent upon the particular printing process and the type of paper being printed on. In general terms however, the problem is that the ink area used in the calculation is related to the dot area as defined by the ink jet volume as first laid down on the substrate, whilst in a practical printing process, there are a number of reasons which cause the dot to spread before it is finally stabilised, a process known as 'dot gain' or tone value increase (TVI).

By rearranging the above formula to give the area a in terms of the other parameters:

$$a = \frac{R_{HT} - R_P}{R_i - R_P} \qquad (23.2)$$

and substituting the measured value for the total reflectance in this formula:

$$a_{eff} = \frac{R_{HT \text{ measured}} - R_P}{R_i - R_P} \qquad (23.3)$$

it is possible for a particular print process to measure the difference in reflectance between the theoretical and measured *effective* values for a number of areas between full coverage and no coverage, and use the values obtained as correction elements in the formula. These values may be as high as 30–60% at 50% calculated reflectance, depending upon the print process.

In addition to the inaccuracy caused by *physical or mechanical* dot gain, there is a further cause of inaccuracy known as optical dot gain, which is of a secondary nature with black ink but plays a more important role with the more transparent coloured inks, as will be shown in the next section.

23.2.3 Scanning the Paper

So we have identified the general approach, but a method needs to be devised for laying down these dots across an image area and the manner of achieving this with printers used in photography is to mechanically scan the image area with a print head comprising a number of ink jets which are capable of emitting a jet of ink in the form of a droplet to form an ink spot, on a repetitive basis. As the print head transverses the paper, a line of ink spots are laid onto the paper and, following the completion of each scan, the paper is stepped forward by the distance equivalent to a scan width and the process is repeated until a complete image is formed.

23.2.4   Forming the Droplets

The mechanism for forming a droplet at the print head exploits the developments in integrated circuit technology. Printed circuit techniques are used to form a droplet-sized reservoir, connected on one side to the ink supply and the other side containing an orifice too small to discharge the ink without pressure being applied to the reservoir. There are two principal methods of applying an intermittent pressure to the reservoir of sufficient strength for the ink to be dispelled in jets which form droplets that settle on the paper. In one method, the reservoir is heated by a resistor lining one side of the reservoir such that when a voltage pulse is applied, the resistor heats up and vaporises a small quantity of the ink which expands in the limited volume of the reservoir building up a pressure, which then expels the remainder of the ink as a jet which forms the droplet. In the alternative approach, the reservoir is formed from a piezoelectric material which, when a voltage is applied across it, constricts into the reservoir, again causing the application of pressure with similar results. In this case, crude control of the droplet size is achieved by varying the voltage applied to obtain several different droplet sizes. These mechanisms are applied at combined rates in the order of several tens of thousands of times a second which enables a travelling print head to cover a paper width in about a second, despite firing spot sizes in the order of picolitres. The reservoir is then replenished from the ink supply at the end of each discharge cycle. These approaches are used by different manufacturers of the commercial printers which are supplied to the photographic fraternity under the generic description of inkjet printers.

Typically, a printer head will contain a number of integrated circuit subsystem printer elements, each element containing up to 200 or so jet-forming reservoirs. The arrangement of the printer elements depends upon the manufacturer and the type of job for which the printer is designed; for example, a mid-range quality colour printer to support a photographic operation might have four printer elements in line at 90 degrees to the direction of scan, providing in the order of 720 jets for each scan, with this layout being repeated for each ink colour such that in a four-colour printer, there might be some 16 integrated circuit printer elements. The manner in which these four sections of the scan are operated is dependent upon the complexity of the image; for simple text-based images, all four elements would be combined to form a 720-jet, 2.5 cm scan, whilst for a complex colour photograph, the ink layers might be built up from four mini scans of 180 jets, each covering about 6 mm of the image.

23.3   Colour Fundamentals in Printing

The foregoing implies that by using a combination of half-tone printing and three inks based upon the subtractive primaries, yellow, cyan and magenta in various combinations, we can produce a colour image. Whilst this is the case, the situation is not as simple as might be construed from the description in Chapter 2 of building a broad gamut of colours from a mixture of these three primaries. In that case, the colours produced were the result of mixing the pigments together in varying quantities; however, when using a half-tone approach the three primaries are laid down as individual dots on the paper at full intensity, there is therefore no variation in the spectral absorption characteristics of each dot as there would be by physically mixing the pigments. Thus assuming for the moment that the ink dots do not overlap, then there is no mixing of the pigments and we are therefore in the unique situation of using the subtractive primaries as effectively three primary light sources, which, together with the unprinted white of the paper, means that the image is formed by four primaries as perceived by the eye. The individual amounts of each primary pigment being dependent upon the area covered by each pigment, whilst the amount of white light reduces as the pigment area increases.

In Chapter 2, the basis of the operation of the subtractive primaries was investigated in some detail and the ideal spectral absorption curves for the three yellow, cyan and magenta primaries were seen to be block rectangular shapes, as illustrated in Figure 2.6, known as *block* dyes, whilst the result of adding any two primaries together was illustrated for actual pigments in Figure 2.7, which showed that roughly equal pair combination amounts produced the additive primaries red, green and blue, respectively. Adding all three primaries in equal proportions produced black.

In printing, the layer of a coloured ink behaves as a filter selectively absorbing light of certain wavelengths as it passes through before being reflected back through the ink by the paper. So now returning to our half-tone printing process, we note that the spectral characteristics of the light leaving the inks results from the reflection of the light from the surface of the paper passing through the ink a second time. If now we assume that the subtractive primary droplets can be arranged to sometimes overlap in a controlled manner, then we will produce at the overlap areas four new colours, red, green, blue and black, for the three pair combinations and the sum of all three pigments, respectively.

Thus, effectively in colour half-tone printing using three primary inks, we have the following eight primary colours:

- white, yellow, cyan, magenta, red, green, blue and black.

This process is illustrated in Figure 23.1, derived from tables in Worksheet 23, which shows the spectral distribution of the three idealised block dye primaries defined in Chapter 2, together with the results of the optical filtering which occurs when pairs of inks are overlaid. When all three inks are overlaid, no light filters through and black is produced.

Since each one of these primaries will contribute elements of the spectrum to the light perceived by the eye, they will effectively act as primary light sources. Thus, the total spectral distribution of the reflected light, $R_{T\lambda,}$ is the sum of the spectral distributions of each of the primary elemental droplets, in accordance with the percentage area of each droplet type, together with the white area, which will be the total area minus the sum of the pigment dot areas. The chromaticities of these primary inks in isolation are plotted in Worksheet 23(a) and illustrated in Figure 23.2.

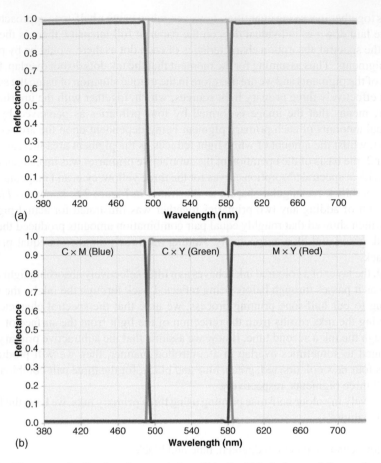

Figure 23.1   The spectral reflection characteristics of idealised block dyes and the result of overlaying them in pairs.

Figure 23.2 includes the sRGB gamut for comparison. It can be seen that although the chromaticities are highly saturated, the gamut of the primaries alone is very limited.

In Figure 23.3, the primaries resulting from the overlays of pairs of the CMY primaries are added to the chart. As can be seen, this results in a chromaticity gamut of comparable size and shape to the sRGB gamut.

    The original CMY primaries are located on a line between the overlaid RGB primaries.

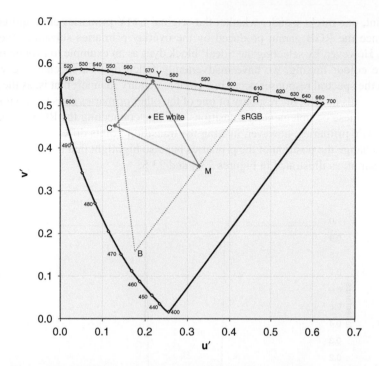

Figure 23.2 Illustrating the chromaticity gamut of the block dye primaries.

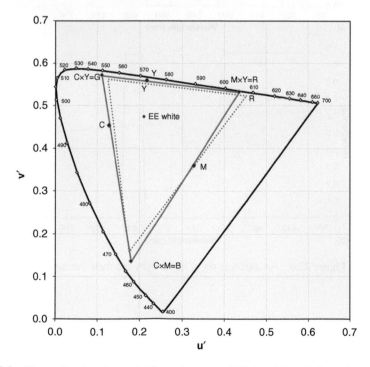

Figure 23.3 Illustrating the chromaticities and gamut of all six of the printing colour primaries.

At this point, one might wonder whether the original CMY primaries are required in their pure form since the RGB gamut produced by the overlay primaries subsumes the original CMY gamut. However, by selecting the 'ideal' block dyes as an example to explain the theory of subtractive colour mixing, we have inadvertently fallen upon a special case where the transitions in the spectral responses occur in a complementary manner; that is, as the response of one primary falls to zero, the response of one of the other primaries rises to maximum. As a consequence of these complementary transitions, the line connecting the RGB primaries will overlay the CMY primaries; however, making the transition points different from each other will not only change the position of the primaries but also highlight that the gamut is indeed a six-primary gamut, as illustrated in Figures 23.4 and 23.5.

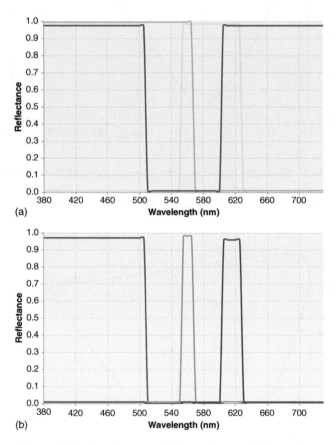

Figure 23.4 An alternate set of block dyes and their overlays.

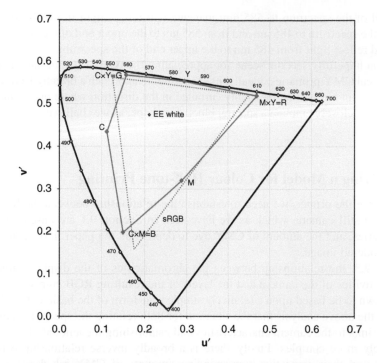

Figure 23.5   The chromaticities of the six alternate primaries.

Figure 23.5 clearly illustrates the contribution made by all six primaries. In Worksheet 23(a), one can change the spectral distribution of the example primaries and see the changes in the gamut appearing in Figure 23.5. It becomes evident that with judicious adjustment, the gamut can be made to match virtually any shape of RGB triangle defined by a set of specified primaries. The reason for this becomes clear if we recall the work on optimal colours in Section 4.7, where it was shown that these colours have very similar characteristics to the block dyes used in these examples.

Thus, to a first degree of approximation, six primaries derived from a block dye approach will always produce a relatively wide chromaticity gamut; however, it should be remembered that the chromaticity gamut represents only two of the three dimensions of a colour space and that these saturated primaries come at a cost, that cost being the relatively narrow band of colours and therefore the low luminance of the primaries. The result is that in these circumstances, we have a large chromaticity gamut but an increasingly small colour gamut as the width of the absorption spectrum of the CMY primaries is increased.

One of the important criteria of the inks for maximum chromaticity gamut is where the cross-overs occur between zero reflectance and maximum reflectance and these wavelengths may be broadly deduced both from an inspection of the chromaticity diagram and, remembering that each of the three primaries must reflect light in two spectral bands, representing two principal adjacent colours. Thus, the cyan primary must reflect both blue and green light, the yellow primary must reflect both red and green light, and the magenta primary must reflect red and blue light. Using the chromaticity diagram as a guide, cyan should therefore reflect light from the lower end of the spectrum to 585 nm; magenta will reflect light in two bands, from the lower end of the spectrum to 485 nm and from 585 nm to the upper end of the spectrum; whilst yellow should reflect light from 485 nm to the upper end of the spectrum.

This section hopefully sets the scene for appreciating the type of gamut one might expect from a set of real CMY primaries; ideally, its shape should approach a broadly RGB triangular chromaticity gamut but will be inevitably curtailed in the direction where one or more of the primaries fails to closely emulate a spectral block dye shape, as we shall see in the last section of this chapter.

23.4   Deriving a Model for Colour Half-tone Printing

In order to drive the printer, we need to establish the relationship between the RGB signals derived from a still camera, which as we have seen in Section 9.1 are based upon known display primaries, and the amount of CMY dye to deposit on the paper in order to obtain an accurately rendered image.

In Section 9.3, the relationship between the chromaticities of the display primaries, the spectral sensitivities of the camera and the levels of the resulting RGB signals were derived and were shown to be based upon a relatively simple transform of the basic CIE XYZ data. It is clear from the above analysis that the situation for both relating the spectral characteristics of the printer inks to the camera characteristics and establishing the amount of ink to deposit is considerably more complex. Firstly, there is a broadly inverse relationship between the RGB signals and their respective corresponding amounts of CMY ink deposited, and in addition, we now have eight primaries which may be used to replicate the desired colour; in effect, unlike with a three-primary system, this means that all but the maximum saturated colours may be rendered by different combinations of the primaries. Finally, as we shall see, there is a non-linear relationship between the amount of ink deposited and the amount of light reflected.

The inverse relationship for ideal 'block' dyes may be broadly but not accurately given by the following simple equations for r, g, b amounts of the linear RGB signals:

$$C = 1 - rR, \; M = 1 - gG \text{ and } Y = 1 - bB.$$

The relationship between the colours represented by the levels of the RGB signals and colours produced by the CMY dye volumes is referred to in the literature, which evolves this relationship as the printer *characterisation function*, which can be defined in two directions. In the forward direction, the function gives the colour values of the printed area obtained from the RGB values, whilst in the inverse direction, the characteristic describes the levels of RGB required to establish known printed colour values. The inverse function may be recognised as identical in concept to deriving the spectral sensitivities of the camera from the chromaticities of the display primaries, as described in Section 9.3, and could be used to provide the data necessary for driving the printer.

The approach to deriving the inverse characterisation function is complex mathematically in terms of the level adopted in this book and, anyway, to a degree, falls back on heuristic approximations as it is not practical for the model to include all the second-degree effects in the printing process. Thus, in what follows, the model will be described in either simplified mathematical terms or in general descriptive terms, which will give the reader an understanding of the physical processes without the detail required to define the mathematical relationships. However, the sources of the work required to fully define the inverse characteristic are given in the references.

23.4.1 Establishing the Forward Printer Characterisation Function

A number of models have been explored to describe the printer forward characterisation function but the one that has received most interest is the Neugebauer model.

23.4.1.1 The Neugebauer Equations for Determining Half-tone Reflectance

In 1937, Neugebauer (Neugebauer, 1937) and, in 1989, Sayanagi (Sayanagi, 1989), in working on a mathematical model to describe the print characterisation function, used the eight primaries described above to express the total reflected light from the half-tone print by wavelength using the sum symbol in the following manner:

$$R_{T\lambda} = \sum {}_i A_i R_{\lambda,i,\max} \tag{23.4}$$

This formula states that the total reflected light at each wavelength λ per unit area is the sum of each of the colour pigment reflectances $R_{\lambda i}$ by wavelength, designated by the ith colour, times the unit area A_i of that colour pigment.

Neugebauer then went on to extend the Murray–Davies equation to incorporate the above reasoning by assuming that if the half-tone dots are printed randomly on the paper, then the Demichel[2] probability equations could be used to express the unit amount of each colour in terms of the effective coverage area of the eight primaries as follows:

$$
\begin{aligned}
A_w &= (1 - a_c)(1 - a_m)(1 - a_y) \\
A_c &= a_c(1 - a_m)(1 - a_y) \\
A_m &= a_m(1 - a_c)(1 - a_y) \\
A_y &= a_y(1 - a_c)(1 - a_m) \\
A_r &= a_m a_y(1 - a_c) \\
A_g &= a_c a_y(1 - a_m) \\
A_b &= a_c a_m(1 - a_y) \\
A_k &= a_c a_m a_y
\end{aligned}
\tag{23.5}
$$

where a_c, a_m, and a_y are the effective fractional coverage areas of cyan, magenta and yellow, respectively; and a_k is the colour represented by black. (In fact, depending upon the type of printer, the half-tone dots vary in the degree of randomness achieved, which causes inaccuracies in this simple equation.)

In 1951 Yule and Neilson undertook detailed work (Yule & Neilsen, 1951) to establish why the simple Murray–Davies model (Murray, 1936) did not provide a match between calculated

---

[2] These equations, known as the Demichel (Demichel, 1924) equations, were first published in 1924 in a now-forgotten French printer's review called *Le Procédé*. (Amidror & Hersch, 2000)

and measured results when using coloured inks. Their work showed that, in addition to the physical dot gain, the effects of light scattering in the paper, caused by some of the light entering a dot exiting via the paper surface surrounding the dot whilst the remainder exiting through the ink, caused a change in the effective area of reflectance of the dot. This effect is referred to as optical dot gain.

It was found that optical dot gain was non-linear and could be described by making the reflected light vary in an exponential manner in the classic Murray–Davies expression developed in equation (23.1):

$$R_\lambda^{1/n} = (1 - a_{\text{eff}})\, R_{\text{p}\lambda}^{1/n} + a_{\text{eff}} R_{\text{i}\lambda}^{1/n} \tag{23.6}$$

where n is a parameter representing the light spreading into the paper and is described as the Yule–Neilson n-value.

The value of n varies depending upon a number of factors: fundamentally, the spread of light in the type of paper used but also other effects, including the varying depth of the ink and its effect in ink overlap areas. Values of n may range from about 1.7 to in excess of 2 as the resolution of the printer increases.

The Yule–Neilson effect modifies equation (23.4) as follows:

$$R_{\text{T}\lambda}^{1/n} = \sum {}_i A_i R_{\lambda,i,\text{max}}^{1/n} \tag{23.7}$$

We are now in a position to establish the relationship between ink dot volumes in digital count terms and the corresponding primary coloured ink reflectances, which is a two-stage process. Equation (23.7) enables a relationship to be derived between the volume of the ink drops in digital count terms and the CMY dot areas. Deriving the second stage is beyond the scope of this book; however, Bala (2003) describes how the Murray–Davies version of equation (23.5) is manipulated to derive the relationship between dot areas and the amount of light reflected from each of the primary inks, using vector-matrix mathematics in an iterative process.

Using this relationship enables an electronic test chart comprising step wedges of the cyan, magenta and yellow dyes between black and white to be used to control the printing of each of the primaries, whose colour parameters may then be measured and used to determine the actual relationship between digital ink level count and the colour of the light reflected by each primary.

23.4.2   Establishing the Inverse Printer Characterisation Function

As indicated in Section 23.4, in order to produce an image with satisfactorily rendered colour using a model-type approach, we need to invert the forward model. This is an exacting and complex task, taking into account that the forward model is already complex, particularly when the secondary effects of the printer process are taken into account. Although a solution was found by Mahy and Delabastita (1998), a more heuristic approach, which is independent of the inverse characterisation function, is normally more generally adopted.

This approach is based upon using the forward model to predict the input necessary to attain a required result and is therefore independent of any particular inverse model. The approach is based upon the following steps:

- The printing of a number of training samples onto the particular type of medium to be used. The samples should explore the full gamut of the printing inks and should aim to take into account the known forward characteristics of the printer such that adjacent samples should, where possible, be perceptibly linearly displaced. (If the RGB samples are derived from a table of equi-spaced L values, for example, this will assist in contributing to making the accuracy of the final interpolation step more likely to produce acceptable results.)
- Derive the inverse function. The colour of each printed sample is measured and used as the basis of a mapping process whereby the RGB input values for a particular sample are matched against the RGB values of the closest printed sample, which in turn are mapped to the values of CMY used to produce that sample. The mapping process is implemented via a look-up table[3] or LUT, whereby the RGB values are used to form the input to a three-dimensional LUT, the output of which are the CMY input values originally used to drive the printer to obtain the sample colours. In an ideal situation, the LUT would hold the values for every defined system colour; however, even in an 8-bit system, this would require the LUT to hold in excess of 16,000,000 values.
- Thus, derive an interpolation procedure to establish the required CMY output values for those colours whose RGB values fall between the training sample values.

The higher the number of training samples, the less interpolation will be required and therefore, taking into account the residual non-linearity of the forward characteristic, the smaller will be any interpolation error, but the larger will be the LUT. Typically between 500 and 2,000 colour samples are used to build the LUT.

23.5 Practical Printer Performance

Having explored the theoretical basis of the printing process, we can now turn our attention to the colour performance of practical ink jet printers. In comparison to the minor limitations of colour gamut in display devices, the printing process is considerably limited by the spectral reflectance characteristics of the inks, which diverge considerably from the ideal block spectral shapes discussed in Section 23.3.

Few printer manufacturers release precise details of the spectral characteristics of the inks used in their printers; thus, in order to provide a description of the colour performance of the printer, we will use the spectral responses measured with the author's spectrophotometer. There are problems with this approach however; ideally to obtain a true response it is necessary to provide areas of print which relate only to the ink of each primary; however, without specialised printer drivers it is difficult to achieve this result precisely. Nevertheless, by producing specialised electronic test charts comprising 100% saturated additive and subtractive primary colours located beyond the gamut of the inks, it is possible to achieve colour patches which inspection with a microscope indicates are very nearly pure colours, that is, there is very little contamination from the other inks. This is the approach adopted for all the results illustrated in this section, and whilst it is considered that the results are representative of the actual ink colour characteristics there may be some variation from the specification of the

---

[3] For those unfamiliar with LUTs, the description at http://www.lightillusion.com/luts.html may be helpful.

manufacturer. The tables of spectral responses and the chromaticity diagrams derived from them are contained in Worksheet 23.

23.5.1 The Three-Colour Inkjet Printer

Whilst very few, if any, inkjet printers currently use only three inks, exploring the characteristics of such a simple printer provides the answers as to why more sophisticated printers use increasing numbers of inks. The spectral reflectance of any specific ink set will be dependent upon the manufacturer, with each manufacturer vying to improve the performance of their inks, not only in terms of their spectral distribution but also in their other characteristics relating to: droplet absorption on the paper, avoidance of damage from abrasion and liquid spills and fading. Thus, the diagrams of spectral distributions appearing in this section, which have been measured[4] by the author, should be considered as representative of the printer type rather than relating to a particular model.

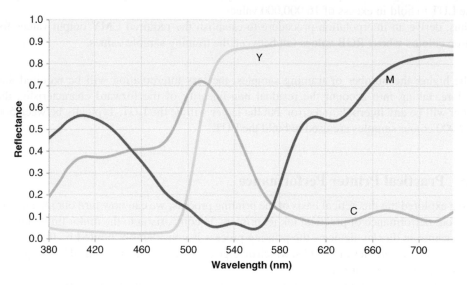

Figure 23.6   Typical spectral reflectance characteristics of CMY inks.

In Figure 23.6, the spectral reflectance of three CMY primaries are illustrated. It is immediately evident that apart from the yellow ink, which conforms reasonably well to the ideal spectral shape, the shapes of the cyan and magenta spectral characteristics diverge considerably from the ideal.

---

[4] Measurements were made on Epson Premium Glossy paper using the X-rite i1 Pro2 spectrophotometer in conjunction with the Spectrashop application. The results at 1 nm intervals were then extracted from Spectrashop into Worksheet 23(z), which samples the data at 5 nm intervals before being passed to Worksheet 23 for processing. Spectrashop: http://www.rmimaging.com/spectrashop.html

Reviewing the responses against the 485 nm and 585 nm cross-over criteria outlined in the previous section, the yellow midpoint response occurs a little away from the ideal at about 510 nm, the magenta midpoint responses occur at about 470 nm and 590 nm, and the cyan at about 550 nm, well away from the ideal of 585 nm.

Furthermore, both the magenta, and particularly cyan, inks have responses which fall off rapidly at the lower end of the spectrum, where ideally they should be fully reflecting; a further illustration of the situation explained in Chapter 2 as to why magenta is always perceived as appearing nearer red than blue in hue. Clearly there are fundamental reasons which prevent an ink being fabricated with characteristics approaching the ideal.

When pairs of coloured inks are overlaid, an approximation of the resulting spectral reflection may be found by convolving their respective response curves as illustrated in Figure 23.7.

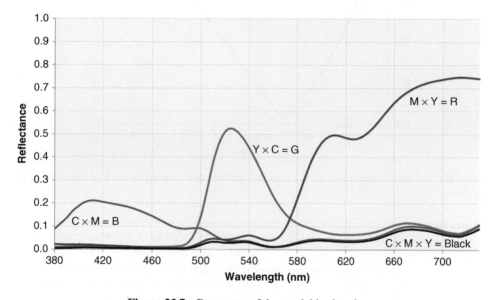

Figure 23.7 Responses of the overlaid primaries.

The limitations in the responses of the CMY primaries are of course multiplied in the responses of the RGB primaries which result from the overlays. The red response does bear some resemblance to the ideal. The green response is broadly constrained within the 485–585 nm limits, indicating that saturated greens will be achieved, albeit only at a relatively low lightness level as the peak response is limited to 50% rather than 100%.

The blue response is very poor, reaching a maximum of only 20% and spreading well across the spectrum as a result of the cyan ink failing to fall to zero at 585 nm, as it ideally should.

Finally, the black is poor, having a Y value of about 3.6 (paper white made equal to 100) and being of a distinctly yellow hue. Such a limited contrast range of 28:1 would be incapable of producing subjectively acceptable renditions of many scenes.

The result of these limitations on the colour gamut is illustrated in Figure 23.8.

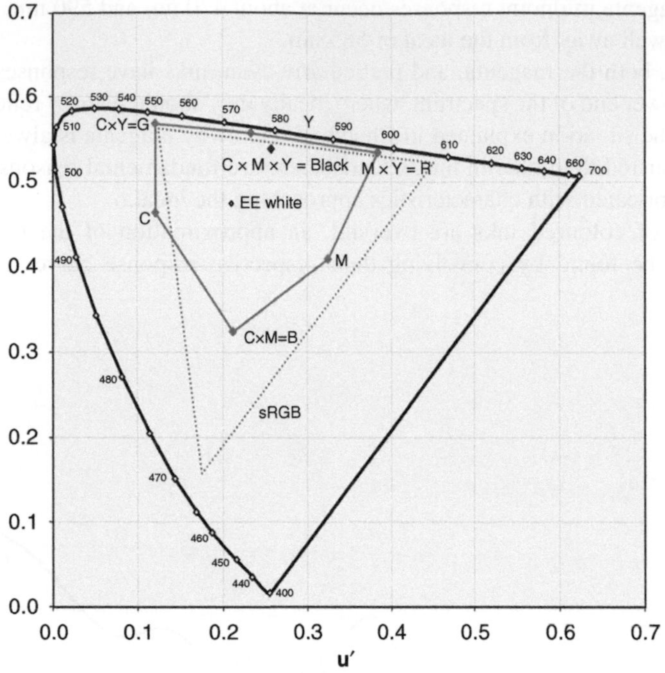

Figure 23.8   The representative gamut of a three-ink printer; also showing the lack of neutrality of the black overlay.

As would be expected from the spectral responses, the location of the blue primary has heavily restricted the extent of the gamut in the blue direction. However, it can be clearly seen that the other primaries do broadly support the shape of the ideal RGB gamut.

Nevertheless, the gamut does cover a substantial area of the colours found in an average scene and is therefore capable of providing a satisfactory rendition of the scene colours unless recognisable saturated colours are present (albeit the darker areas of the scene will be disappointing).

23.5.2   Addressing the Black Problem

In a three-ink printer, black is produced by using maximum concentrations of the three coloured inks, and as we have seen in the previous section, the result is a poor black because of the limited absorption of one or more inks in the spectral bands where their absorption should be total. An obvious solution is to introduce a black ink of high absorption across the full spectrum, and whenever the print driver identifies elements of the scene corresponding to:

- a neutral grey or
- dark tones with colours of low saturation

to replace the coloured inks with a combination of a smaller quantity of the coloured inks with an appropriate amount of black. This process, known as Grey Component Replacement (GCR) or, for commercial presses, Under Colour Removal (UCR), is used universally in inkjet printers in order to bring about very substantial improvements in the rendition of the image. Not only are the near black areas properly rendered but the greyscale neutrality is easier to control.

The black ink is given the abbreviation 'K' in order to avoid any ambiguity with 'B', representing blue. Thus four ink printers are described as CMYK printers.

We have looked at the problem from a colorimetric viewpoint; however, there are other advantages to using black ink as a substitute for the three coloured inks, which include a considerable cost-saving as the coloured inks are considerably more expensive than black ink, and also overcoming problems which can occur as a result of depositing too much ink on the paper, particularly in some types of commercial printing presses.

23.5.3   The Use of Additional Inks

The basic four-ink printer is capable of producing satisfactory prints for the average consumer but does have limitations related to limited colour gamut, gradation of hue and gradation of greyscale. The limitations in colour gamut have been outlined in Section 23.5.1, the limitation in gradation of hue and greyscale relate to the fundamental nature of the half-tone process. As we saw in Section 23.2.4, to lay down a gradual change between zero and maximum of any particular colour, including grey, requires a corresponding variation in the number of ink spots comprising a pixel. Since, even with relatively high resolution printers, there is a limit to the number of spots which make up a pixel, this limits the level of gradation which can be achieved.

In hue terms, the effect is more noticeable in the use of the darker colours cyan and magenta, such that when attempting to reproduce subtle changes in gradation on pale skin, for example, the risk of introducing perceptible contouring between adjacent areas is increased. The solution to this problem is to either increase the resolution of the inkjet process or provide additional inks of pale cyan and possibly pale magenta into the process, whereby a number of spots of the paler colour equate to one spot of the fundamental primary.

The same effect is manifest in the rendition of an even greyscale between black and white, the solution being the addition of a number of additional grey inks with increasing reflectivity. Those printers which claim excellent rendition of monochrome images may have up to three different grey inks.

The limitations in colour gamut were seen to be primarily the result of the divergence of the spectral response of the cyan and magenta inks away from the ideal, leading to a much curtailed area of blue in the chromaticity gamut and a reduced volume in the red area of the colour gamut. It should be appreciated that although the fundamental nature of the restricted responses at the blue end of the spectrum in both the cyan and magenta inks limits their ability to produce red and blue primaries by overlaying, it does not necessary follow that the same would be true of other ink formulations which target these red, green and blue primaries directly. Thus, in order to overcome the limitations of overlay primaries, additional inks may be added to supplement them, the most common being red and/or blue inks. For specialised applications, inks of specific hue may also be included.

Inkjet printers now commonly contain up to six inks, and printers designed for discerning practitioners at the top of the market may contain up to 12 inks, typically a cyan, magenta and yellow, two blacks, three greys, a gloss and a selection of two or three of the following: a pale cyan, pale magenta, and red and blue.

In addition to the ranges of ink colours, printers aimed at the professional market may use pigment inks which can have spectral responses which are closer to the ideal than those illustrated earlier in this section. As an example of the spectral response of the inks in different printers from the range of the same manufacturer, the results measured from two different printers in the range will be described.

23.5.3.1   An Example Mid-Range Printer

The three inks described earlier are part of the ink-set in a mid-range printer which also includes a red and a blue ink with the measured spectral characteristics illustrated in Figure 23.9.

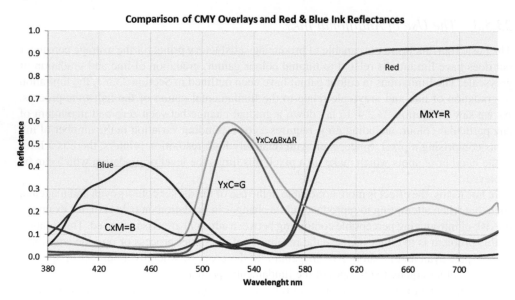

Figure 23.9   A comparison of overlay and dedicated red and blue primaries.

In the figure, the spectral responses of the red- and blue-dedicated primary inks are compared with those primaries derived from the overlay of the CMY primaries.

The closer match to the ideal block primaries is apparent; the blue primary has double the lightness of the overlay primary; it has a spectral shape with a half-amplitude point close to the ideal of 485 nm and furthermore has very little reflectance over the remainder of the spectrum; the red primary has a very similar half-amplitude wavelength to the overlay red, leading to a similar chromaticity but has a much improved lightness response, leading to a significant expansion to the colour gamut in the red area. It is also evident that the printer driver has been adjusted to boost the green lightness by the addition of small amounts of the red and blue primaries.

These observations are supported by the shape of the chromaticity gamut as illustrated by the broken lines in Figure 23.10, where the original CMY gamut is shown for comparison.

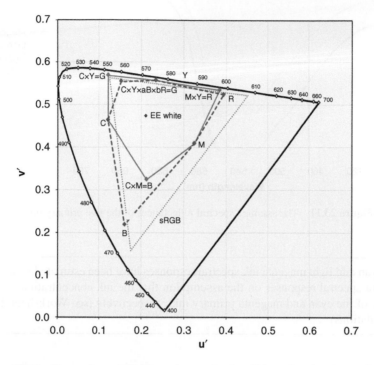

Figure 23.10 Comparison of CMY overlays and red and blue primaries chromaticities.

The addition of the blue ink has extended the gamut considerably, bringing it recognisably closer to the shape of the ideal in terms of an RGB gamut. As expected, the addition of the red ink has had little effect on the chromaticity gamut and the effect of broadening the green primary by the addition of some of the red and blue inks has been to reduce the chromaticity gamut in the green direction. It would appear in this case that there has been a trade-off of chromaticity gamut for an extension of the colour gamut, that is, the saturated colours are lighter but less saturated.

Printers with the capability of producing a gamut as illustrated in Figure 23.10 will render prints approaching the colour range of that achievable by display devices.

23.5.3.2   An Example Professional Printer

The printer to be described uses pigment inks which have spectral responses which are significantly closer to the ideal block responses, as is illustrated in Figure 23.11. The coloured inks in this printer are cyan, pale cyan, magenta, pale magenta and yellow.

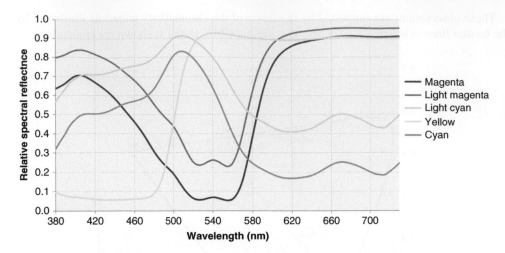

Figure 23.11   The assumed spectral reflectances of the five primary inks.

The light cyan and light magenta ink spectral responses have been estimated from the cyan and magenta spectral responses on the assumption that the ink concentration of the pale inks is 50% of the cyan and magenta primary inks, respectively (see Worksheet 23, Table 12 for calculations).

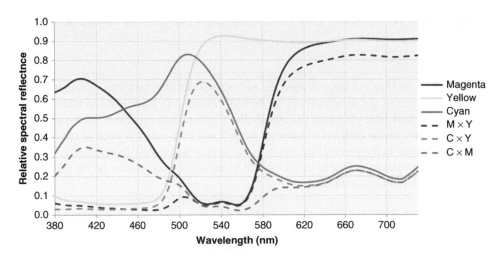

Figure 23.12   Calculated spectral reflectance of the RGB primaries.

When the cyan, magenta and yellow inks are overlaid, the resulting red, green and blue primaries are as illustrated in Figure 23.12, which shows the result of convolving the primary pairs: red = M × Y, green = C × Y and blue = C × M.

The overlaid red and green primaries are as reasonably close to the ideal spectral response as would be expected from the limited match of the original CYM primaries; however, the overlaid blue spectral response is a very poor match to the ideal block dye response.

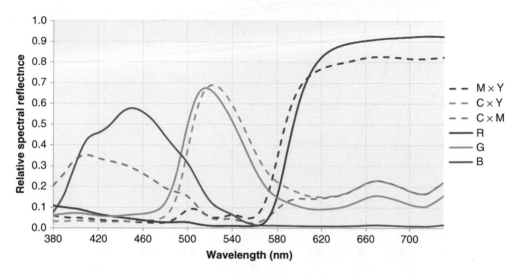

Figure 23.13   Spectral reflectance of the overlaid and measured primaries.

In Figure 23.13, the spectral responses of the derived overlaid primaries are compared with the measured red, green and blue printed primary patches. The close comparison of the red and green responses indicates the validity of the overlay premise; however, the measured blue response is very much better than that predicted from the overlay calculation.

23.5.3.3   Comparison of the Colour Gamuts of the Two Printers

The chromaticity gamuts of the two printers, based upon the measured chromaticities of the patches representing the printed primaries, are illustrated in Figure 23.14.

At first sight, there does not appear to be much to choose between them. The dedicated blue primary of the mid-range printer extends the gamut towards the blue segment of the spectrum locus and the green primary of the mid-range printer is comparatively curtailed.

The reason for the improved chromaticity gamut of the professional printer becomes evident when comparing the measured spectral responses of the two sets of inks, as is illustrated in Figure 23.15. The considerably lighter cyan and magenta inks are responsible for the extended chromaticity gamut.

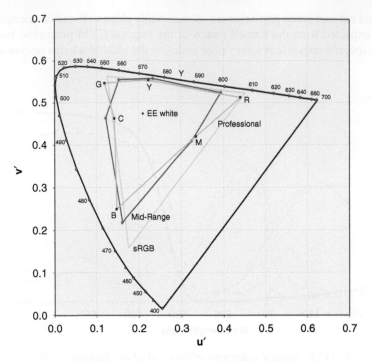

Figure 23.14 Comparison of chromaticity gamuts of the two printers.

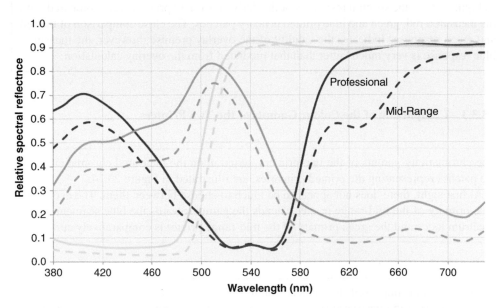

Figure 23.15 Comparison of the spectral responses of the inks of the two printers.

More to the point, although there is a marginal improvement in the chromaticity gamut, the lighter inks will lead to a considerably larger colour gamut, indicating the capacity to render prints with colours of higher chroma.

23.5.4   The Printer Paper

The final step in the printing process is the paper on which the print is rendered since its colour and surface reflection characteristics can significantly affect the rendition of the image.

23.5.4.1   Paper Colour

The light incident on a print travels through the ink layer and is reflected by the paper before returning through the ink and being emitted from the surface of the ink or any protective or gloss coating present. Thus, the spectral reflection characteristic of the paper will be imparted to the light leaving the surface of the image. In particular, for increasingly pale colours, any paper spectral characteristic will increasingly dominate the character of the reflected light.

Evidence indicates that people generally prefer white to appear slightly bluish. This characteristic is so strong, or has become strong by cultural heritage, that very often white paint, white paper and white sheets, etc. contain a fluorescent agent which reacts to any incident ultraviolet light to emit a blue light, causing the paper to have a definite blue cast, an effect often referred to as the result of an optical brightener.

This is no less true for photographic paper and, as a result, very many commonly used papers have a blue bias; this is not a problem if that is what is desired, but it should be borne in mind when critically evaluating a print, that in the highlight areas particularly, the perceived colour can be affected by the colour of the paper white.

In Figures 23.16 and 23.17, the reflectance and chromaticity, respectively, of two white photographic papers are illustrated. The Hahnemuhle Photo Rag has been selected by the

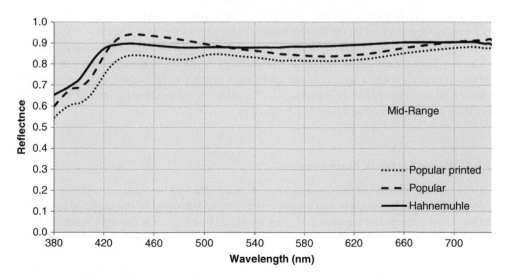

Figure 23.16   The reflectance of a pair of white photographic papers.

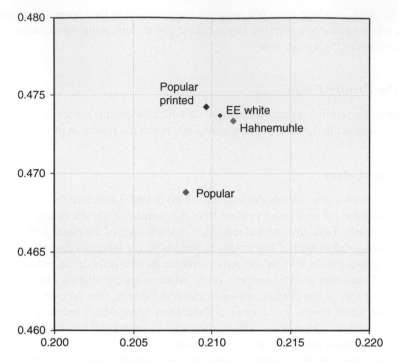

Figure 23.17   The chromaticity of a pair of white photographic papers.

author for its good spectral neutrality, as can be seen, it is very close to equal energy white on the chromaticity diagram. It does however have the perception of a distinct pale cream colour when compared with popular white inkjet papers.

The typical popular white glossy paper shows a distinct rise at the blue end of the spectrum and a dip in the yellow area, leading to its 'white' appearance, albeit it is clearly some way away from equal energy white on the chromaticity diagram in the blue direction.

Of interest is the 'Popular Printed' white which resulted from adjusting the print profile (see Chapter 26) to make white appear as equal-energy white; in so doing, it ensured that on the printer receiving a white signal, a light application of ink was laid down to compensate for the uneven white of the paper. In consequence, the white of the rendered print will be slightly darker than any white surround that may be present.

23.5.4.2   Paper Surface Characteristics

The surface characteristics of the paper can significantly affect its appearance; a glossy surface will reflect that light not absorbed at an angle opposite to but equal to the angle of incidence such that when viewing the paper at any other angle, all that is seen is the light reflected from within the underlying surface. In contrast, a truly matt surface will reflect unabsorbed light equally in all directions, and since this is usually white light, it will mix with the light reflected from the underlying surface, causing any colours generated in the underlying surface to be desaturated.

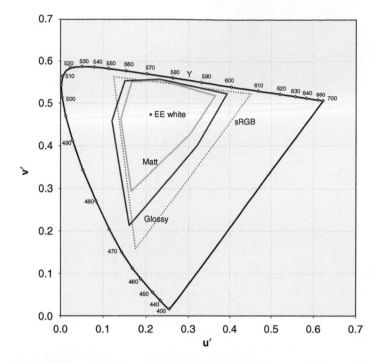

Figure 23.18 Comparison of the same ink gamut on matt and glossy paper.

The result on the chromaticity gamut is illustrated in Figure 23.18, where the gamuts of the inks discussed earlier are plotted when printed on both matt and glossy paper.

Most matt papers have an element of gloss in their surface characteristic, which with care enables the desaturating affect to be minimised by arranging the lighting at 45 degrees to the surface whilst viewing at an angle of 90 degrees.

23.6 Conclusions

The theoretical basis for the half-tone printing process has been reviewed in some detail and the fundamentals of using pigments as primary colours for generating a broad gamut of colours have been charted.

The configuration of modern inkjet printers has been described and their performance in terms of their capability of producing quality colour images has been measured.

Figure 23.18   Comparison of the same ink gamut on mat and glossy paper

The result on the chromaticity gamut is illustrated in Figure 23.18, where the gamuts of the inks discussed earlier are plotted when printed on both mat and glossy paper.

Most mat papers have an element of gloss in their surface characteristic, which with care enables the disturbing effect to be minimized by arranging the lighting at 45 degrees to the surface whilst viewing at an angle of 90 degrees.

23.6   Conclusions

In this section the need has been shown to characterize devices in terms of their colorimetry, and to make use of these sets of data, so that colours can be faithfully portrayed in successive devices.

The paper gamut of a new ink-jet printer has been described, and a new combination of ink set and kind of paper to give quality colour images has been described.

24

Colour Spaces in Photography

24.1 Introduction

A first exposure to the range of colour spaces used in photography is likely to lead to some confusion; there is a plethora of defined colour spaces which are often referred to in the literature without reference to their history or relevance to the situation being described. There are a number of reasons for this, mostly historical as different sectors of the industry acting independently sought to define colour spaces for use in their particular domain. Within Adobe® Photoshop®, the situation is further confused by the availability of a range of legacy colour spaces, some relating to the early days of colour television before an international standard emerged in that area and so are no longer relevant to the majority of photographers. Nevertheless, despite the growing resolve to agree an international standard, it became clear there was also a requirement to define different colour spaces for different purposes.

This chapter sets out to delineate these colour spaces into categories related to the photographic workflow so that the reader may differentiate between those that are currently relevant and those that are only of legacy interest.

24.2 Colour Spaces in Image Capture

Though formally referred to as source colour spaces, those adopted for image capture often continue to be used in subsequent elements of the photographic workflow. The photographer with a 'bridge' or professional camera is offered the option of selecting which of these colour spaces will be used to capture the image of the scene, though often in modern cameras, the image may be captured in two colour spaces simultaneously.

As implied in Chapter 22, the CRT market was driven by the huge domestic television receiver sales which kept the cost of production down and thus made the domestic CRT the practical choice as the display element for consumer computer monitors. The chromaticity coordinates of the primaries of these monitors were therefore those of the television system, and thus, the television chromaticity gamut was incorporated into a photographic colour space

Colour Reproduction in Electronic Imaging Systems: Photography, Television, Cinematography, First Edition. Michael S Tooms.
© 2016 John Wiley & Sons, Ltd. Published 2016 by John Wiley & Sons, Ltd.
Companion Website: www.wiley.com/go/toomscolour

specification. However, as is evident from the considerations outlined in Section 12.2, such a chromaticity gamut is significantly limited in the range of colours it can accurately render.

It was recognised early on that more specialist sectors of the industry would require defined colour spaces which were not limited by the chromaticities of the television gamut.

It should also be recognised that the chromaticity gamut defines only two of the three parameters of a colour space; the lightness parameter defined by the relative luminance signal also forms an important element of the colour space gamut, defining the tonal range which can be rendered. Thus, care should be taken when considering whether a *gamut* is describing a *chromaticity area* or a *colour space* and, where there is the possibility of ambiguity, the description should make it clear which is being described.

There are notionally four colour spaces used in image capture which are named as follows:

- sRGB
- Adobe® RGB
- Adobe Wide Gamut RGB
- RAW RGB

24.2.1   The sRGB Colour Space

In 1996, Microsoft and Hewlett Packard (HP), who were leading players in the evolving digital photographic industry, recognised the requirement for defining a common colour space for use in the embryonic industry. They also recognised that although the International Color Consortium (ICC) (see Section 27.3) were active in addressing the requirement for standardising the approach to specifying colour profiles, which would enable different colour spaces to be used for different elements of the workflow, there were major sectors of the evolving industry that did not need such a level of sophistication but did need effectively a default colour space standard which the industry could adopt.

On the basis of the rationale outlined above, the chromaticity gamut and the white point of the then recently defined television system, as described in Recommendation ITU-R BT.709 (Rec 709), was adopted as a photographic standard, but importantly, a more critical gamma correction characteristic was specified, since the television gamma correction was based upon a different viewing environment than that envisaged for viewing photographic images. This colour space was defined as *the Standard* RGB colour space or sRGB. Subsequently, this specification was adopted as an international standard in IEC61966-2-1.[1]

The approach to defining the transfer function was adopted from the television industry, based upon the combination of a linear and an exponential characteristic (see Section 13.4.3), but the parameters selected for these two elemental characteristics ensured that the characteristic extended over a significantly larger contrast range and was applied more accurately than was the case for the television version. For a description of the gamma correction characteristic of Rec 709, see Chapter 19.

Thus, the parameters of the sRGB colour space are shown in Table 24.1.

---

[1] 'IEC 61966 Multimedia Systems and Equipment – Colour Measurement and Management' is a comprehensive set of international standards comprising 11 parts; Part 2-1 covers the specification of sRGB.

Table 24.1 Parameters of the sRGB colour space

Primaries	x	y	z	u′	v′
sRGB colour space parameters					
Red	0.6400	0.3300	0.0300	0.4507	0.5229
Green	0.3000	0.6000	0.1000	0.1250	0.5625
Blue	0.1500	0.0600	0.7900	0.1754	0.1579
White D65	0.3127	0.3290	0.3583	0.1978	0.4683

Transfer function	R′G′B′
For RGB ≤ 0.00304	$12.92 \times RGB$
For RGB > 0.00304	$1.055 \times (RGB)^{1/2.4} - 0.055$

The chromaticity gamut and associated camera spectral sensitivities of the sRGB colour space are derived in Worksheet 24 and illustrated in Figures 24.1 and 24.2, respectively whilst the gamma correction characteristic, derived in Worksheet 13(b), is shown in Figures 24.3 and 24.4.

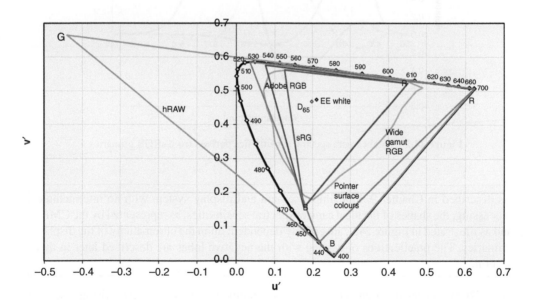

Figure 24.1 Chromaticity gamuts of the image capture colour spaces.

Figure 24.1 illustrates the comparison between the sRGB gamut and the Pointer gamut of real surface colours and highlights the limitation of the sRGB gamut in terms of its inability to accurately render a large range of surface colours of high saturation.

In Worksheet 24, the coefficients required to derive the sRGB CMFs in terms of the XYZ CMFs are calculated by selecting the appropriate button and are shown in Table 24.2.

Table 24.2   Coefficients of XYZ CMFs required to derive the sRGB CMFs

	X	Y	Z
R =	3.4177	−1.6212	−0.5258
G =	−1.0221	1.9783	0.0438
B =	0.0587	−0.2151	1.1146

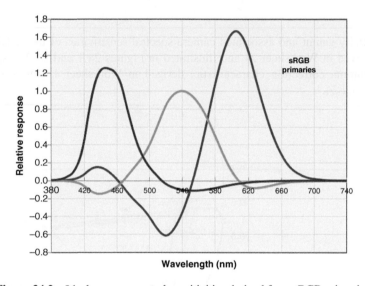

Figure 24.2   Ideal camera spectral sensitivities derived from sRGB primaries.

As described in Chapter 9, in a simple camera and display system with no intermediate processing, the shapes of the ideal camera spectral sensitivities, as represented by the CMFs and as illustrated in Figure 24.2, are entirely dependent upon the chromaticity of the display primaries. The practicalities of dealing with the negative lobes are described later in this section.

The sRGB transfer function is based upon a combined characteristic comprising a section with a linear gain of 12.92 for signal levels at or below 0.304% and a section with a power law characteristic with an exponent of 1/2.4, equal to 0.4167, for signal levels above 0.304%. As illustrated in Figures 24.3 and 24.4, which are derived in Worksheet 13(b), this produces a combined characteristic which is a close match to a single power law characteristic with an exponent of 1/2.2 or 0.4545 for signals in the input range of 10–100% and a reasonably close match in the input range of 4–10%.

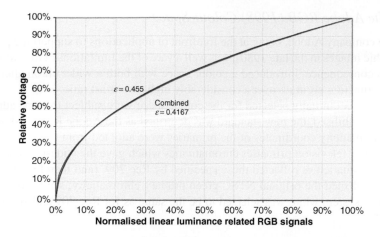

Figure 24.3 Comparison of sRGB gamma correction characteristic and a power law characteristic with an exponent of 0.4545.

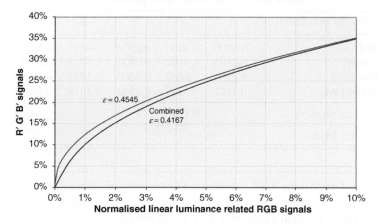

Figure 24.4 As Figure 24.2 but over the input level range of 0–10%.

In contrast to the Rec 709 characteristic, the power law exponent value of 1/2.4 and the higher value of the gain of the linear section of the characteristic of 12.92 ensure the characteristic faithfully matches a power law over a much greater contrast range. As the chart illustrates, there is a good match between the two curves down to the 1% input level, a 100:1 contrast range. The overall system gamma, being the product of the display gamma and the correction gamma, is therefore 2.4 × 0.4545, that is, about 1.1.

The sRGB standard has become ubiquitous throughout not only the photographic industry but also those related to it, such as the graphics and design industries, and most importantly, images on the Internet; it is their default colour space and is regarded as the colour space to be assumed where no identification of the colour space in use has been given. All consumer cameras and printers adopt the sRGB colour space as their default working colour space, enabling colour to be managed without the complication of using profiles to identify the working colour space.

24.2.2   The Adobe RGB (1998) Colour Space

The software company Adobe, being at the forefront of applications to support the processing of photographic images in the late 1980s, was well aware of the limitations of the sRGB colour space and, in consequence, introduced a colour space with both a wider chromaticity gamut and a transfer function accurate over a considerably higher contrast ratio.

As the parameters initially selected for the colour space were subject to some adjustment, once they were stabilised, the new standard was described as the Adobe RGB (1998)[2] colour space. The chromaticity coordinates of the primaries were adopted from those of the various current and legacy television primaries chromaticities which gave the widest gamut; thus, the red and blue chromaticities reflected those specified by Rec 709, (and therefore also sRGB), and the green reflected the original NTSC green primary chromaticity; see Section 17.2 and Table 17.1. The system white was selected to be the same as for Rec 709 and sRGB, that is, Illuminant D65.

Table 24.3   Parameters of the Adobe RGB colour space

Adobe RGB (1998) colour space parameters					
Primaries chromaticities	x	y	z	u'	v'
Red	0.6400	0.3300	0.0300	0.4507	0.5229
Green	0.2100	0.7100	0.0800	0.0757	0.5757
Blue	0.1500	0.0600	0.7900	0.1754	0.1579
White D65	0.3127	0.3290	0.3583	0.1978	0.4683
Transfer function	$R'G'B'$				
For $0 > RGB \leq 1$	$(RGB)^{1/2.19921875}$				

The parameters of the Adobe RGB colour space are detailed in Table 24.3.

The positions of the primaries on the chromaticity chart are illustrated in Figure 24.1, where the gamuts of the source primaries may be compared.

In Worksheet 24, the matrix to derive the CMFs required for the camera spectral sensitivities is calculated and gives the values listed in Table 24.4.

Table 24.4   Matrix values to derive Adobe RGB CMFs from XYZ[3]

	X	Y	Z
R =	2.1529	−0.5958	−0.3635
G =	−1.0221	1.9783	0.0438
B =	0.0142	−0.1248	1.0705

---

[2] http://www.adobe.com/digitalimag/pdfs/AdobeRGB1998.pdf

[3] These values are slightly different in absolute terms to those given in the cited reference for the Adobe RGB (1998) specification; however, the critical relative values are identical.

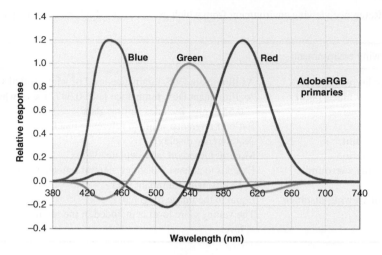

Figure 24.5   Adobe RGB idealised camera spectral sensitivities.

Using the values in Table 24.4, Worksheet 24 calculates the ideal camera spectral sensitivities as illustrated in Figure 24.5. Note that by adopting a more saturated green primary, the negative lobe of the red analysis curve has been significantly diminished by comparison to the sRGB idealised camera spectral sensitivities.

The transfer function characteristic breaks with tradition in being based upon a simple power law function with an exponent or gamma of notionally 1/2.2. However, in order to be compatible with digital encoding and processing elements, the actual value is $1/(2 + 51/256)$ or $1/2.19921875$. Values of RGB outside of the range 0–1.00 are clipped.

The Adobe RGB colour space specification clearly indicates a power law characteristic with no linear section at the lower end of the curve to ameliorate the increasing gain towards infinity as the zero-level input point is approached. However, in practical terms, at some input level below the contrast range of the reference viewing environment, the characteristic must merge into a linear section with an appropriate gain. Measurements carried out in the CS3 version of Photoshop some years ago indicated the gain of the linear section of the Adobe RGB gamma correction characteristic was 32.

In keeping with good practice, the colour space specification already cited was further refined by supplementing it with a subsequently defined compatible viewing environment, though ideally a colour space should be specified to match an already defined display and viewing environment. The essential elements of the reference viewing environment are shown in Table 24.5.

Selecting the Adobe RGB colour space in Worksheet 13(b) with the gain of the linear section set to 32 indicates that the break point of the characteristic between the power law and linear section is at an input level of 0.0422%, roughly one eighth of the reference black level and thus at a level well below the point of perceptibility for the reference viewing environment, indicating a satisfactory match of colour space characteristics with the viewing environment.

Table 24.5 Reference viewing environment from which the Adobe RGB (1998) colour space is derived

Reference viewing environment	
Reference display white point	As Illuminant D65, luminance level of white 160 nits
Reference display black point	As Illuminant D65, luminance level 0.34731% of white = 0.5557 nits. Includes veiling glare
Contrast ratio	Ratio of reference white to reference black = 287.9
Adapted white point	As reference display
Ambient illumination	Monitor turned off, at faceplate, 32 lux
Reference display surround	Extends to at least 2 degrees, neutral reflectance, luminance 32 nits, with Iluminant D65
Image size and viewing distance	Equal to the image diagonal
Glare	The veiling glare level is included in the reference black point level

24.2.3   Adobe Wide Gamut RGB Colour Space

The Adobe Wide Gamut RGB colour space is an extended colour space and has a chromaticity gamut, which as shown in Figure 24.1, is an excellent capture colour space since it is a good compromise between the largest colour gamut available for real colours and a gamut that encompasses very nearly all the Pointer colours on the cyan side of the spectrum.

> The parameters of the Adobe Wide Gamut RGB colour space are detailed in Table 24.6.
> The positions of the primaries on the chromaticity chart are also illustrated in Figure 24.7, where the gamut may be compared with that of the ProPhotoRGB gamut (as described in Section 24.3.1.1).

Table 24.6 Parameters of the Adobe Wide Gamut RGB colour space

Adobe Wide Gamut RGB colour space parameters					
Primaries chromaticities	x	y	z	u'	v'
Red	0.7347	0.2653	0.0000	0.6234	0.5065
Green	0.1152	0.8264	0.0584	0.0363	0.5863
Blue	0.1566	0.0177	0.8257	0.2161	0.0549
White D50	0.3457	0.3585	0.2958	0.2092	0.4881
Transfer function	$R'G'B'$				
For 0 > RGB ≤ 1	$(RGB)^{1/2.19921875}$				

Though not quite as large a gamut as the ProPhotoRGB gamut, this gamut encompasses the cyan colours of the Pointer real surfaces colour gamut, and furthermore, these real colours will utilise more of the available code values. However, if one were being critical, since the green primary is further away from the zero z straight line of the spectrum locus, there will be some highly saturated yellow and yellow-green colours outside of the gamut.

The white point of the gamut is based upon the D50 light source, and the transfer function is identical to the Adobe RGB gamma law, based upon an exponent of notionally 1/2.2.

As with the ProPhotoRGB gamut, it is recommended that one uses 16 bits per channel when using this colour space.

24.2.4   The RAW RGB Source Colour Space

As indicated in Section 22.1, non-consumer cameras provide the option of capturing the image in a non-processed or 'raw' format, that is, using the native colour spectral sensitivities of the camera and recording the RGB signals after only colour balancing, with no gamma correction, matrixing or other processing. The basis of this approach is founded upon the limited processing power of the camera compared with that of the computer, which following image capture is used for processing the image data in such applications as Adobe Photoshop. The higher processing power of the computer provides the photographer with the ability to process the raw RGB data in a manner better able to overcome any limitations in the image capture processing operation.

However, this flexible approach causes a conflict between the requirement to standardise the colour space and the ingenuity of the camera manufacturer to design the camera colour spectral sensitivities for optimum colour response, the details of which will remain proprietary. The shape of the spectral sensitivity characteristics is dependent upon the camera optics, the colour filters and the spectral response of the image sensor, which gives the camera manufacturer a number of degrees of freedom in optimising the design.

Nevertheless, at some point, the raw RGB data will need to be matrixed to match the computer processing colour space and/or the final viewing space, be it the monitor display or the print. There are fundamentally two approaches to resolving this issue: the first is for the camera manufacturer to provide a raw file processing application for the computer which, in addition to providing a range of adjustments to the RGB signals, also provides a matrix which processes the raw colour gamut to that of a recognised colour gamut such as sRGB or Adobe RGB. The second approach, which is both simpler for the photographer with cameras of different manufacturers and is likely to lead to a more consistent rendition of the image, is for the provider of the computer photographic processing application to provide a universal raw file processor which is capable of processing files from any manufacturer which has provided the confidential details of their camera spectral sensitivities. Adobe Camera Raw is a plug-in to the Photoshop application which, on selecting a recognised raw file, automatically loads to provide the appropriate processing before outputting the image in a standard format to the Adobe Photoshop application.

Although the raw colour gamut is proprietary and therefore its chromaticity coordinates are not published, we can consider the form it is likely to take from knowledge of the 'ideal' camera gamuts evolved in Section 12.4. Since the vast majority of stills cameras use a single image sensor, where the light for the red, green and blue pixels is individually filtered rather than shared, one may envisage that the aim would be to design the camera spectral sensitivities to match as closely as possible, for example, the 'Ideal 2' camera spectral sensitivities derived in Section 12.4. Thus, to provide us with a working model of a raw file colour space, we will envisage a hypothetical raw (hRAW) colour space based upon these characteristics as listed in Table 24.7 and whose chromaticity gamut is illustrated in Figure 24.1. It should be appreciated however that, in reality, the difficulty in manufacturing optical filters which precisely complement the response characteristic of the image sensors means the match to this or any other 'ideal' response is likely to be compromised.

Table 24.7 Parameters of the hypothetical RAW colour space

Hypothetical RAW colour space parameters					
Primaries chromaticities	x	y	z	u'	v'
Red	0.7347	0.2653	0.0000	0.6234	0.5065
Green	−2.0578	3.0593	−0.0015	−0.4400	0.6659
Blue	0.1741	0.0050	0.8209	0.2568	0.0166
White D65	0.3126	0.3290	0.3584	0.1978	0.4683
Transfer function	Linear				
For 0 > RGB ≤ 1	RGB				

In Worksheet 24, the chromaticities of the hRAW primaries are used to calculate the matrix which provides the RGB CMFs in terms of the XYZ CMFs as illustrated in Table 24.8.

Table 24.8 Matrix for deriving the hRAW RGB CMFs from the XYZ CMFs

	X	Y	Z
R =	0.7454	0.5013	−0.1611
G =	−0.5090	1.4097	0.0994
B =	−0.0001	0.0004	0.9490

The hRAW CMFs provide the required shapes for the 'ideal' camera spectral sensitivities as illustrated in Figure 24.6.

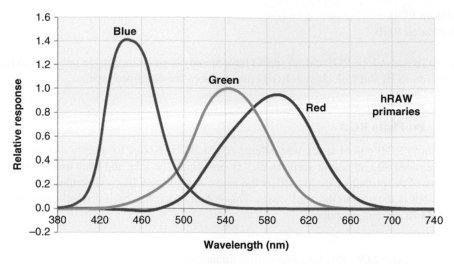

Figure 24.6   The hRAW ideal camera spectral sensitivities.

24.3   Colour Spaces in the Computer

Reference to Figure 22.1 indicates there are four areas within the computer in which the processing of the colour signals requires the definition of a suitable colour space:

- RAW file processor
- Photographic file processor
- Monitor driver
- Printer driver

The colour spaces associated with each of these areas will be described in the context of using Adobe Camera Raw as the raw file processor and Adobe Photoshop as the photographic processor.

24.3.1   Adobe Camera RAW

When an attempt is made to load a raw file recognised by Adobe into Photoshop, the Adobe Camera Raw application loads first to enable the user to select which of four colour spaces the raw file should be converted into from the native camera colour space before adjustments are carried out and the file is opened in Photoshop. The option is also available to save the file in the standard Adobe raw format, that is, as a file with a default .DNG[4] extension.

The four colour spaces available for selection in Adobe RAW are:

- sRGB
- Adobe RGB

---

[4] Adobe uses DNG as an abbreviation for 'Digital Negative' to emphasise the photographic data is stored in a format with a similar range of flexibility of adjustment to that of a film negative from which a print may be taken.

- ProPhoto RGB
- ColorMatch RGB

The sRGB and Adobe RGB colour spaces have already been defined in the previous section; the ProPhoto RGB and ColorMatch RGB colour spaces are described next.

24.3.1.1   ProPhoto RGB

The ProPhoto RGB colour space was defined originally by Kodak to specify a colour space which embraced all real colour surfaces and, in particular, the inks and dyes used in photography. Two of the primaries are located on the zero z-axis of the CIE x,y chromaticity diagram and the other is located on the zero y-axis. This colour space has been formally specified as the ROMM RGB[5] colour space.

Table 24.9   ProPhoto colour space parameters

ProPhoto colour space parameters					
Primaries	x	y	z	u'	v'
Red	0.7347	0.2653	0.0000	0.6234	0.5065
Green	0.1596	0.8404	0.0000	0.0500	0.5925
Blue	0.0366	0.0001	0.9633	0.0500	0.0003
White D50	0.3457	0.3585	0.2958	0.2092	0.4881

Transfer function	$R'G'B'$
For RGB < 0.001953	$16 \times RGB$
For RGB $\geq$ 0.001953	$(RGB)^{1/1.8}$

The ProPhoto RGB colour space parameters are listed in Table 24.9. The white point is specified as that of the D50 source and the transfer function is based upon a power law with an exponent of 1.8, considerably lower than for other colour spaces.

The chromaticity gamut of the ProPhoto RGB colour space is illustrated in Figure 24.7, together with the optimal colours chromaticity gamut (see Section 4.7), which the former embraces.

Thus, the ProPhoto gamut includes all possible non-fluorescing surface colours.

---

[5] US Standard ANSI/I3A IT10.7666:2003.

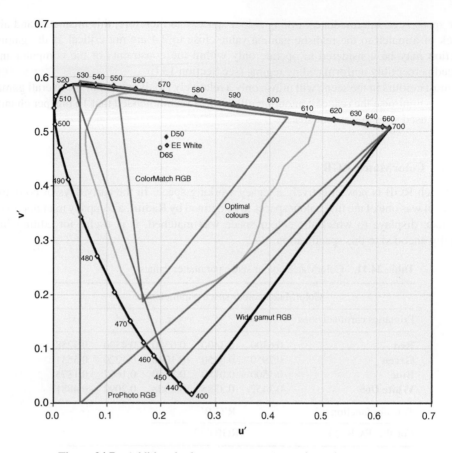

Figure 24.7 Additional relevant computer processing colour gamuts.

Table 24.10 ProPhoto RGB values
from XYZ values

	X	Y	Z
R =	1.3679	−0.2598	−0.0519
G =	−0.5534	1.5325	0.0209
B =	0.0000	0.0000	1.2318

The matrix for deriving ProPhoto RGB values from XYZ values is calculated in Worksheet 24 and shown in Table 24.10.

The gamma correction 'breakpoint' parameter has a value of 0.001953, which is not consistent with the value of 0.000529, the natural breakpoint value of a transition between a combined power law and a linear element characteristic, as calculated in Worksheet 13(b) for the given exponent and linear gain values of 1/1.8 and 16, respectively. However, the ProPhoto colour space is an intermediate working colour space, and therefore, this mismatch and also the lack of a match to the realistic gamma value close to 2.4 are not critical as the gamma correction may be considered to operate only within the constraints of the computer in a balanced perceptibly uniform coding regime (see Section 13.6).

As most colours in the scene will utilise only a relatively small volume of the overall gamut, in order to minimise the risk of contouring effects, it is recommended that 16 bits per channel should be used with this colour space.

24.3.1.2   ColorMatch RGB

ColorMatch RGB is now effectively a legacy working space. In the early days of computer graphics, it was one of the first colour spaces to be defined by Radius, a company manufacturing high grade displays to which the colour space was matched. It is useful for editing files originally encoded to this specification.

Table 24.11   ColorMatch colour space parameter values

ColorMatch colour space parameters					
Primaries chromaticities	x	y	z	u'	v'
Red	0.6300	0.3400	0.0300	0.4330	0.5258
Green	0.2950	0.6050	0.1000	0.1220	0.5631
Blue	0.1500	0.0750	0.7750	0.1667	0.1875
White D65	0.3457	0.3585	0.2958	0.2092	0.4881
Transfer function	$R'G'B'$				
For 0 > RGB ≤ 1	$(RGB)^{1/1.8}$				

The parameters of the ColorMatchRGB colour space are detailed in Table 24.11.

The positions of the primaries on the chromaticity chart are illustrated in Figure 24.7. The ColorMatchRGB gamut is somewhat smaller than the sRGB chromaticity gamut.

24.3.2   Adobe Photoshop

Within Photoshop are two colour spaces, the Profile Connection Space (PCS) and the working space. The relationship between these two colour spaces will become clearer in Section 27.3; however, the forms they take are described in these following sections.

24.3.2.1   The Profile Connection Space

The purpose of the profile connection space is described in some detail in Chapter 27; suffice to indicate at this stage that it is the underlying colour space which drives Adobe Photoshop. Files interchanged between the different colour spaces which interface with Photoshop are first converted from the source colour space into the profile connection space and then converted again into the destination colour space. It is essential therefore that the profile connection colour space is able to accommodate all colours; the most fundamental colour space to meet this requirement is the linear CIE XYZ colour space, complemented by the non-linear CIELAB colour space which more nearly matches the perception characteristics of the eye. Both these colour spaces, it may be recalled, were derived in Chapter 4 and either may be used in the profile connection space.

Encoding in the $L^*a^*b^*$ format has a number of merits:

- Colour is encoded in a manner that is accurately modelled after the human vision system and provides an unambiguous definition of colour without the necessity of additional information such as primary chromaticities, white point and conversion functions.
- Unlike RGB spaces, which are mostly based upon real primary chromaticities, it is not associated with any device.
- Being based upon the CIE XYZ primaries, it encompasses the full gamut of colours perceived by the eye and therefore also all colour spaces which are based upon the chromaticities of real display primaries.
- Because of its visual uniformity, its colours will be recorded relatively compactly for a given perceptual accuracy.
- It is non-proprietary and unambiguous in its structure.
- The advantages for image compression made possible by having separate lightness and chrominance components are very significant.
- Colour spaces such as CIELAB are inherently more compressible than tristimulus spaces such as RGB. The chroma content of an image can be compressed to a greater extent, without objectionable loss, than can the lightness content, making it an ideal candidate for JPEG compression.
- Gamut mapping of an image to match the capabilities of the intended output device is critical but easier to undertake when the chrominance components are in polar form, since the amplitude and phase of its a^*b^* components are closely related to saturation and hue, respectively.

24.3.2.2 Working Colour Spaces

In addition to the Adobe Camera RAW colour spaces defined above, Photoshop offers a very comprehensive range of colour spaces which may be selected as the 'working' colour space for editing and adjusting images. These may be broadly categorised as:

- Good for editing and adjusting for best display
- Good for editing with the display simulating the print which will be rendered
- Good for editing legacy files

The basis for selecting an appropriate colour space will be reviewed in some detail in the chapters on colour management; at this stage, we will limit our interest to the practicality of editing images using the computer display as the criterion; for this purpose, the colour spaces already made available in Camera Raw are an adequate range, possibly supplemented by the 'Adobe Wide Gamut RGB' colour space. The legacy colour spaces may be ignored and the printer-related colour spaces will be reviewed in the final section of this chapter.

Thus, in summary, the working colour space is likely to be selected from amongst the following colour spaces:

- sRGB
- Adobe RGB
- Adobe Wide Gamut RGB
- ProPhoto RGB

The basis of the selection will be reviewed in the chapters on colour management.

24.4   Colour Spaces in Displays

Although the sRGB and Adobe RGB colour spaces are based upon the average characteristics of display devices, they should not be confused with the actual colour space of a particular display, which may, for instance, have primaries with similar characteristics but not identical to the formal specifications; the same is also true of the associated gamma characteristic.

Thus, display profiles are usually finessed by the manufacturer to match precisely the characteristics of the display itself and despatched with the display on an accompanying CD, where they may be transferred to the computer. The operating system then provides the option of selecting either the dedicated profile or one of the generic colour spaces listed in the sections above to drive the display.

Unfortunately, it has become fashionable, possibly for proprietary reasons, for display manufacturers not to publish the chromaticity coordinates of the primaries used in their displays but to classify them in terms as, for example, 'reproducing 92% of the Adobe RGB chromaticity gamut', usually by implication, as referred to the x,y rather than the u',v' chromaticity diagram. This practice prevents users from being aware of precisely the chromaticity gamut a particular display is capable of reproducing.

Generally, the display gamut of a monitor is close to either the sRGB or Adobe RGB gamut, both of which are limited in the range of real surface colours which may be displayed, as Figure 24.1 illustrates.

24.4.1   Future Extended Colour Spaces for Photographic Displays

As noted in Chapter 20 on future television specifications, displays are likely to become available with extended colour spaces when compared with those currently available. The ITU television Recommendation 2020 provides for a colour space with a considerably extended chromaticity gamut; furthermore, pressure is being brought to bear to increase the perceivable contrast range of displays, primarily by increasing the highlight luminance level whilst retaining a low level of perceived black. It is likely therefore that such displays will eventually become available as computer monitors for photographic processing use. Furthermore, as the ability to display photos on television screens becomes ever easier and more popular, it is inevitable that these new television screens will also be used for viewing photographic image files.

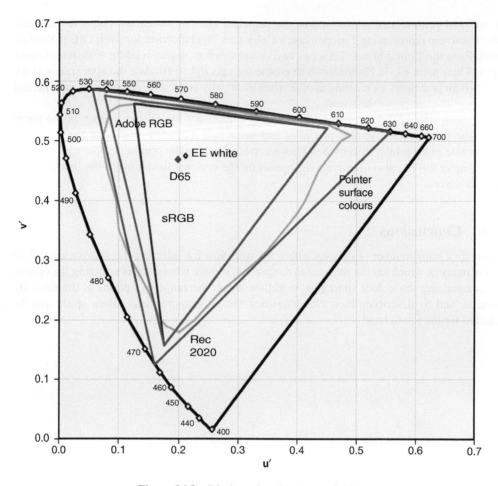

Figure 24.8 Display primaries chromaticities

The chromaticity gamut of Rec 2020 is illustrated in Figure 24.8, together with those of sRGB, Adobe RGB and the Pointer surface colours for comparison.

As can be seen, this gamut covers all but a very few of the real surface colours located in the saturated cyan area of the diagram and is a distinct improvement on the current gamuts.

24.5 Printer Colour Spaces

Printer colour spaces fall into two categories, those associated with any printer connected to the computer on which Photoshop resides and those associated with international print standard organisations, which are invariably CMYK colour spaces.

Although most print-related specified colour spaces are now available as standards of the ISO, nevertheless Photoshop continues to list these spaces as they were originally specified by national print-related organisations, the principal ones of which are Fogra in Germany but effectively representing Europe, Japan Color and 'Specification for Web Offset Printing (SWOP) in the United States. Each has two or more colour spaces relating to different paper types. Their main use in Photoshop is to enable the operator to emulate the appearance of the photograph or artwork as it would appear when eventually printed on one of these commercial presses, that is, to prove their work.

The local inkjet printer colour spaces usually comprise a list related to each of the paper types the printer manufacturer produces and serve two purposes: firstly, as the profile to incorporate in the printer driver when Photoshop is managing the printing colour management (see Chapter 29) but also to provide the option for the operator to soft proof the image on the monitor screen.

24.6   Conclusions

There are a huge number of options within the workflow for selecting the appropriate colour space, many of which are not relevant to the particular stage where one is selecting the option. By categorising the colour spaces by workflow stage, the range of options is dramatically reduced, and by describing their characteristics, the most appropriate colour space may be selected for the job in hand.

25

Component and File Formats

25.1 Introduction

Component[1] and file formats are often inextricably linked, sometimes with only certain component formats allowed for a particular file format, whilst other file formats will support all the various component formats. The aim of this chapter is to review the component formats previously described and defined in Chapter 14 and to provide the background which indicates how these component formats relate to the various file formats in use.

25.2 A Review of Component Formats

At the commencement of the period when consideration was being given to replacing consumer film cameras, one of the principal objectives in designing a digital stills colour camera was to enable the camera to provide a compact file for each of the photos captured in order that these files could be stored efficiently within the camera and exported when required to a printer or computer. As a first step, the then recently developed digital television technology specification ITU-R BT. Rec 601 (Rec 601), briefly described in Section 17.3, was adopted and only in a minor way adapted to the requirements of the stills camera. This approach led to the availability of the digital $YCbCr$[2] components, which were well suited to undergo very significant compression using the discreet cosine transform (DCT), which in turn had recently become available as the result of the work of the Joint Photographic Experts Group (JPEG).

Thus, the camera is required to provide sufficient processing capability to support the following operations:

- Quantisation of the raw geometrically disparate RGB components from the sensor
- Demosaicing of the geometrically disparate RGB components from the Bayer sensor
- Matrixing of the native colour space to a defined colour space

---

[1] In television, the data derived from a sensor is dynamic and constantly changing, which explains why the RGB elements are described as a 'signal', whilst in photography, the RGB elements are static and thus are usually described in terms of 'components'.

[2] In standards documents, the notation in television is YC_BC_R, and in photography, it is YCbCr.

Colour Reproduction in Electronic Imaging Systems: Photography, Television, Cinematography, First Edition. Michael S Tooms.
© 2016 John Wiley & Sons, Ltd. Published 2016 by John Wiley & Sons, Ltd.
Companion Website: www.wiley.com/go/toomscolour

- Gamma correction of the RGB components
- Derivation of the YCbCr components by matrixing and filtering
- DCT compression
- Creation of a standard format file containing the compressed image data

Not surprisingly, these tasks represented a demanding load for the tiny camera computer, which led to compromises being imposed upon the quality of the component processing; compromises which were unlikely to be noticed by the average amateur photographer but which would cause problems for the more discerning user.

In addition, although the camera manufacturer does a remarkable job in designing circuits to automatically compensate for a variety of lighting conditions during image capture; nevertheless, many would prefer to undertake these critical adjustments themselves.

In consequence of these considerations, there was a demand for professional and bridge cameras to make available the raw RGB components externally.

Thus it becomes apparent that there is a requirement for image components in three formats:

- Raw RGB format
- Gamma-corrected R'G'B' format
- YCrCb format

In addition, some file formats have the option of storing the components in the $L^*a^*b^*$ format described in the previous chapter.

25.2.1   The Raw RGB Format

The RGB signals derived from the sensor are, with the possible exception in some cameras of an analogue colour balance adjustment, completely unprocessed; that is, with no matrixing having been applied, their colour characteristics are defined only by the spectral characteristics of the sensor RGB optical filters and the spectral sensitivity of the image sensor; thus, the components are unlikely to relate to a defined colour space.

25.2.2   The R'G'B' Format

The R'G'B' components are produced by the camera from the raw RGB components. They are quantised, demosaiced, matrixed to match a defined colour space and gamma corrected, primarily in preparation for reformatting to YCbCr; however, they can also be made available for external use in some cameras.

25.2.3   The YCbCr Format

In Section 14.3, the reasons for deriving the YCbCr components from the R'G'B' components were explained in some detail. Whilst many authors describe the YCbCr signal as being of a different colour space to the RGB colour space, I prefer to consider the signal as only of a different format. Different colour spaces are of the types described in the previous chapter; reformatting the components in this manner does not change the size or shape of the colour gamut obtained, and therefore, strictly speaking does not change the colour space, which is defined in terms of the chromaticity coordinates of the source device primaries and the component conversion functions.

In adapting the Rec 601 television specification, it was decided to modify the original specification, which provided head and foot room for the components when mapping onto the 255 code levels of the digital signal, by specifying that the YCbCr components utilise the full range of code levels. Thus, for the luma component, black is represented by code level 0 and white by code level 255, and for the chrominance components, black is represented by code level 128 and the maximum negative and positive excursions by code level 0 and code level 255, respectively.

25.2.3.1   CbCr Scaling Factors

It may be recalled from Section 17.3 that scaling factors related to the coefficients of the $R'G'B'$ components comprising the luma or Y' component were applied to the colour difference components so that the peak amplitude of all three signals were made equal to unity. If the scaling factors are x and y and the luminance coefficients of the blue and red components are b and r, respectively, then the scaling factors are:

$$x = 2 - 2b \text{ and } y = 2 - 2r$$

At the time Rec 601 was adopted, the luminance coefficients for blue and red were 0.114 and 0.299, respectively, making $x = 1.772$ and $y = 1.402$.

$$\text{Thus,} \quad Cb = B' - Y'/1.772 \text{ and } Cr = R' - Y'/1.402$$

Since the adoption of Rec 601, the television industry and the monitor industry which supports it have moved on (see Part 5A) with the adoption of new system primaries (Rec 709)[3] and therefore different values of the luminance coefficients of the $R'G'B'$ components comprising the luma or Y' component and, in turn, the values of the scaling factors used to derive Cb and Cr. Since these values are now options for the photographic community, it is clearly important in latter stages of the workflow, when re-deriving the $R'G'B'$ components from YCbCr, to use the correct Cb and Cr scaling factors.

25.3   File Formats

25.3.1   General

Consumer camera image files are usually made available only in JPEG format, sometimes with a choice of compression quality options, whilst professional and prosumer models also include raw format files. Image capture equipment often provides the option of selecting the colour space into which the native RGB components are matrixed before being saved to the image file, often via the selection of a particular 'colour mode', which defines the primaries of the working colour space of the mode selected.

---

[3] The UHDTV specification Rec 2020 utilises primaries with new chromaticity coordinates, which are likely to be adopted by the photographic community as displays incorporating these primaries become available.

At first sight there appears to be a large number of different file formats in use for storing and conveying photographic image data, although, on investigation, many of these formats are effectively either obsolete or obsolescent. In this section we will review the background and specifications of those formats which are still in common use and those in use for ensuring the potential for gaining the optimum quality from the data captured by the camera image sensor. These formats include the various manifestations of the tagged image file format (TIFF) and the two common versions of the JPEG format.

Although there are exceptions, file formats are generally initiated either by an industry leader in the appropriate domain or by groupings of industry leaders who have a strong interest and awareness that unless a specification for a common file format is developed, it will prohibit the growth of the industry.

Once a format has been agreed and begins to show signs that it has industry acceptance through its use by a number of companies, it will be proposed for standardisation by the appropriate international organisation covering the interests of the industry. As indicated in Chapter 22, there are a number of these organisations with overlapping interests and, as a result, the two major organisations with a shared interest in the photographic and printing industry, the International Organisation for Standardisation (ISO) and the International Electro-technical Commission (IEC), often work together through their appropriate committees to agree a standard common to both organisations. Where this work also impinges on the interests of the International Telecommunications Union (ITU), that organisation will sometimes adopt these standards verbatim and allocate an appropriate new title reference for them. Depending upon the status of the specification when it is submitted from industry, will dictate the amount of additional work the international organisations will undertake before a standard emerges. Occasionally these organisations will adopt the specification virtually as is, only adjusting the document format to match their standard approach.

In Figure 25.1, an overview is shown of the relationships between the organisations representing the photographic and printing industries, the international standards organisations and the file formats commonly used in photography and, to a lesser extent, in the related field of graphic arts.

For clarity, the lines which would connect the standards organisations to the file formats have been omitted; however, the organisation responsible for the standard is shown at the top of each box representing a file format.

25.3.2   The File Formats

The file format specifications are comprehensive and detailed; thus, only those parameters directly affecting the colour quality of the rendered image are summarised in the following descriptions.

25.3.2.1   Tagged Image File Format (TIFF)

The Tagged Image File Format (TIFF) was originally formulated by Aldus in 1986, where the specification was processed up to Version 5 before the company was taken over by Adobe®. Adobe modified the format to its final form in Version 6[4] of the specification and published it for general use in 1992, although it remains a proprietary format. By virtue of its open

---

[4] https://partners.adobe.com/public/developer/en/tiff/TIFF6.pdf

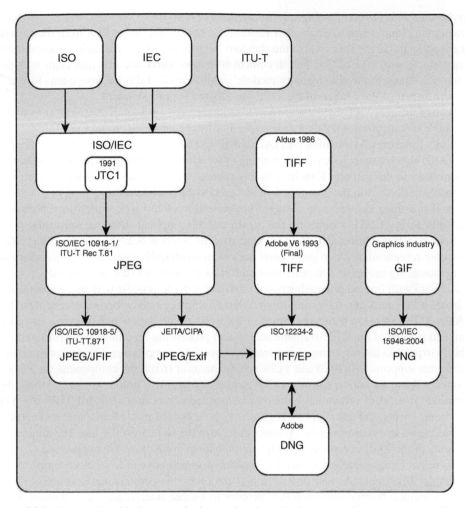

Figure 25.1 The relationship between the international standards organisations, the industry, the companies and the file format specifications.

specification and extensibility, it rapidly became an industry-wide informal standard. It is a comprehensive raster picture format able to accommodate a range of non-compressed and both lossless and lossy compressed files, including JPEG (see Section 25.3.2.3), though the latter is rarely used. A wide range of component formats may be incorporated into the file, including RGB, YCbCr and CIE $L^*a^*b^*$ formats. A large number of tags are available for both specified and private use which extends the usability of the system. Amongst many other parameters, these tags may be used to specify the colour space of the associated raster data; the code range of the components, to indicate whether or not foot room has been allocated; the luminance component R and B values for calculating the coefficients of the chrominance components and the degree of any subsampling of these components. In the latter case, the degree of subsampling, that is, 4:4:4, 4:2:2 or 4:1:0, is the same for both the horizontal and the vertical dimension of the raster (see Section 14.5).

Tagged Image File Format for Digital Photography (TIFF-EP)

Following the requirement to establish a standard for conveying image files from the camera to other applications, the ISO created the standard 'Digital still-picture imaging – Removable memory' document, ISO 12234. This document is in three parts: the first part deals with the format of the 'Basic removable memory module', whilst Part 2 'TIFF-EP Image data format'[5] details the format of the content of the TIFF file adopted for this standard.

The aim of the new TIFF-EP standard was to embrace both a subset of the parameters of the Adobe TIFF specification and the meta data specified by the Japan Electronics and Information Technology Industries Association (JEITA), in their Exchangeable Image File Format (Exif)[6] specification but also to include new parameters to accommodate the characteristics of the raw files generated in the camera. Raw files in this context relate to files which contain the raw components derived from the sensor before any processing is carried out.

Many if not most cameras use a single image sensor of the type which incorporates a colour filter array (CFA) comprising red, green and blue optical filters in sequential pixel resolution in a Bayer mosaic pattern as described in Section 8.2. As the resulting RGB components representing a particular pixel are not geometrically centred, it is necessary to apply demosaicing processing in order to obtain RGB components that are co-sited.

In a camera with limited processing power, the demosaicing process may be compromised; thus, in order to bypass any such compromise, the raw unprocessed components relating to the individual RGB pixels are formatted into the file together with data which describes the positioning and pattern of the CFA to enable subsequent processing on a platform with sufficient capability to process the raw components into RGB co-sited components without compromise.

The format supports both RGB and YCbCr components at 16 bits per component, the YCbCr components being an option from non-CFA sensors. The native colour space is defined by ICC profiles instead of individual colour space parameters as in a standard TIFF file. For YCbCr components, the subsampling regime may be different in the horizontal and vertical directions, again an extension of the options available in the standard TIFF file. The luminance coefficients of the RGB components comprising the luma component are tag defined.

Two types of light source are defined by the camera, usually one of low colour temperature, for example, Illuminant A, and one of a higher colour temperature such as D65. These are complemented by two matching ICC profiles to enable interpolation calculations to be undertaken to match the components to the actual colour temperature of the scene illumination.

Tags are also defined to record the RGB spectral sensitivities and the Optical Electronic Conversion Function (OECF) of the camera.

25.3.2.2   Digital Negative Format (DNG)

Independently, Adobe recognised very early on in the development of digital cameras the same requirements as those addressed by the ISO committee for a non-proprietary format to carry raw RGB components. Furthermore, in identifying the similarity of approach of processing a raw file before it can be used to render an image, to that used in traditional photography in processing a negative before rendering a print, Adobe created the term 'digital negative' to describe the format of the raw RGB components. The aim was to create a digital negative standard which all camera manufacturers could use in the formatting of raw RGB components.

---

[5] http://www.iso.org/iso/iso_catalogue/catalogue_tc/catalogue_detail.htm?csnumber=29377
[6] http://www.cipa.jp/std/std-sec/std-list_e.html

As the owners of the TIFF specification, Adobe followed a very similar approach to the ISO committee in adopting a subset of the TIFF specification as the basis of their new digital negative format[7] (DNG). Naturally, in extending the specification, the same additional parameters were identified but were also supplemented by further parameters which would provide the downstream processor with a finer degree of finesse in the adjustment of the image; these included further ICC profiles, analogue RGB balance data and a reference colour space for individual cameras.

When it became evident that the two groups were following identical goals, Adobe shared their work with the ISO committee responsible for the emerging TIFF-EP format and, as a result, there is a very close correlation between the DNG and the TIFF-EP formats.

The following, which with the permission of Adobe Systems Incorporated, is copied verbatim from the Adobe Digital Negative (DNG) Specification, details the advantages of a standard approach to specifying a raw file format:

- DNG has all the benefits of current camera raw formats, namely increased flexibility and artistic control. In addition, DNG offers several new advantages over proprietary camera raw formats.
- **Self-Contained:** With the current proprietary camera raw formats, software programs wishing to process camera raw files must have specific information about the camera that created the file. As new camera models are released, software manufacturers (and by extension users) must update their software to accommodate the new camera raw formats.
- Because DNG metadata is publicly documented, software readers such as the Adobe® Photoshop® Camera Raw plug-in do not need camera-specific knowledge to decode and process files created by a camera that supports DNG. That means reduced software maintenance and a more self-contained solution for end users.
- **Archival:** Camera manufacturers sometimes drop support for a propriety raw format a few years after a camera is discontinued. Without continued software support, users may not be able to access images stored in proprietary raw formats and the images may be lost forever. Since DNG is publicly documented, it is far more likely that raw images stored as DNG files will be readable by software in the distant future, making DNG a safer choice for archival.
- **TIFF compatible:** DNG is an extension of the TIFF 6.0 format, and is compatible with the TIFF-EP standard. It is possible (but not required) for a DNG file to simultaneously comply with both the Digital Negative specification and the TIFF-EP standard.

Unfortunately, possibly over concerns regarding the protection of the privacy of proprietary information, the DNG format has been adopted by only a few of the camera manufacturers, the remainder continuing to use their own proprietary formats for storing raw RGB components. In consequence, it will be necessary for those manufacturers using proprietary formats, who anticipate their customers will wish to take advantage of the AdobeRAW plug-in to Photoshop, to provide Adobe with details of the structure and content of their files. Where the DNG format does meet the aims of the creator is in the AdobeRAW plug-in to Photoshop, where loaded proprietary files may be saved to the DNG format, thus protecting them for the future against the risk of their camera manufacturer discontinuing support.

---

[7] http://wwwimages.adobe.com/content/dam/Adobe/en/products/photoshop/pdfs/dng_spec_1.4.0.0.pdf

Table 25.1 A summary of the principal image characteristics in each of the file formats described

	Tagged Image File Format	Tagged Image File Format – digital Camera	Digital Negative	Joint Photographic Experts Group	JPEG/JFIF	JPEG/Exif	Portable Network Graphics
File extension	.tif or .TIF	.tif or .TIF	.DNG		JPEG, .JPG, .JPE	JPEG, .JPG, .JPE	.PNG
Component format	RGB YCrCb L*a*b*	RGB YCrCb	RGB	YCbCr	YCbCr	YCbCr	RGB
Compression	None, lossless, lossy (JPG)	None, lossless	None lossless	Lossy (DCT)	Lossy (DCT)	Lossy (DCT)	Lossless
Bits/Component	8/16	8/16	8/16/24/32	8	8	8	8/16
Colour space	As defined	As defined	As defined	N/A	Rec 601	sRGB sYCC	As defined
ICC profiles*	No	Yes	Yes	No	No	No	Yes
Responsible organisation	Adobe V.6	ISO.IEC 12234-2	Adobe	ISO/IEC/ 10918-1	ISO/IEC 10918-5	JEITA-CIPA	ISO/IEC 15948:2004

*See Chapter 27.

25.3.2.3   Joint Photographic Experts Group Format (JPEG)

The exception to the general rule that format specifications are initiated within the industry is the format used for generating and storing compressed images. The importance of image compression was recognised early on by ISO/IEC, who established the permanent Joint Technical Committee 1 to work together with the appropriate ITU committee in this area, with one of its main tasks being to evolve a compression standard for still pictures. This combination of committees, which adopted the name 'Joint Photographic Experts Group' (JPEG), were responsible for creating the compression format based upon the discreet cosine transform (DCT) and giving it the name of the committee responsible for creating it.

The JPEG compression system is a lossy compression system, in that detail in the image to which the eye is less responsive is increasingly removed as higher levels of compression are selected. Relatively low levels of compression are generally not perceived by all but the most experienced of photographers, but as the level of compression is increased, the image will exhibit a loss of sharpness and an increase in artefacts on edges in the image and in areas that are almost, but not quite, of uniform colour.

This specification was extended both by the addition of meta data, that is, data which describes what the primary data is, and by the addition of a file interface format, by two separate organisations in parallel; the Japan Electronics and Information Technology Industries Association (JEITA), who formed the JPEG Exchangeable Image File Format (JPEG-Exif), and ISO/IEC, who formed the JPEG File Interchange Format (JFIF) specification, ISO/IEC 10918.5, known as the JPEG-JFIF format. These two file formats are very similar but not compatible; most cameras use the JPEG-Exif file format and most applications are able to read both formats.

25.3.2.4   Portable Network Graphics (PNG) Format

Until about 1995 the format used by the graphics industry was the Graphics Interchange Format (GIF), which was regarded as less than desirable for a number of reasons, not least because it was a patented format. When the compression patent was due to expire, an industry committee was formed to propose a replacement for the GIF format, the first version of which was released in 1996 as the PNG specification, primarily in support of the use of files on the Internet.

This is a raster graphics file format that supports lossless compression based upon RGB component values at the number of bits per channel appropriate to the number of colours in use but capable of full-picture depth representation of 8 or 16 bits per component or 24 or 48 bits per pixel.

In 2003 the current version of the PNG specification became an ISO/IEC standard and the latest version is ISO/IEC 15948:2004.

25.3.3   File Formats Image Characteristics Summary

In Table 25.1, the principal image characteristics of the various file formats described above are summarised.

In the JPEG specifications there appears to be some ambiguity regarding the colour space used. Considering the original specification drew upon the parameters of the television recommendation ITU-R BT.601, including its colour space, then the initial specification referred to this colour space but the JPEG/JFIF version does not.

26

Appraising the Rendered Image

26.1 Introduction

In photography the manner in which the rendered image is displayed will depend both upon the stage reached in the workflow at the time the appraisal is undertaken and, at the completion of the workflow cycle, whether the appraisal is made when viewing a print or a projected image.

Sometimes the appraisal will be informal, when other factors irrelevant to the quality of the rendered image are being considered; at other times one may wish to critically compare the rendered image produced on a monitor, a projector or a print with the original scene. In each of these cases the viewing environment is likely to be considerably different with all the factors described in Chapter 10 coming into play in affecting the appraised quality of the rendered image.

Since critical subjective adjustments of colour balance and tone scale gradation will be made on the computer monitor, it is imperative that the monitor and the viewing environment are matched to a standard line-up condition in order to provide the most critical appraisal environment consistent with ensuring that when the image is finally rendered on print or projector, possibly for competitive comparison with other images, the result matches expectations.

Using the concepts outlined in Chapter 10 as the basis of our approach, we will consider the performance and set-up of the display device, its associated environmental lighting and the characteristics of the surrounding surfaces in turn for the monitor, the projector and the print. It may be recalled from Chapter 13 in particular, that as the viewing angle the display or print subtends at the eye diminishes, so the influence of the colour of the surrounding surfaces on the chromatic adaptation characteristic of the eye will increase, which in turn will lead to a change in the perception of colour in the image.

26.2 The Monitor and its Environment

Before considering the formal set-up of the monitor it would be expedient to first review its performance and any limitations which might affect the manner in which it is used.

Colour Reproduction in Electronic Imaging Systems: Photography, Television, Cinematography, First Edition. Michael S Tooms.
© 2016 John Wiley & Sons, Ltd. Published 2016 by John Wiley & Sons, Ltd.
Companion Website: www.wiley.com/go/toomscolour

Most modern monitors currently in use (2013/2014) are based upon LCD technology with either cold cathode fluorescent or LED backlighting as described in Section 8.3. Depending on the environmental lighting conditions, the performance of these monitors may sometimes be compromised by their inability to produce a zero light output for a zero signal input, which directly affects the contrast ratio of which the display is capable. We saw in Chapter 13 that the spatial static contrast ratio of the eye in reasonably critical environmental conditions is about 400:1, close to that achievable by LCD displays backlit by LEDs. Furthermore, the electro-optical conversion functions (EOCF) of these displays are fundamentally linear and thus, in emulating the legacy CRT display, they require electronic circuitry to provide the gamma characteristic of the latter to compensate for the inverse characteristic provided by the standard photographic colour spaces (see Chapter 24).

We saw in Section 13.3 how the accommodation and adaptation of the eye is influenced by the viewing environment, particularly in terms of:

- the highlight luminance of the display,
- the angle of view subtended at the eye by the display dimension,
- the luminance and chromaticity of the ambient lighting,
- and the reflection characteristics of the surrounding surfaces.

This accommodation and adaptation level will significantly affect the perception of the displayed image. Furthermore, it was noted in Section 10.6 that the colour space selected at the source should reflect the standard viewing conditions at the location of the display; and it may be recalled that in Section 24.2 the transfer function of both the sRGB and Adobe RGB colour spaces were shown to equate closely to an exponent of 2.2 over most of the perceived contrast range.

In consideration of all the above it becomes clear that if one is to undertake appraisal and adjustment of the rendered image in a consistent fashion it is necessary to specify the set-up of the monitor, the characteristics of the environmental lighting and, when viewing the display at the recommended angle the monitor subtends at the eye, the characteristics of the surfaces in the line of sight.

26.3   Reference Conditions

Clearly the generality of the approach is to provide a set of conditions which provides the best possible perception of the image whilst ensuring that surrounding conditions are not adversely affecting that rendition. Thus the luminance of the display and the level of illumination of the prints should be high enough to ensure that there is no perceptual crushing of the darker tones by any limitations in the accommodation range of the eye, (see Section 13.3). Furthermore, the surrounding surfaces should not impinge in an uncontrolled manner upon the accommodation and adaptation characteristics of the eye, by ensuring relatively low luminance levels and chromaticities to match the white point of the rendered image, respectively.

There are a number of international standards which address the environment for appraising rendered images, including the International Electrotechnical Commission document IEC 61966-2-1, but the most recent and pertinent document is ISO 3664 – 2009 "Graphic technology and photography – Viewing conditions". This document embraces conditions both for the media on which the rendered images are displayed and for their environments, whether in terms of monitor, print or projector, although the latter is limited to projected transparencies. This standard has a comprehensive two-page introduction which explains the background to setting those parameters and their associated values which are important to ensuring critical appraisals, albeit that sometimes it appears that in what follows these are compromised by legacy limitations in equipment and decisions which were made when the original document was drawn up. The supplementary document, ISO 12646 "Graphic technology – Displays for colour proofing – Characteristics and viewing conditions", sets out slightly different parameters and values for the line-up of the monitor and its surroundings for producing proofs.

As an example of the compromises indicated above, one of the decisions in ISO 3664 which appears to cause considerable confusion, is the setting of a different white point for the evaluation of images displayed on the monitor and those appearing in print; the white point for the monitor being consistent with that for television viewing at D65 and that for the print at D50. Though not specifically stated, it would appear that D50 was originally selected as the white point for viewing prints because the average correlated colour temperature (CCT) of the lighting in graphics offices of the day was about 5,000 K; whilst in contrast virtually all monitors sold were set by their manufacturers to a white point of D65. Although the document is clear that the monitor in such a set-up should not be used for proofing, that is, comparisons should not be made between the monitor display and the print, in many studio offices and on many desktops there are no monitors dedicated to proofing, so it is inevitable that comparisons will be made and disappointment experienced in the relatively poor match that results.

The supplementary document, ISO 12646, sets out slightly different parameters and values for the line-up of the monitor and its surroundings to ensure specifically that the rendered image on the monitor, subject to careful colour management procedures, will match the print or hard proof, irrespective of whether the soft proof on the monitor emulates a press or the desktop inkjet printer. This is achieved by setting the white point of the monitor to D50 and implies access to the appropriate set-up equipment to achieve that goal.[1]

Another area of compromise is associated with the highlight level of the monitor and the print. As we saw in Figure 13.7, in order to ensure that the range of luminances of the rendered image is within the optimum contrast range of the eye, the darkest perceivable tones should have a luminance no less than about 1 nit or cd/m^2 and ideally nearer 10 nits in order to ensure they are above the "Barten limit", that is, the luminance level where the contrast range of the eye conforms to the $\Delta L/L$ rule at all perceived levels. Again as we saw in Chapter 13, this implies for images of small field of view a highlight luminance level of at least 100 nits and for large fields of view a level of at least 400 nits. Since the higher of these values was not possible with CRT displays and was also beyond what was generally available in many LCD displays, and furthermore was higher than is usually found for white surfaces in many office environments, the ISO 3664 standard provides a compromise figure of highlight luminance. This compromise figure is preferred for ensuring that a print that would have been judged satisfactory at a higher level of illumination is not disappointing when viewed at office levels. Although the document acknowledges the desirability of a relatively high level of highlight luminance and notes the requirement to do so for perceiving darker tones, it does not justify the reason for this requirement in the terms expressed here.

---

[1] As an expedient, the author has found that using D65 for illuminating the print produces a satisfactory match of the rendered image to that of the D65 display, albeit that ideally the colour components for the printer should have first undergone a transform from the D50 to the D65 white point.

In the following two sections the parameters associated with the two ISO standards are listed and reviewed. However, as there is considerable repetition in the standards in order to embrace sometimes marginally different requirements, the essential elements of the standards are extracted here, leaving behind the legacy figures. Those who do not require the background information to the standards may consider moving directly to the summary of the next three sections in Section 26.7, where ideal values of the parameters for appraising images are summarised.

26.4   Conditions for Appraising and Comparing Images – ISO 3664

Permission to reproduce extracts from ISO standards is delegated to the appropriate national standards institute; in this case BSI Standards Limited (BSI). That permission has been obtained subject to the inclusion in the book of the following paragraph:

Permission to reproduce extracts from British Standards is granted by BSI. No other use of this material is permitted. British Standards can be obtained in PDF or hard copy formats from the BSI online shop: www.bsigroup.com/Shop or by contacting BSI Customer Services for hard copies only: Tel: +44 (0) 845 086 9001, Email: cservices@bsigroup.com.

ISO 3664 sets out conditions for four different viewing requirements:

1. conditions for critical comparison of prints and transparencies, P1 and T1;
2. conditions for practical appraisal of prints, P2;
3. conditions for viewing small transparencies by projection, T2;
4. conditions for appraisal of images displayed on colour monitors.

Note that comparison between the images on the monitor and the print is specifically excluded from these conditions.

In the following, the parameters and values specified in ISO 3664 are provided in summarised form only and thus the reader requiring fuller information is recommended to refer directly to the official document. Furthermore, it should be noted that in order to avoid ambiguity, the style of standards documents tends to be terse; the parameters and their values being given with no associated explanatory text. Informative annexes sometimes provide guidance to understanding the background to the definition of the condition parameters and their values. Thus in the ISO reference conditions which follow, although the parameter values and tolerances (shown in italics) are essentially those that appear in the standard, the style of presentation may be different. Furthermore, where it is considered helpful, material has been added in standard text format to clarify and explain the ramifications of applying the specification when appraising rendered images.

26.4.1   Conditions for Critical Comparison of Prints and Transparencies P1 and T1

The standard defines conditions for prints (P1) and transparencies viewed directly on an illuminated diffuse screen (T1). These conditions are designed with relatively high levels of highlight luminance to ensure the critical comparison of rendered images whether they are print to print, transparency to transparency or print to transparency. The standard also recommends that these conditions are used for the judging and exhibition of photographs.

26.4.1.1 Ambient Conditions

In an open area, as opposed to a preferred viewing booth, surfaces in the field of view should be neutral matt grey with a reflectance of 60% or less.

26.4.1.2 Illumination and Illuminance

The parameters illumination and illuminance are dealt with separately in order that any surfaces adjacent to the plane of the image, which contribute by reflection illumination of a different characteristic to the general illumination, may be embraced in the specification of the *illuminance* of the rendered image.

26.4.1.3 Illuminant

The spectral power distribution (SPD) and the chromaticity coordinates of the reference illuminant shall be D50, with a u'_{10}, v'_{10} chromaticity tolerance of radius 0.005. See the "Illuminants" Worksheet and Section 7.3, respectively for the SPD and chromaticity values; the subscript "10" indicates a chromaticity diagram derived from the CIE observers using the 10-degree observer rather than the more usual 2-degree observer.

The CIE colour rendering index (CRI) of the illuminant, (see Section 7.2), at the viewing surface shall have a general CRI value of 90 or higher and for each of the special CRIs should have a value no less than 80. As noted in Chapter 7, the CIE CRI is a somewhat deprecated method of measuring the chromaticity performance of an illuminant. The visible range metamerism index shall be less than 1.0 and ideally should be less than 0.5.

26.4.1.4 Illuminance at the Print (P1)

The illuminance at the centre of the illuminated viewing surface area shall be 2,000 ± 500 lux and preferably 2,000 ± 250 lux. For viewing areas up to 1 m^2 the illuminance at any point shall not be less than 75% of the illuminance at the centre, for larger areas the lower limit shall be 60%.

Assuming a print reflectance of 89%, this level of illuminance will lead to a print with a highlight luminance of notionally 570 nits.

26.4.1.5 Surround and backing of the Print (P1)

The surround shall extend beyond the image area on all sides by at least a third of the image dimension, except where images are being compared, in which case they may be positioned edge to edge. The surround and backing shall be neutral and matt and have a luminous reflectance of between 10% and 60%; for critical appraisal a mid-grey of 20% reflectance is recommended.

26.4.1.6   Luminance at the Surface of the Transparency Illuminator (T1)

The luminance at the centre of the illuminated surface of the transparency illuminator shall be 1,270 ± 320 nits and ideally should be 1,270 ± 160 nits. Any departures from uniformity over the area should be within 25% of the luminance level at the centre.

26.4.1.7   Transparency Surround (T1)

The surround should be at least 50 mm wide on all sides. It shall appear neutral and have a luminance that is between 5% and 10% of that of the surface of the image plane. This condition may be met by a transparency mounted with an appropriate opaque border.

It is interesting to note that although these conditions are intended to facilitate the comparison of images rendered both in print and in transparency, the standard acknowledges that there will be an approximate 2:1 difference in luminance between the transparency and the print, that is, the print illuminance level is notionally 2,000 lux, which when illuminating a perfect Lambertian reflecting medium will produce a luminance of $2,000/\pi$ nits; as the reflectivity of the print media is likely to be in the order of 89% this translates to a peak luminance of 572 nits, compared to the transparency peak luminance of 1,143 nits, assuming a maximum transparency of 90% on white. Possibly the relative size of the images is influential in making this comparison; however, the standard is silent on this subject.

26.4.2   Condition for Practical Appraisal of Prints P2

It was noted in the early paragraphs of this section that in order to fully perceive the range of dark tones in an image a relatively high level of illuminance is required. Accordingly, the opposite is also true, that is, if the level of illuminance is inadequate it is likely that subtle variations in the dark tones of an image will not be perceived. Since acceptable levels of office illuminance often fall within the range where the differentiation of dark tones is difficult to perceive it follows that prints appraised as satisfactory under the print P1 conditions described above could lead to an unsatisfactory appraisal in less well-illuminated conditions. Thus to ensure that prints which are judged to be satisfactory under the standard conditions are also judged to be acceptable in normal office conditions, a second set of appraisal conditions P2 is specified in the standard for this purpose.

26.4.2.1   Illumination of Prints P2

The illumination shall comply with that described for P1 conditions.

26.4.2.2   Illuminance of Prints P2

The illuminance at the centre of the viewing surface shall be 500 ± 125 lux and the uniformity shall comply with that described for the P1 conditions.

26.4.2.3 Surround and Backing P2

The surround and backing shall be in compliance with the conditions specified for the P1 conditions.

26.4.3 Conditions for Viewing Small Transparencies by Projection (Viewing Conditions T2)

The specifications in these conditions are not to be confused with those normally used for viewing slides in a commercial projector, where the magnification is generally much greater and there is no intent to compare such images with reflection prints.

26.4.3.1 Illumination T2

The Standard indicates that the light emitted from the screen with an empty slide mount in the projector gate shall comply with that described for illumination in the P1 conditions. It would appear that what is being intended here is a description which covers the combined colour characteristics of both the projector illumination and the reflection characteristics of the screen.

26.4.3.2 Luminance T2

The luminance at the screen in the direction of the observer shall be 1,270 ± 320 nits when measured with an empty slide mount in the projector gate.

26.4.3.3 Uniformity of Screen Luminance T2

The screen should be sensibly uniform and conform to the detailed description in the Standard.

26.4.3.4 Surround

The surround shall conform to the conditions specified for T1.

26.4.3.5 Ambient Light and Veiling Glare

Ambient light and veiling glare at the centre of the screen shall not exceed 1% of the maximum screen luminance.

26.4.4 Conditions for Appraisal of Images Displayed on Colour Monitors

ISO 3664 emphasises that these conditions for appraising images on all types of colour monitor are intended purely for that purpose alone and specifically should not be used to compare with a printed image; that is, it is assumed that when using these conditions the appraisal of images on the monitor and the print should take place successively rather than simultaneously. Where it is required to compare the same image simultaneously on a monitor and a print then one should adopt the conditions laid down in ISO 12646, which are summarised in Section 26.5.

26.4.4.1 Display Chromaticity

The display of white on the monitor should have a chromaticity approximating to illuminant D65 within a tolerance of 0.025 of $u'_{10} = 0.1979$, $v'_{10} = 0.4695$ on the CIE 1976 Uniform Chromaticity Scale diagram.

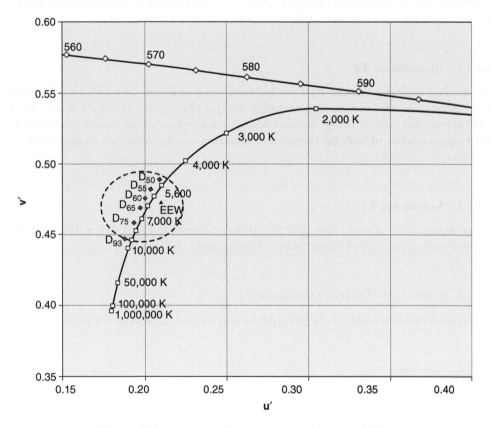

Figure 26.1   Tolerance of display chromaticity set at 0.025.

As acknowledged by the standard, this is an unusually wide tolerance as indicated by the circle on the expanded chromaticity diagram illustrated in Figure 26.1, where it is seen to embrace all the CIE daylight illuminants from D50 to D93.

It would appear that the committee members setting the reference parameters and their values had some difficulty in finding a compromise between that which would be ideal from the perspective of the perceived quality of the rendered image and that which was likely to be achieved in the average graphics office environment of the time. The justification for the large tolerance appears to be based primarily on the well-recognised ability of the eye to adapt to a broad range of illuminants, subject to them having a spectral distribution close to that of the Planckian locus and in the case of a monitor screen, not to be diverted from this condition by high levels of ambient illumination of a different CCT. Furthermore, it tends to be implied that the P2 set of conditions are designed as a supplement to the P1 conditions, primarily to ensure that the perception of the darker tones of a print are not compromised by the lower level of illumination – rather than to critically appraise the colour content.

Nevertheless, in many situations the reality is that only one appraisal environment is likely to be available and the adoption of the large illumination chromaticity tolerance of 0.05 could lead to problems of appraisal in a disciplined colour management situation. Furthermore, despite the warnings of not comparing images rendered on the monitor with those on a print because of the difference in the system white point, this tolerance in fact embraces both white points.

26.4.4.2 Monitor Luminance

The luminance level of the white displayed on the monitor shall be at least 80 nits and ideally should be at least 160 nits.

26.4.4.3 Ambient Illumination

The level of the ambient illumination specified is measured in terms of the luminance it would produce on a perfectly reflecting diffuser located at the position of the faceplate of the monitor. This level will not be greater than 1/4 of the monitor white point luminance and ideally should not be greater than 1/8 of this level.

The CCT of the ambient illumination shall be less than or equal to that of the monitor white point and ideally should be equal to that of the monitor white point.

26.4.4.4 Surround Condition

The area immediately surrounding the displayed image and its border shall be neutral and approximately the same chromaticity as the white point of the monitor. Subject to the following paragraph, the luminance of the border should be no greater than 20% of the white point luminance and preferably not greater than 3% of the white point luminance.

When the monitor is being used to visualise images to be reproduced as hard copy, the border should be of the same colour as the border of the hard copy, that is, white where appropriate for a print and dark for transparencies. However, it is generally preferable that any such border be no more than 1–2 cm wide.

26.4.4.5 Environmental Conditions

The monitor shall be situated so there are no strongly coloured surfaces in the field of view or which may cause reflections from the monitor screen. Ideally all surfaces in the field of view should be neutral.

26.4.4.6   Veiling Glare

Sources of veiling glare should be avoided by situating the monitor such that no sources of extraneous illumination are directly in the field of view or cause discernible reflections from the monitor screen.

26.5   Colour Proofing

Perhaps the most critical appraisal requirements in terms of both precision of monitor performance and the setting of environmental lighting conditions occur for colour proofing, a procedure described in general terms in the latter paragraphs of Section 22.1, and as will be addressed in more detail in Chapter 29. Fundamentally the monitor is switched to an optional condition used to render an image precisely as it will appear when rendered in some other media, that is, either a monitor with different characteristics, a projected image but most usually a print as it would appear printed from a particular printer and paper combination. The printer may be an inkjet, which has a relatively wide colour gamut or one of a range of commercial printers, each with its own, usually relatively limited, colour characteristics. For this appraisal to be fully satisfactory, it is essential that the colour gamut of the monitor fully encompasses that of the device being proofed and as we saw in Chapter 23 the sRGB gamut is likely to be compromised when attempting to proof those printers with a good colour gamut and particularly the inkjet. Currently a monitor with primaries close to Adobe RGB will embrace the gamut of most printer inks.

The ability to emulate the proofing device results from the availability of the appropriate profiles, one to emulate the proofing device and a complementary version for the transformation of the resulting components to match the monitor display profile. In practice even when the monitor gamut is inadequate, the profile will endeavour to provide as good a solution as possible by gamut mapping (Section 12.3).

It is essential when undertaking a proofing task to be fully aware of the ramifications of the human vision, as described in Chapters 10, 13 and in the introduction to this chapter, such that a satisfactory proof is achieved. The test of acceptability being a good match in the direct comparisons of first the hard proof with the soft proof and ultimately the hard proof with the print from the proofing printer.

26.6   Displays and Viewing Conditions for Colour Proofing – ISO 12646:2008

The standard, ISO 12646 "Graphic technology – Displays for colour proofing – Characteristics and viewing conditions" contains both an introduction and an informative annex which provide in useful detail guidance on how to achieve satisfactory proofs. The description of the standard which follows is a summary of the parameters described in detail in the Standard (in italics) together with additional material where appropriate to clarify and comment upon the ramifications of the values and tolerances of these parameters.

The standard provides sets of conditions for both *comparison of monitor and hard copy images* and for *viewing of single images* based upon the P2 conditions of ISO 3664, however since conditions for the latter have already been defined in ISO 3664 they will not be repeated here. In ISO 12646 the option is also provided to view single images using the monitor set to either the D50 or the D65 white point.

26.6.1   ISO 12646 Parameters

The monitor and the viewing booth for the hard proof should be mounted adjacently but care should be taken to ensure light from the viewing booth does not fall on the monitor nor is within the field of view of the observer when viewing the monitor.

26.6.1.1   Image Size

The display shall be capable of displaying an image having a diagonal measurement of at least 43 cm and height of at least 22 cm.

26.6.1.2   Normal Viewing Distance

The normal viewing distance is defined as 500 mm.

   Ideally the viewing distance should be defined in terms of the screen size in order to ensure that the screen presents a constant field of view to the observer, for, as we saw in Chapter 11, the contrast range of the eye, particularly when the surround luminance is very low, is to a degree related to the field of view the image subtends at the eye.

26.6.1.3   Ambient Illumination Falling on the Screen

The ambient illumination falling on the screen should be low. The luminance of a perfectly reflecting diffuser, placed at the position of the faceplate of the monitor shall not be greater than 1/4 of the monitor white point luminance and preferably should be less than 1/8 of this value.

26.6.1.4   Colour Temperature of the Ambient Room Lighting

The colour temperature of the ambient room lighting should be within ±200 K of the colour temperature of the illumination used in the viewing booth.

26.6.1.5   Screen Reflectivity

The luminance of the black level (R = G = B = 0) in the on-state, measured with a spectroradiometer or a colorimeter in a dark room, as specified in 5.6, shall not be greater than 200% of the black level reading in the off-state.

26.6.1.6   Monitor Surround Luminance

The luminance of the area surrounding the monitor shall not exceed 1/10 of the luminance of the monitor reference white.

26.6.1.7   Viewing Booth Conditions

The conditions within the viewing booth shall conform to the viewing condition of P2 in ISO 3664.

26.6.1.8   Black Point of Display

The black point of the display shall have a luminance that is less than 1% of the maximum luminance of the display, that is, a luminance ratio of at least 100:1.

26.6.1.9   Luminance of White Point of Display

The luminance of the white displayed on the monitor shall be at least 80 nits but preferably 160 nits. The luminance of the monitor should be as high as necessary to visually match an unprinted sheet of white paper located close to the monitor having an illuminance of 500 lx (as specified in ISO 3664 for viewing condition P2).

To meet this condition, as noted previously; based upon the white paper having characteristics approaching that of a Lambertian reflector with a reflectance of 89%, the luminance of the display should be about 142 nits.

26.6.1.10   Uniformity of Luminance

The luminance at any point in the display shall be within 10% of the luminance at the centre and ideally within 5% of this value.

26.6.1.11   Chromaticity of Display

The chromaticity of the display shall be set to D_{50}, The chromaticity tolerance shall be within a circle of radius 0.005 on the u′, v′ chromaticity diagram.

26.6.1.12   Uniformity of Chromaticity

For the entire display the chromaticity of every neutral image (defined by equal digital values for R, G, and B) shall be within a radius of 0.01 in u′, v′ from the chromaticity values measured at the centre of the display.

26.6.1.13   Gamma

The value of the target gamma of the display should be chosen, by the vendor, to fall into the range of 1.8–2.4.

The sRGB and Adobe RGB colour spaces used as input to the printer are defined around an effective gamma of 2.2 (Section 24.2), this rather large tolerance on the value of gamma could lead to perceptible differences in tone gradation between the images rendered on the monitor and in print.

26.6.1.14   Point Gamma

The luminance shall be measured for at least 10 neutral colours (R = G = B), approximately equally spaced in lightness, having a luminance greater than 1% of the maximum luminance. The deviation between the normalized measured luminance and the normalized target luminance shall not exceed 10% of the normalized target luminance in every case.

26.6.1.15 Grey Balance (Gamma Tracking)

For at least 10 neutral colours (R = G = B), approximately equally spaced in lightness, having a luminance greater than 1% of the maximum luminance, the tristimulus values shall be measured. For each neutral colour, the colour (chroma) difference, ΔEc, between these measured values and the CIELAB values which are intended to be displayed shall not exceed a value of 3 and preferably not a value of 2.

26.6.1.16 Colorimetric Accuracy

A reference RGB data file comprising at least five equally spaced code values for each channel (e.g., R = 0, 63, 127, 191 and 255, using 8-bit coding) and all combinations among the other channels, having a luminance greater than 1% of the maximum luminance, shall be displayed and measured at the centre of the display. The measured tristimulus values shall be transformed to CIELAB values using the white point chosen by the software application vendor. The average of the colour differences between these values and the LAB values intended to be displayed shall not exceed 5 and preferably not 2. The maximum colour difference shall not exceed 10 and preferably not 4.

NOTE For high quality print work, a deviation of ΔEc < 1 is advisable.

26.7 Summary

This chapter has set out to provide the considerations which should be taken into account when using or establishing an image appraisal facility in terms of both the fundamental issues of perception and how they relate to the recommendations of the relevant ISO standards.

In general terms the two standards reviewed here are relevant and comprehensive; however, in covering a wide range of conditions they can appear complex to understand and in a few areas the tolerances are somewhat lax as a result of the necessity to embrace the legacy situation found in the current practice of some graphics and photographic studios and offices. The different conditions covered for single appraisal and for proof appraisal add to the complexity of the situation.

By simplifying the requirements to only the appraisal of monitor images and prints, both in isolation and in proof comparisons, and exploiting current technology, the opportunity exists to define a work station that would embrace all these conditions of appraisal, whilst still meeting the requirements set out in the two ISO standards. As the adoption of the white point of D50 is a *de facto* standard for appraising prints, such a workstation would be based upon:

- A monitor capable of producing a highlight luminance of not less than160 nits with a white point of D50.
- A viewing booth with a switchable level of illumination to provide an illuminance of either 500 or 2,000 lux at the print surface with a spectral distribution matching D50.

Ideally, by putting the requirement to cover legacy issues to one side, some of the tolerances specified in the ISO standards would be tightened up to help ensure consistency of results between workstations and between those carrying out the appraisals.

These tolerances in abbreviated form would be, for the monitor:

- Viewing distance: 1.5 times the image width.
- Monitor black, (non-dynamic picture control): ≤ 0.25% of white.
- Reflected ambient illumination from screen: ≤ 0.1% of white.
- Luminance of white: to match luminance of print paper in the viewing booth. (Notionally 142 nits when the booth is set for proof (P2) viewing conditions.)
- Display gamma: 2.2 ± 0.05.
- Other parameters generally as defined in ISO 12646.

For the viewing booth:

- Viewing distance: 1.5 times the image width.
- TLCI[2] of the print illumination: ≥ 95.
- When switching illumination levels provision of sufficient time to fully adapt to the lower level of illumination.

For the environment:

- Monitor and print surround surfaces to extend beyond the image to fill 90% or more of the field of view.
- Monitor and print surround surfaces luminance and chromaticity: 20% of screen white ± 20% and D50 within a circle of $\pm$ 0.005 v′, u′.
- Room ambient lighting: D50 with a TLCI ≥ 90. No specular reflection of room source lighting from the screen when observed from the specified viewing distance.
- Room surfaces luminance: ≤ 20% of screen white.
- Surfaces in the occasional field of view and clothing colours: ideally towards neutral chromaticity, no moderate to high saturated colours.

The rationale for these tighter tolerances has been implied in the preceding sections but is summarised here for convenience.

- The viewing distance is specified in terms of the image width to tighten up the angle of view subtended at the eye by the image, which in turn will influence the accommodation and adaptation of the eye, in terms of the area of the surrounding surfaces within the line of sight and their luminance level and chromaticity respectively.
- Monitor black should ideally be zero for RGB code values of 0. This has not yet been possible to achieve with LCD technology without resorting to dynamic picture control and so monitor black level is likely to be a compromise until OLED or some other satisfactory technology is in regular use. (see Section 8.3.) As noted in Section 13.3, experiment indicates that when appraising images under critical conditions the contrast range of the eye can be

---

[2] Television Lighting Consistency Index, see Chapter 18.

in the range of 400:1 and thus the level of black should not be above the level of 0.25% of white if this range of contrast is not to be compromised.

- For the reasons outlined above, in order to preserve the contrast range of the image, it is important that the level representing screen black should not significantly impinge upon the contrast range. Setting the maximum level of reflections from the switched-off screen to 0.1% of screen white will ensure this condition does not exceed 0.35% of white.
- When making proof comparisons it is important that the highlight luminances of both images are a close match.
- As shown in Section 13.7 the effect of small changes in overall system gamma can be substantial, with an average value of $\Delta E_{00}^* = 4.8$ being obtained for the ColorChecker chips for a change in system gamma from unity to 1.2. The aim is to keep the value of ΔE_{00}^* to less than 1.0; however, by specifying a gamma tolerance of 0.05 a compromise average value of $\Delta E_{00}^* = 1.3$ is obtained, as reported by the calculator embedded in Worksheet 13(e).
- The spectral distribution of the source illumination for the print is critical and it was shown in Chapter 18 that the CIE CRI is not necessarily a reliable measure of the capability of the illumination to produce a perceived match to that achieved by a CIE illuminant, whereas the EBU defined TLCI is so. By specifying the value of TLCI to be 95 or greater ensures that the value of $\Delta E_{00}^* \leq 1.0$.
- The effect of the brightness of the surrounding surfaces on the accommodation characteristic of the eye is significant and difficult to quantify, thus in order to ensure a consistent level of accommodation the surrounding surfaces should extend to fill 90% or more of the field of view at the specified viewing distance. For the same reason the luminance and chromaticity of the adjacent surfaces and the ambient room lighting are also specified.
- In order to minimise the risk of other colours in the occasional field of view from influencing the adaptation characteristic of the eye they are specified to be near neutral in colour.

One compromise to this approach to the specification of a workstation would be the loss of an activated screen with a D65 white point, which is useful for appraising images to be used in television. However, with suitable optional monitor profiles this feature could be made available with a change of monitor parameters and a reboot of the operating system.

Adopting the above parameters to complement those defined in the ISO standards for appraising monitor images and prints would assist in ensuring a consistent appraisal, whilst minimising the investment in different workstations for different types of appraisal. Furthermore it would facilitate a convenient and rapid change of conditions to suit the type of appraisal. Nevertheless, it is strongly recommended that the ISO standards should be fully referenced before embarking on a project to establish such an appraisal facility.

27

Colour Management in the Workflow Infrastructure

27.1 Introduction to Colour Management

It is difficult for anybody involved in taking photography seriously not being aware of colour management; its presence is ubiquitous in the books, the magazines and the websites dealing with photography but why, one might ask, is it necessary in a well-designed system which should automatically take care of such matters? By comparing photography with television, the answer to this question becomes clearer. In television there are, in system terms, only variables at the receiver; one colour space is defined for the system and all elements of the workflow are designed to work within the parameters of that colour space, from the camera through post-production and the delivery system to the domestic television set. Whereas in photography, as we have seen, not only is there a plethora of different colour spaces in use but there is also a much more extensive range of equipment available to the photographer. Furthermore in television, the source of capture and post-production is generally in the hands of professionals with only the display being in the hands of viewers who in the main, with modern receivers, are deterred from changing its characteristics. In photography the system is wide open to all users; so despite the incorporation of a colour management system, the flexibility of the operation will leave many opportunities for the untrained to make mistakes during image capture, post and display, which can lead to disappointment when perceiving the rendered image.

In explaining both the need for colour management and how it is effective in resolving problems, we shall build upon the fundamentals covered in Part 4 and in particular the topics covered in Chapters 11, 12 and 13 are pertinent to the application of colour management techniques. In essence, colour management is about managing those aspects of the operation between capturing an image and producing a print which influence the rendering of the colour of the final image and these can be summarised as:

- care in procedural terms in ensuring colour balance and tone scales are retained throughout the process.

Colour Reproduction in Electronic Imaging Systems: Photography, Television, Cinematography, First Edition. Michael S Tooms.
© 2016 John Wiley & Sons, Ltd. Published 2016 by John Wiley & Sons, Ltd.
Companion Website: www.wiley.com/go/toomscolour

- ensuring that whenever there is a transfer of the RGB components from one stage in the workflow to another, which is operating a different colour space, an appropriate transformation of the components takes place to match them to the new colour space.

We have already described these transform processes (Section 12.2) which in general terms fall into two categories. Camera-derived components may be transformed from one colour space to another by first linearising the components when necessary, matrixing them to the chromaticity coordinates of the new primaries and gamma correcting them to match the conversion function characteristic of the new colour space. Moving from an Adobe RGB colour space to an sRGB colour space is an example of the application of this process. Such a transform is straightforward; we know the characteristics of the processes involved and thus can predict the outcome of the transform undertaken.

However, the second category of transform, which is epitomised by the requirement to match the RGB components to the characteristics of the printer, is more complex, as in this case we generally do not know precisely the characteristics of the printer. As we saw in Section 23.3, the solution is to evolve a training process whereby an image file containing a range of known colour samples is sent to the printer and the samples in the resulting print are measured and compared with the originals; the differences between them are used to create a three-dimensional lookup table which is located at the centre of the transform process. Lookup tables may also be used to undertake the change of conversion function or gamma characteristic of the colour space.

Assuming for the purposes of the definition of colour management that the aim is to produce a rendered image which is either a close match to the scene or is perceived to be a close match to the scene, there are three or sometimes four areas where the application of colour management is essential.

1. In system design terms, the establishment of a colour management infrastructure strategy between those elements of the workflow with different colour space characteristics, to ensure the system is capable of supporting the movement of the image file through the workflow without the introduction of impairments.
2. For each session, checks and adjustments where necessary of the scene capture and display equipment to ensure colour balance is achieved and tone scales are properly rendered.
3. For each session, care in the selection and use of the appropriate interface colour space characteristics at each stage of the workflow.
4. On an as-required basis, occasional realignment of one or more of the interfaces in (1) to match any change in the characteristics of a workflow element or to accommodate new elements.

These four areas are critical to colour management and each will be described in some detail in a dedicated chapter; the remainder of this chapter being used to describe how colour management is incorporated within the infrastructure of system design. The subsequent two chapters describe the operational procedures to be carried out whenever critical work is undertaken and the fourth procedure only needs to be undertaken when either the stability of the operating characteristics of a system element is in doubt or a new system element, such as a new paper for printing, has been introduced into the system.

27.2   Establishing the Requirements of a Colour Management Infrastructure Strategy

27.2.1   Transform Options

We have identified the requirement to transform the colour components whenever there is a change of colour space within the workflow. In order to accomplish this, we need the following information for each of the two adjacent stages in the workflow:

- The characteristics of the colour spaces associated with the RGB or CMYK components of each stage, that is the chromaticity of the primaries associated with the derivation of the components,
- the tone scale characteristic,
- for an RGB to CMYK transform also the data required to populate the lookup table in the transform process.

From these parameters the necessary data to undertake the RGB or CMYK transform process can be calculated (see Section 12.2) or implemented respectively; however, where this information is held and how it is used will depend upon which interface in the workflow is being considered and the particular configuration selected for the transform.

27.2.1.1   A Simple Transform Configuration

Figure 27.1 illustrates an example of an image file from a camera driving a monitor; the monitor has no knowledge of the colour space associated with the image file and thus the file must include an embedded *profile*, supplied by the camera vendor, which carries the parameter data associated with both the primary chromaticity coordinates and the gamma law which relates to the RGB components. The monitor will include two related processing elements, the

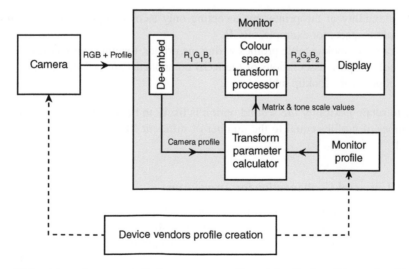

Figure 27.1   A transform process between a camera file and the display of the rendered image.

calculator, which derives both the matrix coefficients and the tone scale transform values, and a transform processor which uses these parameters to undertake the transform of the colour spaces. Such a monitor will also hold a profile, supplied by the manufacturer, which contains the chromaticity coordinates and conversion function parameters for the display. The monitor extracts the embedded data from the incoming component file and loads it into the calculator together with the data from the display profile, from which the parameters necessary to drive the transform processor will be calculated.

The configuration associated with this example of an RGB to RGB transform is a particularly simple case; by including the calculator in the monitor, it is in a position to undertake the appropriate transform irrespective of the nature of the colour space associated with the camera components. In consequence, only the display profile data is required to enable the monitor to match any set of RGB components irrespective of their particular colour space encoding.

Often it is preferred that the camera manufacturer undertakes the calculation required in order to save the cost of implementing that function in the monitor. In such circumstances the profile incorporated in the camera by the manufacturer would contain data relating to the matrix coefficients rather than the chromaticity coordinates, enabling the monitor to load them directly into the transform processor. This approach is however somewhat restrictive in that the monitor display primaries would need to conform to an established standard in order for the camera manufacturer to be able to calculate the required matrix coefficients.

27.2.1.2 A More Complex Transform Configuration

A frequent requirement is the ability to drive a printer with a file derived from a camera. In order to conform to typical configurations most printers are designed with RGB inputs, the transformation to CMYK components taking place within the printer firmware. Such a scenario presents us with a more complex problem since now a lookup table is required within the transform processor and the values required to populate it cannot be simply calculated in the manner of establishing the matrix coefficients in an RGB to RGB transform. Thus in this scenario there are a number of options open to the printer manufacturer:

- Limit the flexibility of the printer by accepting only RGB components conforming to a standard colour space, for example, sRGB.
- Provide a range of lookup table profiles within the printer, together with a means of identifying the colour space associations of the RGB components and selecting the appropriate profile to populate the lookup table.

In order to retain flexibility the second option is likely to be selected which would necessitate a number of profiles equal to the number of different RGB colour spaces likely to be encountered.

27.2.1.3 A Universal Configuration for Transforms

Although printers are available which provide a direct interface to camera image files the more comprehensive configuration is a camera in association with a computer and monitor for image processing, connected to one or more printers and possibly a projector.

In this scenario it is the computer which is required to undertake the transformations necessary and it quickly becomes apparent that although there are default colour space specifications for the camera, in general terms the colour characteristics of the camera, the monitor, the printer and the projector could be one of a number of specifications and furthermore, each model of printer will have a lookup table for each of the paper types supported.

It is evident that a different profile will be required for each pairing of colour spaces and when it is appreciated that across the industry this will include all input types: cameras, scanners, CRT and LCD displays and printers of various types, each with a range of papers with different characteristics, the situation at best becomes extremely complex and is likely to rapidly become unworkable. In the limit for a system comprising n devices, where each device may need to connect with any other device in the system, n^2 profiles will be required, furthermore for each new device or even each new paper type, a further n new profiles; clearly a far from ideal situation.

Profiles of the type described in this section, where their use is dedicated to the connection of two specific devices are referred to as device-dependent profiles.

27.2.2   Requirements of a Colour Management Infrastructure Strategy

The requirements of a colour management infrastructure may be summarised as follows:

- A configuration which minimises the number of different profiles required
- A standardisation of the profile structure and contents
- Profiles able to operate in either direction between two identified colour spaces
- A minimum of processing associated with the transform process
- Device-independent profiles.

27.3   The International Colour Consortium

In the early days of the computer-based photographic and graphics industry it was recognised by the leading vendors that the use of a dedicated profile for each possible combination of colour space was impractical. This recognition led to the establishment of the International Color Consortium (ICC) in 1993, for the purpose of *"creating, promoting and encouraging the standardization and evolution of an open, vendor-neutral, cross-platform color management system architecture and components"*, as their website[1] explains.

The ICC evolved an infrastructure for the universal application of colour management to the photographic workflow and supported it with a specification for the profiles which complements the configuration; the latest version, ICC.1:2010-12, is entitled "Image technology colour management – Architecture, profile format and data structure" and was eventually adopted by the ISO as ISO 15076-1 2005.

It is worth noting that the excellent website hosts not only the specification but also much information relating to the ICC and several white papers describing the background to colour management in the context of the specification and giving guidance on its interpretation.

An understanding of the concepts of the ICC approach to colour management is essential to ensuring that using computer-based photographic applications in day to day operations will lead to the production of prints which are of a satisfactory quality.

---

[1] http://www.color.org/index.xalter

27.3.1 The ICC Profile System of Colour Management

27.3.1.1 System Overview

The basis of the ICC system is the recognition that by splitting the colour space transform operation into two independent steps, to manage the input and output stages respectively, and standardising the parameters of the resultant colour space located between them, would make each transform and its associated profile independent of the colour space of the adjacent device in the workflow path and thus reduce the number of profiles required in a complex operation. The profile in this case contains the data already calculated to undertake the transform rather than the colour space characteristics.

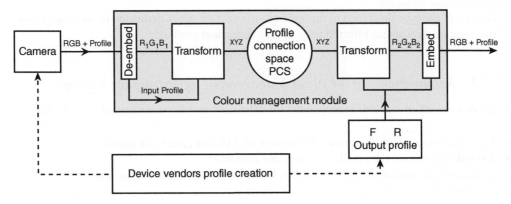

Figure 27.2 A simplified configuration of a colour management module.

In Figure 27.2, the device vendor incorporates the appropriate profile in the camera and possibly a different vendor makes the Output profile available to the colour management module (CMM). In the figure, the essential elements of the ICC framework which undertake the transform task, referred to collectively as the CMM, is illustrated with a colour background. The colour space between the two transform processors is referred to as the Profile Connection Space (PCS). Since this is a standardised colour space with published parameter data, the device vendors are able to undertake the calculations required to establish the data necessary to undertake the transform between the vendor device and the PCS. This transform data is packaged into profiles, as appropriate, both for use within vendor devices and for storage in the profile library of the CMM. Thus the matrix coefficients, LUT population and conversion function data are in a form within the profile which may be used directly to populate the appropriate functions of one of the pair of transform processors in the CMM. For maximum flexibility the profile contains the transform data necessary to undertake the transform in either direction, that is, forward, from the device to the PCS and reverse, that is from the PCS to the device. In the figure this is indicated by the Forward (F) and Reverse (R) ports in the element representing the profile.

It will be noted that for each new device introduced into the system only one additional profile is required, as any combination of device connections can be achieved by linking the appropriate source and destination profiles using the PCS as the interface. Thus the total number of profiles required for a complex system with n devices is n profiles, a considerable advantage on the n^2 number of profiles required for the device-dependent profile solution.

27.3.1.2   Workflow Description

The camera is located on the left of Figure 27.2 where it is populated by one or a number of profiles by the manufacturer to reflect both the colour encoding characteristics of the camera and the desired rendering intent of the image; the default rendering being an sRGB colour space with a perceptual rendering intent. With the exception of the raw file, the profile is embedded with the RGB component data in the file stored on the camera card.

In the CMM the RGB components and the profile in the camera file are separated; the data in the source profile is used to populate the input transform processor, enabling the components to be transformed into the profile colour space. The PCS is a conceptual element only since, except to provide measurement data, the component data it contains are used for no other purpose than to become the input to the destination transform processor, which is populated with the transform data from the destination profile selected from those available in the profile library. The library contains profiles provided by a number of sources: the computer vendor for general use, the photo application vendor, and the vendors of the various devices connected to the computer.

The transform data in the destination profile also accompanies the components to the next stage in the workflow.

27.3.1.3   The Profile Connection Space

The primary requirements of the PCS are that it should be sufficiently large to embrace all colour spaces of real colours and preferably relate to well-established international standards; in addition, it should be device independent. The prime candidates that meet these criteria are those defined by the CIE and described in Chapter 4 as the XYZ colour space of the 1931 Colorimetric Observer using a two-degree field, and the CIE LAB colour space, which is derived from it. In order to strictly delineate the options available in using the CIE colour spaces, those defined in ISO 13655:2009, *Graphic technology – spectral measurement and colorimetric computation for graphic arts images*, based upon a system white of D50 are formally adopted as the PCSXYZ and the PCSLAB colour spaces. Either may be used, the space selected being noted by a tag in the profile.

As the PCSLAB colour space is perceptually more uniform it may be encoded with either 8-bit or 16-bit values; however, the PCSXYZ colour space is encoded only with 16-bit values.

27.3.1.4   Rendering Intents

Rendering intents are fundamental to the PCS and since the PCSXYZ and PCSLAB colour spaces do not accommodate the appearance of the rendered image, the PCS is specified to be used in two different ways: first, in a colorimetric manner and second, in a perceptual manner.

In the case of the colorimetric approach, the measured values relating directly to the colorimetry of the originals and reproduction are retained, subject to the values captured relating to the PCS reference white of D50, otherwise the values are chromatically adapted to the D50 reference white of the PCS. This approach is used for both of the defined colorimetric intents.

Media-Relative Colorimetric Intent

Transformations for this intent shall re-scale the in-gamut, chromatically adapted tristimulus values such that the white point of the actual medium is mapped to the PCS white point (for either input or output).

NOTE: Transforms for the media-relative colorimetric intent represent media-relative measurements of the captured original (for Input profiles), or media-relative colour reproductions produced by the output device (for Output profiles) unless otherwise indicated in the appropriate profile tag.

ICC-Absolute Colorimetric Intent

Transformations for this intent shall leave the chromatically adapted CIEXYZ tristimulus values of the in-gamut colours unchanged.

For the perceptual approach, the intent is based upon the rendering of the image on a standard reference medium under a specified viewing condition in order to provide a target for the source intent, a target which at the printing stage is likely to be amended by re-rendering once the print media has been identified and an appropriate profile selected for the destination transform. This approach forms the basis of both the perceptual intent and optionally the saturation intent options.

Perceptual Intent

In perceptual transforms the PCS values represent hypothetical measurements of a colour reproduction on the reference reflective medium. By extension, for the perceptual intent, the PCS represents the appearance of that reproduction as viewed in the reference viewing environment by a human observer adapted to that environment. The exact colour rendering of the perceptual intent is vendor specific.

Saturation Intent[2]

The exact colour rendering of the saturation intent is vendor specific and involves compromises such as trading off preservation of hue in order to preserve the vividness of pure colours.

27.3.1.5   Reference Perceptual Intent

The reference perceptual intent is the reference print medium on which the perceptual intent calculations are based and is generally in line with the appraisal specifications for print media described in the previous chapter.

The reference medium is defined as a hypothetical print on a paper specified to have a neutral reflectance of 89%, which for an illumination of 500 lux results in a luminance of 142 nits. The darkest printable colour on this medium is assumed to have a neutral reflectance of 0.30911%, that is, 0.34731% of the paper neutral reflectance; giving a media dynamic range

---

[2] NOTE: The subject of rendering intents is considerably more extensive than I have indicated in the above few paragraphs, where I have attempted to summarise sufficient of the topic to provide a working familiarity of the four intents. However for those who require a deeper understanding of the subject, it is strongly recommended that the material on the ICC website be accessed in order to see the extensive coverage provided, both in the annex to the specification and in the associated white papers.

of 287.9:1. The specification does not explain how and on what basis these extremely precise but rather untidy figures for "black" were derived.

The reference colour gamut is provided as an optional target gamut and is defined in terms of a range of colour samples as listed in ISO 12640-3 and tabled in the ICC specification. The table provides the maximum chroma values in terms of C^*_{ab} for each 10 degrees around the a*,b* chromaticity axis and for L* values at intervals of 5 from 5 to 100. However, this target should not be used if by doing so the component values would be clipped in order to locate inside this gamut; if used the fact should be recorded as such in the appropriate tag of the profile.

The reference media viewing conditions are generally in accordance with condition P2 defined in ISO 3664, that is, the average illumination level thought to be found in the office and home, rather than that specified by the more critical appraisal condition P1. The illumination of the environment shall be assumed to be the same as that for the image; the surfaces immediately surrounding the image shall be assumed to be a uniform matt grey with a reflectance of 20%. The reference environment shall be assumed to have a viewing flare equating to 0.75% of the luminance of the reference medium, that is 1.06 nits.

27.3.1.6   The Profiles

Historically, ICC profiles have either an .icc or .icm extension; Windows uses .icm and Apple .icc. There is no difference in the structure or content of these profiles. The ICC website provides a downloadable "Profile Inspector" which enables one to inspect the profile header information and the list of tag tables.

The format of ICC profiles forms the core of the specification since it is crucial that the very many vendors creating and using profiles abide by the rules which are extensively laid out in the specification. The detail of the format is however beyond the scope of this book.

The original ICC profile specification was made available as version V2 and was updated to V4 in 2001. Despite the lengthy period since the introduction of V4, many of the colour space profiles commonly in use, such as sRGB and Adobe RGB are often coded to the V2 specification. Furthermore, users of the Windows Operating System will find that the default viewer of photos, the Windows Photo Viewer, is incompatible with some V4 profiles used to match the display to the operating system; images displayed using this viewer may be rendered incorrectly with all tones appearing too dark.

27.4   The ICC System in Practice

It is instructive to review how the ICC system is implemented in a practical reproduction system, such as the workflow in a computer-based photographic processing system application, with input from a camera file and output to a printer.

Figure 27.3 illustrates a representative system and it can be seen that in this case the CMM comprises not one but three PCSs. The first provides the *working* colour space, the second provides the input to the *output* transform processors and the third is dedicated to the *proofing* system. In this diagram the PCS uses the PCSXYZ colour space, abbreviated to the XYZ space. The configuration of the elements which produce the RGB components in the *working* colour space is identical to that described in detail in the previous section. The *working* colour space provides the environment which supports the range of functionality found in photographic

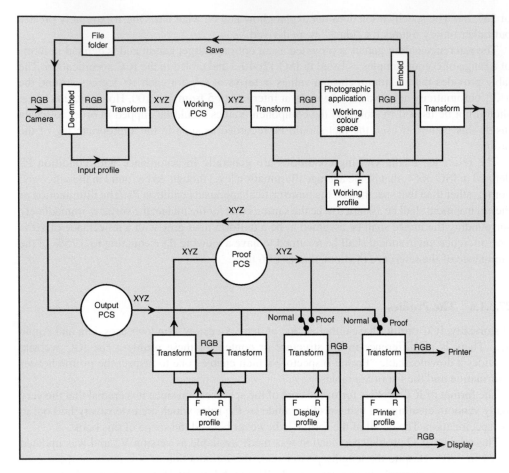

Figure 27.3   Profile workflow in the colour management module.

processing applications – such as Adobe® Photoshop®. The option is usually provided to either use the input profile as the working space or transform it to a dedicated working space. The file may be saved to a folder at any time, embedded with the working space profile.

The working space encoded output from the photographic application is passed to the input transform processor of the *Output* PCS, which is populated with the forward parameter data from the working profile to produce the XYZ components. In the lower part of the diagram, these XYZ components feed three output transform processors in parallel. Assuming that the switches are in the "Normal" position then both the display and printer output transform processors will transform the XYZ components to the display and printer colour spaces respectively.

The XYZ components from the *Output* PCS are also passed to the reverse *proof* output transform processor to provide the components transformed to the proof printer RGB components. Since the gamut of the proof printer is usually smaller than the working RGB gamut this will entail either or both gamut mapping or clipping. The forward parameters of the proof printer profile populate the proof input transform processor to provide the XYZ proofed components which, subject to the switches being in the "Proof" position, drive the display and printer output transform processors to provide an emulation of the proof printer image on the display and printer respectively.

27.5 Summary

In this chapter the general approach to modifying the colour space of a set of RGB components has been reviewed and the difficulty of incorporating the simple concept of device-dependent profiles in a complex operation exposed. The solution to this problem, in the form of the ICC approach to colour management has been explained in the context of dual device-independent profiles for each colour space conversion via a PCS located between them. The flexibility of the PCS in terms of a range of rendering intents available was explored and an example of the use of profiles in a CMM of a complex operation was illustrated in some detail.

28

Colour Management in Equipment and Scene Capture

28.1 Why there is Sometimes a Failure to Match Scene, Display and Print

In the last chapter we reviewed in some detail the ICC-recommended practice of using standard format profiles to ensure that at each stage in the workflow there is an appropriate selection between the colour space of the stage and the coding colour space of the incoming RGB components. Since the use of ICC profiles is now virtually universal, it might therefore be considered that the colour management problem had been satisfactorily dealt with but unfortunately this all too often is not the case.

As we have previously discussed, although in general terms we are looking to match the display and print to the scene,[1] for a number of reasons this is often not the case, either because for aesthetic reasons we are looking to portray a particular mood or because it is recognised that some colours in the scene are outside of the colour gamut of the display or the print. However, in the next two chapters, in order to focus on the fundamentals, we will assume that when the scene is represented by a range of colours which exercise the system but nevertheless are constrained within the display and printer gamuts, the criteria for judging that colour management is working is when we achieve a good, if not perfect, match between the scene, the display and the print.

Nevertheless, even within these constraints, all too frequently that elusive match is difficult to obtain. Often the results are disappointing and sometimes they are a disaster. So what is going wrong? It has become customary to blame one or the other of the profiles in the workflow for no longer reflecting the characteristics of its associated device. Whilst this is occasionally the reason, which we will address in Chapter 30, more often than not it is as the result of one

---

[1] In this context it is assumed that the appraisal conditions are such that they reflect the general requirements of proof appraisal as specified in ISO 12646 and as described in Section 26.5, i.e., the white point of the monitor display and the print illumination are identical.

Colour Reproduction in Electronic Imaging Systems: Photography, Television, Cinematography, First Edition. Michael S Tooms.
© 2016 John Wiley & Sons, Ltd. Published 2016 by John Wiley & Sons, Ltd.
Companion Website: www.wiley.com/go/toomscolour

or more of the following:

- Scene lighting of the incorrect colour temperature, see Section 11.3.
- Poor colour balance in the camera, see Section 11.2.
- A poor colour-balanced image in the workflow prior to assessing a match.
- A poorly set up monitor display.
- A poor viewing environment for the matching assessment.
- A poor gamma match between camera and display.
- A poor gamma match between display and printer.
- Incorrect colour settings within Adobe® Photoshop®.
- Incorrect settings within the printer driver.
- Inappropriate printer paper for the match.

Now at this stage we could use a number of paragraphs to largely repeat from a different perspective much of what has been covered in the fundamental chapters of this book; however, after due consideration I am of the view that the most useful approach is to set ourselves an exercise of attempting to achieve a match of scene, display and print, analysing along the way the correct procedures at each stage and highlighting the traps which are so easy to overlook. This is a more extensive task than one might at first imagine and will therefore be split into two chapters, with the last three of the above list being reserved for the next chapter, that is, in this chapter we will select, illuminate and capture the scene and also ensure the equipment we will be using is adjusted to perform to the specification required to ensure that colour management is able to function as predicted.

With the early versions of Photoshop, establishing the correct colour settings throughout the workflow was far from straightforward. However, with each new version it became evident that considerable thought had been given to assisting the user to make the correct choices by providing explanatory help notes when the mouse was held over a particular option. Nevertheless, for the less experienced user the range of options available and their positioning in various different drop-down menus can still be daunting. So in the following, in order to address this situation, each step in the process will be described with appropriate screenshot images. It must be said therefore that the reader experienced in the use of Photoshop, who is not experiencing colour matching problems of the sort described above, will likely find the pace adopted in these two chapters overly pedestrian and is therefore advised to consider moving directly to Chapter 30. Nevertheless, in discussion with a number of experienced photographers it has become evident that it is not unusual for them to be unaware of some of the pitfalls awaiting them in the workflow process and it might therefore be helpful to at least skim the following material.

28.2   The Exercise of Matching Scene, Display and Print

In the previous chapter we reviewed the *fundamental* basis of avoiding mismatching, now we will turn our attention to reviewing what we may term the *operational* causes of mismatch, that is, the failure to provide in the workflow, the environment required and the correct setting up and adjustment at each stage to ensure that a colour match is achieved between the scene, the display and the print.

In this chapter we will set the conditions for:

- Selecting an appropriate scene for the matching task
- Scene illumination
- Image capture
- Monitor line-up
- Viewing conditions for display and print

In the next chapter we will implement the desktop workflow, including:

- Establishing the desktop working practice colour management parameters
- Previewing the image files
- Colour managing raw files
- Undertaking the Photoshop workflow
- Appraising the match of the scene with the display
- Printing from Photoshop
- Appraising the match of display and print proofs
- Appraising the match of the scene with the print

28.3   The Matching Tests

We have outlined where problems of an operational nature can occur but how do we determine the location of the weak points in the workflow, why they are there and what action do we take to ensure a satisfactory outcome? One solution is to derive a short series of tests which by their diverse nature explore the characteristics of the various stages of the workflow and highlight the means of ensuring that where options of set-up are available the correct options are selected to match the circumstances.

28.3.1   Facilities Required for the Tests

As a first step it is helpful to identify the facilities required to enable the tests to be undertaken:

Primary equipment required:

- Camera
- Scanner
- Computer with Photoshop
- Monitor
- Printer
- Proof viewing station

Supporting items:

- For the scene, a ColorChecker Colour Rendition Chart, suitably illuminated
- An electronic greyscale test chart
- Monitor calibration equipment
- Photographic quality **neutral** white print paper

28.3.2   Review of the Characteristics of the Facilities

28.3.2.1   The Scene

The requirements of the scene are as follows:

- The scene should be capable of direct comparison with the display and the print, the most convenient solution being a chart of some form.
- It should contain samples which test the reproduction capabilities of the system both in terms of chromaticity gamut and tone range.
- The colour characteristics of the samples should be specified such that ideally subjective appraisal of the match between the original and the reproduction can be supplemented by objective measurements.

Figure 28.1   The ColorChecker chart.

The obvious candidate for this task is the ColorChecker Rendition Chart shown in Figure 28.1 which has already been referenced a number of times earlier in this book. Its characteristics may be summarized as follows:

- Critical colours. Representing skin colours, sky and trees, etc. Also difficult colours to reproduce such as purple and dark green.
- Additive and subtractive primaries sufficiently saturated to explore the gamuts of the display and printer whilst being contained within them.[2]
- A true neutral greyscale. For checking colour balance, contrast range and gamma.

The ColorChecker chart is extremely useful since it enables a wide range of critical parameters to be checked.

---

[2] As noted in Section 19.2, the cyan primary is just outside the sRGB gamut.

The top two rows of colour chips represent a range of those colours we have a good record of in our memory and if reproduced incorrectly, will immediately indicate to us that things are not right. Flesh colours are represented by the two top-left chips and others include common scene colours such as sky and trees and colours known to be more difficult than others to reproduce satisfactorily.

The third row contains the six primary colours, first the three additive primaries followed by the three subtractive primaries. The colorants selected were, at the time the chart was designed, representative of the most stable and saturated versions of the primary colours available.

Finally across the bottom is the greyscale, which is perhaps the most critical element of the chart in the context of checking and possibly adjusting controls in Photoshop. The white and greys of the ColorChecker chart are good neutrals, that is, they reflect light evenly at all wavelengths and are reasonably (but not perfectly) matched in terms of chromaticity.

28.3.2.2 Scene Illumination

In order that an image is captured with no bias of illumination across the chart it is imperative that the illumination should be uniform. This is not easy to achieve with any form of artificial lighting so a good alternative is to use daylight.

Although the camera is able to accommodate variations in the colour temperature of the lighting, it will be recalled that major changes in colour temperature will change the colours of the chart, see Section 11.3. Thus it is preferred that the daylight conditions should reasonably approximate to the system design white point, that is, D65. The simplest approach to approximating the ideal conditions is to select a sunny day with scattered white cloud and capture the image when the sun is semi-obscured by cloud. Capturing the chart in the shadows, which, subject to sky conditions could be illuminated only by blue sky light, will cause the colours to be distorted.

It is worth noting that although the ColorChecker chart has a matt surface, nevertheless it is not immune from producing diffuse specular reflections which add an underlying flare to the chart causing the darker colours to be portrayed at very much higher component levels than otherwise would be the case. Thus the chart should be captured at a time when the sun is approximately at 45 degrees to the plane of the chart as a compromise angle which minimises diffuse specular reflection whilst providing a sufficient level of reflectance.

28.3.2.3 Electronic Greyscale

It is worthwhile emphasising the general usefulness and importance of greyscales in checking the white balance of the workflow path in colour reproduction. The critical point to appreciate is that equal amounts of red, green and blue are equal to reference white. A true, that is, a neutral greyscale has all signals equal on each step of the scale, which makes it easy to check that the balance and contrast law shape are matched by checking that the R,G and B components are equal at the output of each stage and that each step equates to an equal step in perceived lightness (see Section 13.9). Thus colour balance is achieved by first setting each of the RGB components of the lightest step to 100% and then ensuring the R,G and B values for each of the individual grey steps are equal.

Although the ColorChecker chart is ideal for checking the balance and contrast laws of the camera, minor irregularities in the chart, the lighting and the camera make it unlikely a

Figure 28.2   The electronic generated greyscale test image.

perfect balance will be achieved on each step of the greyscale. For checking the remainder of
the workflow, that is the computer application, the display and the printer we really need an
electronically generated greyscale to ensure the chart will be neutral.

The greyscale generated in Section 13.9 is ideal for this purpose and is repeated in Figure 28.2
(A version of this greyscale is available on the book website). The steps of the greyscale are set
at levels which should produce equal lightness steps on a monitor with a gamma characteristic
of 2.4. At each end of the greyscale two further ministeps with smaller equal changes in
lightness are incorporated to ensure that, if perceived, there is no serious crushing at either end
of the contrast range. In the lighting environment specified later in this section, one should
just see the darkest ministep but it is unlikely to be seen in the printed version of this page.

*Once we know the greyscale can be reproduced correctly then we can be confident that
all objective adjustments are optimally set and components representing a real scene will be
rendered satisfactorily.*

28.3.2.4   Monitor Calibration

Traditionally, the incorrect set-up of the display together with the quality of the associated
ambient lighting viewing conditions are the major reasons for disappointment when comparing
the scene with the displayed reproduction. Very often, particularly with older displays, the
white point does not match the system white. In recent years however, new displays are usually
well adjusted before they leave the factory.

The aim of calibrating the monitor is to match the peak white of the display, in terms
of luminance level and system white point, with the white of the print when illuminated as
specified below. The reference white of the display in photographic terms is ambiguous, as
was noted in Chapter 26, being D65 for displaying images in isolation and D50 when proof
viewing. Since our aim is to compare print with screen, we are effectively proof viewing, so
the monitor should ideally be adjusted to produce a white of D50,[3] at a luminance level of 142
nits (see Section 26.5).

---

[3] In the past, in the absence of a D50 illuminant for print viewing, I have found that adopting a white point of D65
for both the monitor and the illumination of the print did not detract from providing a critical comparative viewing
environment.

Figure 28.3   A greyscale chart and print viewing area illuminated by a system white source, adjacent to the display.

There are two basic methods of calibrating the display. The traditional subjective method was to "eyeball" the display and a reference white source and use the red, green and blue balance controls of the display to bring about a subjective match to a reference source. Although experienced users could achieve satisfactory results, much practice was required. The alternate approach is to use a spectroradiometer together with calibration software to set up the display.

Depending upon the sophistication of the calibration application, once the application is activated the required parameters of the display may be entered, including highlight luminance, white point and gamma characteristic. The spectroradiometer is hung over the screen (in a similar manner to that illustrated in Figure 28.3) and the software applies a range of test colours in sequence and, in a modern monitor with controls accessible to the application, makes the adjustments necessary to match the requested parameters.

28.3.2.5   The Viewing Station

The requirements of the viewing station in terms of the parameters associated with the illumination of the print and the luminance of the surrounding surfaces were described in some detail in Section 26.5. In summary, the illuminance of the print should be 500 nits, the luminance of the surrounding surfaces should be 20% of the luminance of the print paper and both should match D50. Critically the luminance of the print and the display should match and in practical terms the simplest means of achieving this, following monitor adjustment, is to adjust the print illuminance until the display and print image subjectively match. (As a reminder, the

Figure 28.4   A proof style photo viewing station.

luminance of the print with a paper reflectance of 89% will be $0.89 \times 500/\pi$, which is equal to 142 nits, the display luminance.)

The ISO illumination specifications for print appraisal are very specific; however, for comparing proofs the specification is somewhat vague by comparison, indicating only that the two images should be adjacent. Nevertheless, this comparison is critical, so the closer together are the images, the less the viewer has to avert his or her gaze and so the more critical they can be in judging the match of the comparison.

In the absence of specific recommendations, some years ago I constructed a viewing station around the monitor, which in itself had been located in a corner of the room in close to an ideal viewing position, that is, located such that the centre of the screen is a few degrees below the sitting level line of sight. The viewing distance is about 1.5 times picture width, there are minimal reflections from the room environment and a highly directional script lamp matches the system white.

It is not suggested this is an ideal viewing station configuration but it may be helpful to briefly describe its characteristics in order to place in context later remarks pertaining to the judgement of the match between the images of the display and print when using this configuration.

Figure 28.4 illustrates a long shot of my set-up. The illumination of the print area at the rear is by two system white florescent lamps masked by vertical strips on either side of the monitor. These lamps are under dimmer control such that the peak white of the print can be matched to the peak white of the display. Neutral grey card[4] of about 18% reflectance is

---

[4] Unfortunately I was unable to obtain a true neutral grey card which explains the mismatch between the monitor grey surrounding the image and the marginally green tinted surrounding surfaces.

Figure 28.5   Viewing station from the viewpoint of a person undertaking the appraisal.

used as the background to the viewing area in order to assist the eye to accommodate to the illuminant colour. A system white fluorescent downlight illuminates the keyboard area and the monitor is shielded by an overhanging shelf. Notice the screen is located a few degrees below the horizontal which tests have shown provides an environment with less neck strain during a prolonged session at the computer. Such an approach enables the two images to be seen in adjacent lines of sight.

As illustrated in Figure 28.5, when seated in the working position most of the field of view is taken up with neutral areas illuminated by the standard illuminant, which assists in ensuring the eye is fully adapted and is therefore at its most critical when assessing colour differences.

It cannot be emphasised how important it is to get this comparative viewing environment right. Without a doubt differences in highlight brightness between the screen and the print, and errors in the gamma setting and the colour temperature of the monitor are the cause of much disappointment when comparing the print with the monitor display.

28.3.2.6   Selecting the Print Medium

Most print papers are, by popular demand, on the cool or blue side of neutral. So if we were to print the ColorChecker chart onto such paper it would provide a poor match to the original, since the ColorChecker chart is printed on a paper which is very close to a true neutral. The whites and pale colours would relate to the paper and prevent a match being obtained, irrespective of every other system parameter being correct.

Hahnemühle Photo Rag is a paper which is a very close match to neutral and being a matte finish, its white provides a very close match to the white of the ColorChecker chart. It will be appreciated that since no pigments are deposited at white, the colour of the white chip on the print is set only by the paper colour, so it is essential for these tests to select a paper which matches closely the white chip of the neutral ColorChecker chart.

28.3.3　Aims of the Tests

The aims of the tests are:

- to capture the ColorChecker chart image using a range of colour space settings on the camera and scanner and
- to use the various captured file images to explore and exercise what is happening using the various appropriate settings within both the computer system and Photoshop and thus establish a procedure for using the correct settings during a typical session for producing a print.

28.4　Image Capture

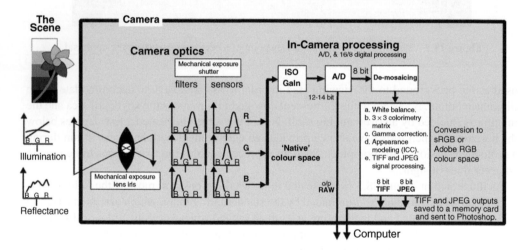

Figure 28.6　Camera configuration.

All image files captured were based upon the ColorChecker chart, which in terms of those shot by the camera, were captured outside on a bright day with no sunlight present. The camera includes the features illustrated in Figure 28.6 and was manually balanced on the white square of the chart but nevertheless prior to correction, images showed small but significant colour balance errors.

The legacy Epson scanner was also used to capture the chart image as it does not embed profiles; however, it does provide the option to select the colour management capture feature which includes a range of "target" colour spaces. This apparently incomplete feature is useful for these tests because they exercise the options of Photoshop when opening a file. For ease of use all files were renamed in a manner which described their source and capture characteristics. The images captured and their characteristics are listed in Table 28.1.

Table 28.1 The exercise image files

Capture	Encoded colour space	Profile	File type
Camera	sRGB	sRGB	JPG
Camera	Adobe RGB	Adobe RGB	JPG
Camera	Native	Untagged	MRW
Scanner	sRGB	Untagged	JPG
Scanner	Adobe RGB	Untagged	JPG
Scanner	Wide gamut RGB	Untagged	JPG

Armed with these files we are now ready to exercise the colour-management-related options in Photoshop.

29

Colour Management in the Desktop Workflow

29.1 Introduction

By using the test files generated as described in the previous chapter, we are now in a position to explore the colour management procedures associated with the desktop workflow, stage by stage based upon the following:

- Establishing the desktop working practice colour management parameters
- Previewing the image files
- Colour managing raw files
- Setting Photoshop workflow colour parameters
- Appraising the match of the scene with the display
- Setting the colour parameters when printing from Photoshop
- Appraising the match of display and print proofs
- Appraising the match of the scene with the print
- Summary of colour management procedures in the workflow.

This chapter describes the detail of following the desktop workflow, from the establishment of the colour management parameters in both the operating system and the photographic processing system, to comparing a match of the print with the original scene. To be useful, such an approach requires reference to the particular menu choices and displays, which though fundamentally similar, have minor differences both within operating systems and within photographic processing systems. With apologies to Mac users, all descriptions of the workflow screen displays relate to using the Microsoft Windows 8© operating system. Furthermore, as the majority of users are committed to Photoshop for their photographic processing system, it is Photoshop CS6 which will be used to illustrate the procedures through the workflow process.

By its nature this chapter includes a number of Adobe® Photoshop® screen shots and the author is grateful to Adobe for the following: "Adobe product screen shot(s) reprinted with permission from Adobe Systems Incorporated."

Colour Reproduction in Electronic Imaging Systems: Photography, Television, Cinematography, First Edition. Michael S Tooms.
© 2016 John Wiley & Sons, Ltd. Published 2016 by John Wiley & Sons, Ltd.
Companion Website: www.wiley.com/go/toomscolour

In the earlier versions of Photoshop, setting colour management-related parameters was somewhat fraught with risk as the selection options were provided with little or no explanation or guidance. With each new version, the underlying logistics of the colour management system have been further rationalised and most importantly at virtually every step, by holding the mouse arrow over the option to be selected, an information panel appears which provides guidance as to how that selection should be used in the situation pertaining at the time.

In the following sections of this chapter, an approach is taken based upon first setting up the colour management parameters of both the operating system and Photoshop, followed by a description of the procedures associated with each stage of the workflow. Where appropriate, at each stage the effect of changing a parameter is evaluated using the range of image files captured as described in the last chapter; the contents of which have parameters designed to exercise the various options within Photoshop.

In order to appreciate what is happening as one loads image files into Photoshop, it is helpful to have this range of image files available, which have been derived at capture using different colour spaces and with and without different embedded profiles. These are the ColorChecker files captured as described in the previous chapter containing an identical critical greyscale and a range of colour samples which will readily show any differences in the display as different parameters are selected in Photoshop.

Once the loading options have been clarified and the displayed images have been reviewed and appraised for matching with the originals, attention is then turned to the printing options and the appraisal of the final prints.

29.1.1   Colour Spaces and Their Conversion in the Computer

Colour management on the desktop is shared between the operating system and Photoshop with the operating system being responsible for matching the RGB components to the display colour space. This colour space is defined by the chromaticity coordinates of the display primaries, the display electro-opto conversion function (EOCF) and the gamma 'correction' incorporated within the monitor. Generically the sRGB colour space is intended to represent the characteristics of a typical display and, in the absence of a dedicated monitor profile, the operating system will supply this profile to the transform processor in the monitor signal path, as illustrated in Figure 29.1. This desktop section of the workflow diagram shows the photo viewing system, the RAW file adjustment and conversion, the Photoshop image adjustment module, the output processing module, the viewing options for normal or proof, the display arrangements and the printing arrangements. The underlying use of the associated profiles is illustrated in more detail in Figure 27.3.

Manufacturers of quality displays provide a profile file (i.e. an .icc or .icm file) which implicitly accurately describes the chromaticities of their particular display primaries. Traditional LCD displays, illuminated by a white backlight, find difficulty in matching the saturation of the sRGB primaries, whilst those LCD displays utilising LEDs as their light source are able to provide primaries of significantly wider gamut, including in some cases the claimed ability to closely match the Adobe® RGB primaries.

In the description of the computer and Photoshop settings associated with the operational workflow which follows, each procedure is supported with a captured screen shot illustrating the procedure being described.

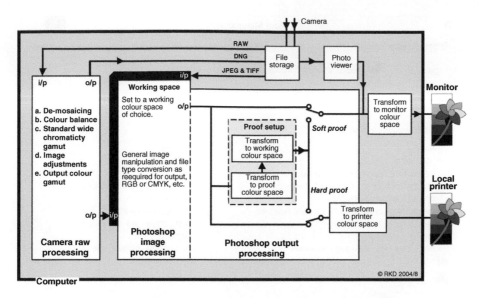

Figure 29.1   Diagram illustrating the summarised desktop workflow.

29.2   Establishing the Desktop Working Practice Colour Management Parameters

Before commencing work on the prepared files it is important to check that the parameters which control the operation of the underlying colour management system are correctly set. These parameters are functional in both the operating system and in Photoshop.

29.2.1   Settings in the Computer Operating System

The operating system settings are limited to selecting the profile which interfaces the coded colour space of the file or working space image to the colour space of the display, as illustrated by the 'Transform to Monitor Colour Space' element in Figure 29.1.

At the time the monitor is installed, an associated display profile file produced by the monitor vendor is usually loaded and stored along with all the other profile files used by both the operating system and Photoshop in the Windows→System32→Spool→Drivers→Color directory.

The colour management settings for the operating system are accessed via the Control Panel settings under the 'Colour Management' tab which will produce a display as shown in Figure 29.2.

If more than one monitor is attached, either may be selected in the options space adjacent to the 'Device' heading to enable the appropriate profile to be selected for each monitor.

A list of all profiles held in the directory whose 'Device Class' is 'Display' will be listed.

Should a dedicated vendor-supplied monitor profile not be available, the sRGB profile can be loaded from the pool of profiles using the 'Add' button to display the list of available profiles for selection.

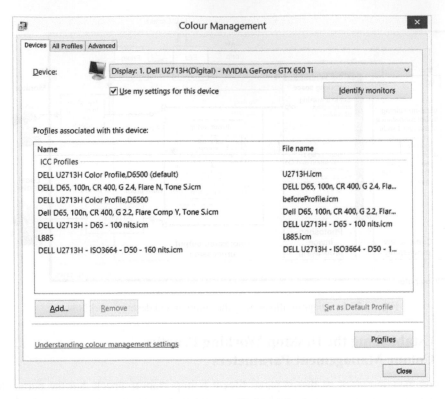

Figure 29.2 Display profile selection.

In addition to the profiles provided by the monitor vendors (Dell and Eizo, respectively in the diagram) are also a range of specific profiles created to match a range of display performance characteristics using the procedures described in Chapter 30 for the creation of profiles.

Selecting the 'All Profiles' tab will display those non-ICC profiles associated with the Windows Colour System (WCS).

If not already the default profile, select the profile appropriate to the monitor and click on 'Set as Default Profile'. *The next time the operating system is rebooted* the selected default profile will drive the display.

29.2.2   Settings in Photoshop

In the context of a colour management system, Photoshop provides a large number of colour spaces to work with from what might loosely be described as system workspaces through to workspaces designated for specific requirements, such as the display profiles, the printer profiles and the print paper profiles.

Unfortunately at every point in the workflow where a profile may be selected, in Photoshop CS and earlier versions, all possible profiles are presented for choice, irrespective of the fact that the majority of those presented are inappropriate for purpose at that point in the workflow, which can cause confusion to the uninitiated. In later versions of Photoshop the options presented at each stage are sometimes tailored to the activity of the particular stage in the workflow.

Photoshop provides for the selection of an appropriate colour space throughout the workflow, from the loading of the image file, through the working space to the print space, and, via optionally the soft proofing space, back to the working space and finally the viewing space. The conversion from one working space to another requires a good deal of calculation, each of the $R'G'B'$ signals must be linearized, matrixed to the PCS XYZ colour space, matrixed to the new colour space and then re-gamma corrected. Where printing is involved, the situation is made more complex by substituting lookup tables for the matrixing process. If gamut mapping is not required, Photoshop undertakes these tasks transparently, though there are limitations to look out for which will be addressed later.

The principal aim of this project is to identify these profile-based transform activities in terms of where they can be accessed in the menu system and to use the captured image files to exercise the various menu options and note the effect of these choices on the displayed images.

29.2.2.1   The Color Settings Panel

The 'Colour Settings' panel is selected under the Edit tab and provides access to setting the parameters for the range of functions which comprise the colour management operation, which include:

- Predefined Settings
- Working Spaces
- Color Management Policies
- Conversion Options
- Advanced Controls

Each of the above functions is assigned a designated area within the Color Settings panel.

> Display the 'Color Settings' panel by selecting Edit→Color Settings from the menu system. The Color Settings panel will appear on screen as illustrated in Figure 29.3.
>
> When the mouse hovers over any of the options of the Color Settings panel a description of the option and, if appropriate a description of the action resulting from the selection, will appear in the 'Description' area at the bottom of the panel. This aid to selecting the most appropriate option has been developed and refined with each new version of Photoshop and is very helpful in ensuring colour management is properly implemented.

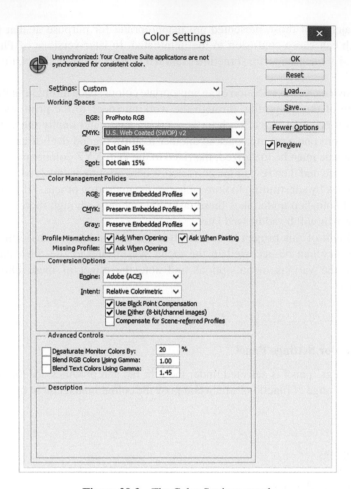

Figure 29.3   The Color Settings panel.

On the right-hand side of the panel, below the 'Save' button, is a button which enables 'Fewer' or 'More' options to be presented. Initially it is apparent that its operation extends the options displayed on the Color Settings panel but it also extends the options presented as each of the boxes in the panel is activated. In general terms the 'Fewer' option presents only those options that for most of the time are pertinent to the area one is operating in and relate directly to the activity associated with the option; whilst the 'More' option provides full flexibility by offering all options irrespective of whether they may or may not be entirely appropriate for the situation.

29.2.2.2   'Predefined' Settings

It will be noted that at the top of the 'Color Settings' panel is the 'Settings' selection, which might less ambiguously be labelled 'Predefined Settings' since the options provided here enable the selection of a file containing the colour settings for each of the parameters illustrated on the panel. This colour settings file (CSF) thus provides a quick means of changing from one set of colour management options to another without the necessity of individually changing the parameters in each of the optional boxes below.

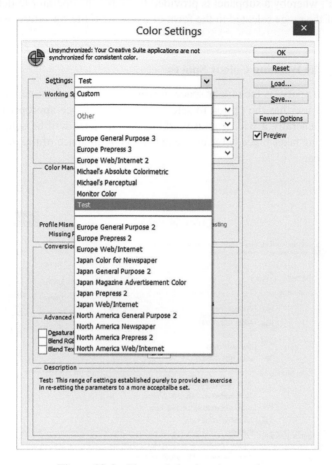

Figure 29.4 The predefined settings options.

Activating the 'Settings' arrow displays over 20 options of predefined settings, primarily associated with the common press standards used in the three dominant printing areas of the world as illustrated in Figure 29.4. If the 'Fewer Options' button is selected then the options are limited to the printing options in the area of the world where the operator is located.

When a 'Settings' option is selected, the range of parameters associated with the particular selection made are placed in each of the boxes within the areas below. For example, if the 'Monitor Color' option is selected, the profile previously selected in setting the display profile for the operating system appears in the RGB working space below. However, since the interest of this project is related to inkjet printing and there is no selection available for this option, selecting 'Test', will load the colour settings previously selected by the author to provide the incentive in the following procedures to select more appropriate parameter values.

As will now be clear, a particular range of colour management settings can be saved, by the user selecting the desired settings in each of the options in the Color Settings panel and clicking on 'Save', whereby a subpanel is provided to enable the operator to define the name and description of the range selected in the form of a new CSF file for future use.

29.2.2.3 Setting the Working Colour Spaces

The 'working colour space' is the colour space in which all image adjustments within Photoshop are made. The user has the option to select the working space appropriate to the tasks to be undertaken. Generally a wide gamut working colour space will be selected to ensure that no gamut clipping takes place between the encoded colour space of the input file and the working colour space.

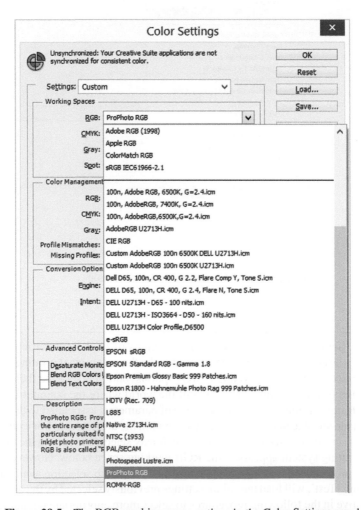

Figure 29.5 The RGB working space options in the Color Settings panel.

The working colour spaces for both RGB and CMYK options are selected in the next area of the Color Settings panel. Our primary interest is the working RGB colour space and, if the 'More' options button is activated, selecting this option will display *all* the RGB colour space profiles on the system including many legacy spaces, as illustrated in Figure 29.5. However, if the 'Fewer' option is active the range of options is restricted to more appropriate colour spaces, with the unfortunate restriction that the ProPhoto colour space is not available whilst the obsolescent ColorMatch RGB colour space is.

The RGB working colour space required may be selected from a choice of some 45 or so spaces as shown in Figure 29.5 (the screenshot has curtailed the list which would otherwise extend below that illustrated) and includes the various print paper profiles associated with any printers connected to the computer.

Since an RGB working colour space which does not inhibit the colour gamut of the input file is usually required, most of those in the 'More' options list are irrelevant to our requirements and the ProPhotoRGB colour space is selected in this instance.

It will be noted that once any of the parameters in 'Working Spaces' are changed, the 'Settings' box will change from the original setting to 'Custom'. Once it is established that a custom configuration will be used on a regular basis it may be saved with a new title for future reference. This title may then be selected from the 'Settings' box list of options, thus providing with one selection the range of colour management options required.

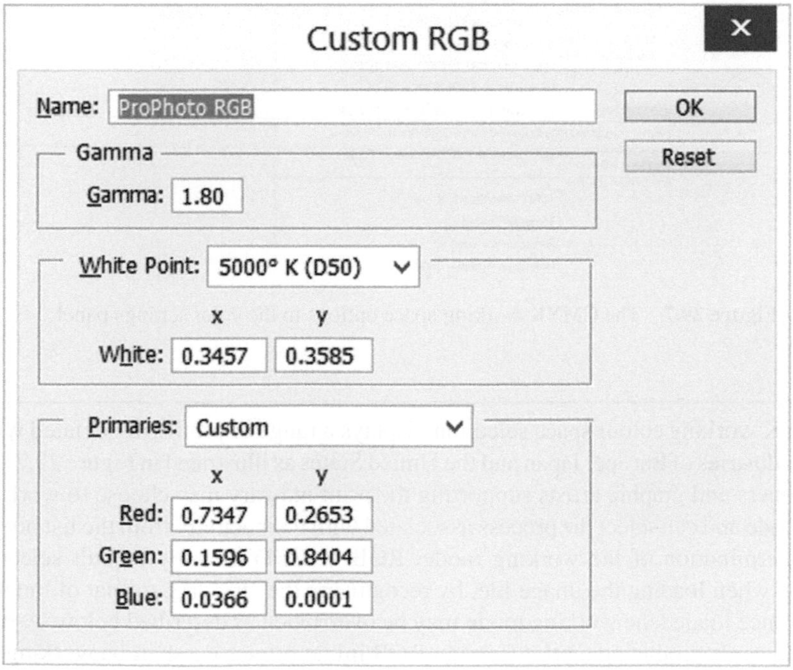

Figure 29.6 Working colour space characteristics.

Photoshop provides a facility for reminding one of the characteristics of the selected working colour space. Ensure that the 'More' option is selected, click on the 'RGB:' option arrow
to display the working colour spaces and then click on Custom at the top of the list, the characteristics of the current working colour space will be displayed as illustrated in Figure 29.6.

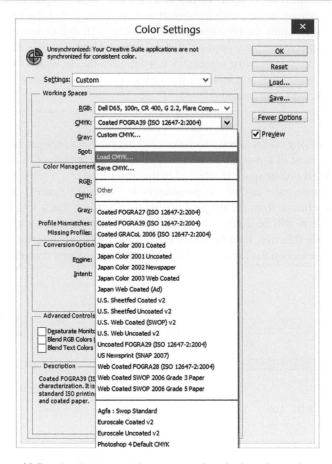

Figure 29.7   The CMYK working space options in the color settings panel.

The CMYK working colour space selection displays a range of profiles associated with the
printing industries of Europe, Japan and the United States as illustrated in Figure 29.7. Those
photographers and graphic artists supporting the print industry may choose to work in the
CMYK mode and can select the process associated with their activity from the list provided.

The determination of the working mode, RGB or CMYK, is by default selected by
Photoshop when loading the image file, by recognising the encoding format of the image;
however, once loaded the working mode may be overridden as described below. As we are
working towards producing a print from an RGB inkjet printer we shall be working in the
RGB mode throughout the workflow.

The 'Gray' and 'Spot' boxes enable dot gain-related parameters to be set. If the predefined 'Settings' option has been set on a recognised group of printer procedures, these boxes will default to the dot gain parameters associated with that group. However, if the CMYK working colour space is changed, the 'Settings' option will default to 'Custom' leaving the original parameters in the 'Gray' and 'Spot' boxes. Therefore if the CMYK colour space is changed, care needs to be taken to ensure the appropriate matching parameter values for the 'Gray' and 'Spot' boxes are in place, since these values do not track the CMYK selection in the same way as is done for a change in the predefined 'Settings' parameters.

29.2.2.4   Colour Management Policies

When the mouse arrow is suspended over the 'Color Management Policies' area of the 'Color Settings' panel a useful summary of the functionality controlled by this area is provided in the 'Description' area as illustrated in Figure 29.8.

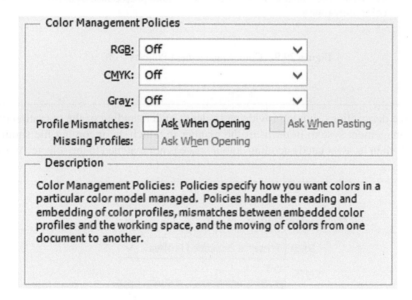

Figure 29.8   The Color Management Policies area of the Color Settings panel.

The options available under each of the three boxes are virtually identical and are as follows:

- Off
- Preserve Embedded Profiles
- Convert to Working RGB

In each of the next three screenshots, Figures 29.9, 29.10 and 29.11, the three Color Management Policies options are displayed together with the 'Descriptions' appropriate to each option.

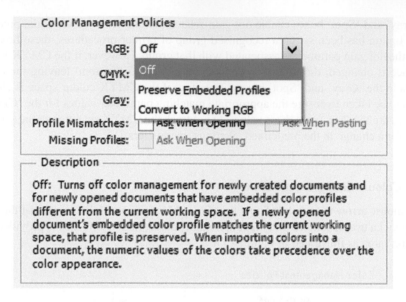

Figure 29.9   Color Management Policies – off.

These three descriptions should provide an in-depth understanding of the operation of the colour management system in terms of the manner in which the incoming file should be processed when its associated encoding colour space is not matched to the selected working colour space.

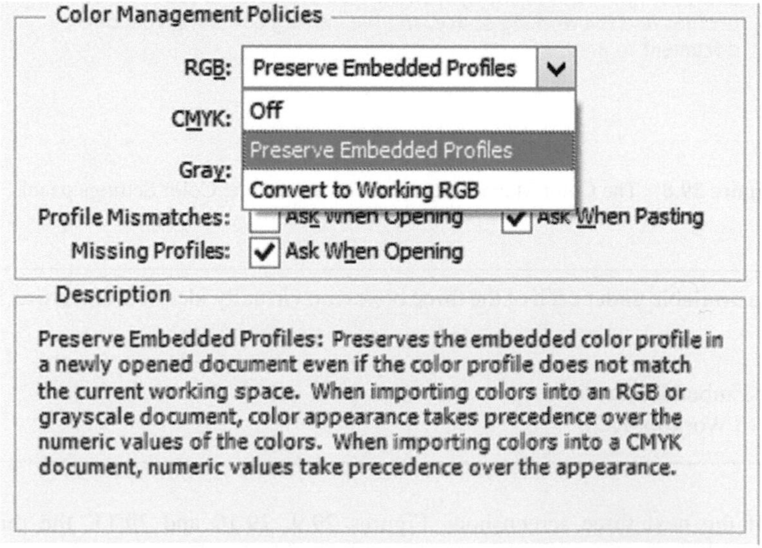

Color Management Policies

RGB:	Preserve Embedded Profiles ▾
CMYK:	Off
Gray:	Preserve Embedded Profiles
	Convert to Working RGB
Profile Mismatches:	☐ Ask When Opening ☑ Ask When Pasting
Missing Profiles:	☑ Ask When Opening

Description

Preserve Embedded Profiles: Preserves the embedded color profile in a newly opened document even if the color profile does not match the current working space. When importing colors into an RGB or grayscale document, color appearance takes precedence over the numeric values of the colors. When importing colors into a CMYK document, numeric values take precedence over the appearance.

Figure 29.10   Colour management policies – preserve embedded profile.

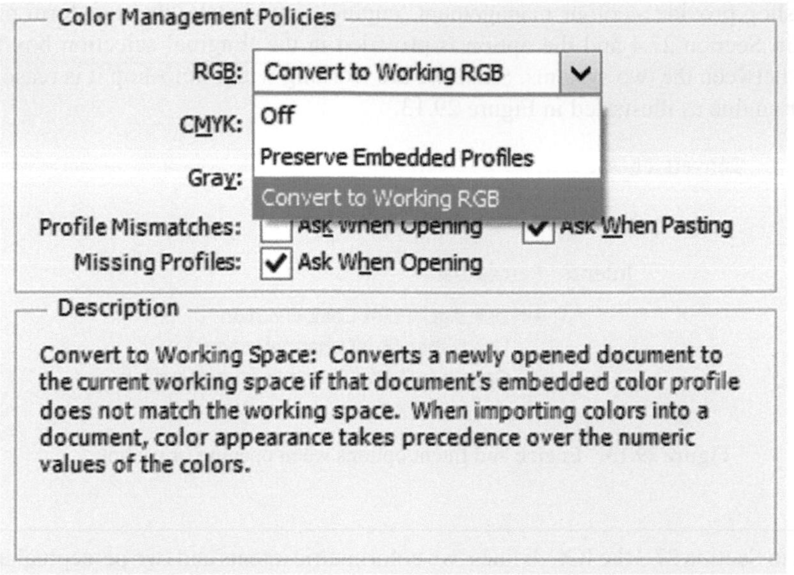

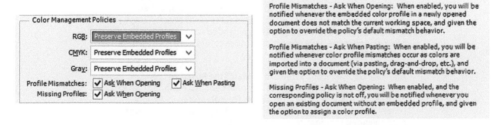

Figure 29.11   Colour management policies – convert to working profile.

In the normal course of events either the 'Preserve Embedded Profiles' or 'Convert to Working Profile' should be selected. My preference is for 'Preserve Embedded Profile', though as we shall see in either case, depending upon subsequent settings, one can make the decision on the options available when loading the file.

The lower section of the Color Management Policies area contains three tick boxes which enables Photoshop to ask the relevant question regarding the preferred option when attempting to open or paste an image that does not match the working colour space. The three descriptions on the right of Figure 29.12 indicate the result which will occur depending upon whether the associated box is ticked or not. In order to minimise the possibility of making a mistake when opening a file, all three boxes should be ticked.

Figure 29.12   Profile Matching when opening or pasting.

29.2.2.5   Conversion Options

This section of the Color Settings panel primarily provides the options for selecting the transform intents associated with the ICC profiles as described in Section 27.3. Both Windows and Photoshop provide a colour management 'engine' to undertake the transform processes described in Section 27.4 and the option is provided in the 'Engine' selection box to make the choice between the two systems. Since we are working with Photoshop it is reasonable to select their engine as illustrated in Figure 29.13.

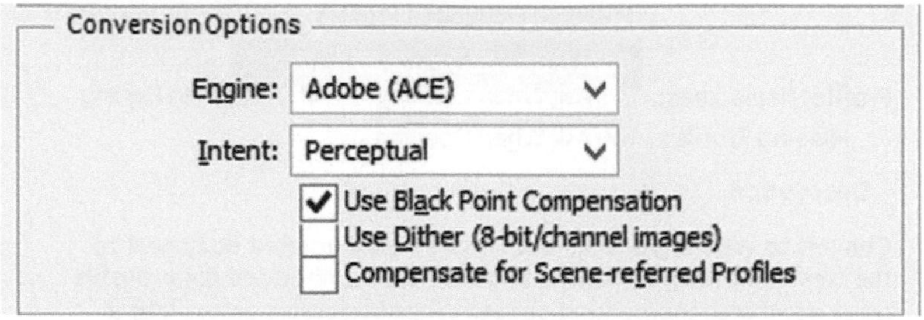

Figure 29.13 Engine and Intent options when opening or pasting.

> As noted in Section 27.3 the ICC defines two colorimetric intents and two perceptual intents. With each option selection Adobe provides a less formal description of these intents than does the ICC, which provides an alternate view of the process; these descriptions are illustrated in Figure 29.14.

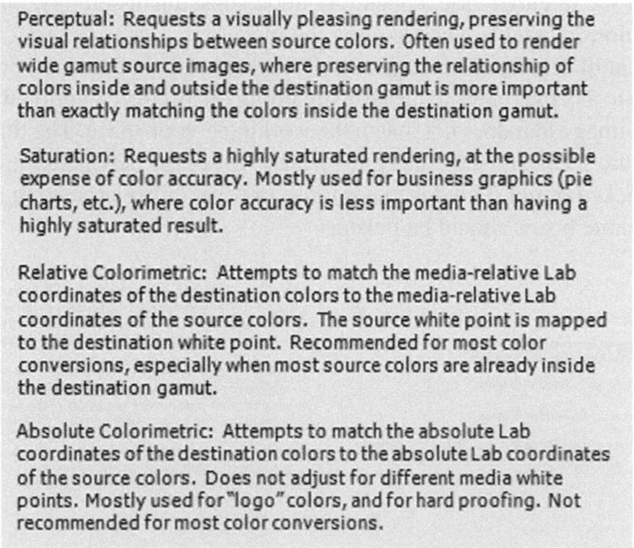

Perceptual: Requests a visually pleasing rendering, preserving the visual relationships between source colors. Often used to render wide gamut source images, where preserving the relationship of colors inside and outside the destination gamut is more important than exactly matching the colors inside the destination gamut.

Saturation: Requests a highly saturated rendering, at the possible expense of color accuracy. Mostly used for business graphics (pie charts, etc.), where color accuracy is less important than having a highly saturated result.

Relative Colorimetric: Attempts to match the media-relative Lab coordinates of the destination colors to the media-relative Lab coordinates of the source colors. The source white point is mapped to the destination white point. Recommended for most color conversions, especially when most source colors are already inside the destination gamut.

Absolute Colorimetric: Attempts to match the absolute Lab coordinates of the destination colors to the absolute Lab coordinates of the source colors. Does not adjust for different media white points. Mostly used for "logo" colors, and for hard proofing. Not recommended for most color conversions.

Figure 29.14 Transform intents descriptions.

As these descriptions imply, the choice of intent is very dependent upon what is being aimed for in terms of achieving the best looking result when transforming between two different colour spaces; if the destination colour space has a relatively limited gamut, then Perceptual will usually produce the most satisfactory result. However, there are times when other alternatives are a better option and there have been occasions when preparing material for this book when the Absolute Colorimetric intent has provided the colorimetric accuracy required.

29.2.2.6   Black Point Compensation

Black point compensation is an Adobe transform procedure which supplements the intents options of the ICC profiles by mapping the lightness or contrast range of the source components to maximise the lightness range in the destination colour space, rather than undertake a mapping which would endeavour to reproduce the original lightness range.

This is a very powerful tool for optimising the appearance of many photographic images and the mathematical basis of its operation is described in an Adobe® white paper[1] from which the following introductory paragraph is extracted:

'The color conversion algorithm consults the ICC profiles of the two devices (the source device and destination device) and the user's rendering intent (or intent) in order to perform the conversion. Although ICC profiles specify how to convert the lightest level of white from the source device to the destination device, the profiles do not specify how black should be converted. The user observes the effect of this missing functionality in ICC profiles when a detailed black or dark space in an image is transformed into an undifferentiated black or dark space in the converted image. The detail in dark regions (called the shadow section) of the image can be lost in standard color conversion.'

In general terms, whenever it is decided that the intent should be 'perceptual', then activating black point compensation is also likely to lead to a more acceptable print. Its use when other forms of intent are selected may or may not improve the situation depending upon the circumstances.

Since for this project we are seeking an objective match between scene, display and print, the intent should be set to 'absolute colorimetric' and black point compensation should not be selected.

29.2.3   Settings Summary

Noting that the aim of this exercise is to produce a reasonable match between the ColorChecker chart, its display on the monitor and its rendition as a print, then the Color Settings panel parameters would be set as illustrated in Figure 29.15.

Note: It is emphasised that the settings illustrated in Figure 29.15 are unlikely to be selected for normal photographic day-to-day operation.

The CMYK working space is arbitrarily chosen since it is not used for this exercise.

---

[1] Adobe Systems' Implementation of Black Point Compensation: http://www.color.org/AdobeBPC.pdf

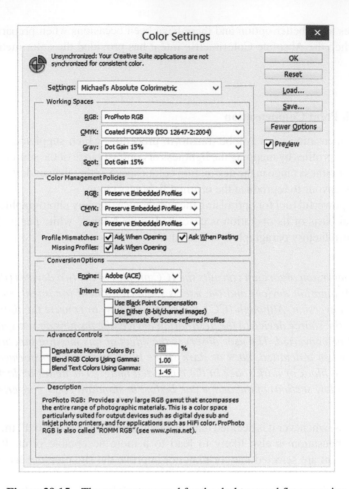

Figure 29.15 The parameters used for the desktop workflow exercise.

29.3   Image Preview

29.3.1   Applications for Previewing

Photoshop is not the ideal application for previewing files, nor is it intended to be. The applications most suited to previewing are:

- The operating system photo viewer
- Adobe Bridge
- Adobe Lightroom

Reference to Figure 29.1 indicates that all paths to the monitor pass through the monitor transform processor, so subject to the application abiding by the ICC colour management system all preview files should be rendered on the display in their correct colours. Unfortunately, this is not always the case; such a situation is summarised in the next section.

29.3.1.1 Windows Photo Viewer

The Windows Photo Viewer is not ideal for critical viewing, one of its principal drawbacks being the bright white surround to the image when viewing images of an aspect ratio different to the monitor display. Ideally an adjustment of surround brightness would be provided.

For those who are using a custom profile for their monitor which is formatted in accordance with the latest (2001) ICC V4 specification, Windows Photo Viewer is likely to display images more darkly than they should be because of the incompatibility between the viewer and some V4 display profiles. In these circumstances it is recommended that either Adobe Lightroom is used for previewing or a new monitor profile is created in accordance with the ICC V2 specification.

29.3.1.2 Adobe Bridge

Adobe Bridge is a very convenient application which satisfies so many previewing requirements but is hugely disappointing in that its behaviour in colour management terms is anomalous. It would appear that unless Adobe Creative Suite is installed the application does not colour manage the images displayed on the monitor, despite being aware of the identity of the image colour space contained in the file. All images for display appear to be tagged with an sRGB profile, irrespective of the native colour space of the file image. This is not a problem if the display has an sRGB gamut but extended gamut displays will render the image gamut mapped to the sRGB gamut.

29.3.1.3 Adobe Lightroom

Adobe Lightroom is another application which is convenient to use as a previewing system; furthermore, it displays all tagged files with the correct colour rendition irrespective of whether the display chromaticities conform to sRGB or to an extended range of chromaticities.

29.3.2 Previewing the Exercise Files

Prior to opening in Photoshop, the images from the exercise JPEG files may be inspected in some detail with the Windows Photo Viewer (using a V2 ICC display profile) in order to appraise the differences in their displayed characteristics without the possible complicating influence of the colour management processes of Photoshop. Untagged images from the same capture source coded into different colour spaces exhibit significant differences in appearance, with as expected, the primary difference being the reduction in displayed saturation as the area of the capture colour gamut is increased. (Some may be surprised at this result, considering it counter-intuitive; however, in simplistic terms, the more saturated are the primaries of the display device, the less is the amplitude required of the components to drive it in order to produce a particular level of saturation. Thus, components relating to wide gamut displays will be lower in amplitude than those relating to those of narrow gamut and so, without the benefit of a tagged profile to inform the display transform processor of the nature of the encoding colour space, the result will be lower saturation for wider gamut components.)

Tagged images derived from the same capture device with different colour space encodings are displayed identically.

No significant differences in the dark steps of the greyscale are perceived, possibly indicating the same gamma law correction had been applied during image capture, irrespective of the colour mode chosen. (One might have expected the difference in the gamma law for the sRGB and Adobe RGB colour space specifications to show up in untagged files as slight differences in the dark steps of the greyscale.)

29.4   Colour Managing Raw Files

Loading Raw files into Photoshop automatically opens the Adobe Camera Raw plugin. This is another powerful application for photographers, providing tools that go beyond those described in the following paragraphs, since these are limited to describing only the colour-related parameters. There are however many excellent books available on the broader use of Camera Raw, one being 'The Digital Negative' (Schewe, 2013).

The RGB components of a raw file have not yet been processed to take on the characteristics of a particular colour space; however, the components will have been shaped by the proprietary characteristics of the capture responses of the optical components of the camera, including those of the image sensor. Furthermore, the RGB components are linear, that is, unlike the JPEG file components they are not gamma corrected. Thus in colour management terms the job of Camera Raw is to transform the proprietary colour space into a defined colour space and embed a matching profile in order that it may be properly colour managed in subsequent stages of the workflow.

Attempting to load a raw file into Photoshop will automatically activate AdobeRaw the image will be converted into the colour space used the last time the application was activated. Figure 29.16 illustrates the screen display when loading the Camera Raw file containing the ColorChecker chart.

The very small blue lettering at the centre bottom of the display gives the current working space of Camera Raw together with the image size. It reads: *'ProPhoto RGB; 16 bit; 2560 by 1920 (4.9MP); 300 ppi'*. Clicking on this information brings up the Workflow Options display illustrated in Figure 29.17.

Figure 29.16   The ColorChecker chart in Camera Raw.

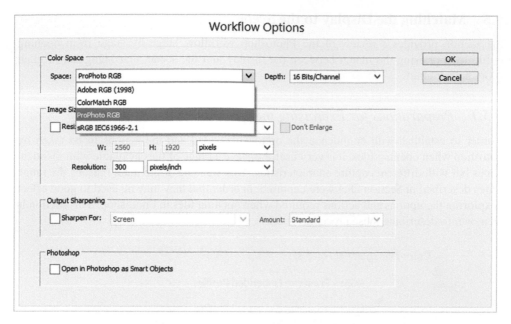

Figure 29.17   Selecting the colour space in Camera Raw.

As Figure 29.17 illustrates Adobe Raw offers the option of four working colour spaces: Adobe RGB, ColorMatch RGB, ProPhoto RGB and sRGB.

Our aim is to minimise any potential gamut clipping or mapping, so the natural choice is ProPhoto RGB.

The raw file carries a good deal of metadata which defines many of the parameter settings of the camera at the time the image was captured, including 'Exposure' and 'Contrast'. All of these adjustable parameters are made available to the user in Camera Raw. Normally for this exercise all adjustments are set at the default zero position when the file is opened and any adjustment required should be limited to the following:

- Set Exposure to just not clip peak levels on the histogram.
- Use the White Balance Tool to white balance the white chip of the greyscale of the ColorChecker chart.

The image file is now ready for loading into Photoshop, which is achieved by selecting the 'Open Copy' button at the lower right of the screen illustrated in Figure 29.16, ('Open Image' in later versions). If the working colour spaces of Camera Raw and Photoshop match, this action will load the image into Photoshop, if there is not a match Photoshop will provide the operator with action options as detailed in the next section.

29.5   Matching the Display to the Scene

This section provides a review of the Photoshop workflow, stage by stage from opening the file to appraising the match between the display and the scene, that being the original ColorChecker chart.

29.5.1   Preparations for Exercising the Workflow

In order to establish with confidence the options provided and the actions to be taken by Photoshop when opening files, it is very useful to have a range of files available with identical images but with different capture characteristics. This was one of the main reasons the range of files described in Section 28.4 were captured, in order that they may be used to good effect in exploring the options and actions required when opening files in Photoshop. This section is given over to describing these procedures.

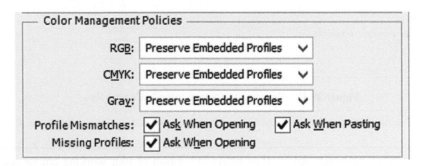

Figure 29.18   The settings in the Color Management Policies panel.

As a reminder Figure 29.18 illustrates the default settings of the 'Color Management Policies' panel described in Section 29.2; however, these settings will be varied in the following descriptions as files are opened in Photoshop to provide the opportunity of becoming familiar with the Photoshop approach to these tasks.

Subject to the appropriate 'Ask When Opening' boxes being ticked, the setting in the RGB box does not alone determine which option panels are displayed when opening files. If the RGB box is set to 'Convert to Working RGB' and the appropriate 'Ask' box is not ticked, then conversion of non-matching files to the working space will take place automatically after the first conversion, subject to the option panel that first appears being appropriately ticked.

29.5.2   Opening Files in Photoshop

Having set the colour management parameters for Photoshop we can start to open our files but before doing so we should be aware of the following:

- Unavoidably, several different situations may present themselves when opening a file in Photoshop which initially can appear somewhat daunting.
- However, Photoshop is very good at assisting the user to make the right choice of settings, whether it is to just remind one of the current working colour space or to guide the uninitiated to the correct choice.

- This is a straightforward matter if the profile space of the input file, that is, from the camera or scanner, and the working space of Photoshop match.
- If they do not match, then the action will depend upon which selection is made and which boxes are ticked in the 'Colour Management Policies' panel.

The following sub-sections describe the result of selecting in turn the various options in the Edit→Color Settings→Color Management Policies→ RGB: settings from the menu system.

29.5.2.1   Color Management Policies →RGB Off

On the Color Management Policies panel select the RGB: to Off. Use File →Open to load the tagged 'Minolta A1 Natural sRGB' file. If the 'Working Space RGB' had been set to sRGB, the file would open with no 'Mismatch' panel being displayed.

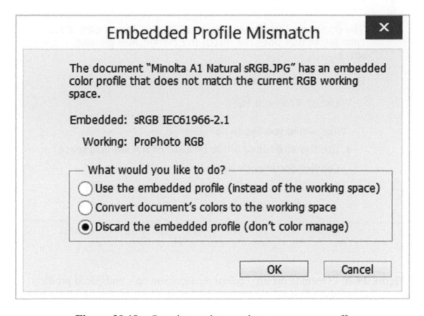

Figure 29.19   Opening option – colour management off.

As there is no match, the Mismatch panel appears as illustrated in Figure 29.19, which always illustrates the embedded profile and the working profile in order to highlight the details of the mismatch.

Since the 'Off' option had been selected, Photoshop assumes colour management is not required and therefore defaults to offering the 'Discard profile' option. Should a file with no embedded profile be opened in this situation, then Photoshop places it straight into the working space with no questions asked.

However, should the user wish to override the 'Off' selection, the option is provided to either use the embedded profile or convert the image to the working space.

29.5.2.2   Color Management Policies →RGB Colour Management Operational – Embedded Profile

If 'Settings' is set to an operational 'Colour Management' configuration, and an attempt is made to open a file with an embedded profile, which matches the selected working colour space, then the file will again open with no questions asked. However, if there is a mismatch between the two environments then Photoshop will present the same panel as the one previously but with a different default option selected, the actual selection depending upon whether Color Management Policies' was set for 'Preserve' or 'Convert' as illustrated in Figure 29.20.

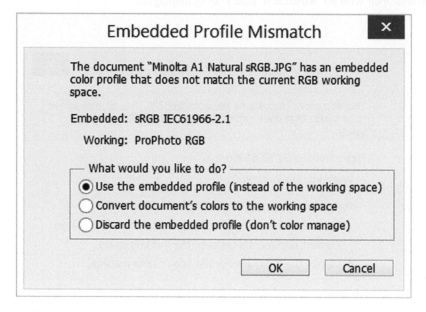

Figure 29.20   Opening option – colour management on – embedded profile.

If one is working simultaneously with a number of files, with different embedded profiles, then it makes sense to convert them all to the chosen working space. However, if this is a single image file there is little advantage in either option, subject to the embedded working space being within the limits of the working colour space. In these circumstances it may be convenient to place the file effectively into a temporary dedicated working space to match the embedded profile.

In converting a file, the embedded profile is amended to match the working space, so care should be taken when saving the file, if one wishes to retain the original file with the original embedded profile.

29.5.2.3   Color Management Policies→RGB Colour Management Operational – no Embedded Profile 1

Open the 'Epson Scanner – Target Adobe RGB' file which has no profile but was known to be encoded to the Adobe RGB specification.

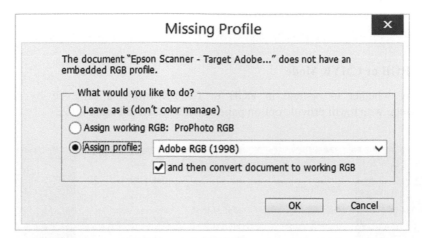

Figure 29.21   Opening option – colour management on – no profile.

> This presents the user with an interesting range of options as shown in Figure 29.21. Assuming that colour management is on, then it is unlikely that the first option would be selected. Often, as in the case illustrated, although there is no embedded profile it is known that the file was created with a specified target colour space, in this case Adobe RGB.

The second option is inappropriate since it is essential to assign a profile which matches the encoded colour space of the file, which is Adobe RGB. Therefore the third option is selected and in addition, the arrow is activated to open the range of colour space options and the Adobe RGB colour space is selected. If the 'and then convert document to working RGB' box is ticked, the image will be converted to the working colour space and a new profile matching the working space will be assigned and will be embedded should the file be saved.

29.5.2.4   Viewing the Open File

The name of the encoded colour space of an open file is displayed on the screen together with the number of bits per digital colour, its positioning depending upon the Window arrangement. For what might be defined as the 'normal' mode with the image of one file only displayed, accessed via Window →Arrange → Consolidate All to Tabs, the colour space name is displayed on the left-hand side of the Widows frame. If more than one image is displayed using Windows →Arrange →Float All in Windows, the name is located in the lower left-hand side of the image frame. If the file is untagged, the name is given as 'Untagged'.

29.5.3   Managing Files and Profiles within Photoshop

So now we know how to deal with opening files into Photoshop, how do we deal with the options available for managing them once they are safely opened?

Photoshop provides the facility to manage files as they are opened in Photoshop but it may subsequently become necessary to change the characteristics, either by changing the working space from RGB to CMYK or assigning or changing a profile.

29.5.3.1   RGB or CMYK Mode

Determining whether to operate in RGB or CMYK mode is achieved by selecting Image →Mode which will provide option panels as illustrated in Figure 29.22.

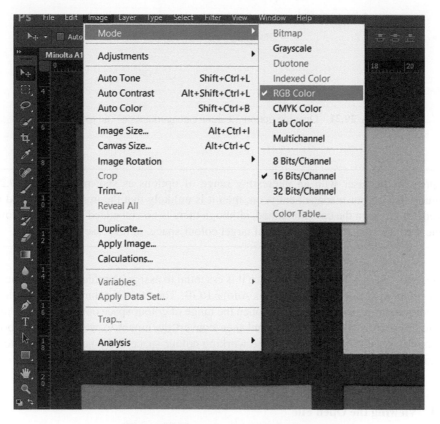

Figure 29.22   Colour mode selection.

The menu selection shown on this screenshot offers three generic types of working space for the file to operate within; 'RGB Color', 'CMYK Color' and 'Lab Color', and provides for a mathematical transform between them.

The particular RGB or CMYK colour encoding space associated with these options was set in the 'Colour Settings' panel as described in Section 29.2.2.3. Image chromaticities which fall within the gamut of all three colour spaces will be displayed identically, irrespective of the working space selected. However if, as with the colour chart, some chromaticities fall outside of the CMYK colour space, then the selection of that space will cause a significant reduction in the displayed saturation of the associated colour patch, caused by the gamut clipping or gamut mapping operation.

This is a permanent loss; switching back to RGB will not restore the original component levels.

The primary working space of interest to us is the default 'RGB Color' space.

29.5.3.2 Managing the Profiles of Open Files

As we have seen, options are available on opening a file to either ignore the option of assigning a profile to an untagged file or adopting or converting the profile associated with the file. These same options are available to colour managing open files.

The 'Assign Profile' and 'Convert to Profile' commands may be found towards the bottom of the 'Edit' menu, appropriately beneath 'Colour Settings'. The following examples illustrate how these commands may be used.

It is assumed for this exercise that we will use the untagged files generated by the Epson scanner as described in Section 28.4, where each of the three files captured were encoded with different image colour spaces as listed in Table 29.1 and described in Section 24.2.

Table 29.1 Epson scanner untagged files

Source	Colour space	Profile	File
Scanner	sRGB	Untagged	JPEG
Scanner	Adobe RGB	Untagged	JPEG
Scanner	Wide gamut RGB	Untagged	JPEG

Non-profiled files with no identification may be properly identified and profiled, and if required converted to the system working space.

The aim of this example is to show that after non-profiled files have been loaded with no identification at open time, they may subsequently be properly identified and profiled and if required converted to the system working space. *The results from undertaking this exercise provide a bonus, in as much as they illustrate in a profound manner the justification of, and exemplary workings of, ICC-based colour management.*

For these tests we will first ensure that sRGB is selected as the Photoshop default working colour space (see Section 29.2.2.3). As the file is opened, Photoshop immediately determines there is no embedded profile and asks how one would wish to proceed.

29.5.3.3 Opening the Untagged Files

Opening an untagged file will cause Photoshop to display the panel illustrated in Figure 29.23.

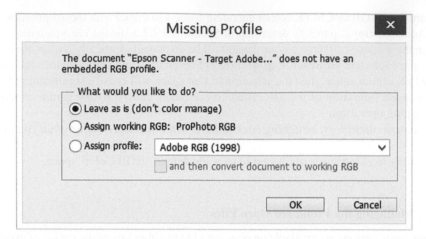

Figure 29.23   Opening untagged files.

Select the 'Leave as is' option and in turn open all three of the untagged files.

When all three untagged files have been loaded, Photoshop is arranged to display the three files simultaneously by using **Window→Arrange→Float All** in Windows. After some juggling and rearranging the three files will be displayed as illustrated in Figure 29.24, a very instructive display.

It will be noted that in terms of saturation three very different looking images appear. Unfortunately in this capture the lettering in the title bar at the top of each image frame is barely legible; however, the files are laid out in order of colour encoding spaces from the top left, with image titles 'Wide Gamut RGB', 'Adobe RGB' and 'sRGB', respectively; also in the title bar, '#' has been added to the file name to indicate there is no confirmation of a match of the capture and the working profiles. Each image is also labelled as 'Untagged RGB (8bpc)' in the lower title bar.

Before progressing further, it may be instructive to discuss why it is that those images captured with the larger gamuts are displayed with decreasing saturation. The component code values, in for example, the Wide Gamut RGB file, are expecting to control saturated primaries and therefore have a lower value than the corresponding values in the sRGB file. However in this display, because there is no profile present to tell Photoshop how to handle the file, these lower values are controlling the working profile sRGB primaries and the result will therefore be a desaturated display. Thus the only one of the three files to be displayed correctly is the sRGB file, since by default that is also the working colour space.

The next step is to assign appropriate matching profiles to each of these files.

Figure 29.24   Display of the three untagged scanner files.

29.5.3.4   Assigning Profiles to the Untagged Files

Using **Edit →Assign Profile**, the screenshot illustrated in Figure 29.25 appears.

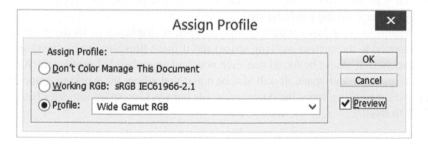

Figure 29.25   Assigning a profile to an untagged file.

We can now assign the appropriate matching profile, respectively to each of the images in turn by highlighting the 'Profile:' button and selecting the appropriate colour space from the drop-down list. As each image is selected, one may click alternately on the 'Preview' box and see the effect of applying the profile to the image.

Figure 29.26   The three originally untagged files with appropriately assigned profiles.

The result of assigning the profiles is illustrated in Figure 29.26.

It will be noted that the sRGB targeted image does not change but the presence of the appropriate profiles causes the other two images to be processed such that they now form a perfect match with the sRGB image. The name of the new profile now appears underneath each displayed image on the left-hand side.

Thus although we now have matching images, each is displayed in its assigned working space (as opposed to the system working space) and if these files were to be saved, and these new files reopened, it would be found that each now had an embedded profile which matched its original targeted colour space. It will also be noted that where the working space does not match the system working space the document's title bar working space indicator has changed from RGB# to RGB*.

29.5.3.5   Converting Profiles

The two non-sRGB image files can be converted into sRGB files by using the 'Edit →Convert to Profile' menu command which will cause the panel illustrated in Figure 29.27 to appear.

Convert both the 'Wide Gamut RGB' and 'Adobe RGB' images to the 'sRGB' working space.

After conversion the display will appear as illustrated in Figure 29.28.

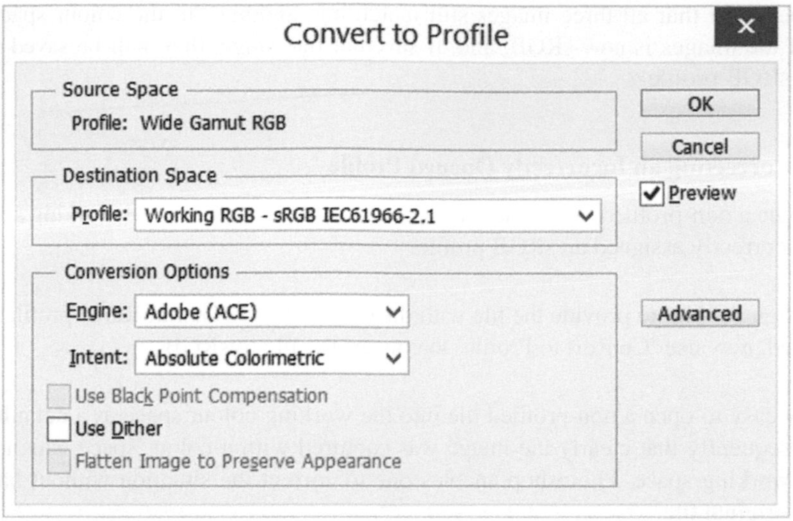

Convert to Profile ✕

Source Space
Profile: Wide Gamut RGB

OK
Cancel

Destination Space
Profile: Working RGB - sRGB IEC61966-2.1 ∨

☑ Preview

Conversion Options
Engine: Adobe (ACE) ∨
Intent: Absolute Colorimetric ∨
☐ Use Black Point Compensation
☐ Use Dither
☐ Flatten Image to Preserve Appearance

Advanced

Figure 29.27 Converting the images to the working colour space.

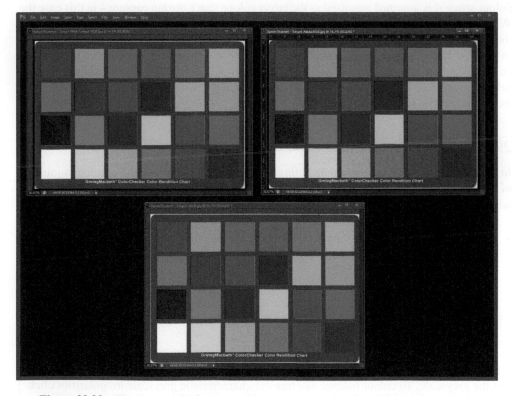

Figure 29.28 The three original un-tagged images converted to the working colour space.

It will be noted that all three images still match one another but the colour space name for each of the images is now sRGB, and, if saved at this stage, they will be saved with an embedded sRGB profile.

29.5.3.6 Correcting an Incorrectly Opened Profile

Let us assume a non-profiled 'Wide Gamut RGB' file is loaded into sRGB working space and has been incorrectly assigned an sRGB profile.

- Use 'Assign Profile' to provide the file with the correct 'Wide Gamut RGB' profile.
- If required, now use 'Convert to Profile' to convert the file to sRGB.

It is very easy to open a non-profiled file into the working colour space as a default – only to find subsequently that clearly the image was captured with a colour space which did not match the working space. Photoshop enables one to correct the situation without having to reopen the original file.

Assuming that the 'Epson Scanner – Target Wide Gamut' file has been inadvertently opened into the sRGB system working space and is now associated with an sRGB profile. The correction procedure is similar to that above: first select 'Assign Profile' and then select the colour space it is assumed was used as the target colour space during capture – in this case the Wide Gamut RGB colour space. Photoshop will now reprocess the image in accordance with the new colour space, edit the profile to Wide Gamut RGB and display the image correctly in its own Wide Gamut RGB working space. If it is now required to convert the image to an sRGB colour space then this may be achieved by selecting 'Convert to Profile' and then selecting the sRGB colour space. Photoshop reprocesses the data to sRGB and also edits the associated profile to sRGB.

29.5.4 Image Adjustment in Photoshop

In order to exercise colour management thoroughly we will undertake the absolute minimum of adjustment of the reference files, by firstly selecting Image→/Adjustments→ Levels which will place the 'Levels' adjustment panel on screen as illustrated in Figure 29.29.

- The 'Set White Point' – sets the colour balance of the white chip of the chart greyscale to neutral and automatically makes it equal to a digital code level 255; prior to adjustment the RGB levels were equal at a level of 251.
- The neutral chips of the ColorChecker chart are not quite neutral. Optionally the 'Set Grey Point' – may be used to set the colour balance of the mid grey.
- Do not adjust the level of the black step of the chart.

To set the white point, click on the 'Set White Point' picker and then click on the white of the greyscale. To colour balance any transfer characteristic greyscale irregularities click on the middle 'Set Grey Point' picker and then click on the third grey chip from zero. That is the limit of any adjustment to the image.

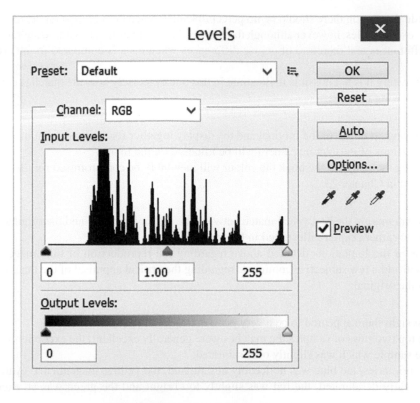

Figure 29.29   White balance adjustment.

There is not always the opportunity to place a white card, or even better a greyscale, in the scene. Nevertheless, more often than not there are whites in the scene which can be used to obtain this very critical white balance using this technique. Most modern cameras track greyscale reasonably well but if it is clear that having set the white point the greys are not neutral then sometimes finding a grey in the scene and using the middle picker can improve the situation. (Care must be taken on whites in shadows since these are lit by what may be a totally blue sky and white balancing on a blue shadow will lead to disastrous results). Setting the white and grey balance is the limit of any adjustment to the image.

The adjusted images of the three files are with minor variations virtually of the same appearance as those of the untagged images illustrated in Figure 29.28 and illustrated in the next section.

29.5.4.1   Appraising the Match of the Scene with the Display

The appraisal of the matches between the scene and the display of the various exercise files was undertaken on a wide gamut display, profiled using the X-rite iPro spectrum photometer and the i1 Profiler application. The screen was set to a highlight luminance of 100 nits, at a white point of D65 and a gamma of 2.2. The ColorChecker chart was evenly illuminated with adjustable level D65 fluorescent lamps and the highlight level set to subjectively match the white chip of the chart with the white chip of the displayed chart.

In an ideal situation there should be no perceptible difference between the rendered images of the six exercise files, however although the images are 'acceptably' close there are relatively minor differences and therefore of course differences when each is compared to the original scene.

Unfortunately it is not possible to illustrate in this book precisely how the matches compare for the following reasons:

- In capturing an image of the original and the display together for an illustration in the book the limitations of the capture camera will be imposed on the image.
- In printing the image in the book the colour will inevitably be compromised for the reasons explained in Chapter 23.

By a small margin the best visual match between the displayed image and the original chart was the raw camera capture illustrated in Figure 29.30.

In view of the limitations detailed above regarding the reproduction of this image in the book, I will add a few subjective comments regarding the critical appraisal of the match of the image to the original:

- At first sight there appeared to be a very good match in all respects.
- Of the top two rows of samples the matches were generally excellent; the exception was the orange sample which was slightly over saturated.
- Of the primaries, the blue was noticeably desaturated, the yellow, magenta and cyan were very slightly desaturated, the red was slightly too bright and the green was an excellent match.

Figure 29.30   Camera Raw screenshot of the ColorChecker chart.

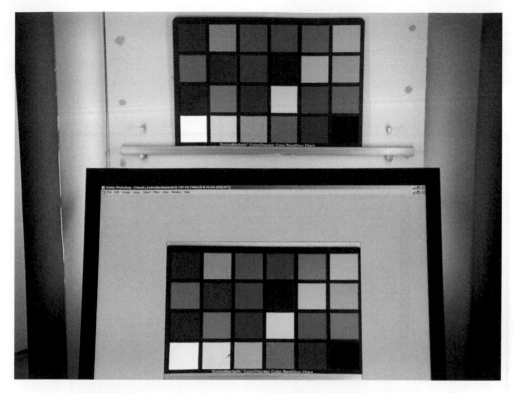

Figure 29.31 The camera match of the original and displayed image.

In summary, on the basis of a subjective appraisal only, the results of this exercise indicate that colour management is working well in producing a very good match between original and display.

Figure 29.31 is an un-retouched photo of the original chart with the screen showing the camera Adobe RGB image. Admittedly this is not a very scientific approach, with the result being open to criticisms of metamerism, etc. Nevertheless, it is considered that this is a satisfactory result for our scene to screen matching criteria.

29.6 Previewing the Soft Proof

Proofing can be carried out on screen using the soft proof feature which is described in detail below. However, before proceeding further it is worth emphasising that the purpose of Photoshop is to prepare images for both desktop printers and for a large number of different press printers of various types. Thus, in order to provide an indication of how the print will appear for these various final destinations, Photoshop provides the facility to switch the display from the 'normal' mode to the 'soft proof' mode. In doing so the dual reversible facility of ICC profiles described in Sections 27.3 and 27.4 is utilised to emulate the proof rendition, that is, the profile provides not only the parameters for transforming the working space image to the print space, but also provides the parameters for the complementary process of converting the print colour space image back to the working colour space. Thus one is provided with the option of displaying either the normal image directly from the working space or selecting for comparison, an image based upon the values in the print colour space; that is, a soft proof of what can be expected from a print.

Before proceeding into the complexities of printing with Photoshop, it may help to clarify the situation by reviewing once more the appropriate elements of the workflow diagram, as illustrated in Figure 29.32.

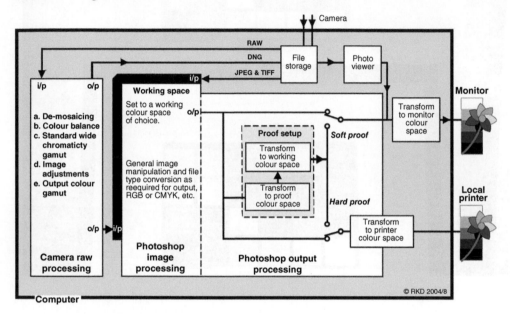

Figure 29.32   The simplified proof switching arrangements in Photoshop.

The complexity in the switching arrangements associated with printing is related to the ability to both view a soft proof and print a hard proof of how the image would appear on a commercial printing press. The switch positions shown relate to the 'normal' viewing and desktop printing positions. The support of proofing enables an appropriate press profile to be utilised in both the reverse and forward directions to enable the appearance of the press print to be emulated individually on both the monitor and the desktop printer.

Setting up the proofing feature is achieved by selecting the menu item 'View/Proof Setup' causing the screenshot illustrated in Figure 29.33 to be displayed.

> The resulting list enables the selection of the printer colour space to be used as the basis of the proof, note the choice of either 'Custom Setup' or 'Working CMYK' as the two options relevant to this task.

The latter option will produce a hard proof based upon the characteristics of the printer system originally selected (see Section 29.2) as the basis for the 'CMYK:' colour space in the 'Color Settings →Working Spaces →CMYK panel' and the profile name will be added to the image title as a reminder that the proofing image is on display.

If the 'Proof Setup' option 'Custom Setup' is selected the 'Customize Proof Condition' panel appears on the display as illustrated in Figure 29.34.

Figure 29.33   The proofing menus selection.

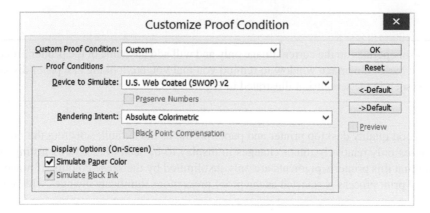

Figure 29.34   The proof customise panel.

Activate the 'Device to Simulate' option and select the alternate printing system required, in this case the 'U.S. Web Coated (SWOP) v2' system.

Select 'Rendering Intent' to 'Absolute Colorimetric'. (Interestingly this option is not provided for the 'Working CMYK' alternative.)

The 'Reset' and the two 'Default' buttons appearing on the screenshot illustration above incorrectly reflect the actual name of these buttons which are 'Cancel', 'Load' and 'Save', respectively. Thus if required, the settings selected may be saved as a template for future use by selecting the 'Save' button and nominating a name for this proof condition. Selecting the 'OK' button will return the display to the 'Color Management' box.

The display may now be switched between the 'normal' display and the soft proof display by selecting the menu item **'View →Proof Colors'** or more conveniently by alternately selecting the keys **'Ctrl+Y'**.

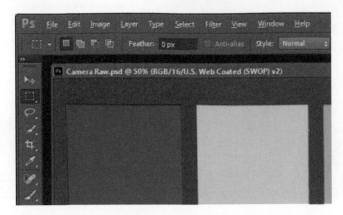

Figure 29.35   Image title includes proofing profile.

This control operates on the current image only and will add the name of the printer process to the title bar of the image window to remind you that you are looking at a proof display, as can be seen in the screenshot illustrated in Figure 29.35.

With a good quality desktop printer and paper profile as the default, selecting the proofing option causes only relatively minor changes in display to be seen even on this testing chart, indicating that this printer's pigments are only just limited by the scene colours. If however a commercial print process is selected as the proof press more significant changes will be seen.

29.7   Matching the Print to the Display and the Scene

29.7.1   Introduction

A workflow process which produces a good match between the scene and display for in-gamut colours has been achieved. The next step is to produce a print to compare with both the display and the original ColorChecker chart.

As the printer and the gamut of its inks or pigments may be incapable of containing the gamut of the display, and furthermore, there is some difficulty in matching the contrast law of the printer with that of the display, it is inevitable that this match is likely to be compromised to some degree. Nevertheless, by selecting the correct options in the workflow between the stage that achieved a good display and that required to produce a print, an optimum rendition of the image can be achieved.

Although inkjet printers operate in a CMYK colour space, they are driven by Photoshop with signals derived from an RGB colour space. So the printer driver will be responsible for the conversion to the appropriate CMYK working space, as described in Chapter 23.

There are colour management settings options in both Photoshop and in the printer driver provided by the printer vendor. However, since both Photoshop and the printer driver offer colour management of the printer, it is important that only one of these options is activated, otherwise a correction profile will be applied twice. Epson indicates that their driver is intended for relatively simple applications, which do not have the sophistication of Photoshop. Thus in this situation colour management is best left to Photoshop and the colour management features of the printer driver should be turned off. The method of achieving this will be described below.

When a printer driver is opened, it displays the range of profiles available for connecting the working space of Photoshop to the printer. For a relatively simple printer only one printer-dedicated generic profile amongst the large range available in Photoshop will be displayed, whilst the more professionally orientated printers will also include profiles for each of a range of related print papers.

29.7.2   Setting the Print Parameters of Photoshop and the Printer

The Photoshop parameters associated with the printer and the printer settings are accessed via the print menu which is activated by the menu item 'File →Print or keys, 'Ctrl+P' resulting in the screenshot illustrated in Figure 29.36.

In legacy terms the print panel was perhaps the area which caused most confusion to the uninitiated, a confusion mostly associated with the option of using either the printer profile supplied by the printer vendor or the Photoshop printer profile. On the release of successive versions of Photoshop, Adobe has provided increasing clarity regarding these options in the print menu, supplemented by helpful comment in the 'Descriptions' box. In this release the most critical comment cannot be missed. Rather than risking it being overlooked, it appears with a warning triangle as the first item in the Color Management box situated on the right-hand side of the main panel. Thus if using Photoshop to manage colours, one is warned: 'Remember to disable the printer's colour management in the print settings dialog box'.

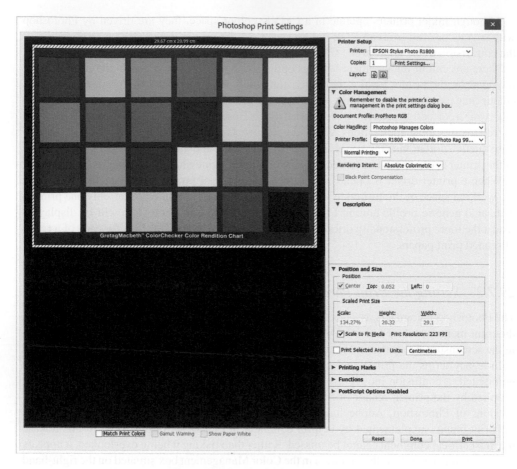

Figure 29.36   The initial print screen.

29.7.2.1   The Printer Settings

In the 'Printer Setup' box, ensure the 'Printer:' selection is set to the appropriate printer. In order to abide by the warning described above, click on 'Print Settings...' this will activate the 'Basic' version of the 'Printer Properties' panel, as illustrated in Figure 29.37.

Set the parameters for the various options and click on 'Advanced' to display the remaining options, illustrated in Figure 29.38. Set the usual parameters as required and in the 'Color Management' box select the 'ICM' option and tick the box beneath for 'Off (No color adjustment)'. These selections ensure that the printer driver does not apply any colour management actions to the image to be printed. Click 'OK' on both panels successively to return to the initial print panel.

Figure 29.37 Printer 'Basic' options.

Figure 29.38 Printer 'Advanced' options.

29.7.2.2   Photoshop Print Settings for the Local Master Print

The colour management menu items of the 'Print' panel are illustrated in Figure 29.39.

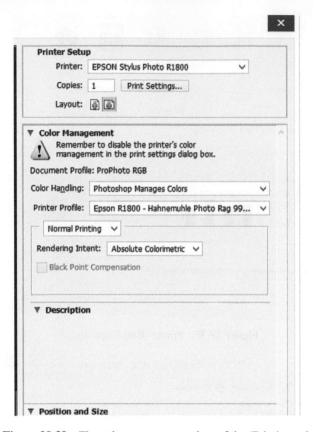

Figure 29.39   The colour management box of the 'Print' panel.

When undertaking the following actions, note in each case the useful descriptions which appear in the 'Description' box.

- Activate the 'Color Handling' options and select 'Photoshop Manages Colors'.
- Activate the 'Printer Profile' options and select the printer and paper profile as appropriate. (The paper of the profile illustrated in Figure 29.34 is a matt paper with a neutral spectral reflection distribution (SRD) to match the media on which the ColorChecker chart is produced.)
- Activate the unnamed printing options and select 'Normal Printing'.
- Activate the 'Rendering Intent' and select 'Absolute Colorimetric'.
- 'Black Point Compensation' should be greyed out when selecting 'Absolute Colorimetric'.

At this final stage the opportunity is available to soft proof the projected rendition of the image as it will appear on the local printer. On the 'Print' menu (Figure 29.36), beneath the image area are three tick boxes:

- Match Print Colors
- Gamut Warning
- Show Paper White

Ticking the 'Match Print Colors' will cause a slight desaturation of the higher saturated chart samples on the displayed image as it emulates the printed image. Ticking the 'Gamut Warning' box will cause all those samples which fall outside of the printer ink gamut to appear as grey.

Finally click on the 'Print' button to produce a normal print of the Photoshop image.

29.7.2.3 Photoshop Print Settings for the Hard Proof

Subject to the colour gamut of the desktop printer fully encompassing the gamut of the commercial printer process, it may be used to produce a hard proof of the latter's performance.

Activate the 'Photoshop Print Settings' panel if not already displayed and set all options as described for producing the local master print as described in the previous section.

In the lower half of the 'Color Management' box, activate the unnamed printing option area to change the setting from 'Normal Printing' to 'Hard Proof'. This will change the remainder of the options in this box as illustrated in Figure 29.40.

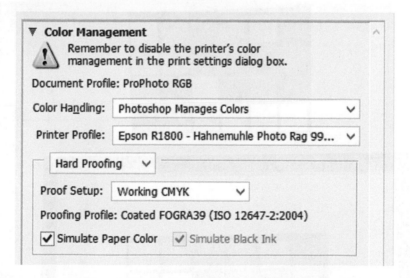

Figure 29.40 Colour management hard proof box.

Activate the 'Proof Setup' options and note the choice of either 'Custom Setup' or 'Working CMYK'. The latter option will produce a hard proof based upon the characteristics of the printer system originally selected as the 'CMYK:' colour space in the 'Color Settings → Working Spaces →CMYK panel' and the profile name will be listed beneath as the 'Proofing Profile'.

If the 'Proof Setup' option 'Custom Setup' is selected the same 'Customize Proof Condition' panel as that previously reviewed when describing the soft proofing arrangements within the main screen of Photoshop appears on the display as illustrated in Figure 29.33. Furthermore, the procedure for selecting the hard proof custom profile is identical to that described in Section 29.6 for the soft proof.

As described previously for the local printer procedure, the tick boxes under the image in the 'Photoshop Print Settings' panel may be ticked to see a soft proof associated with the printing press selected above.

Activating the 'Print' button will produce a hard proof of the image as it would appear on the commercial printing press.

29.7.3   Appraising the Match of Display and Print

As when appraising the match of the display and the original, there are practical reasons why it is not possible to provide a critical objective basis for assessment based upon viewing the image in a printed book. Nevertheless, even with these provisos it can be seen from the photograph of the screen and the print, illustrated in Figure 29.41, that there is a good match.

As with the original capture of the scene, this image of the print and original was captured as a raw file on the same camera, the only adjustments being auto colour balance on the white sample of the original ColorChecker chart.

Figure 29.41   Comparison of print and display.

29.7.4   Appraising the Match of the Scene with the Print

The visual appraisal of the match between ColorChecker chart and the print is subjectively close to excellent and vindicates the use of colour management in ensuring a high quality of colour image reproduction.

The print and the original were photographed together under weak sunlight at approximately 45 degrees and printed to give some indication of the quality achieved as illustrated in Figure 29.42. Such a comparison is a very testing exercise and unfortunately the camera view was not as good as the visual comparison. Nevertheless, it does illustrate that even with this highly critical test the results were reasonably favourable. It is worthy of emphasis that the print appearing here has suffered any inaccuracies that might be present in any of the following processes:

- Original camera capture
- Processing through Photoshop
- Printing

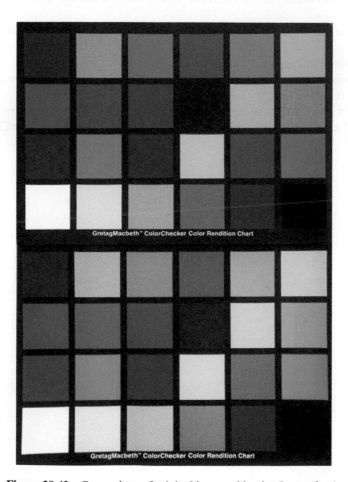

Figure 29.42   Comparison of original image with print (lower chart).

- Secondary camera capture
- More processing through Photoshop
- Printing for the book

Thus although the reader has not seen the original visual comparison, it can be envisaged that without the last three processes the result may be deemed to be very satisfactory.

29.8   Summary of Activities to Assist in Obtaining Good Colour Reproduction

- If possible place a neutral greyscale in the scene.
- To minimise the risk of clipping out of gamut colours, in the camera use in order of preference, RAW, Adobe RGB or sRGB colour mode formats.
- Set up display to standard conditions.
- Provide a matching standard print viewing environment.
- Select Kodak ProPhotoRGB for the RAW viewer and Photoshop; do not change during processing.
- Colour balance the image on a neutral greyscale, standard grey card or a true white in the scene.
- For archive, save the image to a DNG format file.
- Select the option of printer colour management by Photoshop; ensure printer driver colour management is turned off.
- If in any doubt about performance use inks specified by printer manufacturer.
- Select a satisfactory print paper colour of neutral hue if a match to the ColorChecker chart is required.

30

Colour Management by Profile Maintenance

30.1   The Requirement to Incorporate New Profiles

There are occasions when it becomes necessary to incorporate new profiles into the photographic workflow, either because the original no longer reflects the characteristics of the equipment it was designed to complement or because a new item has been added to the workflow without an accompanying profile. Sometimes the vendor develops an improved profile for their product.

In the normal course of events the profiles supplied by equipment vendors adequately reflect the performance of the elements in the workflow they are intended to complement, from the camera to the print media, and by adopting the procedures described in the last two chapters, satisfactory colour reproduction is usually achieved. However, experience indicates that there can come a time when no matter how conscientiously colour management is practised the rendered image compares poorly with the original. This is usually a sign that somewhere in the workflow one of the items of equipment is no longer performing in the manner which dictated the parameter values incorporated in the associated profile, and in consequence some sort of remedial action is required.

The cause of the change in operating characteristics may be either due to instability of performance or, more seriously, an equipment fault. In the latter case either the equipment must be repaired or a decision is made to replace it. However, very often the problem is one of instability caused either by changes within the equipment or in its set-up; assuming the set-up has been checked, then it is reasonable to assume that for one reason or another operating characteristics have changed. In these circumstances the solution is to replace the profile with one matched to the new characteristics. The items which most frequently require attention are the monitor, the printer and the projector; the scanner is a relatively stable item although it can sometimes benefit from being re-profiled.

Occasionally a new item is added to the workflow without the benefit of an accompanying profile; most frequently a new brand of paper for the printer. Since a printer profile is based upon the joint characteristics of the printer and the paper, it is impractical for the vendor of the paper to provide a matching profile for the paper alone. Many paper vendors provide a profile

Colour Reproduction in Electronic Imaging Systems: Photography, Television, Cinematography, First Edition. Michael S Tooms.
© 2016 John Wiley & Sons, Ltd. Published 2016 by John Wiley & Sons, Ltd.
Companion Website: www.wiley.com/go/toomscolour

service whereby they provide a file containing a test set of colour samples from which the user provides a print on the new paper, which is then sent to the vendor who measures the resultant colours and responds with a new profile which is loaded into the computer profile library.

It is usually a relatively straightforward matter to determine which profiles need attention. If after the set-up procedures described in Section 28.3 have been implemented there is a poor match between the original image and the monitor display, then it is likely that the monitor profile should be replaced. However, if there is a good match between the original image and the rendered image on the monitor but a poor match to the printed image, then it is likely that the printer profile requires to be replaced.

The remainder of this chapter provides a brief description of the procedure for generating new profiles, whether that is by the vendor or the user.

30.2   Preparing to Generate a Profile

The theoretical aspects of generating a profile have been covered in earlier chapters (Sections 12.2, 23.4, 27.3 and 27.4.) where fundamentally two types of profiles were identified. A simple matrix-based profile is used when the characteristics of the device are precisely defined and stable, and table-based profiles are used when only an imprecise description of the characteristics is known and/or the characteristics are not stable. In the latter case, there may be a need to generate a new profile each time the characteristics change. For table-based profiles it will be recalled a learning process is required, whereby a set of specified colour samples is used to generate an output from the device and the results are used to build a table which relates the output to the input colours for the profile. It is the latter approach which is usually appropriate to the devices requiring re-profiling; this also includes the new printer paper-printer combination profile.

30.2.1   Items Required in Generating a Profile

From the above description it is evident that in order to generate a profile the following items will be required:

- A file containing colour samples of known RGB values
- A spectrophotometer for measuring the colour samples at the output of the device
- A calculator to calculate the required table values from the measured sample values
- A software application for generating the profile

Taken in isolation the list implies a technically demanding procedure; however, technology has now progressed to the point where the profile software application incorporates all but the spectrophotometer itself. Nevertheless the application does control the spectrophotometer via an interface to the workflow computer in order that the measurements made are captured directly by the application.

A number of vendors supply this type of software application package in kit form with the software guiding the user through a number of steps culminating in the generation of the profile.

In the following, as an example of the procedure, the processes available using the X-Rite 'i1 Photo Pro 2' kit will be briefly described.

30.3   Generating Profiles

30.3.1   The Home Screen

Activating the i1 Profiler causes the 'Home' screen to be displayed as illustrated in Figure 30.1.

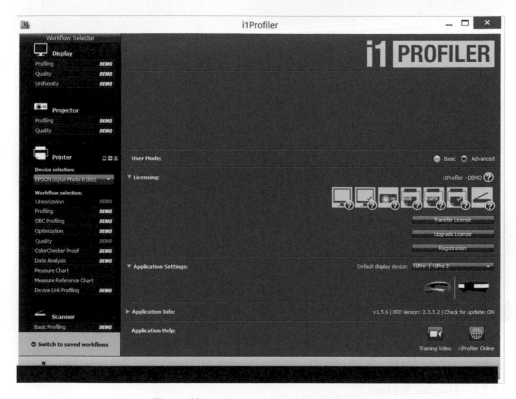

Figure 30.1   The i1Profiler 'Home' display.

On the left of the screen the various profile operations are displayed together with the options available. At the bottom right is the option to display the list of training videos which run through each of the activities associated with producing a profile for each of the items on the left of the screen.

As an example of creating a profile the 'Display Profiling' operation will be described.

30.3.2   Display Profiling

Selecting the Display ▶ Profiling option will cause the Display Settings screen to appear as illustrated in Figure 30.2.

At the bottom of the screen are shown the five stages in producing the profile, with the centre part of the screen providing the options to set the required display settings.

The active stage may be selected via either the ▶ Next icon or the appropriate icon along the bottom of the display.

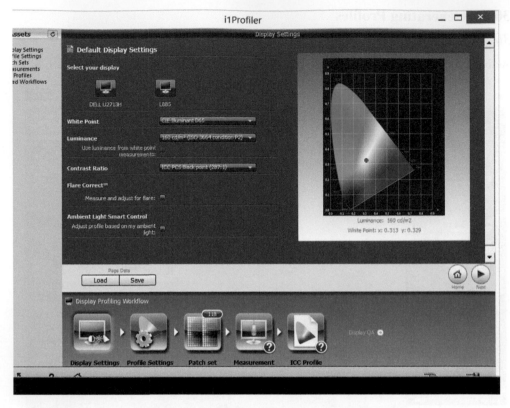

Figure 30.2   Display profile, display settings.

30.3.2.1   Display Settings

The Display Settings offers the option of selecting parameter values from those shown in Table 30.1:

Table 30.1   Display settings

Parameter	Display setting
White point	D50, D55, D65, D75
Luminance	80, 100, 120, 160 (ISO 3664 condition P2), 250 nits
Contrast ratio	Native, Custom 10 – 1000, ICC PCS Black point (287:1), from printer profile
Flare correct	Measure and adjust for flare
Ambient light smart control	Adjust profile based on my ambient light

As can be seen a wide range of settings is available, comprehensively covering a range of normal operations; it is interesting to note that the ISO and ICC recommended settings detailed in Sections 26.3 and 27.3 respectively for highlight luminance and black point are offered as a default.

30.3.2.2 Profile Settings

The Profile Settings offer the options listed in Table 30.2

Table 30.2 Profile settings

Parameter	Profile settings
Chromatic adaptation	Bradford (default), CIECAT02, Sharp, CMCCAT2000
ICC profile version	Version 2, Version 4 (Default)
Tone response curve	Standard (Default), sRGB
Gamma	Variable to 2 decimal places from 1.00 to 3.00
Profile type	Table based, matrix based (Default)

The chromatic adaptation transforms are those described in Section 5.2. Care needs to be taken in selecting the ICC Profile Version; Version 4 (Default) would appear the obvious choice; however, as indicated in Section 27.4, since V4 is not compatible with a number of applications many workers continue to use V2. The tone response curve options are associated with the Gamma settings; if the sRGB tone response (detailed in Section 24.2) is selected the Gamma options are greyed out. Selecting Standard (Default) activates the Gamma selection slider which would normally be set to 2.2.

30.3.2.3 Patch Set

Three patch sets are offered: small – 118 samples, medium – 211 samples and large – 462 samples. Selecting the larger patch set will provide more samples for the software to make increasingly more accurate profiles but will also make the measurement process longer, up to 25 minutes for the large patch set.

If required, patches may also be selected from both the Pantone range, subject to the Pantone Color Manager application being available, and from a JPEG image selectable from the computer folders. In the latter case the software will select up to 20 sample colours from the image and add them to the patch set selected. The various combinations of patch sets may be viewed by selecting the appropriate icon at the top left of the viewing area.

30.3.2.4 Measurement

On selecting the Measurement screen, the patch set is displayed centrally and on the left of the screen two action options are provided. If not already done, the software reminds the user to connect the spectrophotometer to the computer. The first action option is to calibrate the spectrophotometer which is undertaken by placing it on the calibration frame, which incorporates a small white calibration tile, and clicking the calibration button on the display or on the spectrophotometer. The second action is to inform the software whether the monitor will be adjusted via 'Automatic Display Control (ADC)' or manually by the user. ADC is a software adjustment facility incorporated into many current monitors and enables the i1Profile software to manage the monitor adjustments.

Once all the options have been selected the 'Start Measurement' button is activated and the user is guided to set the spectrophotometer against the face of the screen with the aid of the carrier frame provided. Once the software detects the properly located spectrophotometer, it takes control of the screen and commences to display in sequence in full screen mode the patches previously selected. As indicated above this sequence can take up to 25 minutes to complete.

30.3.2.5   ICC Profile

Once the measurements are complete, the ICC Profile icon is selected and the user is invited to provide a name for the new profile. Once named, the new profile is stored in the appropriate folder and also automatically made the new default profile for the monitor.

30.3.3   Other Profiling

The other profiling procedures follow a very similar pattern. Where there are significant differences they are broadly described in the following paragraphs.

30.3.3.1   Printer Profiling

The workflow for producing a printer profile commences with the selection of a suitable patch set, generally the printer patch sets contain more patches than those used for monitor profiling to assist in ensuring a satisfactory level of accuracy from the printer. An additional stage is included in the workflow to format the patches into a suitable test chart in readiness for printing. Figure 30.3 illustrates a print of half the samples from an 800 sample selection. A measurement reading tray is provided to accommodate the resulting print which enables the spectrophotometer to be mounted on a sliding frame above. The user then slowly slides the

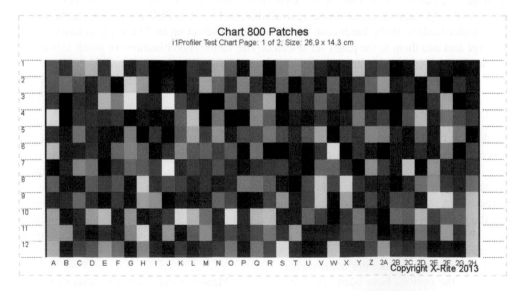

Figure 30.3   Illustration of half the samples from an 800 sample selection print.

spectrophotometer across each row of printed patches in turn enabling the software to measure the colour of each patch. A number of profile parameter values are then set before the profile is created.

30.3.3.2   Scanner Profiling

A recognisable test card is inserted into the scanner, the ColorChecker chart being acceptable. All colour management settings on the scanner should be set to off and a TIFF file (Section 25.3) format selected for recording the file. The contents of the file are then measured and an appropriate profile is added to the file. This profile is then stored in the computer profile folder.

When a file from the scanner is opened in Photoshop, as described in Section 29.5, either the native profile should be replaced with the new profile or, where no native profile exists, the new profile should be assigned to the image file.

Part 5C

Colour Reproduction in Digital Cinematography

Introduction

Cinematography was the last of the reproduction media to commence the adoption of digital technology in place of film during the 1990s/2000s. There were good reasons for this apparent delay; although colour television services had matured for some 50 years, and during that time the quality of reproduction had improved dramatically, nevertheless, when it came to producing motion pictures on large cinema screens with considerably wider angles of view compared to even large screen domestic television, the results compared poorly with what could be achieved by film. The principal limitation was the comparatively poor resolution of television, even the current high definition television system was deemed barely adequate for the larger cinema screen.

Nevertheless, in system technology terms, television, being based on the reproduction of finite sequences of activity, as opposed to digital still photography, is the clear forerunner for digital cinema. Furthermore, unlike photography, which has to service a very extensive range of users each with the freedom to adopt system parameters from a wide range of options, the cinematographic industry had the opportunity to define a single set of worldwide standards. Digital cinema was conceived in an era when international standards in other media were already well established and potentially had a clear stage to set ideal specifications from conception, without the limitations of accommodating legacy electronic practices.

Colour Reproduction in Electronic Imaging Systems: Photography, Television, Cinematography, First Edition. Michael S Tooms.
© 2016 John Wiley & Sons, Ltd. Published 2016 by John Wiley & Sons, Ltd.
Companion Website: www.wiley.com/go/toomscolour

The early serious experiments towards the end of the twentieth century commenced by introducing digital electronics into the post-production (often abbreviated to 'post') processes of some post houses, whilst a few enthusiastic producers went on to use the advanced television cameras of the day to produce digital image tapes. In the wide range of procedures comprising the post workflow these digital images were easier to manipulate and the use of tape based rather than film based clips significantly enhanced the efficiency of the operation. Once post was complete the tape was converted to film for distribution to cinemas; thus establishing a digital workflow pattern to a section of the overall workflow which gave direction to the evolvement of the digital cinema standards.

The following chapters provide a broad outline of the digital cinematographic workflow and a description of how in the 2000s the major production and distribution houses came together to specify the digital cinematography standards for each of the two major workflows: production and post, and distribution and exhibition.

Author's Note. In January of 2015 just as the final chapter of this book was nearing completion, a major review and re-issue of the standards documentation undertaken by the body responsible for deriving these production and post standards became available, providing the author with the incentive to add a further chapter (34) in which the supplementary elements of the specifications are described and in which the status of the industry in the 2010s, in the light of the level of adoption of these industry standards, is reviewed.

Acronyms

Table 31.0   List of acronyms used in the digital cinema image specifications

ACRONYM	Definition
ACES	Academy Color Encoding Specification
ADX	Academy Density Exchange Encoding
AMPAS	Academy of Motion Picture Arts and Sciences
Academy	Academy of Motion Picture Arts and Sciences
APD	Academy Printing Density
ASC	American Society of Cinematographers
CDL	Colour Decision List
CMF	Colour Matching Function
DCDM	Digital Cinema Distribution Master
DCI	Digital Cinema Initiatives
DCP	Digital Cinema Package
DLP	Digital Light Processing
DMPC	Digital Motion Picture Cameras
DSM	Digital Source Master
EDL	Edit Decision List
IDT	Input Device Transform
OCES	Output Colour Encoded Space
PCS	Profile Connection Space
ODT	Output Device Transform
RDD	Reference Display Device
RICD	Reference Input Capture Device
RRT	Reference Rendering Transform
SMPTE	Society of Motion Picture and Television Engineers

Acronyms used in the workplace have been adopted throughout this book, always defined at the first occurrence in a chapter, which is deemed usually sufficient to carry the reader through without reference to a list of acronyms used. However, the specifications associated with the digitisation of the film industry have far more than the usual preponderance of acronyms, particularly those associated with production; and more than many readers will remember when reading through a chapter. Thus although the usual practice of defining the first occurrence will continue, a list of acronyms is also provided below for easy reference whilst reading.

Reference to the system configuration diagram in Figure 32.2 illustrates the functions which relate to many of the acronyms appearing in Table 31.0.

31

The Evolution of Digital Cinema

31.1 Background

Over a period of nearly 100 years the cinematographic industry had developed a well-established, rich and extensive history based on the use of film to capture the image, to edit and manipulate the image in post-production (post), to distribute the film worldwide and to exhibit the film in tens of thousands of cinemas. Furthermore, should they have given it consideration, it would seem likely the cinema audience would have indicated they were very satisfied with the technical quality of the rendered image on screen. Why then should there have been an imperative to change such a successful industry away from film-based to digital-based operations?

In order to respond to this question we first need to be aware of the workflow of the industry from scene capture to exhibition in the local cinema. Figure 31.1 provides an overview of that workflow.

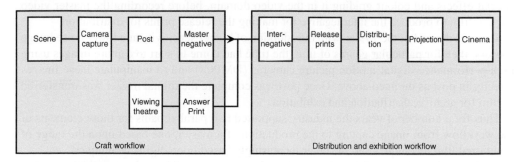

Figure 31.1   A broad indication of the traditional cinematographic workflow.

The cinematographer is the technical craftsman or woman responsible for the appearance of the image, taking in the lighting of the scene, the camera shot and exposure, the artistic adjustments during post and the answer print, which is the culmination of the creative process. From that point on the process is objective, with care necessary to ensure that the release prints when distributed, projected and exhibited reflect the intent of the cinematographer and director as manifest in the answer print.

Colour Reproduction in Electronic Imaging Systems: Photography, Television, Cinematography, First Edition. Michael S Tooms.
© 2016 John Wiley & Sons, Ltd. Published 2016 by John Wiley & Sons, Ltd.
Companion Website: www.wiley.com/go/toomscolour

From the early days of cinematography the basis of operations were such that the logistics of shooting were orientated around one camera, which implied a sequential shooting plan based upon ensuring that irrespective of the time sequence of the plot, all scenes related to a particular location or set were shot before moving on to the next location, which in a major production were possibly weeks or months apart. Such a practice implied the support of a comparatively sophisticated logistical capture and post-production operation, complete with various intermediate film processes to address the negative-to-positive film transfers, special effects, editing, dubbing and colour grading. Once the film master was available, a large number of release prints were required to service each cinema circuit which in turn were distributed around the world, an intensive logistic and expensive exercise.

As early as the 1950s, enthusiastic cinematographic craftsmen with some experience in television production techniques saw the potential for introducing these techniques into the film world, where the much simpler logistics of shooting and flexibility in post-production could save on the costs of the various film transfers and grading sequences required, and companies were established to exploit these savings. Once the television-based video tape was mastered in post it was transferred to film using a telerecording technique whereby a film camera was arranged to view a special high-definition television (HDTV) display. These early attempts were not successful. The technology of the time was incapable of providing the degree of resolution, dynamic tone range and lack of electronic noise accepted as normal in the film world.

By the early 1990s, with the advances in television technology which had led to the move from analogue to digital technology, the setting of the HDTV standard and the availability of equipment to support the standard, it became practical to transfer the processed negative film to video. During that decade an increasing number of post houses offered the service of transferring the film negative-based images to tape, editing and conforming it, applying special effects and colour grading it in the video domain, before recording the master video back to film to produce the internegative for making the release prints for distribution.

Towards the end of the 1990s, with the adoption of higher sensitivity and lower noise image sensors, then for particular genres of film, the time had come to start to capture images using high performance digital motion picture cameras (DMPCs) and to manipulate these images directly in post as outlined above. Once post was complete the digital master was transferred to film for archive, distribution and exhibition.

Thus for a number of years the industry supported two parallel paths for those elements of the workflow from image capture to the production of a master, one based upon the range of traditional film techniques and the other increasingly based upon digital techniques.

As the number of productions adopting digital techniques proliferated so too did the number of colour space specifications used for scene capture, leading to confusion and often poor results when it became necessary to use different post houses for different aspects of post-production. Furthermore, there was a widespread concern in the industry relating to the poor match between the accepted 'film look' of several decades and the look of the digital material, which often manifested itself as a strong resistance to adopting digital methods of production.

These concerns led to the various representative organisations of the craftsmen, engineers, technicians and operators in the industry to appoint committees to review the situation with the aim of establishing projects to evolve the technical specifications to cover every aspect of the digital workflow from lighting the scene to exhibiting in the cinema.

In considering the extent of this project, the work-flow naturally split into two clearly identifiable sets of activities, the production and the cinema operations. The production operation comprises scene capture, post, and the rendering of the master; and the cinema operation comprises the production of the release prints, their distribution, the cinema environment and projection. In 2004, under the auspices of The Science and Technology Council of the Academy of Motion Picture Arts and Sciences (the Academy or AMPAS) some 50 leading technologists and practitioners, with contributions from the American Society of Cinematographers (ASC), addressed the role for generating a specification for the former; and the major studios came together in 2002 to form the digital cinema initiatives (DCI) group to address the latter. The DCI relied upon the society of motion picture and television engineers (SMPTE) to establish a technology committee to specify the image aspects of the specification for the cinema operation. These three organisations called upon experts from across the film and television industries to form the various committees which undertook this very considerable range of tasks. The SMPTE also took on the responsibility for transforming the resulting Academy specification into an international standard.

Although these groups responsible for the two main elements of the workflow worked independently of each other, each was aware of the interface point in the workflow between them and for the sake of technical coherence we will consider the work of the two groups as a single project; a project with the objective to take the mixed operations of film and digital to a fully digital workflow from scene capture to exhibition, with options to ingest film-derived material at appropriate points in the workflow.

31.2 Workflow at Project Commencement

The mixed workflows which had evolved at the time of project commencement can be represented by the paths illustrated in Figure 31.2.

The workflows for which each organisation was responsible are indicated by the shaded backgrounds of the diagram.

It was recognised that the traditional film workflow indicated in the top line of the Academy workflow would continue for several years, albeit with slowly diminishing relevance.

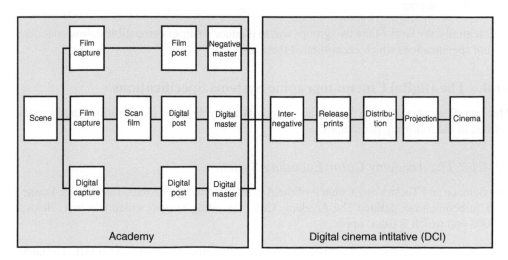

Figure 31.2 A representation of the principal workflows at project inception.

The intermediate line of the Academy workflow represented the growing number of post houses capable of transferring the image of the scene captured on film to a digital format, where the ongoing post operations can be handled more flexibly, accurately and cost-effectively. The resulting digital master is transferred to film for initiating the traditional release prints and distribution process.

The bottom line of the Academy workflow illustrates the path taken by those relatively few cinematographers at that time who had already embraced digital capture techniques, enabling the captured material to be used directly in the digital post operation.

The workflow path for which the DCI would be responsible, that is the release prints, distribution, the cinema and the projector is illustrated on the right of the diagram, which with very few exceptions, was still based upon traditional film methods.

31.3   Common Goals of the Specifications

There was an understanding from commencement of the project that film-based operations had achieved a high standard of technical excellence and a strong belief that it should not be compromised in any manner by the specifications to be evolved; on the contrary it was considered that as a result of what had been learned through the evolving specifications for television and photography there was now an enhanced appreciation of the parameters and their associated values which could contribute to the fidelity of reproduction, such that parameters could be specified which would ensure that future advances in technology could be exploited within the specifications. Between image capture and display, for the first time in media reproduction systems, these specifications would be *device independent*.

Some of the relevant project goals embraced by the specification organisations were:

- A colour gamut embracing all colours
- A dynamic range exceeding the dynamic range of the eye using 16-bit floating-point numbers for colour encoding
- A resolution capability associated with 4K pixels
- The capability of a seamless interchange of material between the various elements of the workflows in post

Essentially, the work of the two groups was to produce a pair of compatible colour encoding system specifications which encapsulated these aims.

31.4   The Digital Cinematographic Systems Specifications

The complementary system specifications produced by the two organisations are broadly defined in the following paragraphs.

31.4.1   The Academy Color Encoding System

The Science and Technology Council of the Academy, through its Image Interchange Framework Subcommittee, defined The Academy Color Encoding System which was first released in 2008 and which is based upon:

- The Academy Color Encoding Specification (ACES) – Specification S-2008-001. The latest Version 1.0.1 was released in August of 2011.

This principal specification was supplemented by a number of complementary specifications to cover interfacing within the image interchange framework:

- Specification for Logarithmic Encoding of ACES Data for use within legacy Color Grading Systems
- Specification S-2013-001 ACESproxy, an Integer Log Encoding of ACES Image Data
- Specification S-2008-002 Academy Density Exchange Encoding (ADX) and the Spectral Responsivities Defining Academy Printing Density (APD)
- Draft Procedure P-2013-001 Recommended Procedures for the Creation and Use of Digital Camera System Input Device Transforms (IDTs)

With the exception of the draft procedure, these specifications were submitted to the SMPTE for processing to international standards formats, which resulted in the following SMPTE documents:

- SMPTE ST 2065-1 Academy Color Encoding Specification (ACES)
- SMPTE ST 2065-2 APD Spectral Responsivities, Reference Measurement Device and Spectral Calculation
- SMPTE ST 2065-3 ADX– Encoding APD Values
- SMPTE ST 2065-4 ACES Image Container File Layout

The specifications in the SMPTE ST 2065-1 standard do not extend to cover the full range of workflow elements specified in the ACADEMY Specification S-2008-001. Furthermore, the SMPTE documents are formal standards which contain very little of a descriptive nature, thus the description of this system in Chapter 32 is based upon the ACADEMY ACES system supplemented by references corresponding to where the SMPTE document is silent.

31.4.2 The Digital Cinema System

The Digital Cinema Initiatives organisation approved the Digital Cinema System Specification[1] in 2005 with Version 1.2 appearing in 2012. This is a comprehensive specification which sets out to cover every aspect of the workflow from production of the distribution master, through packaging, transport, the theatre systems, projection and security.

The defining of those aspects of the specification relating to reproduction, that is the image encoding characteristics, the cinema viewing environment and the projector characteristics, was a role undertaken by the SMPTE on behalf of the DCI, which evolved the following standards and recommended practice:

- SMPTE Standard 428-1-2006 Digital Cinema Distribution Master
- SMPTE Standard 431-1-2006 D-Cinema Exhibition Screen Luminance Level, Chromaticity and Uniformity
- SMPTE Recommend Practice 431-2 -2007 Reference Projector and Environment for Display of DCDM in Review Rooms and Theaters

---

[1] http://dcimovies.com/specification/DCI_DCSS_v12_with_errata_2012-1010.pdf

These SMPTE standard documents contain very little descriptive material; however they are supplemented by the SMPTE Engineering Guideline 432-1-2010 which provides an excellent description of the reasoning behind the selection of the parameters and their values which appear in these standards. In addition Thomas Maier, the author of the Guideline, has published much of this material in a number of issues of the SMPTE Journal. (Maier, 2007-1), (Maier, 2007-2), (Maier, 2007-3), (Maier, 2007-4), (Maier, 2008-5) (Maier, 2008-6). Much of the description of the Digital Cinema System appearing in Chapter 33 is derived from the material appearing in these journals.

32

Colour in Cinematic Production – The Academy Color Encoding System

32.1 Introduction

This chapter describes the Academy Color Encoding System specified by the Subcommittee of the Science and Technology Council of the Academy of Motion Pictures Arts and Sciences.

Excerpts from the Academy Color Encoding System (ACES) specifications are used with the permission of the Academy of Motion Picture Arts and Sciences.

32.2 System Definition

The first task of the Image Interchange Framework Subcommittee was to define those system elements to be covered by the specification. As indicated in the previous chapter, it was recognised that digital workflows were already established within the film industry and it would be necessary therefore to embrace these activities within the image interchange framework (IIF). Essentially, two workflow paths were identified to be covered by the specification, which effectively were the two paths already containing digital elements as broadly illustrated by the lower two paths in Figure 31.1 and which are detailed in Figure 32.1.

The specification embraces all activities of the workflow from the scene to the file to be used as the input to the distribution and exhibition system, defined as the digital source master (DSM). Thus, in the words of the specification:

'The Academy Color Encoding System (ACES) is a set of components that facilitates a wide range of motion picture workflows while eliminating the ambiguity of today's file formats. The framework is designed to support both all-digital and hybrid film-digital motion picture workflows'.

Colour Reproduction in Electronic Imaging Systems: Photography, Television, Cinematography, First Edition. Michael S Tooms.
© 2016 John Wiley & Sons, Ltd. Published 2016 by John Wiley & Sons, Ltd.
Companion Website: www.wiley.com/go/toomscolour

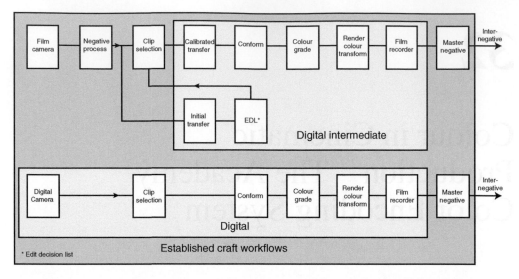

Figure 32.1   The workflows as originally addressed by the Academy Color Encoding System (ACES).

The basic ACES components are:

- Color encoding and metric specifications, file format specifications, color transformations, and an open source reference implementation.
- A set of reference images and calibration targets for film scanners and recorders
- Documentation on the architecture and software tools
- This toolkit is intended to serve as a distribution mechanism for key components of the framework including the reference implementation transforms, reference images, and documentation.

Some of the original ACES documents have now been adopted by the Society of Motion Picture and Television Engineers *(SMPTE) as SMPTE Standards and others are available as Academy Technical Bulletins.*[1]

In essence, the philosophy adopted for the system configuration of the IIF builds upon the same philosophy adopted for colour management in photography, as described in Section 27.3; that is, at the centre of the configuration is a universal colour space which embraces all colours and the full spatial dynamic contrast ratio of the eye (see Section 13.3.3) and transforms the output to an agreed set of reference viewing conditions. This approach supports the coupling of the input and output devices to the IIF via interfaces with a suitable transform characteristic between the ACES colour space and the colour space of the connected device. The main difference between the photographic approach and the ACES approach is that the ACES colour space becomes the post-working colour space to be used throughout the entire image creation workflow, whereas in the photographic workflow, the profile connection space (PCS) acts merely as an interface to the core of the configuration, whilst Photoshop, for example, provides a separate colour space as the working colour space.

---

[1] https://github.com/ampas/aces-dev/tree/v1.0/documents /ZIP

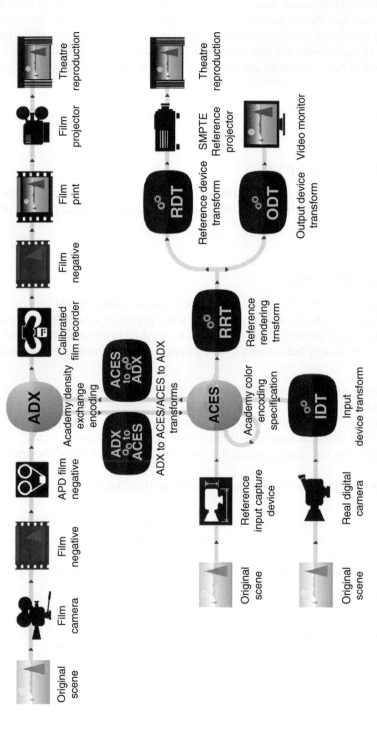

Figure 32.2 The IIF workflow (from the ACES PowerPoint presentation).[2]

2 http://www.oscars.org/sites/default/files/acesoverview.pdf

The resulting IIF configuration is illustrated in Figure 32.2: the workflow based upon film capture is shown at the top of the diagram, the reference capture device, that is, the ideal capture device, is shown on the central workflow and the real all-digital approach is shown in the lower workflow. The interfaces between the film capture elements and the ACES, and between the ACES and the film distribution elements are dependent upon the Academy Density Exchange Encoding (ADX) as specified in S-2008-002.

The remainder of this chapter describes the digital workflows in the central and lower lines of the diagram as defined in the ACES Specification S-2008-001, Version 1.0.1, August 2011.

32.3   The ACES Colour Space

In defining the ACES colour space, the objective of the specification was that the space should embrace all colours at all relative luminance levels within the spatial dynamic contrast range of the eye.

32.3.1   Chromaticity Gamut

In order to capture all colours, the ACES chromaticity gamut must entirely embrace the CIE 1931 spectrum locus on the chromaticity diagram. The criterion for establishing the ACES chromaticity gamut is based on using the x,y chromaticity diagram to construct the gamut and would appear to be very simply but sensibly evolved from setting the following conditions:

1. Set the red-to-green straight line section of the spectrum locus as the major element of one side of the colour space triangle.
2. Set the green primary where the line defined in (1) projects to intersect with the y-axis
3. Set the red primary on the line defined in (1) at the red end of the spectrum locus.
4. Extend a line from the red primary along the line joining the two ends of the spectrum locus.
5. Set the blue primary where the line defined in (4) intersects the y-axis.

Applying these criteria results in the chromaticity gamut illustrated in the CIE x,y chromaticity diagram in Figure 32.3a. (In Figure 32.3b the same gamut is portrayed in the CIE u',v' chromaticity diagram).

The chromaticity coordinates of the ACES colour primaries resulting from this construction are shown in Table 32.1.

Table 32.1   Chromaticity coordinates of the ACES colour space

ACES	x	y	z	u'	v'
Red	0.7347	0.2653	0.0000	0.6234	0.5065
Green	0.0000	1.0000	0.0000	0.0000	0.6000
Blue	0.0001	−0.0770	1.0769	0.0002	−0.3338
System white D60	0.3217	0.3377	0.3407	0.2008	0.4742

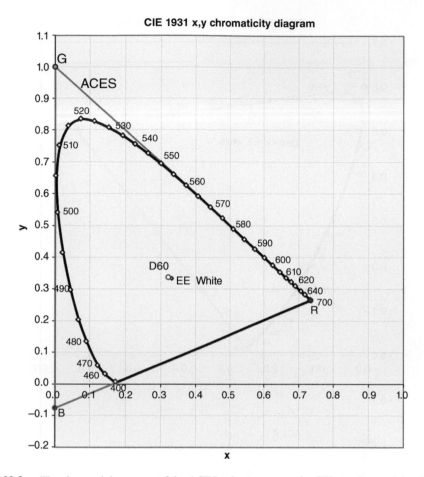

Figure 32.3a   The chromaticity gamut of the ACES colour space on the CIE x,y chromaticity diagram.

It will be noted that the blue primary is not located precisely at the point indicated by the criteria specified at the beginning of this section; that is, it is located slightly off the x-axis with a value of $x = 0.0001$ rather than zero. The reason for this 'offset' appears to be somewhat arcane, with the author failing to discover the basis for it, despite email correspondence with two of the IIF Subcommittee members.

The system white point appearing in Table 32.1 is daylight, specified as the CIE Illuminant D60 and appears to have been selected as the nearest CIE illuminant to the average chromaticity of the xenon-based projectors used for theatre reproduction. This places the ACES system white between the photographic system white of D50 and the television system white of D65.

32.3.2   *Relative Luminance Contrast Range*

In Section 13.3, a dynamic range of 5,000:1 was suggested as approaching the limit of the contrast range of the eye; thus, if the ACES colour space is to accommodate the full dynamic

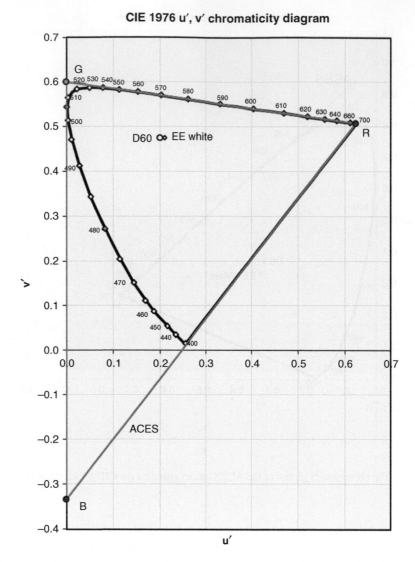

Figure 32.3b   The chromaticity gamut of the ACES colour space on the CIE u′,v′ chromaticity diagram.

range the eye is capable of perceiving, then it would be wise to regard this range as the minimum to be embraced.

The ACES colour space is scene referred; that is, it is a linear colour space, which means it does not have the protection from digital contouring artefacts afforded by perceptual uniform coding. By using the relationships defined in Section 13.6 as the criteria to establish the number of bits required to ensure that no digital contouring of the image is perceived, we can use Worksheet 13(c) to obtain the graphs illustrated in Figure 32.4, which show the perceptibility of bit contouring against the human visual modulation threshold (HVMT) over the luminance range of the display using a binary fixed-point 16-bit digital encoding regime.

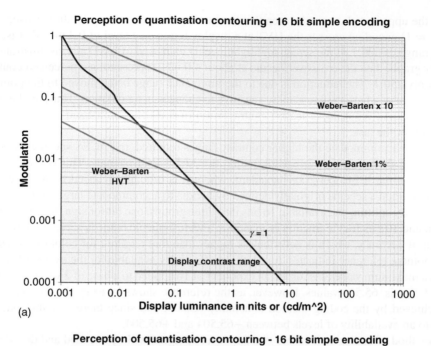

(a)

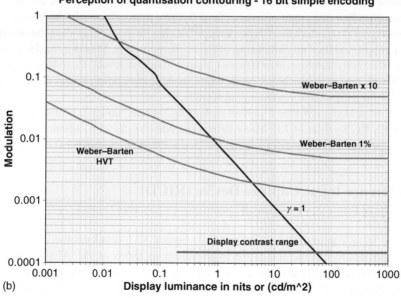

(b)

Figure 32.4   Graphs illustrating the equivalent modulation level of a single binary bit change with a display contrast range of 5,000:1. In the upper graph, the display highlight is set to 100 nits and in the lower graph to 1,000 nits.

From the upper graph, it can be seen that the line representing the modulation caused by a linear ($\gamma = 1$) encoding crosses the HVT at a display luminance level of 0.2 nits, that is, at a contrast range of 500:1. If the screen highlight level is increased to 1,000 nits, as illustrated in the lower graph, then the crossover occurs at about 4.0 nits, representing a perceived contrast ratio of only 250:1. The reduction is due to the luminance range being moved to the point on the HVMT curve where the eye is more sensitive to changes in modulation. Thus, on the basis of the criterion of requiring at least a contrast ratio of 5,000:1, 16-bit fixed-point coding is inadequate for the task.

In order to accommodate the higher contrast ranges required, the ACES specifies a 16-bit half-precision floating-point format for coding the ACES RGB signals. This coding format is specified in the IEEE 754-2008[3] standard, where it is officially referred to as 'binary16' and is based upon a scheme originally defined by Nvidia to accommodate large contrast range image files. By splitting the available 16 bits into three groups, one bit for sign, five bits for the exponent and 10 bits for the fraction or mantissa, a wide dynamic range is achieved at the cost of a level of accuracy, which is nevertheless higher than that required for colour processing. As the footnote reference indicates, four bits are effectively available for the exponent, which thus has a maximum value of 2^4 or 16: an exponent of 16 applied to a base of 2 will provide for up to 2^{16}, that is, 65,536 values. However, as the reference shows by example, the maximum value achieved by the coding method is actually 65,504, and since there is a sign bit, this equates to an availability of levels between $-65,504$ and $+65,504$.

This method of coding will ensure that any level of signal will be encoded and decoded to a level of accuracy to a minimum of five decimal figures; thus, effectively the accuracy of the level of input signal is retained; there is no perceptible change in level between the original and the decoded values and thus no contouring of the rendered image. In addition, since this form of coding is effectively transparent in terms of input and output levels, there is no requirement to map the levels representing 'black' and 'white' to specific coding levels as there is when digitally coding television and photographic signals.

The ACES is scaled such that a perfect reflecting diffuser under a particular illuminant produces ACES RGB values of 1.0. Many scenes include objects with radiance values greater than that of a perfect reflecting diffuser; hence, ACES values well above 1.0 can be expected.

32.4   Reference Input Capture Device (RICD)

32.4.1   RICD Spectral Sensitivities

The reference input capture device illustrated in Figure 32.2 is a hypothetical device with ideal characteristics to capture the scene RGB values in terms of the ACES colour space characteristics. Thus, the RICD scene colour spectral sensitivities or spectral sensitivities in terms of the standard CIE XYZ colour matching functions (CMFs) can be found using the relationship derived in Appendix F and applied in Worksheet 32(a) which produces the matrix illustrated in Table 32.2.

In Worksheet 32(a), the matrix illustrated in Table 32.2 is used to produce the RICD scene colour spectral sensitivities from the CIE CMFs; these spectral sensitivities are illustrated in Figure 32.5.

---

[3] This format is described at http://en.wikipedia.org/wiki/Half-precision_floating-point_format

Table 32.2 The matrix for deriving the ACES RGB capture spectral sensitivities from the XYZ colour matching functions (CMFs)[4,5]

	X	Y	Z
R =	1.0760118426	0.0000000000	−0.0000999175
G =	−0.5082796015	1.4075877241	0.1006918774
B =	0.0000000000	0.0000000000	1.0159913478

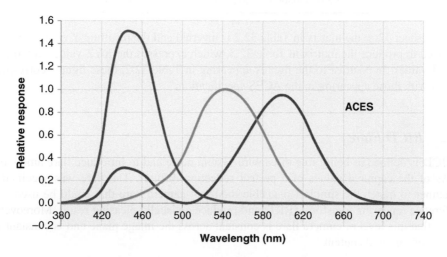

Figure 32.5 The spectral sensitivities of the Reference Input Capture Device (RICD).

Since the green and blue ACES primaries either match or are located close to the CIE Y and Z primaries and the red primary is on the extreme of the spectrum locus, the resulting spectral sensitivities have a marked similarity to the CIE CMFs illustrated in Figure 4.4. However, the ACES minor red lobe is slightly smaller than the equivalent x lobe due to the red primary being located well within the XYZ gamut.

In Worksheet 32(a), the values appearing in WS Table 1 are convolved with the spectral power distribution (SPD) of CIE Illuminant D60 in WS Table 3 and the results for each convolution summed to show that equal values of RGB signal are obtained, that is, a white

---

[4] The values in the matrix are given to 10 decimal places to be consistent with the accuracy given for the values in the ACES specification. However, there is a scaling difference between the figures in Table 32.2 and those in the specification because the worksheet normalises the G coefficients by adjusting the summation of the centre line coefficients of the matrix to be equal to 1.0.

[5] If the criteria for locating the primaries on the CIE x,y diagram had been adhered to, the Z value of the red primary would have also been zero, making the matrix interestingly symmetrical, with four of the nine values equal to zero, the inevitable result of locating two of the primaries on the y-axis of the chromaticity chart.

balance for a scene neutral illuminated by a D60 source is achieved by the RICD with no balance adjustments being required.

Table 32.3 Matrix for deriving XYZ values from RGB values

	R	G	B
X =	0.9525523959	0.0000000000	0.0000936786
Y =	0.3439664498	0.7281660966	−0.0721325464
Z =	0.0000000000	0.0000000000	1.0088251844

In Worksheet 32(a) the matrix in Table 32.2 is inverted and the resulting Y coefficients are normalised to produce the matrix in Table 32.3, which expresses the XYZ values in terms of the RGB values. In contrast to the figures appearing in Table 32.2, these figures correspond precisely with those appearing in the ACES specification.

32.4.2 RICD Flare

The RICD is defined as being free of capture system noise and to introduce flare amounting to 0.5% of the captured values of a perfect reflecting diffuser. The flare was specified in this manner to match the simplest plausible model of camera flare that could be used with the reference rendering transform (RRT) and produce a visually pleasing result. Moreover, it assumes that this level of camera flare is constant across the image plane and is constant for all values of captured content.

32.5   The Input Device Transform

Since it is unlikely that the native spectral sensitivities of a real camera can be made to match those of the RICD, the purpose of the input device transform (IDT) is to transform the colour space of the real camera to that of the RICD.

As described in the previous section, the ideal characteristics of the capture device are the spectral sensitivities defined by the CMFs of the system primaries, and it was shown in Section 17.2 that there are practical difficulties in shaping the characteristics of the optical colour filters and the image sensor(s) to match precisely the shape of any ideal set of CMFs and the same is equally true of the ACES primaries CMFs, as illustrated in Figure 32.5.

If it were assumed that the capture device had responses approaching the shape of the blue and green CMFs and the primary response of the red CMF, then the secondary red response in the 380–500 nm range could be emulated by inverting a fraction of the blue signal and adding it to the red signal via a suitable matrix.

A simpler approach would be to aim for the characteristics of an ideal camera response as described in Section 12.4, where the input device gamut is defined by selecting specific locations of primaries on the u′,v′ diagram, which leads, in the two examples given, to diminishing amplitudes of the red secondary lobe required. In the second of the examples, the secondary lobe is fast approaching zero level amplitude; thus, with such spectral sensitivities only a very low level of matrix coefficients would be required.

Once the optimum combination of spectral sensitivities and matrixing has been determined, any further improvement in matching the ideal curves would be achieved by establishing suitable look-up tables as discussed in Section 23.4.2. The Academy has produced a draft procedure, 'P-2013-001 Recommended Procedures for the Creation and Use of Digital Camera System Input Device Transforms (IDTs)', to provide guidance on the design of the IDT.

32.6   An IIF System Configuration for Viewing the Graded Signals Defined in the ACES Colour Space

In order to avoid much repetition, it is assumed that the reader is familiar with the material related to appraising the rendered image contained in the first three sections of Chapter 26 and with the concepts of applying transforms to encoded images to match them to a reference viewing environment, as discussed in Sections 27.3.1.4 and 27.3.1.5.

As noted earlier, it is intended that the ACES colour space is the adopted working colour space throughout the post-processes; however, it is evident that during processing and subsequently, these processed signals will require transformation to a colour space or spaces matched to their display and its environment, in particular, the grading display and the cinema projector. It was also noted in Section 32.2 that the colour gamuts for both the post and the distribution workflows would encompass all colours; thus, when considering the required transformations, it is well to keep in mind that both these displays will be required to accommodate signals encoded in a colour space whose chromaticity coordinates extend beyond the spectrum locus.

32.6.1   Objective versus Preferred Rendition

Assuming for the moment a display with a fundamentally linear transfer characteristic, then in colorimetric terms, a simple matrix to transform from ACES RGB coding to the display RGB coding would ensure rendered images with minimal colorimetric errors. Such a solution would provide excellent results in terms of colour fidelity; however, that is not necessarily what is required, since the intent is for the material to be appreciated by a cinema audience inculcated into viewing excellent-quality pictures resulting from the experience of decades of finessing the images derived from film with its logarithmic characteristics. The means of addressing the ACES method of extracting the essence of the desirable characteristics of the traditional film path characteristics and applying it to the ACES master is dealt with in the next section on the RRT.

32.6.2   Matching Reference and Cinema Displays

A representative workflow of an ACES system is illustrated in Figure 32.6. The Digital Cinema Initiative (DCI) have adopted a naming regime for the master file prepared by post for distribution as the DSM and the format of the file used for distribution as the digital cinema distribution master (DCDM); this nomenclature has been used in the figure, which, in order to close the grading reference display matching loop, also includes the exhibition cinema and projector. A scene captured by the camera sensors and encoded in the native colour gamut of the camera is transformed by the IDT into the ACES colour space, which, with the exception of the displays, is the colour space then retained throughout the post-processes.

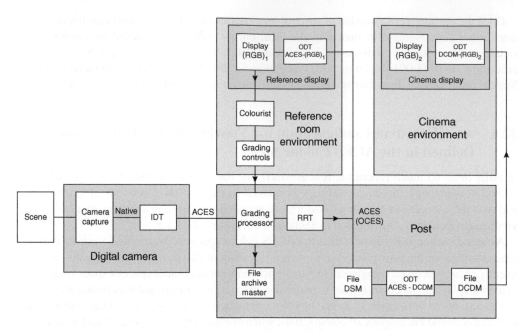

Figure 32.6   Matching graded material on the reference and cinema displays.

It must be emphasised that the configuration illustrated in Figure 32.6 is simplistic; it is based upon a simple colorimetric approach where the display device is fundamentally of a linear nature[6] and the problems of distributing the ACES 16-bit digital signal (McElavain et al., 2012) are not addressed.

The grading loop includes the RRT (sometimes located in the reference display), which imposes desirable filmic characteristics on the image. It is defined in the specification as 'the signal-processing transform that maps an ACES representation of an image to an Output Colour Encoding Space (OCES) representation appropriate for viewing on the Reference Display Device'. The OCES is defined as 'the color image encoding used by the Image Interchange Framework to represent images to be displayed on the Reference Display Device (RDD)'; however, since the RDD is in turn defined as 'an ideal output device with an unlimited color gamut and a dynamic range exceeding any current or anticipated real output device', it would appear that the OCES is synonymous with the ACES colour space. The RRT encoding is transformed by the output device transform (ODT)[7] in the display unit to the chromaticity gamut and the contrast range of the grading reference display.

When considering the colour characteristics of the grading reference display, it might be assumed as an initial consideration that it should match the characteristics of the cinema display such that a satisfactory grading achieved on the grading reference display would

---

[6] Which fundamentally they are; however, as we shall see in Section 32.8, a projector designed for the cinema is likely to include a transform to complement the DCDM perceptibly uniform encoding, and a flat-panel display may contain a gamma circuit to complement signals originating in a television or photographic environment.

[7] The term ODT is retained in order to provide consistency with Figure 32.2; however, as can be seen from the diagram, ODTs are often located at the input of the output device rather than in the post-processing operation.

produce a matching rendition in the cinema. However, such an approach could result in a situation where the industry became permanently tied to an increasingly obsolescent standard as improved colour gamut projectors were introduced into the system but the requirement to continue to support legacy equipment remained. To avoid this situation, the *ideal* RDD is defined by the ACES to complement the RICD, that is, as indicated above, 'the RDD is an ideal output device with an unlimited color gamut and a dynamic range exceeding any current or anticipated real output device'; however, whereas a capture device is theoretically capable of *capturing all colours*, a three-primary display is theoretically incapable of *displaying all colours*.

Nevertheless, since both the cinema projector and the reference display are provided with signals encoded in a colour space, the primaries of which have chromaticity coordinates which are located external to the spectrum locus, then irrespective of the content of the graded material on the reference display, and subject to appropriate transforms and compromises of any gamut mapping, that content will be matched in chromaticity terms on the cinema display. Thus, there is not the requirement to match the chromaticity gamuts of the cinema and grading reference displays, and both may independently be replaced by displays of wider gamuts, with only the secondary compromise of any gamut mapping required to be considered.

In evaluating the reference display image, the colourist adjusts the grading controls to achieve the desired results; once adjustments are complete, the file is saved as the DSM. The contents of the DSM file are also transformed into the DCDM colour space in readiness for distribution; as will be described in the next chapter, the DCDM colour space is, like the ACES colour space, also based upon a colour gamut with chromaticity coordinates external to the spectrum locus and therefore does not impart any colour restrictions during encoding. Furthermore, the perceptible uniform encoding introduced by the DCDM encoder is precisely complemented by the cinema display ODT, making the distribution path transparent to colour information. The cinema display ODT transforms the DCDM colour space signal to a linear signal with RGB coding and contrast range mapping appropriate to the display and environment characteristics, ensuring the displayed image will represent an ideal perceptible match to the grading reference display. As the white points of the two displays are marginally dissimilar, there will be very small differences in the colorimetric match of the displays, subject, that is, to the cinema display chromaticity gamut being identical to or larger than the reference display. Should the cinema display chromaticity coordinates fall inside of the chromaticity gamut of the grading reference display, then there will be a requirement to provide colour gamut mapping (Frohlich et al., 2014) to accommodate those colours which fall outside the cinema display gamut.

In practical terms, perhaps the nearest set of primaries to that of the theoretical RDD are those defined as the ideal display primaries in Section 9.2, together with a very similar slightly compromised set which were subsequently specified in the ultrahigh definition television (UHDTV) system described in Section 20.4. As the latter set is likely to become the standard for extended gamut displays, it is also likely to be adopted at some point for the grading and cinema displays. Such displays would be capable of accurately rendering nearly all the Pointer surface colours, and in the intermediate period before such displays are widely adopted for review rooms and the cinema projector, their gradual and uneven introduction would be achieved without significant compromise to the rendered image.

In order to ensure however that a minimum chromaticity gamut is achieved in the cinema, the SMPTE have defined a reference projector specification based upon the colour characteristics of the xenon-illuminated digital light processing (DLP) projector in 'SMPTE RP 431.2.2007

D-Cinema Quality – Reference Projector and Environment'. This recommendation specifically indicates that the reference gamut is the minimum gamut required and that extended gamuts may be used. At the time of writing, laser-based projectors with a colour gamut approaching the UHDTV Rec 2020 gamut are starting to be introduced.

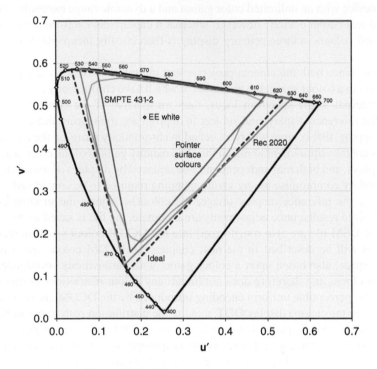

Figure 32.7   Display chromaticity gamuts.

The gamuts of the displays discussed are compared in Figure 32.7. As can be seen, the Rec 2020 gamut is very close to the ideal gamut, the latter being illustrated with a dotted line. The Rec 2020 gamut encompasses all but a few of the Pointer surface colours, and both gamuts significantly exceed the SMPTE 431-2 gamut based upon the xenon-illuminated DLP projector.

32.6.3   Which File for Archive?

Whether the DSM or the file untransformed by the RRT is archived is a matter of some consideration, in as much as, at the time of writing, the characteristics of the RRT are not stable and it is not clear whether it will be defined in terms which enable it to be bidirectional.

In philosophical terms, an archive file should be the closest representation possible to the original material, and as the ACES system has been defined to ensure that its colour space fully represents the original scene colours, then this is the version which should be archived.

However, if the RRT were defined in terms which enabled it to be used in either direction, then a case could be made for the convenience of storing the RRT version as the archive, on the basis that if the original coding was required, it could be obtained by applying the reverse RRT. A further point to consider is that as yet only candidate versions of the RRT have been published, so 'baking in' a transform for the archive, when in the future possibly improved versions of the RRT may become available, is a short sighted philosophy.

32.7   The Reference Rendering Transform

The RRT is defined in the specification as: 'the signal-processing transform that maps an ACES representation of an image to an Output Color Encoding Space (OCES) representation appropriate for viewing on the Reference Display Device'. However, although this defines its location in the workflow, it does not define either its purpose or its characteristics, and although a number of candidate versions of the RRT have been evolved by the IIF Subcommittee, none to date (2014) appear to have been judged to be entirely satisfactory by the post industry.

To address the purpose and characteristics of the RRT in any detail is beyond the scope of this book; however, in fundamental terms, its purpose is to render the scene-referred captured image with its unlimited chromaticity gamut and contrast range in such a manner that, in conjunction with the appropriate ODT, it produces an accurately perceived and pleasing image on the defined display.

Much of the difficulty in achieving an acceptable candidate RRT appears to be in establishing agreement on interpreting the subjective description of a 'pleasing image' and the manner in which such an image is achieved by the RRT. The premise (Fielding and Maier, 2009) appears to be that since the filmic appearance has been popularly accepted over many decades, then the rendition of the 'pleasing' picture should be based upon a generic version of the film 'look'. In the cited article, the authors lay down a clear path of definitions associated with the RRT and provide a fundamental view of its required characteristics, based upon the film tone curves, to provide the pleasing rendered image. Work continues in this area and further papers on the topic are available (McElavain et al., 2012) and (Iwaki and Uchida, 2013) which describe RRTs with varying levels of complexity based upon a combination of 1D and 3D LUTS and chromaticity transform matrices.

32.8   The Reference Display and Review Room

The ACES specification states: 'The image interchange framework reference projector and associated viewing environment are equivalent to those defined in SMPTE RP 431-2-2007'. Since both projectors and other displays are used as the reference display in some colour grading suites, it will be assumed that the specifications in the SMPTE recommended practice apply to any display used for grading. The recommended practice also refers to SMPTE RP 431-1-2006, which defines the characteristics of the screen associated with the projector.

RP 431 has an identical set of conditions to cover both the review room and the theatre or cinema; however, the tolerances associated with the review room are generally tighter than those specified for the cinema. In the following, emphasis is given to the review room, its display and the interfacing of the latter to the IIF.

32.8.1   The Reference Display

32.8.1.1   Pixel Count

Although digital cinema uses the same nomenclature as television to describe the sampling structure of the displayed picture, that is, in terms of multiples of 'k', in cinematic terms, the value of k in the horizontal direction is 1,024 compared with 960 for television. The minimum sampling structure for the reference display should be 2,048 horizontal and 1,080 vertical pixels, that is, 2k.

32.8.1.2   The Reference Display Colour Space

RP 431 defines a minimum colour gamut for the reference display defined by the chromaticity coordinates of the display primaries shown in Table 32.4 but notes that, in practice, the reference display may have a larger gamut. The display white point represents the average values of the chromaticity coordinates of a number of xenon lamp sources whose characteristics change slightly with running time.

Table 32.4   The chromaticity coordinates and white point of the reference display

SMPTE RP 431-2	x	y	z	u'	v'
Red	0.680	0.320	0.000	0.4964	0.5255
Green	0.265	0.690	0.045	0.0986	0.5777
Blue	0.150	0.060	0.790	0.1754	0.1579
White: average xenon	0.314	0.351	0.335	0.1908	0.4798

In early work associated with selecting the chromaticity coordinates of the cinema display, these primary chromaticities were referred to as the 'P3' set of primaries, and although that nomenclature was never, as far as the author is aware, used in any SMPTE or ACES specification, its use has become widespread within the media industry.

The chromaticity gamut represented by these primaries is illustrated in Figure 32.7. All colours falling within the minimum colour gamut must be rendered within an accuracy of $\Delta E^*_{ab} = 4$.

32.8.1.3   Transfer Function

RP 431 does not differentiate between the cinema projector and the review room display with regard to the transfer function parameters. As indicated in Section 32.5.2, the cinema projector is driven by a DCDM file which incorporates perceptibly uniform fixed-point encoding with an exponent of 1/2.6 and which RP 431 addresses by defining a projector complementary transfer characteristic based upon an exponent of 2.6. Since the projector itself is a linear device, this is achieved by including a suitable exponential transform in the projector. However, depending upon the actual IIF configuration adopted to serve the purpose of the simple configuration illustrated in Figure 32.6, it will often not be convenient to feed the review room display with a DCDM encoded signal.

Furthermore, since other coding regimes have been proposed to service the IIF infrastructure (McElavain et al., 2012), it would complicate the required transforms unnecessarily to include the 2.6 exponent transfer function component in the review room display device. Thus, although some post-operations may well include a reference display with a built-in transfer function to match the requirements of RP 431, together with the infrastructure to match, for the purposes of this description, it will be assumed that this is not the case.

32.8.1.4 Display Luminance

The screen luminance is defined in RP 431-1 as 48 nits and a chromaticity of $x = 0.314$, $y = 0.351$. RP 431 also sets tolerances for these parameters and for areas away from the screen centre.

32.8.1.5 Sequential Contrast

RP 431 defines sequential contrast to be measured to include the contribution of ambient light resulting from the review room environment defined in Section 32.8.2. The nominal value of the sequential contrast should be at least 2000:1.

32.8.1.6 Intra-frame Contrast

Intra-frame contrast is measured using a 4×4 checkerboard pattern and should include the ambient light. It should have a minimum value of 100:1.

32.8.2 Review Room Environment

32.8.2.1 Ambient Light Level

Stray light reflected from the screen should be minimized. The use of black, non-reflective surfaces on all surfaces other than the screen, along with recessed lighting, is recommended.

With the display turned off, the ambient light level reflected by the screen should be less than 0.01 nits.

32.8.2.2 Reference Viewing Position for Colour Grading

The reference viewing position for colour grading shall be at a distance of 1.5–3.5 screen heights. Lighting on work surfaces or consoles should be masked and filtered to eliminate any spill onto the screen.

32.9 The IIF Output Device Transforms (ODT)

The ODTs serve the purpose of matching the generic OCES of the RRT to the colour spaces of the display devices or the distribution system. In general, this will require the ODT to provide two or more of the following transform elements:

- Colour space transform matrix
- Perceptual uniform fixed-point encoding
- Gamma correction
- Contrast law mapping

The OCES colour space is based upon the chromaticities of the ACES primaries, a linear encoding and a comparatively large contrast range. Thus, contrast law mapping will be required for all practical ODTs.

At this time, there are no specific definitions for the characteristics of the ODTs, which are therefore likely to be proprietary in implementation; however, in general terms, they will follow the pattern described in the following examples.

32.9.1   The Reference Display ODT

For the simplified IIF configuration illustrated in Figure 32.6, the reference display ODT would be a simple matrix to convert between the ACES (OCES) and the display colour space; in Worksheet 32(b), the required matrix is calculated for the chromaticities of both the RP 431 and the Rec 2020 sets of primaries and tabled in Tables 32.5 and 32.6, respectively.

Table 32.5   Matrix for converting ACES to SMPTE 431-2 chromaticity space

	R_{ACES}	G_{ACES}	B_{ACES}
R_{431}	2.0877	−0.7413	−0.3464
G_{431}	−0.1301	1.2304	−0.1003
B_{431}	0.0067	−0.0638	1.0571

Table 32.6   Matrix for converting ACES to Rec 2020 chromaticity space

	R_{ACES}	G_{ACES}	B_{ACES}
R_{2020}	1.5091	−0.2590	−0.2501
G_{2020}	−0.0776	1.1771	−0.0995
B_{2020}	0.0021	−0.0311	1.0291

In reality, the current (2014) lack of a specified coding regime for transporting 16-bit digital signals around the IIF almost certainly means the OCES will require converting to a 12-bit digital encoding, which in turn will require some form of perceptible uniform fixed-point encoding in order to avoid contouring effects; see Section 13.8. In these circumstances, it will be necessary for the ODT to include LUTs to provide the complementary decoding characteristic in addition to the appropriate matrix.

32.9.2   The Digital Cinema Distribution Master ODT

The DCDM colour space uses the CIE XYZ primaries and a perceptually uniform fixed point encoding which is based upon an exponential law with an exponent of 1/2.6. Thus, the DCDM ODT will comprise principally of a matrix with the characteristics shown in Table 32.7, which was calculated using Worksheet 32(b), followed by look-up tables to provide the exponential response.

Table 32.7   Matrix for converting from ACES to DCDM colour space

	R_{ACES}	G_{ACES}	B_{ACES}
R_{DCDM}	0.9999	0.0000	0.0001
G_{DCDM}	0.3440	0.7282	−0.0721
B_{DCDM}	0.0000	0.0000	1.0000

32.9.3   The Television ODT

The current HD television system is defined by Rec 709, see Chapter 19, and to match this specification, the television ODT will require a conversion from ACES to the Rec 709 colour space which will include the appropriate gamma correction. The required matrix is calculated in Worksheet 32(b) and is detailed in Table 32.8.

Table 32.8   Matrix for converting from ACES to Rec 709 colour space

	R_{ACES}	G_{ACES}	B_{ACES}
R_{709}	2.5513	−1.1195	−0.4318
G_{709}	−0.2759	1.3660	−0.0902
B_{709}	−0.0173	−0.1485	1.1658

Following matrixing and appropriate contrast law mapping, the gamma law to be applied to obtain the desired Rec 709 colour space is as defined in Section 19.3.2.1.

32.10   Colour Management in Production and Post

32.10.1   General Considerations

The relative simplicity of the ACES system ensures that when fully implemented and properly managed, then when compared with those of photography, colour management problems are dramatically reduced. The definition of one colour space embracing all colours for capture and post, together with the detailed and mostly tight specifications for the grading/review environment, reduces the opportunity for mismanagement of colour to occur.

32.10.2   Potential Problem Area

In specification terms, there appears to be only one area of comparative laxity and that is in specifying the field of view of the colourist. Since the only area of luminance in the field of view is the image, then the accommodation level of the eye will to a degree be dependent upon the area the displayed image subtends at the eye. Since this area is in turn dependent upon the viewing distance, which is specified as 1.5–3.5 screen heights, representing a change in field of view area of greater than 5:1, the level of adaptation between these two extremes could be significantly influenced, thus modifying the spatial dynamic contrast ratio of the eye (see Section 13.3.3). It is likely that this would result in a colourist located in turn at these extreme positions applying different settings to the rendered image of those scenes containing significant areas of dark detail.

32.10.3   Preserving the Look

One area of the production and post-operation which has traditionally been problematic is the preservation of the 'look' of the scene, set by the cinematographer during shooting or the early stages of post, when the material may be subject to subsequent grading on different equipment in different post houses. The problem in essence is one of communication; how can the settings arrived at during shooting be preserved and interpreted in the later stages of post? Grading equipment from different vendors often uses different nomenclature for the adjustments and often implements adjustments in a different mathematical manner.

This problem was addressed by the expert cinematographers, colourists and engineers of the American Society of Cinematographers (ASC), who have evolved a simple but effective means of describing the adjustments made to the captured scene after notional corrections for colour balance (and the application of the camera IDT and the appropriate display ODT) and recording them in a colour decision list (CDL).

Three mathematically defined adjustments are specified for each of the red, green and blue colour signals; these are defined as slope, offset and power. These unique names are given in order to differentiate them from the sometimes mathematical ambiguous names associated with the corresponding classic adjustments of gain, lift and gamma, respectively. Although these adjustments may be made available directly to the operator, the intention is that the usual proprietary adjustments are used to obtain the desired look and the vendor will interpret the adjusted values in terms of the ASC values to compile the CDL, where the ASC numbers are recorded for each scene. In addition to the original nine parameters defined above, a further parameter, saturation is provided as the tenth adjustment.

The intent is that the CDL will be used by the colourist at the various grading stages to implement the previous adjustment decisions as the basis for the final adjustments which may then be 'baked in' to the signal.

Clearly for this scheme to work successfully it is essential that all vendors implementing the scheme translate their adjustments to the ASC-defined parameters in the same manner. Furthermore, the set-up of the monitor and the viewing conditions associated with first adjustment must emulate as far as is practically possible the conditions of the grading suite.

33

Colour in the Cinema – The Digital Cinema System

33.1 Introduction

This chapter describes the D-Cinema System, a system which is designed to ensure that the rendering of the image captured by the digital source master (DSM) in cinematic production is retained by the digital cinema distribution master (DCDM) during distribution and exhibition in possibly several thousand cinemas. This is the system instigated by the Digital Cinema Initiatives (DCI) organisation, an overview of which was given in Section 31.4.2 and which is specified in 'SMPTE Standard 428-1 Digital Cinema Distribution Master – Image Characteristics' (ST 428-1).

The description of the system which follows is consistent with the approach adopted in describing the television and photographic systems in the preceding chapters; that is, the system is described from the bottom up. For a complementary description, where the system is described from the top down, which may lead to a more fully rounded understanding of the reasoning behind the adoption of some of the parameters and their values, the reader is recommended to the material by Maier cited in Chapter 31. These references also contain much fundamental material, which in this book is covered in earlier chapters.

33.2 System Requirements

The DCDM encoding format was derived to be as far as practically possible *device independent* but also to acknowledge the requirements of the cinema projector and its viewing environment as first defined in Section 32.5.2 by the document 'SMPTE RP 431-2 D-Cinema Quality – Reference Projector and Environment' (RP 431-2). (It was noted in Section 15.2.1 that an output-referred system specification is dependent upon the parameters of the viewing environment.)

Colour Reproduction in Electronic Imaging Systems: Photography, Television, Cinematography, First Edition. Michael S Tooms.
© 2016 John Wiley & Sons, Ltd. Published 2016 by John Wiley & Sons, Ltd.
Companion Website: www.wiley.com/go/toomscolour

The criteria for evolving the parameters and their values for the DCDM encoding format appear to have included the following:

- A chromaticity gamut to encompass all colours
- A colour space to encompass both the contrast range of any foreseeable display technology and any foreseeable system white point chromaticity
- Constant luminance encoding
- No perceived contouring of the image at any luminance level
- Capability of ensuring the rendered image in all cinemas shall match within the tolerances specified in RP 431-2
- The encoding to be defined in absolute display terms, rather than in relative display terms, as is the case for television and photography

The following description of the system, in describing how these criteria were met, leads to the definition of the system which is formally captured as a specification in SMPTE ST 428-1.

33.3 Image Structure

The image structure is based upon the cinema 'K' definition of pixel numbers described in Section 32.6, where 1K is 1024 horizontal pixels by 540 vertical pixels; pixels are square; that is, they have an aspect ratio of 1:1. Three operational levels have been specified as shown in Table 33.1.

Table 33.1 D-Cinema image structure and operational levels

Operational level	K level	Horizontal pixels	Vertical pixels	Frames rate
1	4	4,096	2,160	24
2	2	2,048	1,080	48
3	2	2,048	1,080	24

The number of pixels at each operational level shall not exceed the numbers shown in Table 33.1.

33.4 The D-Cinema Encoding Colour Space

33.4.1 The Chromaticity Gamut

In order that the chromaticity gamut encompasses all colours, the system primaries will be located externally to the spectrum locus. A number of proposals were considered and finally a decision was made to adopt the CIE XYZ primaries as the location for the DCDM RGB primaries. Such an approach has the advantage that constant luminance encoding follows automatically.

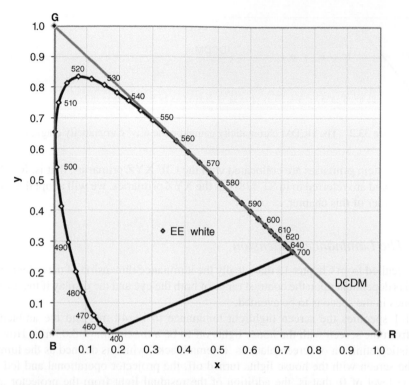

Figure 33.1   The DCDM chromaticity gamut on the *x,y* chromaticity diagram.

The DCDM chromaticity gamut is illustrated on the CIE *x,y* chromaticity diagram in Figure 33.1. The RGB primaries are co-located with the XYZ primaries.

Although there were originally concerns that, as much of the space enclosed by the gamut represented non-real colours, there would be an apparent inefficient use of the code bits, it transpired this was not a relevant issue.

The chromaticity coordinates of the primaries are shown in Table 33.2, and the chromaticity gamut is also shown on the u′,v′ chromaticity diagram in Figure 33.2.

Table 33.2   Chromaticity coordinates of the DCDM primaries

CIE XYZ	*x*	*y*	*z*	u′	v′
Red	1.0000	0.0000	0.0000	4.0000	0.0000
Green	0.0000	1.0000	0.0000	0.0000	0.6000
Blue	0.0000	0.0000	1.0000	0.0000	0.0000
EE white	0.3333	0.3333	0.3333	0.2105	0.4737

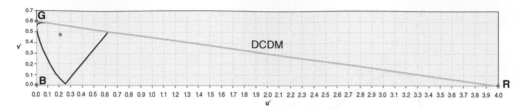

Figure 33.2   The DCDM chromaticity gamut on the u′,v′ chromaticity diagram.

Since the system primaries are collocated with the CIE XYZ primaries, that is, R = X, G = Y and B = Z, and are referred to in ST 428-1 as the XYZ primaries, we will adopt that notation for the remainder of this chapter.

33.4.2   The Luminance Dimension

It may be recalled from Chapter 13 that ideally the luminance dimension of the colour space to be defined is dependent upon the contrast range of both the eye and the display if the 'crushing' of dark tones in the scene is to be avoided.

RP 431-1 specifies the screen highlight luminance to be 48 nits and the ambient light reflected from the screen with the house lights off to be a maximum of 0.03 nits. However, a more realistic definition of screen black is 'cinema black', which is defined as the luminance level of the screen with the house lights turned off, the projector operational and fed with a signal code level of 0, that is, the addition of the residual light from the projector and the ambient light reflected from the screen. RP 431-2 defines a minimum sequential contrast ratio for the projector of 1,200:1.

It is well recognised that the higher the contrast ratio achieved, the more realistic and pleasing is the perceived image, and in terms of defining the luminance dimension of the colour space, the specification calls for a contrast range of 5,000:1 to be accommodated; that is, an implied assumption that a cinema black luminance level of less than 0.01 nits may eventually be achieved.

33.4.3   The Digital Coding Requirement

The relationships between the contrast ranges of the eye and the viewing environment, and the perceptibility of digital contouring artefacts, was explored in some detail in Sections 13.3 and 13.6, respectively, where it was shown that by adopting perceptible uniform encoding with appropriate parameter values, digital contouring artefacts could be made to be imperceptible.

Worksheet 13(c) plots the perceptibility curves of the eye and the luminance changes of fixed-point perceptible coding curves for any quantisation bit range between 8 and 16 and any exponent value between 1 and 3 using the highlighted green parameters in the area shaded pink. Noting that in addition to the requirement to produce a quantisation curve below the human visual modulation threshold (HVMT), the other criterion is to select the minimum quantisation bits to minimise storage and transport capacity requirements. Manipulating the number of encoding bits and the exponent or gamma value in the worksheet leads to a compromise of 12 bits and an exponent value of 2.6, as illustrated in Figure 33.3. Thus, 12 bits will provide for 4,095 quantisation levels.

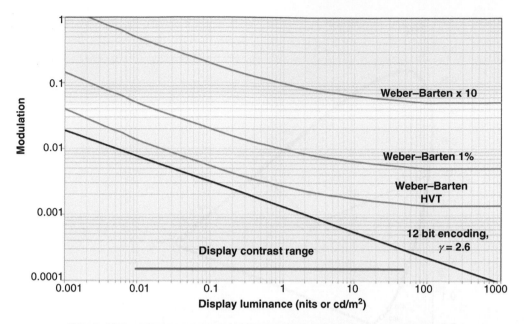

Figure 33.3 Luminance ΔL curve for 12-bit encoding with an exponent of 2.6.

In Figure 33.3, the straight line curve for a 12-bit encoded signal using a perceptible uniform fixed-point encoding exponent of 2.6 is shown to be below the HVT curve throughout the dynamic range of the system. Thus, such a system will not introduce any perceptible quantisation contouring.

Adopting the parameters derived above and using the same format as that used for television and photography, the relative relationship between the X,Y,Z tristimulus signals and those encoded by the system is therefore given by:

$$X' = \text{INT}[4,095 \times (K \times X)^{1/2.6}] \text{ and similarly for Y' and Z'} \tag{33.1}$$

where INT is the nearest integer value and the value of K, representing the coding white point level, has yet to be defined.

33.4.4 Accommodating the System White Points

It was noted in the system criteria listed at the beginning of this chapter that the system should accommodate any foreseeable system white chromaticities. The notional system white, as indicated in Table 33.2, is equal-energy white (EEW), the white normally complementing the CIE XYZ primaries and, by definition, the white obtained when normalised X = Y = Z = 1, and therefore when $x = 0.3333$, $y = 0.3333$ and $z = 0.3333$. Other whites, particularly those further towards blue or red than EEW on the chromaticity diagram, will, for the same luminance, have a value of Z or X exceeding the value 1.0, because of the fall-off in the

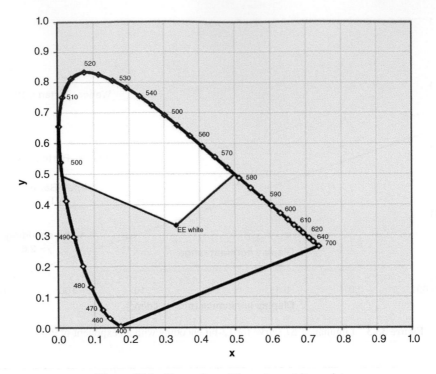

Figure 33.4   Plane of possible system whites points.

luminous efficiency function of the eye away from the peak of the Y response. Illuminants redder than EEW are unlikely to be considered as system white points, but those illuminants towards blue on the CIE daylight locus are likely to fall into this category and the encoding system must be scaled to accommodate them. By selecting two values of X and Z in turn below a maximum of 1.0, whilst keeping the Y value at a level of 1.0, the resulting x,y values may be calculated and plotted on the CIE x,y chromaticity chart (Worksheet 33(b) WS Table 1). The resulting two lines may then be extrapolated in both directions to meet and to intercept the spectrum locus respectively, producing a plane of possible system white chromaticity values, for XYZ normalised values not exceeding the value of 1.0, as illustrated in Figure 33.4.

In Figure 33.4, the plane of possible system white chromaticities, representing X, Z values which do not exceed 1.0, is illustrated in white. As anticipated from the rationale outlined above, this is the Y-dominated sector of the spectrum; both the X and Z sectors contain no system whites. Figure 33.5 illustrates the chromaticity plots of the required system whites on an expanded chromaticity scale.

In order to scale the system to accommodate the required system whites, it is necessary to determine the normalised XYZ values for the range of system whites under consideration:

- The ACES system white
- The Cinema system white, defined by RP 432-1
- The CIE-defined system whites between D50 and D65

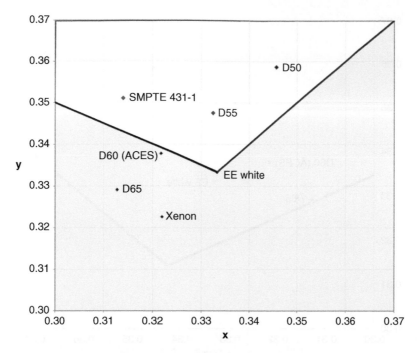

Figure 33.5 Plot of system white points.

These whites, together with their values after perceptible uniform encoding, that is, raised to the power of 1/2.6, are calculated in Worksheet 33(b) and shown in Table 33.3.

Table 33.3 System whites and their $X'Y'Z'$ values for a constant luminance value of 1.0

System white	x	y	z	X	Y	Z	X'	Y'	Z'
EEW	0.3333	0.3333	0.3333	1.0000	1.0000	1.0000	1.0000	1.0000	1.0000
SMPTE 431-1	0.3140	0.3510	0.3350	0.8946	1.0000	0.9544	0.9581	1.0000	0.9822
ACES	0.3217	0.3377	0.3406	0.9526	1.0000	1.0086	0.9815	1.0000	1.0033
D50	0.3457	0.3586	0.2957	0.9640	1.0000	0.8246	0.9860	1.0000	0.9285
D55	0.3325	0.3475	0.3200	0.9568	1.0000	0.9209	0.9832	1.0000	0.9688
D60 (ACES)	0.3217	0.3377	0.3406	0.9526	1.0000	1.0086	0.9815	1.0000	1.0033
D65	0.3128	0.3291	0.3581	0.9505	1.0000	1.0881	0.9807	1.0000	1.0330

From Table 33.3, it can be seen that D65 is the system white with the highest Z' value of 1.033, a value which must be accommodated in the coding range of $0 - 4,095$. Let us assume initially that the code value of 4,095 takes on the Z' signal value of 1.033; since the Y value remains at 1.0, it would have a code value of 4,095/1.033 or 3,964. However, in order to ensure that the D65 system white point is well within the plane of available white points, the Y maximum code value is set to 3,960. The ratio of the maximum code value to the maximum Y code value is thus 4,095/3,960, a factor of 1.034, which when transferred to the linear signals by raising it to the power of 2.6 takes on a value of 1.0911. This is the factor by which the linear signals must be reduced in order to accommodate a D65 system white point within the coding range.

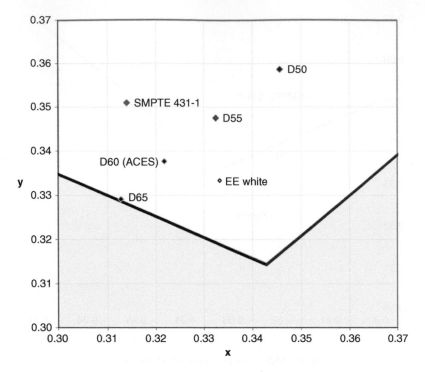

Figure 33.6   Plot of enhanced system white points.

In Worksheet 33(b) WS Table 3, the chromaticity coordinates are calculated for the revised code levels, resulting in a broader plane of chromaticities with a maximum relative luminance of 1.0 and which now embraces the D65 white point, as illustrated in Figure 33.6.

When this scaling factor is applied to equation (33.1), the following set of equations will result:

$$X' = \text{INT}[4,095 \times (X/1.0911)^{1/2.6}] \text{ and similarly for Y' and Z'} \qquad (33.2)$$

33.4.5   The Absolute Encoding Equations

A further system requirement is that the encoding equations should express the display colour in absolute terms rather than the traditional approach of expressing them in relative or normalised terms. Thus, the projector screen maximum luminance defined in RP 431-1 should be applied to equation (33.2); this is specified for the reference projector as a luminance $L = 48$ nits. Thus, applying the factor 1.095 to this luminance level provides an absolute factor of 52.37, leading to the following equation as defined in ST 428-1:

$$X' = \text{INT}\left[4,095 \times \left(\frac{L \times X}{52.37}\right)^{1/2.6}\right] \text{ and similarly for Y' and Z'} \qquad (33.3)$$

where $L = 48$.

Transposing this equation enables code levels to be equated directly with the absolute levels of X, Y, Z at the screen.

$$X = \left(\frac{52.37}{L}\right) \times \left(\frac{X'}{4,095}\right)^{2.6} \text{ and similarly for } Y' \text{ and } Z' \tag{33.4}$$

33.4.6 Signal Excursion Code Levels

Signal excursion code levels will be discussed in terms of the luminance signal, Y, and the white and black levels.

White Level

White has a level of 1.0, so substituting values in the parentheses of equation (33.3) leads to:

$$\left(\frac{48 \times 1}{52.37}\right) = 0.9166$$

Thus, the code value of white on Y is: $\quad CV_{Y'} = \text{INT}[4,095 \times 0.9166^{1/2.6}] = 3,960.$

This is to be expected since the factor 52.37 was chosen to make this so. Although the factor was chosen to leave head room for a change to a system white of D65; nevertheless, it also provides a small margin of head room for signal excursions which may extend beyond white.

Black Level

As explained in Chapter 13, black level has many meanings depending upon the level to which the eye is accommodated, which is in itself dependent upon the average display luminance, the luminance of the relatively small area at the centre of attention and the luminance of the surrounding surfaces. Nevertheless, it was indicated earlier in this section that the system should accommodate a contrast range of 5,000:1, which if used would provide a subjective black level represented by a level of 0.0002 of white level. Using equation (33.3) to obtain the code value for this black level leads to a code value of 362; since this level is regarded as approaching the limit of perception, there is clearly significant foot room for excursions below this level.

Absolute or Relative Colorimetric Encoding

It was noted that the maximum level of cinema black is related to the specified minimum cinema contrast range of 1,200:1, that is, a level of 0.00083, well above the notional limit of 0.0002 noted above; therefore, it is apparent from the discussion in Section 13.9, that in this case, the contrast range is limited by the cinema black level rather than the encoding parameters.

Figures 33.7 and 33.8 illustrate the manner in which dark tones in the scene are compressed for the range of contrast ratios currently specified and those possibly achieved in the future. Interestingly, assuming a spatial dynamic contrast ratio of the eye of 5,000:1, a theoretical improvement in the contrast range from 5,000:1 to 10,000:1 will extend the perceivable range of dark tones, since the level of cinema black will effectively raise the otherwise unseen changes in projector dark tones to the level of perceptibility.

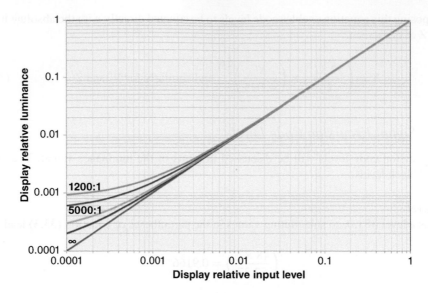

Figure 33.7 Effect of screen contrast range on tone reproduction.

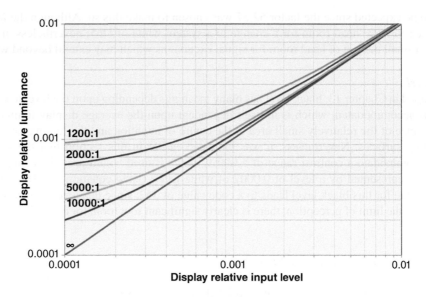

Figure 33.8 Expanded version of Figure 33.7

If all cinemas had an identical cinema black level, it would be possible by adopting absolute colorimetric encoding to compensate for its existence by adjusting the dark tones at source to provide compensation but eventually a code level would be reached where changes were imperceptible and clipping would occur abruptly. A better approach is to allow the addition of the cinema black to the black represented by the dark tone code levels, ensuring there will always be a perceived change in dark tone level from the screen, albeit the change will be less than that of an absolute colorimetric encoding system. The latter would provide larger, more accurate changes down to the cinema black level but would clip all signals below this level.

In practice, the situation is resolved subjectively and satisfactorily by the colourist, who, being located in a very similar lighting environment to the cinema, adjusts the black level for subjectively optimum pictures.

33.5 DCDM Interfaces

33.5.1 The Input to the DCDM Encoding System

By the nature of the digital cinema configuration, it would be inconvenient to undertake the DCDM encoding other than at the post-house following completion of the DSM. Thus, the DCDM encoding is likely to be incorporated within the image interchange framework (IIF) of the post-house as one of the output device transforms (ODTs), as described in Section 32.7.2.

33.5.2 The Output of the DCDM Encoding to the Projector

Current projectors are fundamentally linear devices and thus require to be fed by linear signals; in consequence, each projector will require transform circuitry to be installed which matches the perceptually encoded DCDM signal to the characteristics of the projector.

The transform circuitry comprises two stages: linearization of the perceptually uniform encoded signal and a transform matrix to match the encoded primaries to the primaries of the projector. The linearization will be accomplished by three look-up tables embodying the relationships:

$$X = \left(\frac{52.37}{L} \right) \times \left(\frac{X'}{4,095} \right) 2.6 \text{ and similarly for } Y' \text{ and } Z'$$

Once in the linear domain, the signal will require 16-bit encoding to avoid the perception of contours resulting from digital encoding, as explained in Section 32.2.

The parameters of the reference projector colour gamut are specified by RP 431-2 and are incorporated in Table 33.4; it will be noted that this is a full-colour space specification.

Table 33.4 The SMPTE reference projector primaries chromaticities and colour gamut

SMPTE 431-2	x	y	Y	z	u′	v′
Red	0.6800	0.3200	10.1	0.0000	0.4964	0.5255
Green	0.2650	0.6900	34.6	0.0450	0.0986	0.5777
Blue	0.1500	0.0600	3.3	0.7900	0.1754	0.1579
White Av xenon	0.3140	0.3510	48.0	0.3350	0.1908	0.4798

The chromaticity gamuts of the CIE XYZ, RP 431-2 and Rec 2020 primaries are illustrated in Figure 33.9.

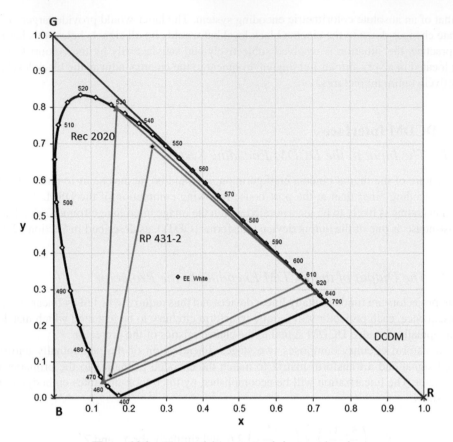

Figure 33.9   Chromaticity gamuts of cinema primaries.

In Worksheet 33(a), the parameters of the transform matrix for matching the DCDM primaries to the reference projector primaries are calculated and the coefficients are shown in Table 33.5.

Table 33.5   DCDM to reference projector primaries matrix

	X	Y	Z
$R_{RP431\text{-}2}$	2.4381	−1.0180	−0.4201
$G_{RP431\text{-}2}$	−0.7113	1.6897	0.0216
$B_{RP431\text{-}2}$	0.0369	−0.0876	1.0507

As laser projectors are introduced, it will be necessary to provide an alternative transform matrix to match the DCDM primaries to the primaries of the laser projector. In the event that the laser primaries match those specified by Rec 2020, as listed in Table 33.6 and illustrated in Figure 33.9, the matrix shown in Table 33.7 would be required.

Table 33.6 The chromaticity coordinates of the Rec 2020 primaries

Rec 2020	x	y	z	u'	v'
Red	0.7080	0.2920	0.0000	0.5566	0.5165
Green	0.1700	0.7970	0.0330	0.0556	0.5868
Blue	0.1310	0.0460	0.8230	0.1593	0.1258
White D65	0.3127	0.3290	0.3583	0.1978	0.4683

Table 33.7 DCDM to Rec 2020 primaries matrix

	X	Y	Z
$R_{Rec2020}$	1.6316	−0.3557	−0.2759
$G_{Rec2020}$	−0.6337	1.6165	0.0172
$B_{Rec2020}$	0.0168	−0.0428	1.0260

In terms of colour fidelity, the parameters of the digital production and cinema systems have been defined in such a manner that there is no theoretical limitation to perceiving a perfect rendition of the original scene.

33.6 Distribution

Once the DCDM is complete, it is ready for packaging into the DCI digital cinema package (DCP), together with the other programme-related streams of data, such as the audio and subtitles, which are referred to generically as 'essence' data, where it is then in a format suitable for distribution to the cinemas. The description of the DCP is beyond the scope of this book.

34

Colour in Cinematography in the 2010s

34.1 Progress in Adopting the Digital Specifications

The specifications for digital cinematography[1] were published in the latter half of the 2000s, so mid-way through the 2010s is an appropriate time to review their status and the progress being made towards their adoption in the day-to-day working practices of the movie industry.

The changes required in these working practices in moving from a film to a digital based operation were outlined in the Introduction to Part 5C, and are now virtually complete; the use of film for scene capture is now relatively rare. Furthermore, the demands for greater sophistication in high production value television programmes have led to these productions being shot on digital cine cameras and thus the requirement for many post-production or 'post' houses to accommodate both cine- and television-based productions in their workflows.

The adoption of procedures encapsulating new specifications is dependent upon two principal factors: the availability of the core equipment which incorporates them and the inclination of those employed in the craft and technology of the industry to adopt the procedures which embrace them.

The first of these criteria, the availability of Academy Color Encoding System (ACES) compatible equipment in the workflow, has to a large extent been met. Over the period from the publishing of the ACES specifications in 2008, the vendors of equipment have responded by ensuring options are available to select ACES-based procedures. Digital cine cameras have been developed which provide the option of working to the specifications and camera vendors have provided the complementary input transforms to their camera spectral sensitivities to the vendors of grading equipment. Virtually all grading systems now incorporate a plethora of input transforms to cover the range of cameras employed in production and offer the option to the colourist of selecting from a number of colour spaces, which include the ACES colour space, the working colour space they prefer. The current candidate version of the reference

---

[1] The Academy Color Encoding System (ACES) by the Science and Technology Council of the Academy of Motion Picture Arts and Sciences, and the Digital Cinema Initiatives (DCI) consortium system standardised by the SMPTE.

Colour Reproduction in Electronic Imaging Systems: Photography, Television, Cinematography, First Edition. Michael S Tooms.
© 2016 John Wiley & Sons, Ltd. Published 2016 by John Wiley & Sons, Ltd.
Companion Website: www.wiley.com/go/toomscolour

rendering transform (RRT) and a number of output transforms are provided to match the working colour space to the colour spaces of the grading display and to the range of media formats. Grading displays may be either projector or LCD/LED based, and although the latter are very much improved in terms of overcoming the fundamental limitation of the technology to provide a contrast ratio matching the specification, nevertheless without some compromise in dynamic contrast response (see Section 8.3.1.3), there is still room for improvement. Laser projectors are being introduced with chromaticity gamuts approaching that of Rec 2020, in line with the option provided for in SMPTE ST 431-1-2006.

The level of adoption of the ACES specifications into the working practices of production and post naturally depends upon the inclination of the individuals and companies involved to do so and it is apparent that large sections of the industry have not yet embraced the specifications in their day-to-day operations. Nevertheless, there is now a broad awareness of the trend to adopt the specifications and also of the working options available to adopt procedures based upon their use in the cameras and systems which support the production and post operations. The background to this situation will be explored further in Section 34.3.

Since the publication of the ACES specifications in 2008, the Academy Subcommittee responsible has continued its work on evolving a generic RRT acceptable to broad sectors of the industry and a number of candidate versions have been released. The Subcommittee has also been receptive to feedback from industry regarding the adoption of its specifications. This work has led to a full review of the documentation and an augmentation of the specifications, which in December 2014 resulted in a re-issue of the supporting documentation. Where appropriate, this documentation has adopted a style designed to be less intimidating to those working in the field that do not have a specialisation in colour science. The augmented specifications and the elements of the new documents which pertain to colour reproduction are described in the next section and the ramifications of their adoption are further addressed in the section on systems and workflows (Section 34.3).

Virtually all cinemas have now converted to digital projector–based screenings and are supported by the digital cinema package (DCP), which incorporates the JPEG 2000 compressed and encrypted version of the digital cinema distribution master (DCDM) standardised by the SMPTE and described in Chapter 33.

34.2   The ACES in the 2010s

This section provides a description of the status of the ACES resulting from the publication[2] in December 2014 of the package of documents which details the ACES specifications and describes their use. The Academy describes these documents as the first official release of version 1.0; the original documents covered in Chapter 32 are now described as 'pre-release' documents. Nevertheless, all the technical specifications appearing in the pre-release documents are incorporated in the current release.

Excerpts from the ACES specifications are used with the permission of the Academy of Motion Picture Arts and Sciences.

---

[2] https://github.com/ampas/aces-dev/tree/v1.0/documents/ZIP

34.2.1 A Brief Review of the December 2014 ACES Documentation

34.2.1.1 Overview

The ACES Subcommittee of the Academy's Science and Technology Council has undertaken a major review of the ACES and the level of its adoption in the cine industry, which has resulted in a clearer distinction between its work on the specifications and the work of the SMPTE on the standards which have evolved from these specifications. A consequence of which is that specifications which are 'internal' and relate to what might be seen as peripheral to the core ACES are retained as ACES specifications, whilst those which are of a more universal nature are adopted by the SMPTE and published as standards. In ACES terms, these latter specifications are described only in their technical bulletins. The full list of the current documents is shown in Table 34.1.

Table 34.1 The ACES December 2014 documents

Document	Title	Version/Date
SPECIFICATION		
S-2013-001	ACESproxy, an Integer Log Encoding of ACES Image Data	V2.0 Dec 2014
S-2014-002	Academy Color Encoding System – Versioning System	V1.0 Dec 2014
S-2014-003	ACEScc, a Logarithmic Encoding of ACES Data for Use within Color Grading Systems	V1.0 Dec 2014
TECHNICAL BULLETIN		
TB-2014-001	Academy Color Encoding System (ACES) Version 1.0 Documentation Guide	V1.0 Dec 2014
TB-2014-002	Academy Color Encoding System Version 1.0 User Experience Guidelines	V1.0 Dec 2014
TB-2014-004	Informative Notes on SMPTE ST 2065-1 – Academy Color Encoding Specification (ACES)	V1.0 Dec 2014
TB-2014-005	Informative Notes on SMPTE ST 2065-2 – Academy Printing Density (APD) – Spectral Responsivities, Reference Measurement Device and Spectral Calculation and SMPTE ST 2065-3 Academy Density Exchange Encoding (ADX) – Encoding Academy Printing Density (APD) Values	V1.0 Dec 2014
TB-2014-006	Informative Notes on SMPTE ST 2065-4 – ACES Image Container File Layout	V1.0 Dec 2014
TB-3014-007	Informative Notes on SMPTE ST 268M:2003 Am1 – File Format for Digital Moving Picture Exchange (DPX) – Amendment 1	V1.0 Dec 2014
TB-2014-009	Academy Color Encoding System (ACES) Clip-level Metadata File Format Definition and Usage	V1.0 Dec 2014
TB-2014-010	Design, Integration and Use of ACES Look Modification Transforms	V1.0 Dec 2014
TB-2014-012	Academy Color Encoding System Version 1.0 Component Names	V1.0 Dec 2014

Where the documents in Table 34.1 relate to new or augmented ACES specifications, they are described in Section 34.2.2; the remainder of the documents fall broadly into three groups:

- those that provide a more descriptive explanation of the specifications than was originally provided;
- those that merely point to the corresponding SMPTE standard;
- those that describe the augmented specifications.

Those falling into these categories are briefly described in the following; the reader who requires a more extensive exposure of their contents is referred to the original documents. The contents and style of TB-2014-001 and TB-2014-002 particularly provide an indication of the importance the Subcommittee attaches to making the aims of the ACES known to a broader representative sector of the industry.

It is difficult to improve upon the summary descriptions which appear in each of the technical bulletins; thus where appropriate, they have been extracted and placed in the descriptions which follow below, indicating as such by using italics for the extracted material.

34.2.1.2   TB-2014-001

In TB-2014-001, the application of the ACES to the production and post system is described as follows:

'The key components of the ACES system are ACES encodings, ACES image files, ACES transforms and associated files, and an ACES clip-level metadata container that describes how the ACES image files were viewed when created or modified. ACES Version 1.0 is the first official release of these components. These components may be enhanced in subsequent releases based on industry requirements. Feedback from ACES Product Partners and end users made it clear that such a dynamic environment requires a clear system for version-control and naming of ACES components'.

34.2.1.3   TB-2014-002

In TB-2004-002 'User Experience Guidelines', the policies of the committee are enunciated at some length, as the extract which follows indicates. The document provides very useful guidelines to those new to ACES and those who previously may have considered that because of the style of presentation, the effort required to understand the specification was beyond what they were prepared to commit:

'A goal of ACES 1.0 is to enable widespread adoption by encouraging consistent implementations in production and post-production tools throughout the complete film and television product ecosystem spanning capture to archiving. This is a very diverse set of tools, each used by professionals with different sets of skills. Furthermore, each manufacturer has established their own set of conventions for how to structure their user experience to best serve their market. Clearly, it is neither feasible nor appropriate for

these guidelines to specify in minute detail every aspect of a user interface (e.g. "all products must use a set of vertical drop-down menus labeled in 10-point Helvetica").

That said, the feedback from users on the first wave of products implementing the pre-release versions of ACES has been clear in the need for guidelines. One common comment is that the implementations are so different, figuring out how to configure ACES in one product is of little help when configuring the next. For example, naming conventions are different for no apparent reason.

Another common concern is that the system is too reliant on acronyms and uses unfamiliar concepts (e.g., what is a "reference rendering transform"?). Although some of these acronyms have become familiar within the inner circle of ACES product partners and early adopters, it must be acknowledged that the tolerance for these terms is much lower amongst the general population of industry professionals (e.g. how would one explain what an RRT is to an editor, CG animator, or anyone else without some color science background).

As the ACES project transitions from technical development to wider industry deployment and the release of version 1, it is appropriate that we take a fresh look at how to portray the system to an audience that includes end-users in addition to engineers and color scientists. Although the technical terms and acronyms will continue to be used within the engineering community, these guidelines introduce a new set of terms intended to be simpler and more familiar to a wider set of users'.

As an indication of the move away from acronyms where possible, the IDT is now referred to as the 'input transform' and the RRT and the ODT acting in series combination as the 'output transform', a transform which may be preceded by a user 'look transform'.

It is recommended that any reader intent upon a complete understanding of the means of implementing the ACES in his or her workflow should, following the completion of this chapter, access and read the Guidelines document particularly and review the advisability of accessing the other documents in the current release which are relevant to his or her needs.

The Guidelines document also refers to the new version of the ACESproxy specification and the new ACEScc specification, which are described in Section 32.2.2 and referred to further in subsequent sections.

34.2.1.4   Technical Bulletins Which Relate Directly to SMPTE Specifications

These technical bulletins primarily describe only the relationship between the original ACES specification and the almost identical SMPTE standard.

TB-2014-004 includes the original (2008) ACES specification and additional informative material which explains how to calculate the values appearing in the table associated with the original Annex B, which lists the RGB values of the ColorChecker chart when taking into account the flare characteristic defined in the reference input capture device (RICD).

TB-2014-005 describes the method for interfacing film captured images into and out of the ACES colour space.

TB-2014-006 basically refers only to the associated SMPTE ST 2065-4 standard for the ACES Image Container File Layout specification, which is intended to be compatible with software and hardware capable of reading and writing the OpenEXR format.

TB-2014-007 basically refers only to SMPTE ST 268M:2003 Am1 – File Format for Digital Moving Picture Exchange (DPX) — Amendment 1, primarily for the purpose of specifying a container for images in the Academy Density Exchange Encoding (ADX).

34.2.1.5 TB-2014-009

TB-2014-009 specifies the ACES clip-level metadata file ('ACESclip'), which is a 'sidecar' XML file intended to assist in configuring ACES viewing pipelines and to enable portability of ACES transforms in production. It is likely to become a useful tool for colour management and to minimise errors in the setting up of the workflow in post.

34.2.1.6 TB-2014-010

TB-2014-010 specifies the 'Look Modification Transform (LMT) which imparts an image-wide creative "look" to the appearance of ACES images. It is a component of the ACES viewing pipeline that precedes the Reference Rendering Transform (RRT) and a selected Output Device Transform (ODT). LMTs exist because some color manipulations can be complex, and having a pre-set for a complex look makes a colorist's work more efficient. In addition, emulation of traditional color reproduction methods such as the projection of film print requires complex interactions of colors that are better modeled in a systematic transform than by requiring a colorist to match "by eye". The LMT is intended to supplement—not replace—a colorist's traditional tools for grading and manipulating images'.

The document *'describes the use of ACES Look Modification Transforms (LMTs) for ACES-based color management. It provides several use cases for LMTs, defines how LMTs are expressed and are carried along with clips and projects, discusses LMT use in the context of a workflow employing ACES-based color management, and concludes with design guidelines for LMTs. This document also describes optimal use of LMTs and suggests several ways in which an LMT may be designed to support flexible mastering and archiving workflows'.*

34.2.1.7 TB-2014-012

TB-2014-012 entitled 'Component Names' notes that *'ACES component names have technical names that emerged from the engineering and development process. While the names make sense to the scientists, engineers and early adopters that "grew up" with the system, the larger adoption community targeted for adoption by ACES Version 1.0 does not have the historical knowledge and context of the ACES pioneers and a large majority of that community does not have the technical training needed to understand many of the existing names. This Technical Bulletin documents the ACES component naming conventions as agreed to by the ACES'.*

The new names listed in the document are shown in Table 34.2, together with, where appropriate, the original acronym as defined in Table 31.0, which appears in the introduction to Part 5C.

Where appropriate, the recommended nomenclature will be used in the remainder of this chapter.

Table 34.2 Recommended nomenclature for ACES terms

Original nomenclature	Recommended nomenclature
Colour primary sets	
SMPTE 2065-1:2012 primaries, a.k.a. 'ACES primaries'	ACES primaries 0 or AP0
ACES 'working space' primaries, a.k.a. 'Rec.2020+'	ACES primaries 1 or AP1
Encodings	
SMPTE 2065-1:2012, a.k.a. 'ACES'	ACES2065-1
'ACES wire format', a.k.a. 'ACESproxy', 'ACESproxy10', 'ACESproxy12'	ACESproxy
SMPTE 2065-1:2012 with Rec.2020+ primaries, log encoding, floating point encoding, a.k.a. 'ACES working space'	ACEScc
VFX-friendly encoding, i.e. integer version of 'ACES working space', with ACESproxy transfer function[3]	ACEScg
Transforms	
Input device transform (IDT)	Input transform
Look modification transform (LMT)	Look transform
'RRT plus ODT' a.k.a. 'ACES viewing transform'	Output transform
Containers	
Clip-level metadata file	ACESclip file. Alternate: ACES xml
Academy-ASC common LUT format file, a.k.a. 'CLF file'	Academy-ASC common LUT format. Alternates: Common LUT format, clf file

34.2.1.8 The Reference Rendering Transform

Although the RRT is mentioned several times in the documentation, its only definition appears in Annex A to TB-2014-002 on 'User Experience Guidelines' as follows:

'RRT (Reference Rendering Transform) — Converts the scene-referred ACES2065-1 colors into colorimetry for an idealized cinema projector with no dynamic range or gamut limitations'.

Although a number of candidate versions of the RRT have been introduced since the publication of the original specification, 2014 marks the first official release of the RRT[4] in versions for both the forward and the inverse directions.

34.2.2 The Augmented Specifications

The December 2014 package of documentation includes a modified version of the ACESproxy specification, S-2013-001, and a new specification, S-2014-003, which describes a new working space for use in colour grading systems. In general terms both these colour space specifications share much in common, including a chromaticity gamut based upon a new set of primaries and a logarithmic encoding, albeit the characteristics of the logarithmic encodings are different.

To differentiate the new primaries from the original ACES primaries, the original primaries are now designated as the AP0 primary set and the new primaries as the AP1 primary set.

---

[3] It has been pointed out that this description is anomalous as ACEScg has a linear light transfer function and AP1 primaries, and is stored as a floating-point encoding.

[4] https://github.com/ampas/aces-dev/tree/v1.0/transforms/ctl/rrt

34.2.2.1   The AP1 Primaries and the ACES-AP1 Colour Space

The AP1 primaries are based upon the Rec 2020 primaries, adjusted to slightly increase the saturation of each primary in order to minimise the risk of highly saturated Pointer surface colours falling outside of the resulting chromaticity gamut. The advantages of adopting the colour space based upon these primaries will be discussed in Section 34.3.3.3.

Table 34.3   Chromaticity coordinates of the AP1 primaries

	x	y	z	u'	v'
Red	0.7130	0.2930	−0.0060	0.5603	0.5181
Green	0.1650	0.8300	0.0050	0.0523	0.5914
Blue	0.1280	0.0440	0.8280	0.1565	0.1210
White D60	0.3217	0.3377	0.3407	0.2008	0.4742

The chromaticity coordinates of the AP1 primaries are shown in Table 34.3, together with the chromaticity of the system white point, which is D60, the same as that of the AP0 primaries set.

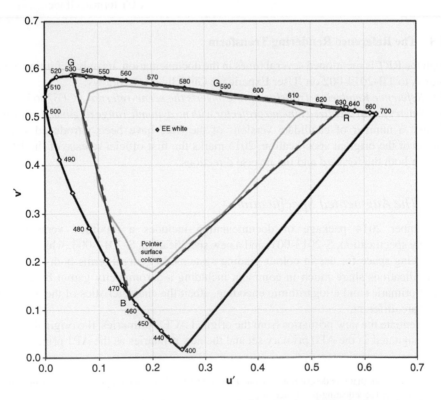

Figure 34.1   The AP1 primaries and the Rec 2020 primaries, the latter shown dotted.

The AP1 primaries gamut and the Rec 2020 primaries gamut for comparison are plotted in the chromaticity diagram derived in Worksheet 34(a) and illustrated in Figure 34.1. They are so similar that it is difficult to separate them on this full-scale chromaticity diagram. However, there are important differences: The red and green AP1 primaries are located just outside the spectrum locus, thus ensuring all maximum saturated red to yellowish-green hues are within the chromaticity gamut; similarly, the blue primary is also located just outside the spectrum locus, leading to slightly fewer saturated cyan hues being outside the chromaticity gamut than would otherwise be the case for the Rec 2020 primaries.

References to the ACES colour space continue to be used without the use of an AP0 suffix but when it is transformed to the AP1 primaries it is identified here as the ACES-AP1 colour space. This colour space is the basis of both the ACESproxy and ACEScc colour spaces described below but confusingly, in the Academy specifications it is given a different name in each case, possibly because different groups defined these two colour spaces. The matrix required for the transform from ACES to ACES-AP1 is calculated in Worksheet 34(b) and shown in Table 34.4.

Table 34.4   Matrix for transforming ACES signals to ACES-AP1 signals

	R_{AP0}	G_{AP0}	B_{AP0}
R_{AP1}	1.4514	−0.2365	−0.2149
G_{AP1}	−0.0766	1.1762	−0.0997
B_{AP1}	0.0083	−0.0060	0.9977

34.2.2.2   The S-2013-001 ACESproxy Colour Space

As the ACES-AP1 colour encoding system uses a 16-bit half-precision floating-point colour encoding method, it is not directly compatible with current 10-bit and 12-bit transport systems; thus, the ACESproxy encoding system, which is required to transport the encoded image to a local monitor, specifies a 10-bit and a 12-bit encoding system, which as we have seen in Section 13.6.3 require a perceptible uniform encoding strategy if contouring artefacts are to be avoided. As logarithmic encoding is relatively common place and familiar, in digital cine cameras and grading systems, in order to emulate the characteristics of film, a logarithmic-type integer encoding is defined for the ACESproxy encoding system to transport a representation of the ACES-AP1 floating-point image.

The contrast range of the ACES-AP1 signal is too large to be accommodated in the contrast range available to 10-bit and 12-bit fixed-point encoding systems; thus, it is limited by clipping the signal level below a point representing subjective black in the environment in which it is intended the proxy signal will be viewed. In consequence, the specification emphasises this encoding should not be used for any other purpose but on-set monitoring.

The transcoding formula in the specification is given in a slightly different and simplified form (ignoring the rounding rules) as:

$$\text{ACESproxy10 CV} = 64 \qquad\qquad\qquad \text{for ACES-AP1 CV} \leq 2^{-9.72}$$

$$\text{ACESproxy10 CV} = (\log_2(\text{ACES-AP1 CV}) + 2.5) \times 50 + 425 \quad \text{for ACES-AP1 CV} > 2^{-9.72}$$

$$(34.1)$$

where CV = code value. (In S-2013-001 ACES-AP1 is referred to as 'ACESproxyLin')

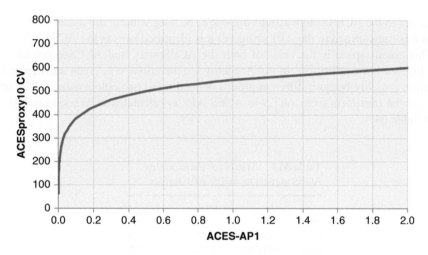

Figure 34.2   The transcoding characteristic of the transform between ACES-AP1 and ACESproxy10.

The characteristic for an exposure range one stop above a fully reflecting white surface used in the transcoding transform is plotted in Worksheet 34(c) and is illustrated in Figure 34.2. The code values on the y-axis result from a 10-bit coding.

The constants in equation (34.1) are designed to place the (dark tones) clipped code values of ACES-AP1 in the range of code values defined in Rec 709, thus enabling established transport systems to be used for conveying the signal from, for example, a digital cine camera to an on-set monitor. If the monitor is set to white on a signal level of 0.9 (with the signal level of a perfect reflector being 1.0) and the clip level is set at $2^{-9.72}$, that is, a level of 0.001184, a contrast range of approximately 760:1 will result. Taking the $\log_2$ of $2^{-9.72}$ gives a figure of -9.72; thus, in equation (34.1), the addition of 2.5 provides a black level of -7.22; the multiplier of 50 provides the number of code values for each value produced by the log expression, and the addition of the 425 term places black level at the code value of 64, as is the case for Rec 709.

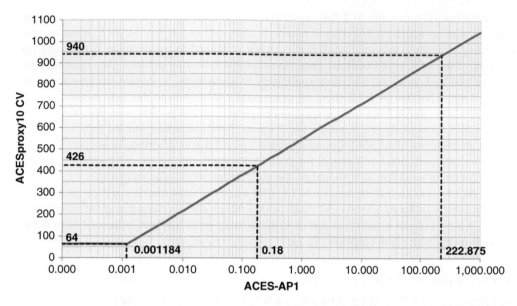

Figure 34.3 As Figure 34.2 but illustrated on a log/lin plot.

Figure 34.2 is useful for illustrating the shape of the log characteristic, but to see the full contrast range of image values, it is necessary to use a logarithmic chart as illustrated in Figure 34.3. Some of the signal levels of interest and their code values are also shown. It will be noted that the characteristic requires an input of 222.875 to equate to the notional white code value of 940, ensuring the coding scheme will not clip the highlights of any realistic signal level.

The specification also provides a formula for 12-bit coding which delivers the same shape of characteristic but the constants in the formula are changed to ensure that the black code level is at 256 and the white code value is at 3,760.

It can be seen from an inspection of the graph in Figure 34.3 that the range of ACES-AP1 signal levels between 0 and 1.0 will be represented only by ACESproxy code values of between 64 and 550; even more crucially, the levels between 0.18 and white at 1.0 are constrained to a range of code values between 426 and 550, which means that when displayed on a standard Rec 709 monitor without an inverse transform in place, the images will be severely lacking in contrast but nevertheless useful for judging image composition.

34.2.2.3 The S-2014-003 ACEScc Colour Space

Specification S-2014-003 is a new colour space introduced by ACES in recognition that in using its primary colour space (ACES-APO) for grading, it can in certain circumstances lead to problems and can be counter-intuitive in terms of the manner in which adjustment of the grading controls affects the appearance of the image.

In colorimetric terms, the ACEScc colour space is based upon the same ACES-AP1 colour space as that on which ACESproxy is based, as described in Section 34.2.2.2; however, the logarithmic encoding scheme is different in that it is defined to ensure there is generally no clipping of the ACES-AP1 contrast range, in order that following grading, the signal may be transformed back to the ACES colour space with no impairment. In certain circumstances, as we shall see in Section 34.3.3.3, there are occasionally exceptions to this general rule.

Nevertheless, the ACEScc colour space is designed to be compatible with the ACESproxy colour space, particularly in terms of the use of on-set 'look' metadata using the American Society of Cinematographers (ASC) Colour Decision List (CDL), as described in Section 32.10.3.

The use of the ACEScc is transient for grading purposes only, and in consequence, there is no file container specified, as it should not be used for interchange or archiving.

The ACES signal is transformed to ACES-AP1 using the same matrix coefficients as those listed in Table 34.4 for ACESproxy. (In S-2014-003 ACES-AP1 is referred to as 'ACESccLin'.) The ACES-AP1 signal is then encoded as 32-bit floating-point numbers as described in IEEE P754 using the formula in equation (34.2).

$$\text{ACEScc} = (\log_2(2^{-15} \times 0.5) + 9.72)/17.52 \qquad\qquad \text{for ACES-AP1} \leq 0$$
$$\text{ACEScc} = (\log_2(2^{-16} + \text{ACES-AP1} \times 0.5) + 9.72)/17.52 \quad \text{for ACES-AP1} < 2^{-15} \qquad (34.2)$$
$$\text{ACEScc} = (\log_2(\text{ACES-AP1}) + 9.72)/17.52 \qquad\qquad \text{for ACES-AP1} \geq 2^{-15}$$

The transcoding characteristic resulting from equation (34.2) is plotted in Worksheet 34(c) and illustrated in Figure 34.4.

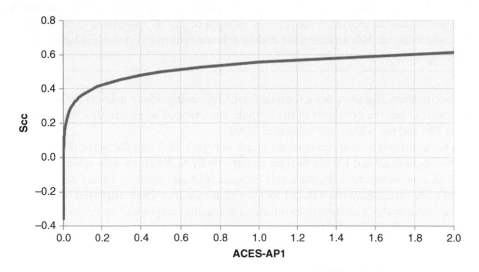

Figure 34.4   The transcoding characteristic ACES-AP1 to ACEScc.

Although the log/lin graph in Figure 34.5 looks similar to the ACESproxy graph, it can be seen on inspection that the characteristic covers a much increased range of contrast.

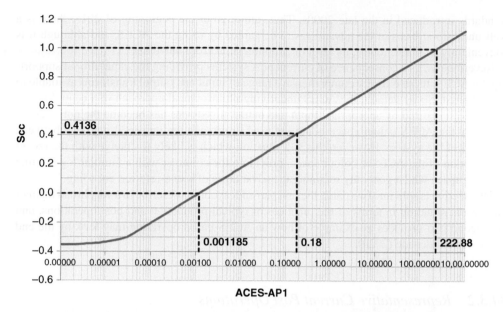

Figure 34.5 The log/lin presentation of the same data portrayed in Figure 34.4.

One drawback of the ACEScc scheme is that there will be circumstances when it cuts across the ACES philosophy of preserving all scene-referred colours; out-of-AP1-gamut colours will produce negative signals which cannot be encoded by a log characteristic as logarithms of negative numbers are not possible. In consequence, any negative values which would otherwise result from the ACES to the ACES-AP1 matrix will require mapping or clipping before the ACEScc logarithmic element of the transform. It is understood the Academy is addressing this issue.

34.3 Production and Post — System Configuration and Workflows

34.3.1 Introduction

This final section differs from the remainder of the book in as much as it attempts to describe not so much the technology of colour reproduction but the manner in which that technology is being adopted within the cinematographic industries and how it may develop in the foreseeable future. In consequence, it is to a degree subjective; it is based to a large extent on contacts between the author and a small number of practitioners within the industry, both post house colourists and technologists concerned with the equipment that supports them.

As indicated in the introduction to this part, the post industry in particular evolved a range of working practices during the change from film to digital operations, practices which were supported by manufacturers of grading systems who were prepared to provide different solutions for different post houses; this led to many different established patterns of work in the period before the Academy specifications were formulated.

The post operation is complex and the successful practitioners within it are often reluctant to put aside their hard-learned individual procedures to adopt new ones based upon the standards introduced in the late 2000s. Thus, although at the beginning of 2015, there is a well-understood basis of the advantages to be gained by using the ACES, and although it is conveniently available to use, its level of adoption is far from universal.

Nevertheless, as a consequence of current post systems providing the capability of supporting the ACES, there is a very comprehensive range of input and output transforms available to the colourist. So by default, the ACES underlying philosophical approach of a central working space supported by appropriate input and output transforms has become the norm, albeit generally the ACES colour space is not yet used to provide the universal interface between elements of the system. However, this is a situation which is likely to change with the recent introduction of the ACEScc colour space, providing the colourist with a familiar logarithmic working space whilst generally retaining the advantages of scene-referred encoding.

This section is split into two parts: the first part endeavours to describe in a very broad manner the current situation in terms of the practices of typical production and post operations, and the second part provides an example of how an ACES-based operation may look by the end of the current decade (2020).

34.3.2 Representative Current Post Operations

The diverse requirements of post ensure there is not a 'typical' operation; the operation of each post house tends to reflect not only the market niche it was established to support but also the view of its principal colourists in specifying its equipment and its mode of operation. Thus, the descriptions which follow are merely representative; they do not purport to describe any particular operation.

Figure 34.6 illustrates a post house grading room, configured in this instance for dealing with high end programmes and commercials, principally for television. When material for

Figure 34.6 A post-production grading room.

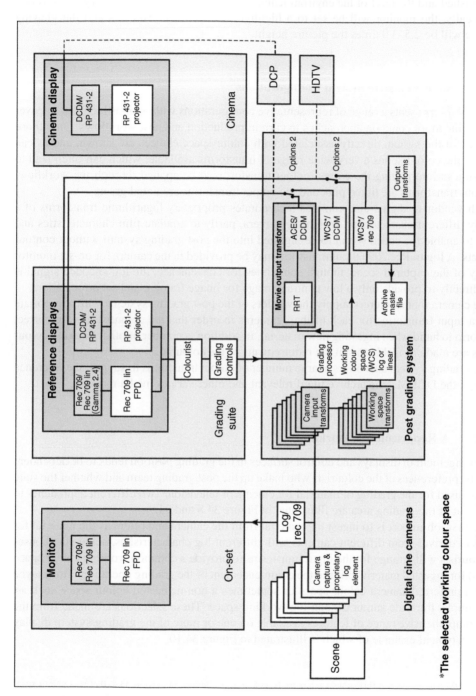

Figure 34.7 A representative 2015 production and post system configuration.

*The selected working colour space

the cinema is being graded, the monitor is replaced by a projector, the surround lighting is extinguished and the level of the environmental lighting is brought down to very low levels.

Typically, the monitor will be set to a highlight luminance of 100 nits and the viewing distance will be 2.5–3.0 times the picture height.

34.3.2.1   Representative System Configurations

Figure 34.7 represents a range of representative configurations with options included to cover some of the more common approaches to current production and post workflows. Only those elements of the system directly associated with colour space choices are shown, and as can be seen, the colourist has a very wide range of transforms available, which can often lead to confusion in interpreting the correct colour transforms to be applied through the workflow; the input transforms are tinted pink and the output transforms are tinted green.

Each vendor of digital cine cameras incorporates proprietary logarithmic transforms of a slightly different characteristic within the camera, partly to emulate film characteristics and partly to enable the coded signal to be ingested into the post grading system without contour artefacts. A log to Rec 709 output transform may be provided in the camera for on-set monitor display of the captured scene, though sometimes for convenience, the log encoded signal is used directly to provide only a low contrast image for image framing and set arrangement.

The camera manufacturers supply the vendors of the post grading systems with appropriate camera input transforms for each of their cameras in order that the colourist has the correct transform to hand when ingesting new material. In addition, a number of other working colour spaces are made available to suit the requirements of the colourist.

The grading system also offers a large number of output transforms to support the reference displays, the DCDM,[5] the archive file if relevant and other various media markets.

34.3.2.2   A Representative Workflow

The arrangement of displays and control surfaces in the grading position tends to be dependent upon the preferences of the colourists who make up the post grading team and whether the suite is being used for the grading of material for cinema or television. Two different approaches to the layout of the grading area are illustrated in Figure 34.8 and 34.9.

The first job in post is to ingest the material from the camera and often, as the material for a project derives from different cameras with different log characteristics, some post houses or colourists will arrange for the input transforms to provide a common working colour space (WCS) for ingested material prior to the commencement of the grading operation. This maybe either a preferred camera colour space or sometimes a non-dedicated colour space such as, for example, the wide gamut Cineon-log working space. These selections are made from the very comprehensive range of features available on one or more of the grading system display panels, a typical example of which is illustrated in Figure 34.10.

---

[5] The DCDM colour space in this context refers to the colour space defined in Section 33.4, that is, a colour space comprised of an XYZ chromaticity gamut, a tone response characterised by a power law with an exponent of 1/2.6 and a contrast range defined by 12-bit perceptible uniform encoding. Within the post operation, this colour space is sometimes referred to as the DCI XYZ colour space.

Figure 34.8 A monitor grading view.

The configuration of the colour space transform elements following the working space varies considerably depending upon the practices of the post house and the individual colourist and whether grading is being undertaken for the cinema or DVD/television. These variations in configuration are accommodated by the 'Option' connections in the Figure 34.7.

In the following, initially the ideal 'default' configuration is described, which uses colour management practices to match the working colour space to the various colour spaces of the

Figure 34.9 A grading view in a high end post set-up position; the projection screen is not shown.

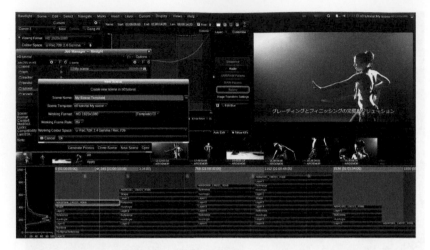

Figure 34.10 The detail in a typical grading set-up screen.

environment. In the subsequent description, the alternative custom and practice approach still used by many in the industry is outlined.

The Colour-Managed Workflow

With reference to Figure 34.7, the output from the grading processor is in the form of the WCS, which in turn will be the favoured working space of the colourist, usually one of the logarithmic working spaces. This working space will be transformed to match the reference display working space, either the projector or the flat-panel display (FPD). Appropriate transforms are available to undertake this match, labelled as the WCS/DCDM or the WCS/Rec709, respectively. In this situation, the Options 1 and 2 selections will be connected to the output of the transforms. At the reference displays, the input transforms convert the characteristics of the signal to match that of the appropriate display. Thus, although there may be limited or no support of the ACES, this approach is based upon sound colour management principles.

The path to the projector maybe configured to include the movie output transform, which includes the RRT and the ACES/DCDM transforms. The RRT has taken several years to stabilise and is not yet used universally within the post house environment.

Despite the availability of the means to do so, indications are that few post houses have adopted a colour management approach to grading.

The Custom and Practice Workflow

It appears that the custom and practice evolved in the era of the digital intermediate process continues to be widespread in the post industry, that is, to use no matching transform between the logarithmic coding associated with the grading processor working colour space and the reference displays. In consequence, prior to the commencement of the grading operation, there will be a very significant mismatch of colour spaces in the workflow, which will be manifest by a very low contrast, low saturation rendered image on the reference display.

The colourist compensates for this mismatch by using the grading controls to increase the contrast and saturation until a pleasing and acceptable image is rendered on the reference

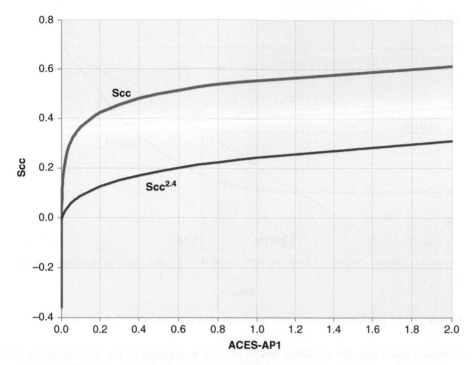

Figure 34.11 Rendering characteristic of a 'custom and practice' workflow on linear plot.

display. However, it is unlikely that this compensatory adjustment is capable of achieving the same results as that achieved by a colour-managed workflow; the wide colour gamut of the working colour space will not match the display primaries, and it is unlikely that the tone adjustments will mimic the precise characteristics of the missing colour-managing transform.

Effectively, the grading processor controls are compensating for the combined mismatched characteristics of the logarithmic working space and the reference display characteristics, in the case of the FPD, the emulated display gamma of 2.4. In Worksheet 34(c), a combined characteristic based upon a representative ACEScc logarithmic characteristic and the gamma of a Rec 709 display is calculated and plotted on to both linear and log/lin graphs, which are illustrated in Figures 34.11 and 34.12, respectively.

Although the display gamma compensates to a degree for the high gain of the logarithmic characteristic at low signal levels, providing an improved point gamma, nevertheless, the shape of the resulting curves clearly shows why the rendered display is so lacking in contrast; for a signal input level from the camera equating to 1.0, the combined characteristic will produce a relative light output from the display of only 0.24. In addition, as was shown in Section 29.5.3.3, rendering a large chromaticity gamut signal onto a display with a smaller chromaticity gamut will lead inevitably to a desaturated display.

It is to be hoped that the new initiatives from the Academy will precipitate a stronger movement towards embracing a colour management approach to the grading operation.

Some of the major post houses configure display rendering transforms (DRT), which are effectively proprietary versions of the ACES RRT to convey both a particular 'look', which

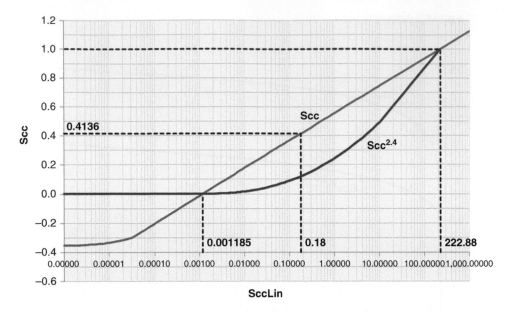

Figure 34.12   Rendering characteristic of 'custom and practice' workflow on lin/log plot.

may be incorporated into the workflow and to provide a mapping of the working space tone range to the tone range of current practical displays.

Keeping track of the colour spaces through the workflow is often a complex task, and some vendors of colour grading systems provide a dedicated panel within one of their grading screens to list the detail mapping of the path from ingest to output in terms of the colour spaces the encoded image encounters at each stage, as is illustrated in Figure 34.13.

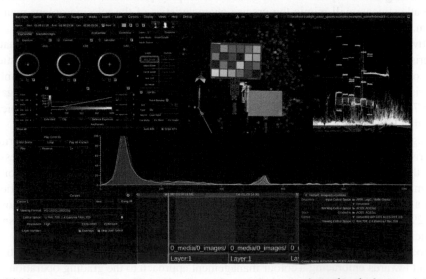

Figure 34.13   A grading display illustrating the 'colour space journey' panel at the bottom right-hand corner.

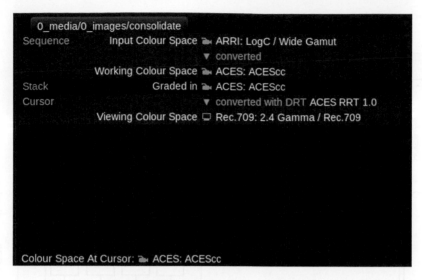

Figure 34.14 Colour space journey detail.

Figure 34.14 illustrates a cropped view of the lower right-hand corner of the full-screen display shown in Figure 34.13. This path through the system is called the 'colour space journey' by Filmlight, the vendor of the Baselight colour grading system, and illustrates that the new ACEScc working space has already been incorporated into the system.

34.3.3 An ACES System Configuration

34.3.3.1 Introduction

It may be helpful to envisage the form a system might take which is configured entirely around the ACES specification. As will become apparent, there are many ways such a system could be configured and Figure 34.15 is merely a representative form; however, it does provide a platform on which the issues which need to be addressed at various points in the system can be explored. In recognition of the trend to use the same cameras and post facilities for both movie and high production value television productions, the grading suite is shown equipped with the facility to use either a cine projector or an FPD grading monitor. It is assumed the projector configuration matches the configuration of a cinema projector, in that it contains a transform to accommodate a DCDM colour space configured input. In such a suite, the display surround and environmental lighting would be switchable between the requirements of the two media types.

In order not to overcomplicate the diagram in Figure 34.15, where common components of different elements in the workflow exist, a single component is used to illustrate both elements of the workflow. For example, the Rec 2020 output transform is shown feeding both the reference FPD and the ultrahigh definition television (UHDTV) distribution chain simultaneously, whereas in reality, this would form a sequential work pattern. The diagram is primarily intended to illustrate only the colour-related processing of the system and, in

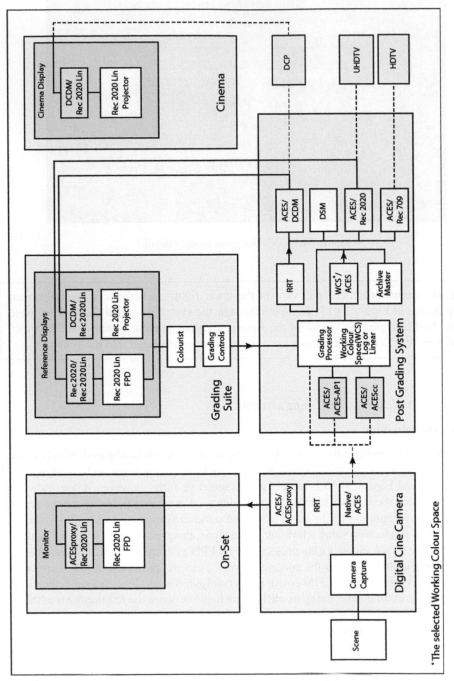

Figure 34.15 A representative system configured entirely on the ACES specification.

*The selected Working Colour Space

particular, the input and output transforms, which are shown tinted pink and green, respectively. The file structures required for transporting the signal between system elements are not addressed in this review.

34.3.3.2 Production

Within the camera, the native or raw images from the sensor(s) are processed by a proprietary transform to provide a standard ACES format signal in preparation for ingesting into the post-grading system. In addition, in order to provide a signal which can be transported to a local on-set monitor, the ACES signal is routed via an RRT and an ACES to ACESproxy transform to the on-set monitor. By including the RRT in the signal path, the rendered image will have a similar appearance to the un-graded image on the reference display in the grading suite, subject to the on-set monitor being located in an area with appropriate subdued lighting. Such an approach facilitates the imposition of a certain 'look' on the encoding during shooting which, with the support of the CDL, can be accurately replicated in post; see Section 32.10.3.

The on-set monitor incorporates a linear FPD with Rec 2020 primaries and, depending upon the sophistication of the display, a 2.4 gamma transform and, possibly, an ACESproxy to Rec 2020 transform. By the time Rec 2020 is fully implemented, the 2.4 gamma transform may be regarded as an unnecessary legacy item and therefore not included; its inclusion or otherwise will naturally influence the configuration of the ACESproxy to Rec 2020 transform, which may be integrated into the monitor or be an external module.

Without the input transform, the log-encoded ACESproxy signal will be displayed as a very flat image with low contrast but with enough detail to view items in the scene. To view the scene critically, a transform is required with dual characteristics, firstly, a transform with the inverse of the ACESproxy log characteristic, followed by a matrix to transform the signal from an AP1 chromaticity gamut to a Rec 2020 primaries chromaticity gamut.

Table 34.5 The AP1 to Rec 2020 conversion matrix

	R_{AP1}	G_{AP1}	B_{AP1}
R_{2020}	1.0393	−0.0114	−0.0279
G_{2020}	−0.0007	1.0006	0.0001
B_{2020}	−0.0057	−0.0223	1.0280

The required matrix is calculated in Worksheet 34(b) and its coefficients appear in Table 34.5. As will be shown in the next section, the chromaticities of the two sets of primaries are very close, which results in matrix coefficients very close to either 1.00 or zero; thus, unless very high colour accuracy of the rendered image is required on set; in practical terms, this matrix may not be necessary.

34.3.3.3 Post

In the manner in which the post-grading system is portrayed in Figure 34.15, it is assumed that the vendor has incorporated the option for the colourist to work in either logarithmic,

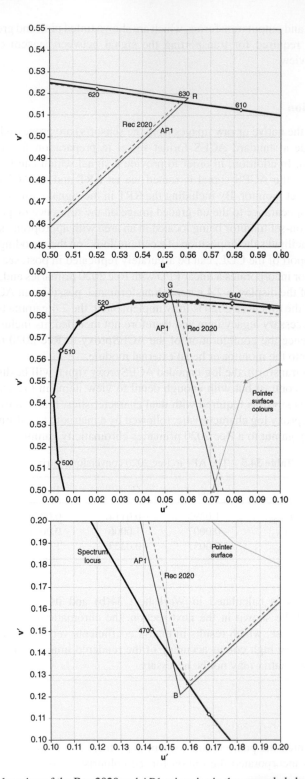

Figure 34.16 The location of the Rec 2020 and AP1 primaries in the expanded chromaticity diagrams.

that is, ACEScc colour space, or linear, that is an ACES-AP1 (ACESccLin) colour space, and thus has provided during ingest, the option of an ACES to ACEScc input transform or an ACES to ACES-AP1 input transform. In addition, the colourist can choose to work directly in ACES colour space, an important occasional option for very wide chromaticity gamut material, as we shall see. The ACES to ACES-AP1 matrix transform characteristics are defined in Section 34.2.2.2.

Generally speaking, it is assumed that the colourist will prefer to work in either the ACES-AP1 or ACEScc formats as this has the advantage that the colour controls are operating in a colour space virtually identical to the Rec 2020 reference display. The reason this is so is highlighted in the three expanded versions of the chromaticity diagram illustrated in Figure 34.16, where both the ACEScc-associated AP1 primaries and Rec 2020 primaries are plotted.

It can be seen that, from the viewpoint of a practical adjustment of colour controls, the two sets of primaries are virtually identical, with the advantage that the AP1 set encompasses all the saturated red to yellow-green colours. Thus, adjustment of any of the primary colour controls will be intuitive since only that colour value will change at the display, whereas when working with the AP0 primary set, adjustment of a primary colour will also cause secondary adjustment to the other primaries at the display.

The output from the grading processor is transformed from the selected working colour space back to the ACES working space; thus, subject to none of the scene colours being outside of the AP1 colour gamut, the full range of the scene-referred encoding is preserved. In the event that a gamut warning indication is given for a particular scene that some colours are outside of the AP1 gamut, the colourist has the option to switch to the ACES working space to grade that scene, which will require the reference display feed to then include either a gamut clipping or a gamut mapping element but will ensure that the full range of scene-referred colours are preserved for the archive file. It is important that this is a switchable option for the limited scenes with out-of-AP1 chromaticity gamut colours, as otherwise, the gamut mapping will unnecessarily limit the saturation of in-gamut colours for all other scenes.

Table 34.6 AP1 to AP0 primaries matrix

	R_{AP1}	G_{AP1}	B_{AP1}
R_{AP0}	0.6955	0.1407	0.1639
G_{AP0}	0.0448	0.8597	0.0955
B_{AP0}	−0.0055	0.0040	1.0015

The WCS/ACES output transform has a bypass when the selected working space is ACES; for other options, the transform includes an appropriate antilog transform characteristic and an AP1 to AP0 matrix, the values of which are derived in Worksheet 34(b) and shown in Table 34.6.

Following the recovery of the ACES encoding, the signal is in an ideal scene-referred format in readiness to service a number of outputs, all of which, with the exception of the archive file, in one way or another, terminate in displays which, by comparison with the ACES encoding, are currently of limited contrast range. Thus, there is the requirement to match the tone scale of the ACES encoding to the limited tone scale of the displays.

This was one of the requirements of the RRT introduced as part of the ACES, together with the requirement, perhaps overly simply put, to match the rendered image to the type of images enjoyed by cinema-goers over recent decades. Since its introduction in 2008, the RRT has evolved through several candidate iterations, each with its critics, until the full release of V1.0 in December 2014. As far as the author is aware, the Academy has not released any documentation to the public to date which describes the RRT, but it would appear that each subsequent candidate release has given more emphasis to the need to match the tonal scale and less to emulate the film appearance.

In consequence, the RRT now appears to be viewed as a generic transform required in the workflow of all grading procedures, whether for the cinema or for DVD/television, in order to provide at least an approximate mapping of the ACES tonal range to the tonal range of the display. It seems likely that additional output transforms will be developed to use in series with the RRT to provide a more precise match for specific viewing environments. As a consequence of the above rationale, the RRT in the diagram is shown in all but the archive master signal paths.

The amended ACES signal from the RRT is required to service a wide range of facilities, including the reference monitoring system, the DSM file and the various media markets.

There are two feeds to the reference monitoring system, though only one will be used at any one time depending upon whether the current grading operation is for cinema or television viewing. Though not essentially so, it is assumed in this case that for cinema viewing emulation, the display will be a laser projector with Rec 2020 primaries and will incorporate the same DCDM[6] to linear Rec 2020 transform as is found in a cinema laser projector. Thus, the path to the reference display will incorporate the same functionality as the path to the DCP agency, that is, an ACES to DCDM output transform as described in Section 32.9.2.

The alternate feed to the reference displays is designed to feed an FPD, primarily for use in grading material for the DVD/television market. Since some sort of compression will be required for the monitor path in order to avoid contouring artefacts, it is assumed that in this case, the monitor will incorporate a classic television inverse gamma correction characteristic Rec 2020 to Rec 2020 linear transform. Thus, an ACES to Rec 2020 output transform will be required in the post-grading system with characteristics as described in Section 32.9.1, which comprises a matrix as defined in Table 32.6, followed by a classic Rec 2020 gamma corrector. (As noted earlier, the Rec 2020 gamma corrector characteristics are likely to be amended.)

The characteristics of the reference FPD input transform is a classic Rec 2020 gamma emulation element only as no chromaticity space matrix is required. For the reference projector display, the characteristics of the transform have already been described in Section 33.5.2 and Table 33.7.

In order to preserve the characteristics of the scene-referred encoding, the archive master is formed from the output of the WCS/ACES transform. In this manner, if it is later required to generate a new DSM, this may be achieved by a simple transfer which includes the RRT in the signal path, and possibly in the future, an RRT with characteristics related to displays of greater contrast range.

One possible design implementation of a fully ACES system has been described and some of the issues facing the designer have been explored. However, it is emphasised that as the ACES system begins to be adopted in the future, quite different approaches to the one reviewed here are likely to unfold.

---

[6] Also referred to as the DCI XYZ format.

Appendices

Appendices

A

Photometric Units

A.1 The Physical Aspects of Light

Light is that range of frequencies in the electromagnetic spectrum which may be perceived by the human eye. Its spectrum in comparison with other electromagnetic forms of energy is illustrated in Figure A.1 where it can be seen that it occupies a frequency band between about 360 and 830 THz. A Terahertz is equivalent to 10^{12} Hz and a Hertz is defined as a cycle per second.

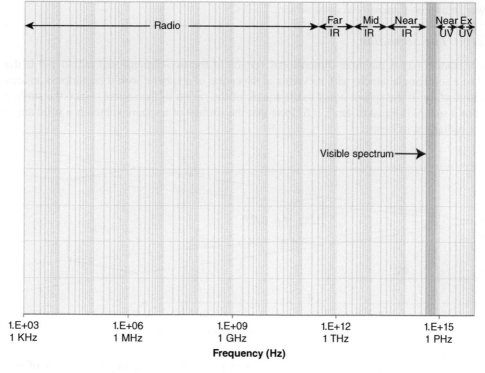

Figure A.1 Lower frequency electromagnetic spectrum.

Colour Reproduction in Electronic Imaging Systems: Photography, Television, Cinematography, First Edition. Michael S Tooms.
© 2016 John Wiley & Sons, Ltd. Published 2016 by John Wiley & Sons, Ltd.
Companion Website: www.wiley.com/go/toomscolour

The spectrum is divided into bands which each occupy frequencies between a factor of 3 and a factor of ten times 3; for example, 30–300 MHz is designated the VHF band. The radio spectrum is divided into nine bands which extend beyond domestic radio services at the lower end and above satellite television services at the upper end. The infrared bands fall between the radio bands and the visible spectrum, and the ultraviolet (UV) bands commence above the visible band and include the extreme UV band. Beyond UV are the x-ray and gamma ray bands.

Note that the light band occupies only about 17% of the lower end of the formerly designated *near-UV* band, which stretches from 300 THz to 3 PHz; thus it represents a truly diminutive part of the complete electromagnetic spectrum.

Since light is a form of electromagnetic energy it may be measured in watts, although as we shall see, since the response of the eye to differing frequencies varies significantly, watts of light is a concept which is somewhat limited in its usefulness to describe the effect which is evoked in the eye. Nevertheless, one's familiarity with physical units may make an initial review using a physical approach to measuring electromagnetic energy easier to comprehend than by using physiological units.

A.2   Power in a Three-Dimensional Environment

It may be useful at this point to review the geometric unit associated with the definition of power in a three-dimensional environment.

A.2.1   The Steradian

The steradian (sr) is the three-dimensional version of the two-dimensional radian and is the unit of a solid angle which, with its vertex at the centre of a sphere, would encompass an area on the surface of the sphere equal to the square of the radius of the sphere. Since the surface area of a sphere is equal to $4\pi r^2$, it follows that the surface of a sphere subtends a solid angle of 4π radians at its centre.

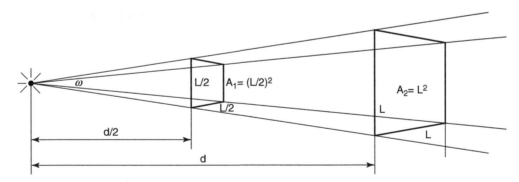

Figure A.2   Geometry for the steradian.

In Figure A.2 the two surfaces may be considered to be located on the surfaces of two spheres, respectively whose centre is located at the apex of the triangle.

The figure illustrates that for two surfaces, one at twice the distance from the apex of the pyramid formed by the solid angle ω, the ratio of the area of the surfaces to the square of the distance from the apex is always a constant which is equal to the value of the solid angle.

Thus:

$$\omega = \frac{\text{Area}}{(\text{Distance})^2} = \frac{A_2}{d^2} = \frac{L^2}{d^2} \text{ and also } \omega = \frac{A_1}{\left[\frac{d}{2}\right]^2} = \frac{\left[\frac{L}{2}\right]^2}{\frac{d^2}{4}} = \frac{L^2}{d^2} \text{ sr}$$

A.2.2   Radiation Intensity, I

Imagine now a point source of electromagnetic energy radiating a power of P watts uniformly in all directions as illustrated in Figure A.3.

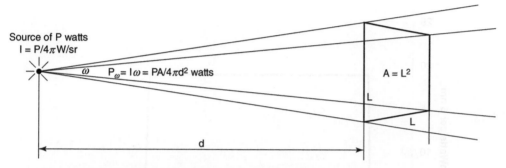

Figure A.3   Indicating the power radiated into a solid angle ω by a point source of power P.

Then since the source may be imagined at the centre of a sphere, its radiation intensity I is:

$$I = P/4\pi \text{ watts/steradian or W/sr}$$

If a surface of area A is placed at right angles to the direction of power flow then no matter at what distance d it is from the source, it may be imagined to form a very small part of the area of an imaginary sphere at that distance and will therefore subtend a solid angle ω at the source, where

$$\omega = A/d^2 \text{ sr}$$

Thus the power in the solid angle ω is:

$$P_\omega = I\omega = (P/4\pi)\omega = PA/4\pi d^2 \text{ W}$$

Since the surface intercepts all the energy in the solid angle it subtends and since we know that the power per steradian is $P/4\pi$ W, we can therefore deduce that the power incident upon the surface is also:

$$P_\omega = PA/4\pi d^2 \text{ W}$$

By dividing this expression by the area A, we can obtain a general expression for the power falling on a unit area. Thus power per unit area is:

$$PA/4\pi d^2 A = P/4\pi d^2 \text{ W/m}^2$$

If the units of distance and area are in metres then the power intensity of irradiation is $P/4\pi d^2$ watts per square metre.

Thus we have developed the well-known relationship that the level of electromagnetic energy from a point source falls off with the square of the distance between source and object.

A.3   A Useful Theoretical Source of White Light

As a tool to assist us in exploring the relationship between the power of an electromagnetic source of energy in the visible spectrum and its corresponding perception as a source of light with a calculated luminous intensity, we will define a simple theoretical (but not a practical) light source, S_t.

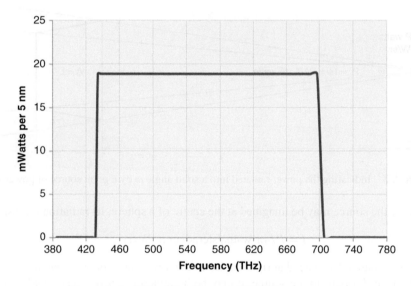

Figure A.4   SPD of the theoretical 1W light source tool.

Thus this point source of light, S_t, will have equal spectral power distribution (SPD) across only the extent of the visible spectrum from 435–700 THz and have a total emitted electromagnetic power of 1 W. The SPD of this source tool is illustrated in Figure A.4.

A.4   The Physiological Aspects of Light

Having established the basic physical units for describing a point source of electromagnetic energy, let us now turn our attention to the particular electromagnetic energy to which the eye is sensitive; by definition this is light energy.

The response of the eye to light of different frequencies varies very slightly from person to person, even for people with normal colour vision. However, based upon statistical experimental work, a standard response for the eye was agreed by the CIE in 1924 and is formally referred to as the CIE Photopic Spectral Luminous Efficiency Function or more generally just as the luminosity function or the V_f or V_λ curve, depending upon whether frequency or

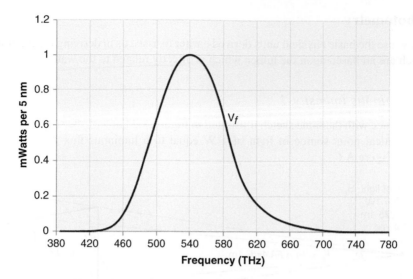

Figure A.5   The luminosity function of the human eye.

wavelength, respectively is being used to measure the spectral characteristics. This curve is illustrated in Figure A.5.

The absolute limits for the response of the eye for measurement purposes are taken as 360–830 THz. However, the 1% response points of 435–700 THz give a more practical representation for day-to-day experiences.

Since the spectra of light sources are usually far from flat, it can be seen from the V_f curve that measuring the light energy in purely physical terms would be irrelevant as a means of evaluating the subjective response of the eye.

Therefore, we need to define a physiological unit for measuring the effect of electromagnetic energy upon the eye, that is, a light unit which corresponds directly to watts at a specific frequency. That unit is defined as the lumen (lm) and the frequency chosen is 540 THz, the peak of the luminosity function.

Thus the eye may be said to have a maximum luminous efficiency of K_m lm/W at the peak of the luminosity function and the lumen is defined as *the luminous flux provided by a monochromatic light source of a power of 1/683 watts at 540 THz*[1], that is, the value of K_m is 683 lm/W. The reason for the value of K_m being made equal to 683 rather than some convenient factor such as 1 or 1,000 is historic and relates to the luminosity of a standard candle.

By convolving: the SPD of the theoretical light source illustrated in Figure A.4, the luminosity function illustrated in Figure A.5 and K_m, the luminous efficiency of the eye, we can calculate the total luminous flux of the 1 W theoretical light source tool. This calculation is carried out in Worksheet A1 and is found to be approximately 275 lm.

---

[1] In order to provide a clear and intuitive relationship between the fundamental concept of power and its *direct* equivalent in photopic terms, I have taken the liberty of amending the SI definition of the relationship by defining it in terms of the lumen rather than the candela. The latter, being a measure of lumens per steradian, I consider it to be a derived unit, which is defined in the next section.

A.5   Photometry

We will now use the basic physical units derived earlier to assist us in deriving the photometric units, which are all based upon the lumen which is directly related to the watt.

A.5.1   Luminous Intensity, I

Let us commence with the same sketch illustrated in Figure A.3 but with the source exchanged for the theoretical point source of light of 1 W equal to a luminous flux F of 275 lm, as illustrated in Figure A.6.

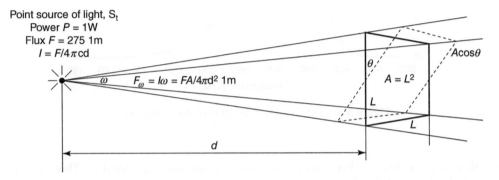

Figure A.6   Indicating the luminous flux emitted into a solid angle ω by a point light source of F lm.

In photometry, light power is defined in terms of luminous flux F and is a measure of the energy causing a sensation of brightness or lightness in the eye.

Thus our point source of power 1 W will have a flux of F lm and the luminous intensity I will be:

$$I = F/4\pi \text{ lm/sr or candela}$$

The unit of luminous intensity is the candela, abbreviated to cd.

The luminous intensity is the parameter which enables us to judge the brightness of a remote source of light which subtends a small solid angle at the eye. One may envisage in Figure A.6 the pupil of the eye replacing the surface area A, whereupon it becomes apparent that all the light within the solid angle ω is incident upon the pupil.

Here we have considered a point source of light for ease of deriving the units involved; however, any geometrically small source of light, which may have a directional emission, will be subject to the same rules; once the intensity of the source in a particular direction is known, it can be specified in terms of candela. Thus lamps are often specified in terms of a polar diagram which indicates the candela value in the direction of interest.

A.5.2   Illumination and Illuminance, E

In photometry, illumination is not a defined term but is used in a general manner to describe the character of light falling upon a scene in terms of its SPD or its colour temperature. Nevertheless, in general usage it is often used to describe the intensity of illumination where more accurately the term illuminance should be used.

Illuminance describes the intensity of illumination of a particular surface within a scene. Thus the illuminance E of a surface is defined as the amount of luminous flux falling upon unit area and is measured in terms of lumens per square metre or lux.

Returning to the point source, if we assume a source of $F = 4\pi$ lm, the luminous intensity I will be 1 lm/sr or 1 cd. At a distance d the solid angle subtended by a surface area of A square metres will be:

$$\omega = A/d^2 \text{ sr}$$

The flux emitted into the solid angle ω will be:

$$F_\omega = I\omega = F/4\pi.A/d^2 = FA/4\pi d^2 \text{lm}$$

This is the flux falling upon or illuminating, the area A.
Thus the level of illuminance E of the surface A is:

$$E = F_\omega/A = I\omega/A = IA/Ad^2 = I/d^2 \text{lm/m}^2 \text{ or lux}$$

Thus at a distance 1 m away from a source of intensity 1 cd, that is, 1 lm/sr, the level of illuminance will be 1 lux.

When the surface is at an angle θ to the normal, the level of illuminance will be

$$E = I \cos \theta/d^2$$

A.5.3   Surface Luminance, L

A.5.3.1   Surface Characteristics – The Lambertian Surface

A surface may be characterised by the means in which it reflects light. A totally matt surface reflects incident light at all angles to the normal and is known as a perfect diffuser, whilst a glossy surface reflects most of the light into the angle complementary to the angle of incidence; these reflections are termed specular reflections. Most surfaces have a characteristic which is a mixture of these two extremes, though in the general case there is an inclination towards a matt characteristic with little of the light being specularly reflected.

A Lambert surface is a special case of a matt surface which is a perfect diffuser and thus has no specular reflection and when evenly illuminated appears equally bright in all directions.

How is this appearance of equal brightness achieved? An incremental area of the surface may be defined as δs and will have a luminous intensity into a solid angle ω normal to the surface of $F\delta s/\omega$ cd. The same incremental surface viewed at an angle θ to the normal will have a projected area of $\delta s.\cos\theta$ and will subtend a smaller solid angle of $\omega\cos\theta$ sr and thus provide a luminous intensity of

$$F\delta s \cos \theta/\omega \cos \theta = F\delta s/\omega$$

Thus although the luminance intensity falls off with the cosine of the angle to the normal to the surface, so does the projected area from which it emanates, indicating that the brightness of the surface of a Lambert radiator remains a constant for any angle of view.

A.5.3.2   The Luminance of a Surface

A surface will have luminance as a result of emitting or reflecting light from its surface.

Consider a small surface ds_1 emitting light towards a receptor with a fixed aperture A as illustrated in Figure A.7. If the area is small enough it may be envisaged as a point source and treated as such. The aperture forms a solid angle ω with the surface ds_1 which emits a luminous flux F_ω into that solid angle.

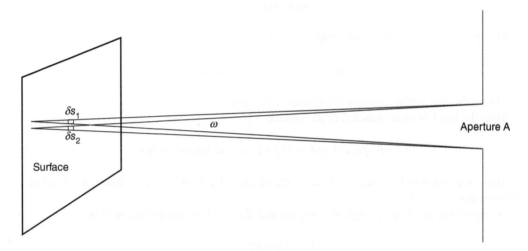

Figure A.7   Indicating how adjacent very small areas contribute to the luminous intensity.

The surface ds_1 would then have a luminous intensity of $I = F_\omega / \omega$ lm/sr or cd in the direction towards the aperture.

If the area ds_1 is very small, it may be assumed that a further identical area ds_2 immediately adjacent to ds_1 and radiating the same number of lumens/steradian will also contribute F_ω lumens to the aperture, giving a total of two F_ω lumens from a surface of two ds.

It is evident therefore that the luminance seen from the aperture is the luminous intensity per $ds = I_\omega / ds$, and therefore, when considering a larger area the luminance is proportional to the integration of the flux from all areas ds, which illuminate the aperture, that is, by the unit area.

Thus luminance L is measured in terms of candela/metre^2 or nits.

The unit of luminance is the nit, a convenient derived expression for *lumens per steradian per square metre* or just *candela per square metre* (cd/m^2); nevertheless, it is common for the latter to be used when specifying the luminance of a surface.

It should be particularly noted that the area which appears in this definition is not the area of the aperture but the area presented to the aperture.

A.5.3.3   Reflected Light from a Surface

Let us assume that a lambert surface with a reflection factor of 1 is normal to a light source which is providing an illuminance of E lux. Then since the entire luminous flux incident on the surface is reflected it has a value of EA lm or E lm/m^2.

The unit for describing reflected light in terms of lumens per square metre is the apostilb (asb) though its use is deprecated. If the reflection factor is ρ, then the luminance is ρE asb.

Though easy to comprehend, a moment's reflection will indicate this is not a very useful description of the luminance of a surface. As we have seen above, the appearance of the surface is dependent upon the intensity of the luminous flux leaving the surface, that is, the luminance in terms of candela per square metre. Furthermore, it is important that we are in a position to simply measure the luminances of a surface and this becomes impractical if it is required to measure *all* the luminous flux from the surface before the level could be specified. Thus the apostilb is a unit more suited to the laboratory than to the scene to be captured.

Referring to Figure A.7 again, it can be envisaged that if the area of the aperture A were to be considered an area on the inside of a hemisphere centred on the surface, then it would be possible by a knowledge of the geometry of the hemisphere to calculate the total flux from the surface in terms of the intensity of the luminous flux leaving the surface. Since we know the total flux leaving the surface is E asb, we would then have the relationship between luminance in apostilbs and luminance in candela per square metre or nits. Such calculation requires two sets of integrations based upon the radius and the circumference of the hemisphere, respectively and is beyond the scope of this book; however, the result is a simple relationship based upon π.

Thus, for a Lambert surface the luminance may be expressed in two ways: either as ρE asb or $\rho E/\pi$ nits or cd/m^2, where E is the level of illuminance in lux and ρ is the reflection factor.

This result provides us with a simple relationship between the level of illuminance, or in practical terms the level of illumination, in lux and the luminance of a surface in nits or candela per square metre.

In recent years there has been an impetus to move away from the use of derived units in photometry to the point where they are sometimes described as deprecated. It may be that there is concern that in using derived units one may become in conceptual terms a step away from the physical reality of the basic units. This is true of the apostilb and to a lesser extent, the nit. In consequence, the most common method of expressing luminance is in terms of candela per square metre, though if one were to extend the approach of using basic units to the extreme then luminance would be expressed in terms of lumens per steradian per square metre, not a particularly helpful step. Thus in this book, to be consistent with other derived units such as the amp, which is actually 1 coulomb per second, the nit will be used as the measure of luminance, together with reminders in appropriate places that this corresponds to candelas per square metre.

The photometric units are summarised in Table A.1.

Table A.1 Relationships between photometric units

Quantity	Symbol	Basic unit	Derived	Abbrv
Luminous flux	F	Lumen	Lumen	lm
Luminous intensity	I	Lumen/steradian	Candela	cd
Illuminance	E	Lumen/square metre	Lux	lx
Luminance (any surface)	L	Lumen/steradian/square metre	Candela/square metre or Nit	cd/m^2 or nt
Luminance (Lambert radiator)	L	Lumen/steradian/square metre Lumen/square metre	Nit Apostilb	nt asb

B

The CIE XYZ Primaries

B.1 Deriving the Chromaticities of the CIE XYZ Primaries from CIE RGB Primaries

This appendix also appears embedded with the actual calculations in Worksheet 4(b).

Based upon the criteria for the location of the XYZ primaries laid out in Section 4.4, primaries will appear on the r,g chromaticity diagram as shown in Figure B.1.

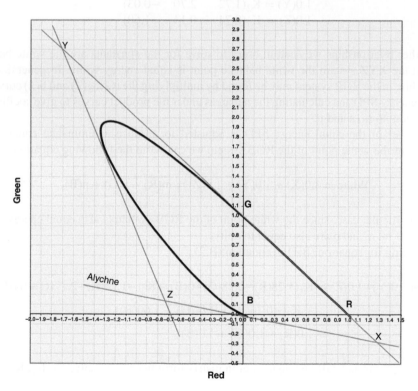

Figure B.1 Locating the XYZ primaries on the CIE Standard Observer r,g chromaticity diagram.

Colour Reproduction in Electronic Imaging Systems: Photography, Television, Cinematography, First Edition. Michael S Tooms.
© 2016 John Wiley & Sons, Ltd. Published 2016 by John Wiley & Sons, Ltd.
Companion Website: www.wiley.com/go/toomscolour

From the CIE Standard Observer Primaries r,g chromaticity diagram in Figure B.1, by inspection the approximate coordinates of the XYZ primaries are:

	r	g
1.0(X)	1.28	−0.28
1.0(Y)	−1.72	2.70
1.0(Z)	−0.74	0.14

In order to establish the relationship between the RGB and XYZ primaries we first need to establish the XYZ primaries in terms of the amounts of the RGB primaries rather than in terms of chromaticity coordinates, that is, we need to carry out the reverse of the normalization process.

Since we do not know the original amounts of the RGB primaries, we must introduce the constants K_1, K_2 and K_3 to establish the relationship; thus, remembering that $b = 1 - r - g$, we can construct a matrix as follows:

$$
\begin{array}{cccc}
 & R & G & B \\
1.0(X) = K_1(1.28 & -0.28 & 0.00) \\
1.0(Y) = K_2(1.72 & 2.70 & -0.03) \\
1.0(Z) = K_3(0.74 & 0.14 & 1.60)
\end{array}
\qquad \text{(B.1)}
$$

The other relationships we need in order to solve for the constants are the white balance points. In the XYZ system the white balance point is equal energy white. In general, equal energy white in the RGB system may be found by multiplying the $r(\lambda)$, $g(\lambda)$ and $b(\lambda)$ curves by an equal energy SPD and calculating the net areas under the product curves to give coefficients of RGB equal to m, n and p.

If the level of the equal energy SPD is adjusted to produce a luminance equal to the luminance of the white point of the RGB system we can write the following equation:

$$\text{White} = 1.0(X) + 1.0(Y) + 1.0(Z) = m(R) + n(G) + p(B) \qquad \text{(B.2)}$$

In this particular case, since the white point of the RGB system is also equal energy white, the values of m, n, and p are also equal to unity.

Substituting for equation (B.1) in equation (B.2)

$$1.0(X)+1.0(Y)+1.0(Z) = K_1(1.28(R)-0.28(G)+0(B))+K_2(-1.74(R)+2.77(G)-0.03(B))$$
$$+ K_3(-0.74(R) + 0.14(G) + 1.60(B))$$
$$= 1.0(R) + 1.0(G) + 1.0(B)$$

Collecting terms in RGB and laying out in matrix form in order to solve for K_1, K_2, K_3.

	K_1	K_2	K_3
1.00(R) =	1.28	−1.74	−0.74
1.00(G) =	−0.28	2.77	−0.14
1.00(B) =	0.00	0.03	1.60

Any of the usual algebraic or determinant methods of solving simultaneous equations may be used to determine the values of the constants. The author chose to use the matrix functions of Excel[1] in Worksheet 4(b) for solving all equations of this type appearing in this book.

Solving the matrix gives:

$$K_1 = 1.8513$$

$$K_2 = 0.5303$$

$$K_3 = 0.6184$$

Substituting in equation (B.1) for K_1, K_2, K_3

	R	G	B	
1.00000X	2.3697	−0.5184	0.0000	
1.00000Y	−0.9121	1.4318	0.0106	(B.3)
1.00000Z	−0.4576	0.0866	0.9894	

We now have unit amounts of the XYZ primaries in terms of the RGB primaries. What we require however, in order to establish the basis of a new colorimetric system, are unit amounts of RGB in terms of XYZ. We therefore need in matrix terms to invert it.

Inverting using the worksheet:

	X	Y	Z	
1.0000R	0.4899	0.1775	−0.0019	
1.0000G	0.3106	0.8114	−0.0087	(B.4)
1.0000B	0.1994	0.0111	1.0106	

The approximate figures taken from the graph inevitably lead to small 'errors' in the derived figures. The values for X and Z are relatively arbitrary within the constraints of the criteria laid down in Section 4.4 but the values for Y must accurately represent the luminosity coefficients of the RGB primaries if the coefficient of Y is to be a measure of the luminance of the colour. These values have already been established to five-figure accuracy by using the method referred to in Section 4.3.

Since the definition of the values of X and Z are not critical to five-figure accuracy they are rounded to two significant figures in order to simplify calculations. (In the days before digital calculation methods were available this was an important accuracy and efficiency consideration.)

If now equation (B.4) is amended to incorporate the rationale outlined above, the equations specified by the CIE result, as shown in the following table.

	X	Y	Z	
1.00000R	0.4900	0.1770	0.00000	
1.00000G	0.3100	0.8124	0.01000	(B.5)
1.00000B	0.2000	0.0106	0.99000	

---

[1] It is the material in this worksheet from which this appendix is derived and the active matrix functions contained there which will enable the reader to more easily follow in depth the procedure described here.

or in equation terms:

$$(R) = 0.4900(X) + 0.1770(Y) + 0.0000(Z)$$
$$(G) = 0.3100(X) + 0.8124(Y) + 0.0100(Z) \qquad (B.6)$$
$$(B) = 0.2000(X) + 0.0106(Y) + 0.9900(Z)$$

Any colour c which was matched by $m(R) + n(G) + p(B)$ may be matched by $i(X) + j(Y) + k(Z)$.

Replacing RGB in this match with the values given in equation (B.6)

$$0.490m(X) + 0.177m(Y) + 0.0m(Z) + 0.310n(X) + 0.8124n(Y) + 0.010n(Z)$$
$$+ 0.200p(X) + 0.0106p(Y) + 0.990(Z) = i(X) + j(Y) + k(Z)$$

and collecting similar terms gives the following set of equations to three significant figures:

$$i = 0.490m + 0.310n + 0.200p$$
$$j = 0.177m + 0.812n + 0.011p \qquad (B.7)$$
$$k = 0.000m + 0.010n + 0.990p$$

Since the $r(\lambda)$, $g(\lambda)$, $b(\lambda)$ curves represent a set of spectrum colours matched in the rgb system then they may be used with equation (B.7) to derive the $x(\lambda)$, $y(\lambda)$, $z(\lambda)$ colour matching functions.

Thus:

$$x(\lambda) = 0.490r(\lambda) + 0.310g(\lambda) + 0.200b(\lambda)$$
$$y(\lambda) = 0.177r(\lambda) + 0.812g(\lambda) + 0.011b(\lambda) \qquad (B.8)$$
$$z(\lambda) = 0.000r(\lambda) + 0.010g(\lambda) + 0.990b(\lambda)$$

B.2   The XYZ Primaries Located on the CIE RGB Primaries Chromaticity Diagram

For the sake of completeness we can now use the same procedure in reverse to establish the precise values of the XYZ primaries on the r,g chromaticity diagram.

Inverting equation (B.5):

	R	G	B
1.0000X	2.2346	−0.5152	0.0052
1.0000Y	−0.8965	1.4264	−0.0144
1.0000Z	−0.4681	0.0888	1.0092

Normalizing to establish the accurate values of r and g

	r	g	
X	1.2750	−0.2778	
Y	−1.7393	2.7673	(B.9)
Z	−0.7431	0.1409	

Equation (B.9) represents the precise co-ordinates of the XYZ primaries on the rg chromaticity diagram and may be compared with the approximate values at the beginning of this appendix that were derived from an inspection of the diagram.

C

The Bradford Colour Adaptation Transform

C.1 The Standard Bradford Transform

The Bradford transform (Luo et al., 1998), so named since it resulted from work carried out at the University of Leeds under the sponsorship of the UK Society of Dyers and Colourists based in the nearby city of Bradford, has become accepted as a sound basis for predicting the effects of adaptation to a reasonable degree of accuracy.

The transform uses the same structure as all other transforms which followed in the steps of the von Kries transform, that is:

1. Measure the XYZ values of: the sample colour under the reference illuminant, white under the reference illuminant and white under the test illuminant.
2. Transform these XYZ values to RGB values using a transform matrix.
3. Apply correction factors to the RGB values for the sample colour.
4. Transform the corrected RGB values for the colour sample to XYZ values.

These steps are detailed in the following:

Step 1. The values of XYZ for the three sets of data measured are given as follows:

For the sample colour: X_c, Y_c, Z_c

For the white under the reference illuminant: X_{wr}, Y_{wr}, Z_{wr}

For white under the test illuminant: X_{wt}, Y_{wt}, Z_{wt}

Colour Reproduction in Electronic Imaging Systems: Photography, Television, Cinematography, First Edition. Michael S Tooms.
© 2016 John Wiley & Sons, Ltd. Published 2016 by John Wiley & Sons, Ltd.
Companion Website: www.wiley.com/go/toomscolour

Step 2. Use the Bradford transform to transform from X,Y,Z to R,G,B for all values of X,Y,Z in step 1.

$$\begin{bmatrix} R \\ G \\ B \end{bmatrix} = M_{BFD} * \begin{bmatrix} X/Y \\ Y/Y \\ X/Y \end{bmatrix}$$

where

$$M_{BFD} = \begin{bmatrix} 0.8951 & 0.2664 & -0.1614 \\ -0.7502 & 1.7135 & 0.0367 \\ 0.0389 & -0.0685 & 1.0296 \end{bmatrix}$$

Obtaining R_c, R_{wr}, R_{wt}, etc. for all the measurements and where the subscripts are c = sample *colour*, wr = *white* under *reference* illuminant and wt = *white* under *test* illuminant.

Step 3. Apply the adaptation correction to the RGB colour sample values

$$R_a = R_{wt} \left(R_c/R_{wr} \right)$$
$$G_a = G_{wt} \left(G_c/G_{wr} \right)$$
$$B_a = B_{wt} \left(B_c/B_{wr} \right)^p$$

where $p = \left(B_{wr}/B_{wt} \right)^{0.0834}$

Step 4. Transform from R_a, G_a, B_a to X_a, Y_a, Z_a

$$\begin{bmatrix} X_a \\ Y_a \\ Z_a \end{bmatrix} = [M_{BDF}]^{-1} * \begin{bmatrix} R_a Y \\ G_a Y \\ B_a Y \end{bmatrix}$$

where

$$[M_{BDF}]^{-1} = \begin{bmatrix} 0.9870 & -0.1471 & 0.1600 \\ 0.4323 & 0.5184 & 0.0493 \\ -0.0085 & 0.0400 & 0.9685 \end{bmatrix}$$

C.2   The Linear or Simplified Bradford Transform

Often the non-linear element associated with the blue value is considered as inessential and a simplified version of the Bradford transform results. In consequence it then becomes possible to concatenate the above steps into one equation:

$$
\begin{bmatrix} X_a \\ Y_a \\ Z_a \end{bmatrix} = [M_{BDF}]^{-1} * [M_{ADT}] * [M_{BDF}] * \begin{bmatrix} X \\ Y \\ Z \end{bmatrix}
$$

where

$$
[M_{ADT}] = \begin{bmatrix} R_{wt}/R_{wr} & 0 & 0 \\ 0 & G_{wt}/G_{wr} & 0 \\ 0 & 0 & B_{wt}/B_{wr} \end{bmatrix}
$$

In Workbook 5 the Bradford 3×3 Transfer formula is laid out and used to calculate the transfer parameters to be used between any two illuminants. The XYZ values of a range of illuminants are supplied and available for inserting into the appropriate cells of the calculation.

The workbook matrix functionality is used to calculate the linear Bradford Transform Matrix between a reference illuminant of D65 and a test Illuminant A.

$$
\begin{bmatrix} X_a \\ Y_a \\ Z_a \end{bmatrix} = \begin{bmatrix} 1.2191 & -0.0489 & 0.4138 \\ -0.2952 & 0.9106 & -0.0676 \\ 0.0202 & -0.0251 & 0.3168 \end{bmatrix} * \begin{bmatrix} X \\ Y \\ Z \end{bmatrix}
$$

D

The Semiconductor Junction

The outer energy band of an atom in which electrons reside at a temperature of zero degrees kelvin is referred to as the valence band and the number of electrons found in the valence band determines the characteristics of the material as illustrated in the periodic table. When the valence band is full the material is chemically inactive, whilst if only one electron is present or one missing from a full band the material is extremely active.

Semiconductor diodes are formed when two pieces of semiconductor material such as silicon or germanium are doped with p and n type material, respectively at very low levels of concentration and brought together to form a p-n junction.

When a p-n junction is first created, conduction band (mobile) electrons from the N-doped region diffuse into the P-doped region where there is a large population of holes (vacant places for electrons) with which the electrons 'recombine'. When a mobile electron recombines with a hole, both hole and electron vanish, leaving behind an immobile positively charged donor (dopant) on the N-side and negatively charged acceptor (dopant) on the P-side. The region around the p-n junction becomes depleted of charge carriers and thus behaves as an insulator.

However, the width of the depletion region (called the depletion width) cannot grow without limit. For each electron–hole pair that recombines, a positively charged dopant ion is left behind in the N-doped region and a negatively charged dopant ion is left behind in the P-doped region. As recombination proceeds more ions are created, an increasing electric field develops through the depletion zone which acts to slow and then finally stop recombination. At this point, there is a 'built-in' potential across the depletion zone.

If an external voltage is placed across the diode with the same polarity as the built-in potential, the depletion zone continues to act as an insulator, preventing any significant electric current. This is the reverse bias phenomenon. However, if the polarity of the external voltage opposes the built-in potential, recombination can once again proceed, resulting in substantial electric current through the p-n junction (i.e. substantial numbers of electrons and holes recombine at the junction). For silicon diodes, the built-in potential is approximately 0.7 V (0.3 V for Germanium and 0.2 V for Schottky). Thus, if an external current is passed through the diode, about 0.7 V will be developed across the diode such that the P-doped region is positive with respect to the N-doped region and the diode is said to be 'turned on' as it has a forward bias.

Colour Reproduction in Electronic Imaging Systems: Photography, Television, Cinematography, First Edition. Michael S Tooms.
© 2016 John Wiley & Sons, Ltd. Published 2016 by John Wiley & Sons, Ltd.
Companion Website: www.wiley.com/go/toomscolour

E

Light Amplification in Lasers

This material is a lightly edited version of the excellent description of the amplification of light in a laser found in Wikipedia at http://en.wikipedia.org/wiki/Population_inversion.

In physics, specifically statistical mechanics, a *population inversion* occurs when a system (such as a group of atoms or molecules) exists in a state with more members in an excited state than in lower energy states. The concept is of fundamental importance in laser science because the production of a population inversion is a necessary step in the workings of a standard laser.

E.1 Boltzmann Distributions and Thermal Equilibrium

To understand the concept of a population inversion, it is necessary to understand some thermodynamics and the way that light interacts with matter. To do so, it is useful to consider a very simple assembly of atoms forming a laser medium.

Assume there are a group of N atoms, each of which is capable of being in one of two energy states, either

1. The *ground state*, with energy E_1; or
2. The *excited state*, with energy E_2, with $E_2 > E_1$.

The number of these atoms which are in the ground state is given by N_1, and the number in the excited state N_2. Since there are N atoms in total,

$$N_1 + N_2 = N$$

The energy difference between the two states, given by

$$\Delta E_{12} = E_2 - E_1,$$

determines the characteristic frequency v_{12} of light which will interact with the atoms; This is given by the relation

$$E_2 - E_1 = \Delta E = h \vee_{12},$$

h being Planck's constant.

Colour Reproduction in Electronic Imaging Systems: Photography, Television, Cinematography, First Edition. Michael S Tooms.
© 2016 John Wiley & Sons, Ltd. Published 2016 by John Wiley & Sons, Ltd.
Companion Website: www.wiley.com/go/toomscolour

If the group of atoms is in thermal equilibrium, it can be shown from thermodynamics that the ratio of the number of atoms in each state is given by a Boltzmann distribution:

$$\frac{N_2}{N_1} = e^{-(E_2 - E_1)/kT}$$

where T is the thermodynamic temperature of the group of atoms, and k is Boltzmann's constant.

We may calculate the ratio of the populations of the two states at room temperature ($T \approx 300\ K$) for an energy difference ΔE that corresponds to light of a frequency corresponding to visible light ($f \approx 5 \times 10^{14}$ Hz). In this case $\Delta E = E_2 - E_1 \approx 2.07$ eV, and $kT \approx 0.026$ eV. Since $E_2 - E_1 \gg kT$, it follows that the argument of the exponential in the equation above is a large negative number, and as such N_2/N_1 is vanishingly small; that is, there are almost no atoms in the excited state. When in thermal equilibrium, then it is seen that the lower energy state is more populated than the higher energy state, and this is the normal state of the system. As T increases, the number of electrons in the high-energy state (N_2) increases, but N_2 never exceeds N_1 for a system at thermal equilibrium; rather, at infinite temperature, the populations N_2 and N_1 become equal. In other words, a population inversion ($N_2/N_1 > 1$) can never exist for a system at thermal equilibrium. To achieve population inversion therefore requires pushing the system into a non-equilibrated state.

E.2   The Interaction of Light with Matter

There are three types of possible interactions between a system of atoms and light that are of interest:

E.2.1   Absorption

If light photons of frequency f_{12} pass through the group of atoms, there is a possibility of the light being absorbed by atoms which are in the ground state, which will cause them to be excited to the higher energy state. The probability of absorption is proportional to the radiation intensity of the light, and also to the number of atoms currently in the ground state, N_1.

E.2.2   Spontaneous Emission

If a collection of atoms are in the excited state, spontaneous decay events to the ground state will occur at a rate proportional to N_2, the number of atoms in the excited state. The energy difference between the two states ΔE_{21} is emitted from the atom as a photon of frequency f_{21} as given by the frequency–energy relation above.

The photons are emitted stochastically, and there is no fixed phase relationship between photons emitted from a group of excited atoms; in other words, spontaneous emission is incoherent. In the absence of other processes, the number of atoms in the excited state at time t, is given by

$$N_2(t) = N_2(0)\, e^{\frac{-t}{\tau_{21}}}$$

where $N_2(0)$ is the number of excited atoms at time $t = 0$, and τ_{21} is the lifetime of the transition between the two states.

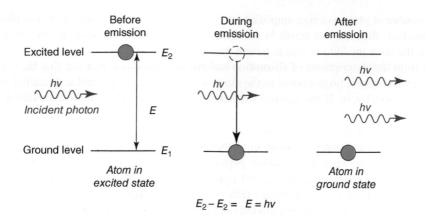

Figure E.1 Illustrating stimulated emission.

E.2.3 Stimulated Emission

If an atom is already in the excited state, it may be perturbed by the passage of a photon that has a frequency f_{21} corresponding to the energy gap ΔE of the excited state to ground state transition. In this case, the excited atom relaxes to the ground state, and is induced to produce a second photon of frequency f_{21}. The original photon is not absorbed by the atom, and so the result is two photons of the same frequency. This process is known as *stimulated emission* and is portrayed in Figure E.1.

Specifically, an excited atom will act like a small electric dipole which will oscillate with the external field provided. One of the consequences of this oscillation is that it encourages electrons to decay to the lowest energy state. When this happens due to the presence of the electromagnetic field from a photon, a photon is released in the same phase and direction as the 'stimulating' photon, and is called stimulated emission.

The rate at which stimulated emission occurs is proportional to the number of atoms N_2 in the excited state, and the radiation density of the light. The base probability of a photon causing stimulated emission in a single excited atom was shown by Albert Einstein to be exactly equal to the probability of a photon being absorbed by an atom in the ground state. Therefore, when the numbers of atoms in the ground and excited states are equal, the rate of stimulated emission is equal to the rate of absorption for a given radiation density.

The critical detail of stimulated emission is that the induced photon has the same frequency and phase as the incident photon. In other words, the two photons are coherent. It is this property that allows optical amplification, and the production of a laser system. During the operation of a laser, all three light-matter interactions described above are taking place. Initially, atoms are energized from the ground state to the excited state by a process called pumping, described below. Some of these atoms decay via spontaneous emission, releasing incoherent light as photons of frequency, v. These photons are fed back into the laser medium, usually by an optical resonator. Some of these photons are absorbed by the atoms in the ground state and the photons are lost to the laser process. However, some photons cause stimulated emission in excited-state atoms, releasing another coherent photon. In effect, this results in optical amplification.

If the number of photons being amplified per unit time is greater than the number of photons being absorbed, then the net result is a continuously increasing number of photons being produced; the laser medium is said to have a gain of greater than unity.

Recall from the descriptions of absorption and stimulated emission above that the rates of these two processes are proportional to the number of atoms in the ground and excited states, N_1 and N_2, respectively. If the ground state has a higher population than the excited state $(N_1 > N_2)$, the process of absorption dominates and there is a net attenuation of photons. If the populations of the two states are the same $(N_1 = N_2)$, the rate of absorption of light exactly balances the rate of emission; the medium is then said to be optically transparent.

If the higher energy state has a greater population than the lower energy state $(N_1 < N_2)$, then the emission process dominates, and light in the system undergoes a net increase in intensity. It is thus clear that to produce a faster rate of stimulated emissions than absorptions, it is required that the ratio of the populations of the two states is such that $N_2/N_1 > 1$. In other words, a population inversion is required for laser operation.

E.3   Selection Rules

Many transitions involving electromagnetic radiation are strictly forbidden under quantum mechanics. The allowed transitions are described by so-called selection rules, which describe the conditions under which a radiative transition is allowed. For instance, transitions are only allowed if $\Delta S = 0$, S being the total spin angular momentum of the system. In real materials other effects, such as interactions with the crystal lattice, intervene to circumvent the formal rules. In these systems the forbidden transitions can occur, but usually at slower rates than allowed transitions. A classic example is phosphorescence where a material has a ground state with $S = 0$, an excited state with $S = 0$, and an intermediate state with $S = 1$. The transition from the intermediate state to the ground state by emission of light is slow because of the selection rules. Thus emission may continue after the external illumination is removed. In contrast fluorescence in materials is characterized by emission which ceases when the external illumination is removed.

Transitions which do not involve the absorption or emission of radiation are not affected by selection rules. Radiationless transition between levels, such as between the excited $S = 0$ and $S = 1$ states, may proceed quickly enough to siphon off a portion of the $S = 0$ population before it spontaneously returns to the ground state.

The existence of intermediate states in materials, as we will see, is essential to the technique of optical pumping of lasers.

E.4   Creating a Population Inversion

As described above, a population inversion is required for laser operation, but cannot be achieved in our theoretical group of atoms with two energy levels when they are in thermal equilibrium. In fact, any method by which the atoms are directly and continuously excited from the ground state to the excited state (such as optical absorption) will eventually reach equilibrium with the de-exciting processes of spontaneous and stimulated emission. At best, an equal population of the two states, $N_1 = N_2 = N/2$, can be achieved, resulting in optical transparency but no net optical gain.

E.5 Three-Level Lasers

To achieve non-equilibrium conditions, an indirect method of populating the excited state must be used. To understand how this is done, we may use a slightly more realistic model, that of a *three-level laser*. Again consider a group of N atoms, this time with each atom able to exist in any of three energy states, levels 1, 2 and 3, with energies E_1, E_2 and E_3, and populations N_1, N_2 and N_3, respectively.

Note that $E_1 < E_2 < E_3$; that is, the energy of level 2 lies between that of the ground state and level 3.

Initially, the system of atoms is at thermal equilibrium, and the majority of the atoms will be in the ground state, that is, $N_1 \approx N$, $N_2 \approx N_3 \approx 0$. If we now subject the atoms to light of a frequency, $f_{13} = \frac{1}{h}(E_3 - E_1)$ the process of optical absorption will excite the atoms from the ground state to level 3. This process is called *pumping* and does not necessarily always directly involve light absorption; other methods of exciting the laser medium, such as electrical discharge or chemical reactions, may be used. The level 3 is sometimes referred to as the *pump level* or *pump band*, and the energy transition $E_1 \rightarrow E_3$ as the *pump transition*, which is shown as the arrow marked **P** in Figure E.2.

If we continue pumping the atoms, we will excite an appreciable number of them into level 3, such that $N_3 > 0$. In a medium suitable for laser operation, we require these excited atoms to quickly decay to level 2. The energy released in this transition may be emitted as a photon (spontaneous emission); however, in practice the 3→2 transition (labeled **R** in the diagram) is usually *radiationless*, with the energy being transferred to vibrational motion (heat) of the host material surrounding the atoms, without the generation of a photon.

An atom in level 2 may decay by spontaneous emission to the ground state, releasing a photon of frequency f_{12} (given by $E_2 - E_1 = hf_{12}$), which is shown as the transition **L**, called the *laser transition* in the diagram. If the lifetime of this transition, τ_{21} is much longer than the lifetime of the radiationless $3 \rightarrow 2$ transition τ_{32} (if $\tau_{21} \gg \tau_{32}$, known as a *favourable lifetime ratio*), the population of the E_3 will be essentially zero ($N_3 \approx 0$) and a population of excited state atoms will accumulate in level 2 ($N_2 > 0$). If over half the N atoms can be accumulated in this state, this will exceed the population of the ground state N_1. A population inversion ($N_2 > N_1$) has thus been achieved between level 1 and 2, and optical amplification at the frequency f_{21} can be obtained.

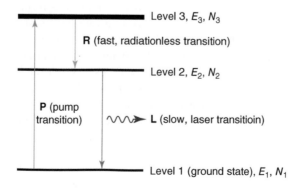

Figure E.2 A three-level laser energy diagram.

Because at least half the population of atoms must be excited from the ground state to obtain a population inversion, the laser medium must be very strongly pumped. This makes three-level lasers rather inefficient, despite being the first type of laser to be discovered (based on a ruby laser medium, by Theodore Maiman in 1960). A three-level system could also have a radiative transition between levels 3 and 2, and a non-radiative transition between levels 2 and 1. In this case, the pumping requirements are weaker. In practice, most lasers are *four-level lasers*, as described in the next section.

E.6 Four-Level Lasers

As illustrated in Figure E.3 there are four energy levels, energies E_1, E_2, E_3, E_4, and populations N_1, N_2, N_3, N_4, respectively. The energies of each level are such that $E_1 < E_2 < E_3 < E_4$.

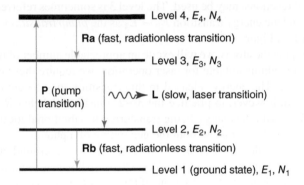

Figure E.3 A four-level laser energy diagram.

In this system, the pumping transition **P** excites the atoms in the ground state (level 1) into the pump band (level 4). From level 4, the atoms again decay by a fast, non-radiative transition **Ra** into the level 3. Since the lifetime of the laser transition **L** is long compared to that of **Ra** ($\tau_{32} \gg \tau_{43}$), a population accumulates in level 3 (the *upper laser level*), which may relax by spontaneous or stimulated emission into level 2 (the *lower laser level*). This level likewise has a fast, non-radiative decay **Rb** into the ground state.

As before, the presence of a fast, radiationless decay transitions results in the population of the pump band being quickly depleted ($N_4 \approx 0$). In a four-level system, any atom in the lower laser level E_2 is also quickly de-excited, leading to a negligible population in that state ($N_2 \approx 0$). This is important, since any appreciable population accumulating in level 3, the upper laser level, will form a population inversion with respect to level 2. That is, as long as $N_3 > 0$, then $N_3 > N_2$ and a population inversion is achieved. Thus optical amplification, and laser operation, can take place at a frequency of f_{32} ($E_3 - E_2 = hf_{32}$).

Since only a few atoms must be excited into the upper laser level to form a population inversion, a four-level laser is much more efficient than a three-level one, and most practical lasers are of this type. In reality, many more than four energy levels may be involved in the laser process, with complex excitation and relaxation processes involved between these levels. In particular, the pump band may consist of several distinct energy levels, or a continuum of levels, which allow optical pumping of the medium over a wide range of wavelengths.

Note that in both three- and four-level lasers, the energy of the pumping transition is greater than that of the laser transition. This means that, if the laser is optically pumped, the frequency of the pumping light must be greater than that of the resulting laser light. In other words, the pump wavelength is shorter than the laser wavelength. It is possible in some media to use multiple photon absorptions between multiple lower-energy transitions to reach the pump level; such lasers are called *up-conversion* lasers.

While in many lasers the laser process involves the transition of atoms between different electronic energy states, as described in the model above, this is not the only mechanism that can result in laser action. For example, there are many common lasers (e.g., dye lasers, carbon dioxide lasers) where the laser medium consists of complete molecules, and energy states correspond to vibrational and rotational modes of oscillation of the molecules. This is the case with water masers that occur in nature.

In some media it is possible, by imposing an additional optical or microwave field, to use quantum coherence effects to reduce the likelihood of an excited-state to ground-state transition. This technique, known as lasing without inversion, allows optical amplification to take place without producing a population inversion between the two states.

F

Deriving Camera Spectral Sensitivities

F.1 General Solution for Deriving the Camera Spectral Sensitivities from the Chromaticity Coordinates of the Display Primaries in Terms of the CIE Colour Matching Functions

In general terms the approach is as follows.

The chromaticity coordinates of the RGB primaries and the system white in the CIE x,y system are known but the tristimulus values are not since they have been lost in the normalisation process; therefore to express the RGB primaries in terms of the XYZ primaries three constants representing those used in the normalisation process need to be applied. The equations may then be expressed as follows:

$$
\begin{aligned}
1.0(\mathrm{R}) &= k_1(x_\mathrm{r}(\mathrm{X}) + y_\mathrm{r}(\mathrm{Y}) + z_\mathrm{r}(\mathrm{Z})) \\
1.0(\mathrm{G}) &= k_2(x_\mathrm{g}(\mathrm{X}) + y_\mathrm{g}(\mathrm{Y}) + z_\mathrm{g}(\mathrm{Z})) \\
1.0(\mathrm{B}) &= k_3(x_\mathrm{b}(\mathrm{X}) + y_\mathrm{b}(\mathrm{Y}) + z_\mathrm{b}(\mathrm{Z})) \\
\mathrm{W_s} &= x_\mathrm{w}(\mathrm{X}) + y_\mathrm{w}(\mathrm{Y}) + z_\mathrm{w}(\mathrm{Z})
\end{aligned}
\tag{F.1}
$$

where x, y, z are the chromaticity coordinates of the R, G, B, primaries and system white point, respectively, R,G,B and X,Y,Z, are unit amounts of the six primaries and $\mathrm{W_s}$ is the system white point.

Since the drive of the display device is always adjusted so that an equal peak signal on all drives produces the system white $\mathrm{W_s}$, we can define a further relationship:

$$
\mathrm{W_s} = 1.0(\mathrm{R}) + 1.0(\mathrm{G}) + 1.0(\mathrm{B})
$$

but the system white is also defined in terms of its chromaticity coordinates, therefore

$$
1.0(\mathrm{R}) + 1.0(\mathrm{G}) + 1.0(\mathrm{B}) = \mathrm{W_s} = x_\mathrm{w}\mathrm{X} + y_\mathrm{w}\mathrm{Y} + z_\mathrm{w}\mathrm{Z}
\tag{F.2}
$$

Colour Reproduction in Electronic Imaging Systems: Photography, Television, Cinematography, First Edition. Michael S Tooms.
© 2016 John Wiley & Sons, Ltd. Published 2016 by John Wiley & Sons, Ltd.
Companion Website: www.wiley.com/go/toomscolour

Substituting in equation (F.2) for RGB, from equation (F.1) enables a set of three equations to be established which, using the techniques detailed in Appendix 2, may be solved simultaneously with a second set of a basically similar nature to produce a generic set of equations of the type:

$$R = k_2 k_3 [(y_g z_b - z_g y_b)X + (z_g x_b - x_g z_b)Y + (x_g y_b - y_g x_b)Z]$$
$$G = k_1 k_3 [(y_b z_r - z_b y_r)X + (z_b x_r - x_b z_r)Y + (x_b y_r - y_b x_r)Z] \qquad (F.3)$$
$$B = k_1 k_2 [(y_r z_g - z_r y_g)X + (z_r x_g - x_r z_g)Y + (x_r y_g - y_r x_g)Z]$$

and where

$$k_1 = (y_g z_b - z_g y_b)x_w + (z_g x_b - x_g z_b)y_w + (x_g y_b - y_g x_b)z_w$$
$$k_2 = (y_b z_r - z_b y_r)x_w + (z_b x_r - x_b z_r)y_w + (x_b y_r - y_b x_r)z_w \qquad (F.4)$$
$$k_3 = (y_r z_g - z_r y_g)x_w + (z_r x_g - x_r z_g)y_w + (x_r y_g - y_r x_g)z_w$$

These equations are used in Worksheet 9 to derive the spectral sensitivities of a camera matched to the chromaticity coordinates of a set of three primary colours.

G

Chromaticity Gamut Transformation

G.1 Introduction

Chromaticity gamut transformation is the process by which RGB signals derived from a 'source' camera designed to feed a display with a defined set of primaries, are processed to appear as if they were derived from a 'target' camera designed to feed a display with a different defined set of primaries. Thus RGB signals derived for one display gamut may be converted to a different set of RGB signals for a display with a different chromaticity gamut.

The general approach to gamut transformation is to establish the XYZ values of the scene from the RGB values of the source camera and then use these scene XYZ values with the RGB matrix of the target camera to establish the RGB values of the target camera.

These relationships are expressed in terms of matrices of the type developed in Appendix F and illustrated in Worksheet 9.

Thus the operations necessary are to establish the XYZ tristimulus values of a colour in the scene from a knowledge of the RGB tristimulus values produced by the source camera using the inverse of the camera matrix and then apply the matrix of the target camera to these XYZ values of the scene derived from the source camera to obtain the RGB values as if derived from the target camera.

G.2 Procedure

In the event that the source and target systems use the same system white point, this becomes a simple matrix operation where the inverse of the source camera matrix is multiplied by the target camera matrix as follows:

$$\text{If the source camera is represented by } \begin{bmatrix} R_1 \\ G_1 \\ B_1 \end{bmatrix} = M_1 \begin{bmatrix} X \\ Y \\ Z \end{bmatrix}$$

$$\text{and the target camera by } \begin{bmatrix} R_2 \\ G_2 \\ B_2 \end{bmatrix} = M_2 \begin{bmatrix} X \\ Y \\ Z \end{bmatrix}$$

Colour Reproduction in Electronic Imaging Systems: Photography, Television, Cinematography, First Edition. Michael S Tooms.
© 2016 John Wiley & Sons, Ltd. Published 2016 by John Wiley & Sons, Ltd.
Companion Website: www.wiley.com/go/toomscolour

Then the inverse of M_1 is given by $\begin{bmatrix} X \\ Y \\ Z \end{bmatrix} = M_1^{-1} \begin{bmatrix} R_1 \\ G_1 \\ B_1 \end{bmatrix}$ and multiplying the inverse of the

source matrix by the target matrix leads to:

$$\begin{bmatrix} R_2 \\ G_2 \\ B_2 \end{bmatrix} = M_2 M_1^{-1} \begin{bmatrix} R_1 \\ G_1 \\ B_1 \end{bmatrix}$$

In Worksheet 12(a), M_1 and M_2 above are represented by matrix 1 and matrix 6, respectively and the inverse of M_1 by matrix 11. As a means of checking the veracity of the approach, if matrix 6 is multiplied by matrix 11 and the resulting matrix used to calculate the RGB values from camera1 the results obtained relate precisely with the RGB values obtained by convolution of the target camera 2 spectral sensitivities and the chosen sample surface colour.[1]

However, if this simple relationship is used when the system white point of the two systems are different the result is better than no correction but does produce significant errors. It is therefore necessary to include the appropriate system white chromaticities in the matrix calculations.

These operations may be detailed as follows:

1. Apply the inverse of the matrix which describes the source camera spectral sensitivities to the RGB values to obtain the XYZ values under the camera 1 system white illumination
2. Apply the correction for the source system illumination to obtain the XYZ values that would have been obtained under an equal energy illuminant
3. Apply the correction for the target system illumination to obtain what the XYZ values would be under the target system white illumination
4. Apply the matrix which describes the camera spectral sensitivity of the target camera to obtain the RGB values from the target camera
5. Apply a colour balance factor to ensure that the reference 100% signal is achieved on white in the scene from the emulated Camera 2.

In mathematical terms these operations may be defined as follows:

First, to obtain the scene XYZ values from the RGB tristimulus values of the source camera it is necessary to apply a correction for the source illumination to the inverse of the source camera matrix.

The RGB values representing the source illumination of a neutral in the scene at the output of camera 1 are described by its scene XYZ chromaticity coordinates:

$$\begin{bmatrix} R_{1i} \\ G_{1i} \\ B_{1i} \end{bmatrix} = M_3 \begin{bmatrix} X \\ Y \\ Z \end{bmatrix} \text{ and its inverse is } \begin{bmatrix} X \\ Y \\ Z \end{bmatrix} = M_3^{-1} \begin{bmatrix} R_{1i} \\ G_{1i} \\ B_{1i} \end{bmatrix}$$

---

[1] The 'Guide to the Colour Reproduction Worksheet' provides detailed guidance on using the worksheets.

Thus the scene 1 colours as they would be illuminated by an equal energy illuminant will have scene tristimulus values in terms of the output of camera 1 which may be given as follows:

$$\begin{bmatrix} X \\ Y \\ Z \end{bmatrix} = M_1^{-1} \begin{bmatrix} R_1 \\ G_1 \\ B_1 \end{bmatrix} M_3^{-1} \begin{bmatrix} R_{1i} \\ G_{1i} \\ B_{1i} \end{bmatrix}$$

(G.1)

The camera 2 relationship when capturing a scene illuminated by illuminant 2 will be as follows:

$$\begin{bmatrix} R_2 \\ G_2 \\ B_2 \end{bmatrix} = M_2 M_4 \begin{bmatrix} X \\ Y \\ Z \end{bmatrix}$$

(G.2)

where M4 is the camera 2 illuminant matrix,

$$\begin{bmatrix} R_{2i} \\ G_{2i} \\ B_{2i} \end{bmatrix} = M_4 \begin{bmatrix} X \\ Y \\ Z \end{bmatrix}$$

Thus if camera 2 were to capture scene 1 the result can be found by replacing XYZ in equation (G.2) by equation (G.1) which describes the XYZ values of the surfaces in scene 1.

$$\begin{bmatrix} R_2 \\ G_2 \\ B_2 \end{bmatrix} = M_2 M_4 M_1^{-1} \begin{bmatrix} R_1 \\ G_1 \\ B_1 \end{bmatrix} M_3^{-1} \begin{bmatrix} R_{1i} \\ G_{1i} \\ B_{1i} \end{bmatrix}$$

(G.3)

The reader is reminded that care must be taken when multiplying matrices, it does not follow that $M_1 \times M_2$ will give the same result as $M_2 \times M_1$.

When the chromaticity coordinates of the system primaries and white points are known then equation (G.3) can be concatenated into a single matrix, as is done in Worksheet 12(a).

In Figure G.1 the arrangement of matrices which reflects the formula above is illustrated. The dotted lines are intended to show the manner in which the output of one matrix is used by another. The constants in matrix 22 are derived from the reciprocal of the sum of the central row of constants in matrix 21 in order to ensure the values of each of the rows in matrix 23 sum to a value of 1.0.

The numbers in the bottom left hand corner of each matrix refers to the number given to the matrices in Worksheet 12(a), where the matrix layout broadly replicates that illustrated in the figure.

In Worksheet 12(a), gamut characteristics for the two camera systems may be copied from the 'Primaries' worksheet and pasted into the appropriate coloured cell ranges in the worksheet. (This operation is automated for some primaries by selection of 'control buttons' at the top left of the worksheet.) The worksheet calculates the various matrix operations illustrated in Figure G.1 and concatenates the result into matrix 23 which for convenience is copied to a position immediately beneath the cell blocks containing the primaries chromaticity coordinates at the top left of the sheet.

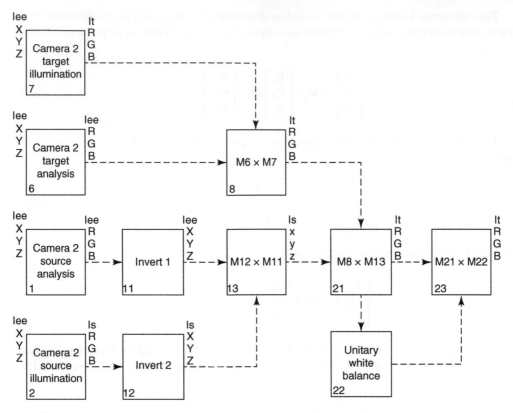

Figure G.1   Gamut transformation: matrices arrangement.

The 'CIE', 'Illuminants' and 'Surfaces' worksheets are used to provide the underlying data to calculate the RGB tristimulus values from the spectral sensitivities of the two cameras. These two sets of data are compared with the values derived from the various matrices to check the veracity of the matrices formulae.

When the gamuts of the two cameras share the same illuminant, then the arrangement shown in Figure G.1 produces tristimulus levels which match precisely with those obtained from the convolution and integration of the illuminant SPD and the camera spectral sensitivities. When the two system white points are different the results of the comparison are completely accurate on neutral colours but deviate slightly on coloured surfaces with increasing deviation with increasing saturation and diminishing luminance of the colour sample. Nevertheless, even on saturated colours the errors are small to negligible on the largest of the three RGB values and small to insignificant on the other two values.

For the example given in Chapter 12, based upon converting sRGB values to Adobe RGB values, the worksheet produces the following matrix values (The 'a' subscript refers to Adobe RGB and the 's' subscript to sRGB.):

$$
\begin{bmatrix} R_a \\ G_a \\ B_a \end{bmatrix} = \begin{bmatrix} 0.7152 & 0.2849 & -0.0001 \\ 0.0000 & 1.0000 & 0.0000 \\ 0.0000 & 0.0142 & 0.9588 \end{bmatrix} \begin{bmatrix} R_s \\ G_s \\ B_s \end{bmatrix}
$$

This matrix produced a match of values to a range of sample colour values calculated using convolution and integration of the spectral sensitivities, the light source and the colour samples.

When the worksheet was used to produce a matrix for converting the RGB values from an NTSC television camera to those from a Rec 709 camera, which has a different system white, the matrix produced matching results on neutral colours. On the Gretag chart also a very good match on flesh colours and non-highly saturated colours. On the saturated primary and complementary primaries chips the maximum error in the matches was 1%.

H

Deriving the Standard Formula for Gamma Correction

H.1 General

The curve representing the inverse of a CRT transfer characteristic is a power law function given by $V = L^{1/\gamma}$ where L is the representation of the image luminance of the scene and γ is the exponent of the CRT transfer function, approximately equal to 2.5 and V is the output voltage to drive the display. For convenience we will express the exponent of the gamma corrector, $1/\gamma$, as ε equal to 0.40.

$$\text{Thus} \quad V = L^{\varepsilon} \tag{H.1}$$

The plot of this curve is illustrated in the following figure:

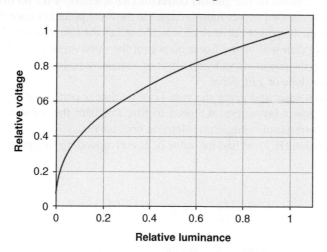

Figure H.1 Gamma correction characteristic.

An inspection of the curve indicates that the slope, which represents the gain, increases as the level of the input luminance diminishes, approaching an infinite gain at very low levels.

Colour Reproduction in Electronic Imaging Systems: Photography, Television, Cinematography, First Edition. Michael S Tooms.
© 2016 John Wiley & Sons, Ltd. Published 2016 by John Wiley & Sons, Ltd.
Companion Website: www.wiley.com/go/toomscolour

The gain at any input luminance level is obtained by differentiating the expression for V from (H.1) above:

$$G = \frac{\delta V}{\delta L} = \varepsilon L^{\varepsilon-1} = \frac{\varepsilon}{L^{1-\varepsilon}} \tag{H.2}$$

At a luminance level of 1% the gain from the above formula is 6.34.

Practical gamma correction for the CRT characteristic is a compromise because of the source noise which would be amplified to very high levels when the signal strength is low and the increasing amounts of gain required at diminishingly lower signal levels.

The usual compromise is to agree the parameters of a gamma correction characteristic which is a combination of two curves: an idealised power law curve, broadly representing the inverse of the CRT power law, and a fixed straight line maximum gain curve that will be applied below a certain breakpoint signal level. As will be seen below, once these two parameters, the exponent of the power law and the gain of the linear section, are fixed, if a smooth overall characteristic is to be achieved with no dislocations, it then follows that the gain of the power law portion of the characteristic and the breakpoint between the linear portion of the curve and the power law characteristic also have fixed values. The higher is the gain selected for the linear portion of the corrector; the lower is the breakpoint between the two curves.

H.2 Establishing the Gamma Correction Parameters for the General Situation

To match the two elements of the gamma corrector characteristic with no dislocations, the slopes or the gains of the two curves must be equal at the breakpoint between the two curves. However, the point on the power curve slope which matches the linear slope will always be at a higher output level than will be the linear output for the same input level. Therefore it will be necessary to drop the power curve by an offset or a negative pedestal p to bring the curves together at the same slope or gain value.

Taking the general case, and given the gain G of the linear portion of the curve and the exponent ε of the power law curve, we need to first calculate the value of L_B, the input luminance, which corresponds to the same slope on both curves.

Transposing equation (H.2), to find the value of L, corresponding to a particular gain G on the power curve:

$$L^{1-\varepsilon} = \frac{\varepsilon}{G} \qquad \text{and} \qquad L = \left(\frac{\varepsilon}{G}\right)^{\frac{1}{1-\varepsilon}} \tag{H.3}$$

Let L_B and G_B be the luminance level and the gain, respectively at the breakpoint between the two curves. The gain as defined above is the gain of the linear expression G_B.

$$\text{Then } L_B = \left(\frac{\varepsilon}{G_B}\right)^{\frac{1}{1-\varepsilon}} \tag{H.4}$$

We are now in a position to calculate the offset value.

Substituting for L_B from (H.4), the electrical signal output, V_l at the breakpoint from the linear part of the characteristic will be:

$$V_l = G_B \times L_B = G_B \left(\frac{\varepsilon}{G_B} \right)^{\frac{1}{1-\varepsilon}} \tag{H.5}$$

The output from the power curve part of the characteristic will be:

$$V_p = (L_B)^{\varepsilon} = \left(\frac{\varepsilon}{G_B} \right)^{\frac{\varepsilon}{1-\varepsilon}} \tag{H.6}$$

Thus the offset required of the power curve to bring it down to match the slope of the linear curve is the difference of these two values:

$$p = V_p - V_l = \left(\frac{\varepsilon}{G_B} \right)^{\frac{\varepsilon}{1-\varepsilon}} - G_B \left(\frac{\varepsilon}{G_B} \right)^{\frac{1}{1-\varepsilon}}$$

Having created effectively a negative pedestal to be applied to the power section of the characteristic, then its output corresponding to a maximum input of 1.00 will no longer be 1.00, but will be:

$$V_x = 1 - (V_p - V_l) \tag{H.7}$$

Thus it will be necessary to apply a gain to the power characteristic to ensure that for an input value of 1 there is a corresponding output level of 1.

Now, as we have seen, the power section of the characteristic extends from a level of V_l to V_x and thus has an amplitude of $V_x - V_l$. In order to ensure that the curve fills the space between V_l and 1.00, that is, an amplitude of $1 - V_l$, it will be necessary to provide amplification.

Thus the amplification or gain m required is:

$m = \frac{1-V_l}{V_x - V_l}$ and substituting for V_x from (H.7) above and simplifying:

$m = \frac{1-V_l}{1-V_p}$ and substituting for V_l and V_p from (H.5) and (H.6) above:

$$m = \frac{1 - G_B \left(\frac{\varepsilon}{G_B} \right)^{\frac{1}{1-\varepsilon}}}{1 - \left(\frac{\varepsilon}{G_B} \right)^{\frac{\varepsilon}{1-\varepsilon}}} \tag{H.8}$$

Now in reality the negative pedestal or offset p is applied *after* the signal has been amplified and since the general expression required for the correcting gain to obtain unity, when a negative pedestal or offset of p is applied, is of the form $V = (1 + p)L - p$ and since $m = 1 + p$ then:

$V = mL - p$ and since from above $p = m - 1$,

then $V = mL - m + 1$

So building the gain m and offset $m - 1$ into the simple power law element of the characteristic we arrive at the general equation:

$$V = mL^{\gamma} - m + 1 \tag{H.9}$$

Finally, noting that the slope of the power curve for a particular input L has been modified by the factor m, we need to recalculate the breakpoint between the power and linear curves comprising the gamma correction curve.

The criterion remains the same, that is, the slope of the curves must match at the breakpoint. Thus we need to establish the value of L at which the slope or gain equates to that of the linear curve.

Differentiating (H.9) to find the gain G for any input value:

$\frac{\delta V}{\delta L} = G = m\gamma L^{\gamma-1}$ and transposing to find L, $L = (G_B/m\varepsilon)^{1/\varepsilon-1}$

Thus the final breakpoint value of luminance L_b for a linear gain of G_B is given by:

$$L_b = \left(\frac{G_B}{m\varepsilon}\right)^{\frac{1}{\varepsilon-1}} = \left(\frac{m\varepsilon}{G_B}\right)^{\frac{1}{1-\varepsilon}} \tag{H.10}$$

H.3 Calculating the Gamma Correction Parameters for a Particular Situation

The approach is to commence with the two independent variables, the exponent ε of the power element of the characteristic and the gain G of the linear element of the characteristic.

These two parameters are all that is necessary to calculate the two dependent variables, the gain m of the power equation and the level of the input luminance L_b at the crossover point. The offset p is simply related to the gain m.

As an example let us assume for a particular gamma corrector an exponent of the power equation element of the characteristic $\varepsilon = 0.41667$ and the gain of the linear element of the characteristic $G_B = 12.92$. These are in fact the independent variables of the parameters of the sRGB specification used in photography.

First we need to calculate m the gain of the power element of the characteristic using the equation derived in (H.8) and substituting for G_B and ε:

$$m = \frac{1 - G_B\left(\frac{\varepsilon}{G_B}\right)^{\frac{1}{1-\varepsilon}}}{1 - \left(\frac{\varepsilon}{G_B}\right)^{\frac{\varepsilon}{1-\varepsilon}}} = \frac{1 - 12.92\left(\frac{0.41667}{12.92}\right)^{\frac{1}{1-0.41667}}}{1 - \left(\frac{0.41667}{12.92}\right)^{\frac{0.41667}{1-0.41667}}} = \frac{0.9642}{0.9140} = 1.0549$$

and the offset p is given by $p = m - 1 = 1.0549 - 1 = 0.0549$.

We are now in a position to calculate L_b the crossover point. From (H.10):

$$L_b = \left(\frac{m\varepsilon}{G_B}\right)^{\frac{1}{1-\varepsilon}} = \left(\frac{1.0549 * 0.41667}{12.92}\right)^{\frac{1}{1-0.41667}} = 0.00304 \text{ or } 0.304\%$$

H.4 Specifying the Opto-Digital Transfer Characteristic of a Colour Reproduction System

The five parameters required to specify the opto-digital transfer characteristic of a colour reproduction system are usually specified in the following manner, where the results obtained above are used as an example.

For the sRGB colour reproduction system to three significant figures:

$$V = 1.055L^{0.4167} - 0.055 \quad \text{for} \quad 1 \geq L \geq 0.00304$$
$$V = 12.920L \qquad\qquad\qquad \text{for} \quad 0.00304 > L \geq 0$$

where L is the luminance of the image $0 \leq L \leq 1$ and V is the corresponding electrical signal.

The first line of the specification indicates that for a luminance signal equal to or above the 0.304% level and not greater than the 100% level, the power law equation should be used. The second line indicates that for a luminance signal level below the 0.304% level and equal to or above the zero level, the linear gain equation should be used.

H.5 Practical Calculations

The procedure outlined above to establish the parameters of the transfer characteristic is somewhat tedious and therefore the equations have been built into the gamma correction Worksheet 13(b). The worksheet enables one to enter the two independent parameters of any opto-digital characteristic and the resulting dependent variables are calculated and displayed in the associated graphs.

For convenience the graphs for the sRGB system are copied from Worksheet 13(b) into Figures H.2 and H.3 below. Figure H.2 illustrates the characteristic for the full range of input luminance.

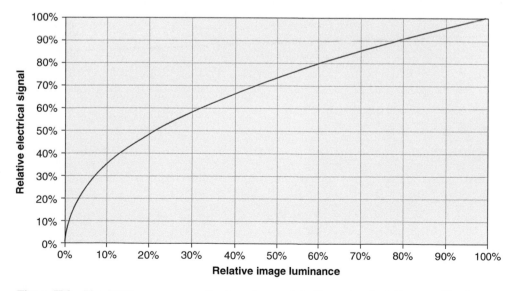

Figure H.2 The sRGB gamma correction law characteristic illustrating the full range of luminance.

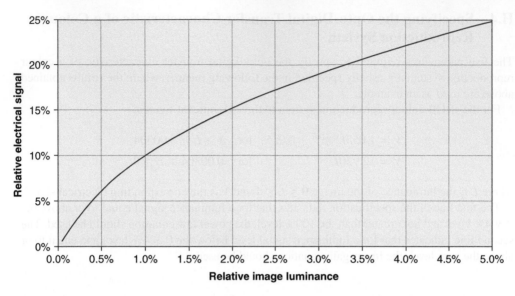

Figure H.3   The sRGB gamma correction characteristic illustrating the crossover range of luminance.

Figure H.3 illustrates the characteristic for the luminance range 0–5% in order to illustrate the smooth crossover from one equation to the other at the break point luminance level of 0.3%.

I

CIE Colour Matching Functions

I.1   Values for a 2 Degree Field

nm	$x(\lambda)$	$y(\lambda)$	$z(\lambda)$
380	0.0014	0.0000	0.0065
385	0.0022	0.0001	0.0105
390	0.0042	0.0001	0.0201
395	0.0076	0.0002	0.0362
400	0.0143	0.0004	0.0679
405	0.0232	0.0006	0.1102
410	0.0435	0.0012	0.2074
415	0.0776	0.0022	0.3713
420	0.1344	0.0040	0.6456
425	0.2148	0.0073	1.0391
430	0.2839	0.0116	1.3856
435	0.3285	0.0168	1.6230
440	0.3483	0.0230	1.7471
445	0.3481	0.0298	1.7825
450	0.3362	0.0380	1.7721
455	0.3187	0.0480	1.7441
460	0.2908	0.0600	1.6692
465	0.2511	0.0739	1.5281
470	0.1954	0.0910	1.2876
475	0.1421	0.1126	1.0419
480	0.0956	0.1390	0.8130
485	0.0580	0.1693	0.6162
490	0.0320	0.2080	0.4652
495	0.0147	0.2586	0.3533
500	0.0049	0.3230	0.2720
505	0.0024	0.4073	0.2123

(*continued*)

Colour Reproduction in Electronic Imaging Systems: Photography, Television, Cinematography, First Edition. Michael S Tooms.
© 2016 John Wiley & Sons, Ltd. Published 2016 by John Wiley & Sons, Ltd.
Companion Website: www.wiley.com/go/toomscolour

nm	$x(\lambda)$	$y(\lambda)$	$z(\lambda)$
510	0.0093	0.5030	0.1582
515	0.0291	0.6082	0.1117
520	0.0633	0.7100	0.0782
525	0.1096	0.7932	0.0573
530	0.1655	0.8620	0.0422
535	0.2257	0.9149	0.0298
540	0.2904	0.9540	0.0203
545	0.3597	0.9803	0.0134
550	0.4333	0.9950	0.0087
555	0.5121	1.0000	0.0057
560	0.5945	0.9950	0.0039
565	0.6784	0.9786	0.0027
570	0.7621	0.9520	0.0021
575	0.8425	0.9154	0.0018
580	0.9163	0.8700	0.0017
585	0.9786	0.8163	0.0014
590	1.0263	0.7570	0.0011
595	1.0567	0.6949	0.0010
600	1.0622	0.6310	0.0008
605	1.0456	0.5668	0.0006
610	1.0026	0.5030	0.0003
615	0.9384	0.4412	0.0002
620	0.8544	0.3810	0.0002
625	0.7514	0.3210	0.0000
630	0.6424	0.2650	0.0000
635	0.5419	0.2170	0.0000
640	0.4479	0.1750	0.0000
645	0.3608	0.1382	0.0000
650	0.2835	0.1070	0.0000
655	0.2187	0.0816	0.0000
660	0.1649	0.0610	0.0000
665	0.1212	0.0446	0.0000
670	0.0874	0.0320	0.0000
675	0.0636	0.0232	0.0000
680	0.0468	0.0170	0.0000
685	0.0329	0.0119	0.0000
690	0.0227	0.0082	0.0000
695	0.0158	0.0057	0.0000
700	0.0114	0.0041	0.0000
705	0.0081	0.0029	0.0000
710	0.0058	0.0021	0.0000
715	0.0041	0.0015	0.0000
720	0.0029	0.0010	0.0000
725	0.0020	0.0007	0.0000
730	0.0014	0.0005	0.0000
735	0.0010	0.0003	0.0000
740	0.0007	0.0003	0.0000
745	0.0005	0.0002	0.0000

nm	$x(\lambda)$	$y(\lambda)$	$z(\lambda)$
750	0.0003	0.0001	0.0000
755	0.0003	0.0001	0.0000
760	0.0002	0.0001	0.0000
765	0.0002	0.0001	0.0000
770	0.0001	0.0001	0.0000
775	0.0001	0.0000	0.0000
780	0.0000	0.0000	0.0000

J

Guide to the 'Colour Reproduction Workbook'

J.1   Introduction

The workbook is provided primarily for those who wish to become fully familiar with the mathematics associated with colour and its reproduction. Whereas in the text of the book the basis and results from equations are provided in sufficient detail to satisfy the average reader, the worksheets clearly expose the mathematics behind each solution. For those whose interest goes beyond that which is explained in the text alone but find the mathematics somewhat daunting, the more complex worksheets are provided with macros. These macros are controlled by 'buttons' with underlying simple instructions, which when activated by a click, cause one set of parameters in a formula to be replaced with another set, in order to see dynamically the effect of the changes in the accompanying charts derived from the formulae.

As an example, in Worksheet 9, which deals with the calculations to establish both the position of the primaries on the chromaticity chart and the drawing of the charts which illustrate the camera red, green and blue spectral sensitivities, the system primary set can be changed by activating the appropriate button and the changes in the position of the primaries on the chromaticity diagram and the change in the illustration of the camera spectral sensitivities can be observed.

The worksheets cover each stage in the reproduction workflow, which has provided the opportunity to bring the calculations of each relevant worksheet together in Worksheet 19. Thus this worksheet is effectively a mathematical model of the complete reproduction process. The parameter values in the model may be changed by macros to illustrate the performance of the system under different specification criteria, in terms of the colour difference values between the original ColorChecker chart values and the displayed reproduced values.

J.2   Structure of the Workbook

The Workbook is comprised of some 50 worksheets arranged in order to correspond with the chapter order through the book, with each sheet title commencing with the chapter number

Colour Reproduction in Electronic Imaging Systems: Photography, Television, Cinematography, First Edition. Michael S Tooms.
© 2016 John Wiley & Sons, Ltd. Published 2016 by John Wiley & Sons, Ltd.
Companion Website: www.wiley.com/go/toomscolour

and, if more than one worksheet is required for a chapter, versions (a), (b) etc. are included. The first worksheet is the Contents page which contains a list of all the worksheets, access to which can be achieved by clicking the left mouse button on the worksheet title. One can return to the contents page from any other worksheet by activating Cntrl + C on the keyboard.

Rather than produce huge worksheets, which are difficult to negotiate with limited screen sizes, a new worksheet is provided for each variant of a similar topic, albeit that sometimes this leads to a repetition in the presentation of data.

Occasionally two apparently identical charts appear in adjacent locations. One is the original chart still dynamically associated with the data cells, whilst usually the second chart is a copy taken for a specific set of data which relates to the narrative in the text of the book.

Much of the basic data is used in many of the worksheets and in order to avoid repetition and inconsistency each set of basic data is stored in separate worksheets at the end of the Workbook. Thus there are worksheets for CIE data, Surface Reflectances, Illuminant SPDs and Primaries Chromaticities Coordinates data sets. When required this data is imported into the relevant chapter related worksheet, usually under the control of a macro.

The worksheets vary considerably in complexity; the simple sheets require no description, other sheets may be described in the book or in a text box immediately under the sheet title and for the more complex sheets the description is contained in Section 5 of this Guide.

J.3   Some General Guidance on Using the Worksheets

J.3.1   Brief Instruction for Using the Worksheet

If any instructions are required to review the use of a particular worksheet, they are located adjacent to cell A1 at the top left of the displayed sheet.

J.3.2   Range Naming

Data ranges which are frequently accessed by a formula are often given appropriate names in order that when inspecting a formula it is easier to comprehend to which parameters the range relates. For example the data range associated with the chromaticities of the sRGB primaries located in the 'Primaries' worksheet at the cell range A144:F148 is identified by the name 'sRGB'.

J.3.3   Macro Activation

Activation of the macros is by selection of a coloured button from a group which is generally located in the top left section of the worksheet to enable one to rapidly negotiate to them from anywhere in the worksheet. Occasionally however the macros are hidden below the list of the titles of a set of parameters; in either case there are instructions as to their operation at the top left of the worksheet. Where a formula with a number of variables is displayed, the variables under user control will have a green background.

For those familiar with the use of Excel, one may enter other values into the parameter set which the macro normally accesses in order to produce results for particular requirements. In addition of course one may copy appropriate data from the data set worksheets at the end of the workbook directly into the source position for the calculation.

J.3.4   Common Calculations – Convolution, Integration and Matrix Functions

One of the most common requirements for calculation in colour work is convolution, that is to multiply together two sets of data relating to values throughout the spectrum, for example the SPD of a source of illumination and the reflectance of a surface being illuminated. More often than not these data sets are listed in the appropriate basic data set worksheets and rather than bring these into the relevant worksheet the formula refers to the appropriate data set in the remote worksheet by name, leading to only one column of calculated values against wavelength in the primary worksheet.

An extension of the above calculation is integration, that is to sum the result in order to establish the area under the multiplied curves or effectively the integration value. When only the integration value rather than the response at each wavelength is required, then as an alternative to this space-filling approach the Excel SUMPRODUCT (convolve and integrate) function is frequently used to obtain the same result far more efficiently.

To process colour data between sets of data based upon different primaries requires the use of matrix mathematics, which to undertake manually is relatively complex and time consuming, thus another extremely useful Excel set of functions are those that enable matrix calculations to be undertaken very efficiently. These functions are used extensively in the appropriate worksheets and it is worth noting for the uninitiated that to complete a matrix instruction it is required, after selection of the required range(s), to press Cntrl, Shift and Enter together to complete the instruction.

J.3.5   Copying Data from the Data Worksheets

Each column of data in the data worksheets is headed with a data title and it is important when copying such a data column into one of the chapter worksheets to include the data title, which then provides a reference as to which data is current.

J.4   The Data Worksheets

There are five data worksheets which are located in the workbook following the chapter worksheets with these abbreviated titles:

- CIE
- Surfaces
- Illuminants
- Primaries
- Camera
- Matrices

Each of these worksheets contain named data ranges appropriate to the title of the worksheet and which are described briefly in the following sub-sections.

J.4.1 CIE Related Data

The worksheet contains four data tables:

- The CIE standard 2 degree Colour Matching Functions (CMFs)
- The CIE x,y chromaticity co-ordinates of the CMFs
- The CIE u′,v′ chromaticity co-ordinates of the CMFs
- The CIE 'D' Illuminant component Spectral Power Distributions (SPDs)

In addition a diagram has been plotted for each of the x,y and u′,v′ sets of chromaticity coordinates which are used as templates for wherever they are required in each of the chapter worksheets.

J.4.2 Surfaces Spectral Reflection Characteristics

This worksheet contains a number of tables relating to well established groups of colour surface samples:

- CIE CRI Test Colour Samples
- ColorChecker Chart – BBC measurements
- ColorChecker Chart – National Physical Laboratory (NPL) measurements
- Lucideon CERAM standard tiles
- A set of simulated highly saturated primary surface colours
- Pointer surface colours – derived chromaticity coordinates from Pointer chromaticity chart
- Pointer surface colours h_{uv} and L^* values
- Pointer surface colours – intermediate calculation tables
- Pointer surface colours – calculated maximum chromaticity coordinates

J.4.2.1 ColorChecker Chart

Two tables of reflectance values for the ColorChecker chart are provided; one set measured by the NPL and the other by the BBC. Since including these tables in the worksheet, the EBU has standardised the Television Lighting Consistency Index, which in turn uses the BBC measured values to obtain the index values. Thus unless there are reasons to do otherwise the BBC values should be used where there is an option to do so.

The chapter worksheets which use the ColorChecker reflectance values refer to the values located in the Surfaces Reflectance Data worksheet at the cell range R5:AP87, thus check that the BBC values at cell range CU5:DS87 have been copied into cell range R5:AP87 before proceeding.

J.4.2.2 Pointer Surface Chromaticities

The chromaticity coordinates of the Pointer surface colours are not available as values in Pointer's paper, however, as they are an important set of reference data which are used frequently against which to compare other data, two approaches have been used in this worksheet to derive the chromaticity coordinates:

- By measuring the coordinates directly from the chromaticity diagram appearing in Pointer's paper
- By calculating the coordinates from the h_{uv} and L^* values which are tabled in the original paper

The values obtained from these two approaches are plotted on a u',v' chromaticity diagram for comparison in the worksheet. Generally the values obtained from measuring those obtained from the Pointer chromaticity diagram are used in the chapter worksheets, since these, when plotted, more nearly reflect the appearance of the original Pointer diagram.

J.4.3   Illuminant SPDs

The Illuminant SPDs worksheet contains a wide range of illuminant SPDs, both CIE specified illuminants and practical illuminants of various types, arranged in tables of similar functionality. The tables are as follows:

- CIE SA, SC and EE illuminants
- CIE Illuminants as calculated using CIE formula
- CIE Illuminants as tabled in the reference literature
- CIE 'F' range of Illuminants
- CIE FL3 range of illuminants
- Chapter 7 related illuminants
- Illuminant data provided by Alan Roberts
- Other illuminants

J.4.4   Primaries – Chromaticity Coordinates

The Primaries worksheet contains the chromaticity coordinates of most of the primaries and their corresponding system illuminant used in colour measurement and colour reproduction. In all 29 primary sets are listed, 3 are associated with colour measurement and 26 are associated with colour reproduction.

Each primary table contains the chromaticity coordinates in both the x,y,z and u',v' coordinate diagrams and its worksheet location named to make it easier to quickly determine which table is being referred to in any macro which accesses the table.

J.4.5   EBU Standard Camera Spectral Responsivities

Since the EBU standard camera spectral sensitivities data sets are required in a number of worksheets they are provided within their own data worksheet.

J.4.6   Correction Matrices

For convenience the matrices required to correct for camera spectral sensitivities based upon the primary positive lobes only of the ideal CMFs of the system primaries are given in this worksheet.

J.5   The Chapter Worksheets

The chapter worksheets vary considerably in complexity and features which leads to different approaches to describing the use of each worksheet. Where the calculations and resulting diagrams are self-evident no description is provided; where only a paragraph of description is required, it is provided in a text box under the heading of the worksheet in the top left hand corner of the sheet and where a fuller description is required it is included in this section under the following paragraph headings where each heading includes the number of the worksheet title.

Often the more complex worksheets contain macros, controlled by buttons, which when activated replace the data on which the calculations are based with data representing a different set of conditions; a common example being the chromaticity coordinates of the display primaries. As the associated diagrams are based upon these calculations, this feature is a powerful tool in indicating dynamically the effect of a change in the basic data on the characteristics of the reproduction system.

The resulting diagrams are labelled 'Dynamic', whilst when a data copy of the resulting calculation or a 'Picture' copy of a dynamic diagram is taken to provide the basis of a figure in the book, they are often referred to as 'Static' to indicated that changing the base data will not change either the set of data or the diagrams.

The general layout of the worksheets places the macro buttons near the top left of the sheet and the resulting calculations to their right, whilst the dynamic diagrams appear below the buttons. Any static diagrams appear below the dynamic diagrams. The spectral data resulting from the application of the calculations appears in the third column to the right of the calculations.

Thus generally the user can access the worksheet, operate the macro buttons and see the resulting effect on the dynamic diagrams without changing worksheet view.

J.5.1   Worksheet 1, Colour Perception – Responses of the Eye

This is a static worksheet.

The cone responses of the eye are plotted from both the Thomson – Wright data and from the best fit to this data derived from the CIE $\bar{x}(\lambda), \bar{y}(\lambda), \bar{z}(\lambda)$ CMFs, which are themselves generated from different data representing the responses of the cones. These two sets of data are used to derive the figures for Chapter 1 which illustrate the cone spectral sensitivities.

J.5.2   Worksheet 7(a), CIE Colour Rendering Index

In Section 7.2 of the Book the method of establishing the CRI is described as follows:

The general approach to formulating a method of measuring the ability of a light source to become a satisfactory source for illuminating a scene for colour reproduction is to test the source against a reference source of known even spectral distribution when both in turn illuminate a range of specified test colours. The reference illuminants used are theoretical spectral distributions based upon either tungsten or daylight sources at a colour temperature matched to the correlated colour temperature (CCT) of the test source.

The methodology adopted is based upon using the spectral distributions of the reference illuminant and test sources, the spectral reflectivity of the test colours and colour response curves of the cones or the camera to calculate the overall response to the stimuli in terms of the values of the red, green and blue signals generated for both the reference illuminant and the test source. In the case of a camera the signal levels are converted first into XYZ values and subsequently into one of the CIEmetrics for measuring colour difference in order that an index based upon the range of colour samples can be generated to reflect the suitability of the test source as an illuminant for colour evaluation or reproduction.

Establishing the SPD of the Reference Illuminant

Worksheet 7(a) is split into two sections vertically, on the left are the calculations associated with establishing the CCT of the test illuminant and for deriving the SPD of the reference illuminant. On the right are the calculations associated with chromatic adaptation and on the right of this section are the charts illustrating the colour of the test samples together with their spectral reflectances.

This worksheet uses several formula some of which appear in the text of the book and others which have been established by workers in the field and also yet others specified by the CIE. These formulae are outlined in text boxes located adjacent to the area of the worksheet where they are used.

It is necessary to first establish the CCT of the test illuminant since this figure is used to calculate the SPD of the reference illuminants, see cell range B6:G88. The x,y chromaticities of the test illuminant are first derived from its SPD using the CIE XYZ curves and then the McCamy formula is used to ascertain the CCT from the x,y chromaticities.

The reference illuminant SPD is a Planckian radiator for the test illuminant with a CCT below 5000K and a CIE defined daylight radiator for other temperatures. Thus it is necessary to have calculations which support the derivation of the SPD for either of these two situations and then select the required SPD for the chromatic adaptation calculations. The SPD of the Planckian radiator is calculated in column 'J' and that of the daylight D illumination in column 'K' using the formula listed in the text boxes to the right of these columns.

The required reference illuminant SPD figures are normalised in column 'L' to enable a chart to be drawn which compares the spectral distributions of the test and reference SPDs. As a check on the method, the CCT of the reference SPD is calculated at cell 'O21' in order that it can be compared with the test illuminant CCT.

To use the worksheet, the SPD of the test illuminant together with its column heading illuminant name should be copied into the worksheet commencing at location C7. This column is given a green background to ease the copying process when rapid changes are required. Sample test illuminant SPDs are available for copying/paste values into the 'CEI CRI' worksheet' from the 'Illuminants' worksheet.

At the top of the LHS of the worksheet the CCT and CRI figures derived from the calculations are shown so that the worksheet can be used for sequentially calculating these figures in a rapid manner.

Establishing the CRI of the Test Illuminant

On the right of the vertical grey separating divide, the chromatic adaptation figures derived from the SPDs of the two illuminants are illustrated and the formulae for deriving the correction figures are outlined in the text box beneath the figures.

The x,y and u,vchromaticities of the CIE Test Colour Samples(TCS) are calculated for both illuminants in the cell ranges V7:AJ20 and V24:AJ33 respectively, the adaptation corrections are applied to those obtained from the test illuminant at cell range V37:AJ44 using the formula in the text box below and the colour differences are calculated for each colour sample in V47:AJ47. From these colour difference figures the R_i for each sample is calculated at V49:AJ49 and the average taken to provide the figure for R_a V50.

Below the calculations two charts are located where the chromaticity changes are plotted from the figures derived at V54:AJ58. The LH chart illustrates the 8 standard CIE colour samples and the RH chart shows the extended range of 14 colour samples.

J.5.3   Worksheet 7(b), MCC Colour Rendering Index

Worksheet 7(b) is an amended version of Worksheet 7(a) and thus laid out in an identical manner, the main difference being that the colour samples are taken from the ColorChecker (CC) chart and the colour adaptation correction uses the CAT02 transform.

The explanatory formula text boxes have not been included on this worksheet. In other respects the LHS of the worksheet is identical to the CIE CRI worksheet.

The CIE adaptation calculations are repeated but for the CC samples rather than for the CIE samples.

In addition the CAT02 calculations are performed and colour difference values based upon the a*,b* colour space are calculated.

Charts are illustrated of the adaptation transforms and the values of the colour differences.

J.5.4   Worksheet 9, Deriving Idealised Camera Spectral sensitivities

Worksheet 9 has all the fundamental mathematics built in to enable the colour gamut and the spectral sensitivities of a camera to be calculated for any set of primaries in terms of their x,y chromaticity values and the system white point. As such its core elements are used as a template in many other worksheets. The x,y primary chromaticity coordinates are located at cell range N6:P11and from these the u',v' coordinates are calculated and used to provide the data for displaying the chromaticity gamut on the dynamic chromaticity chart. In addition the matrix formula derived in Appendix 6 of the book is used to derive the relationship between the $\bar{x}(\lambda), \bar{y}(\lambda), \bar{z}(\lambda)$ CMFs and the r,g,b CMFs in the cell range N15:R53. The relationship derived in Matrix 3 is used to derive the idealised camera spectral sensitivity data at cell range W7:Z87 from which the camera spectral sensitivities characteristics are illustrated in the dynamic chart.

The three primaries relevant to Chapter 9 are made available to the calculations by the selection of 'buttons' at the top left of the sheet, these are 'ITU BT.709 (Rec 709)', 'Widest' and 'Ideal'.

Click on one of the buttons to copy the appropriate primaries set chromaticity coordinates from the Primaries worksheet into the calculation section of the worksheet.

The resulting chromaticity gamut and idealised camera spectral sensitivity characteristics are immediately displayed in the dynamic charts below the macro selection buttons.

J.5.5   Worksheet 12(a), Colour Gamut Transformation – Matrix Derivation

The purpose of this worksheet is to derive a single matrix for converting the values of RGB from a camera with a defined set of spectral sensitivity characteristics associated with one set of system primaries and white point to those values which would have been obtained if the scene had been shot with a further camera with characteristics associated with a different set of primaries and system white point.

In addition it independently calculates the values of RGB obtained from these two cameras from first principles using the cameras spectral sensitivities, the scene illumination SPDs and a selected sample colour and compares these values with those obtained from the matrix calculations in order to check the veracity of the matrix coefficients.

The matrix arithmetic for deriving the camera spectral sensitivities for the two cameras with different primaries is identical to that derived for a single camera in Worksheet 11, albeit displayed in an abbreviated form.

The source primaries data for the camera spectral sensitivities matrix derivations are the two coloured cell ranges at the top of the worksheet. Different primaries data may be copied to these cell ranges from the 'Primaries' worksheet using the named primaries buttons on the left of the worksheet. The resulting matrices, which form the basis of the calculations for the conversion matrix, are located at Matrix 1 and Matrix 6 respectively in the 'Conversion Matrix Development' section of the worksheet located at A53. The gamuts of the two sets of primaries are illustrated to the right of the primaries data blocks and the ideal camera spectral sensitivities are illustrated to the right of the chromaticity diagram.

The basis of the approach to deriving the conversion matrix is to first work back from the RGB values from Camera 1 and its scene illumination, to ascertain the XYZ values of the surface colours in the scene and then apply the characteristics of Camera 2 and its scene illumination to these surface colours to obtain the RGB values from Camera 2. The detailed description of this approach is described in Appendix 7 of the book and the associated diagram is illustrated in Figure A.7.1 which is repeated below for convenience.

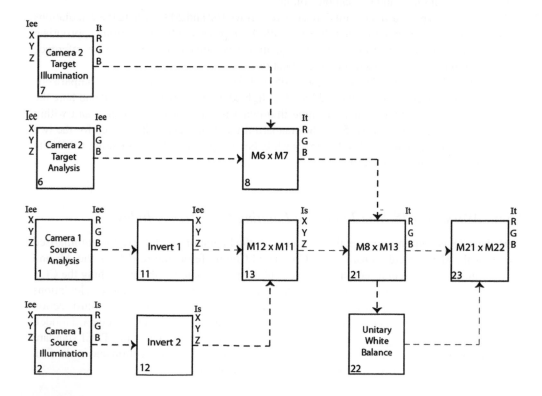

The matrix numbers in the worksheet correspond to the matrix numbers in the lower left corner of each of the matrices illustrated in the diagram.

For convenience, the conversion matrix from the calculations is copied onto the opening area against a grey background in order that as different primaries and white points are selected the resulting conversion matrix is immediately seen without the need to change the view.

Checking the Matrices Function Correctly

The remainder of the worksheet is given over to checking that the conversion matrix provides the same values as would have been obtained directly from Camera 2.

Three relatively critical colour samples are made available for the comparison checks, dark blue, orange and equal reflectance white; either of which may be selected by the appropriate button at row 104. This macro copies the appropriate reflectance characteristic of the sample from the Surfaces worksheet to AM7:AM87.

The primary sets selection macros also copy the appropriate illuminant SPD from the Illuminants worksheet to the cell ranges AJ7:AK87.

In the 'Calculation of the RGB Values' area between rows130 and 160 the 'integration' calculations are undertaken for a number of conditions by multiplying together the camera spectral sensitivity characteristic, the illumination SPD and the spectral reflectance of the sample colour at each wavelength through the spectrum. These values are then summed to give the RGB values before white balancing. For each camera three calculations are undertaken each with the camera spectral sensitivity characteristic: firstly with the sample colour, then with the illuminant – which should give the same value for each of the RGB levels and finally with both the illuminant and the sample colour.

The 'Comparison of Partial Results'area between rows 162 and 203 is where the calculations of RGB values above are compared directly with those projected by the various appropriate matrices. The matrix number allocated in the matrix development area is given for reference as to which matrix is being used for the calculation.

In the Full Results area between rows 205 and 217, the calculations and comparisons associated with the conversion matrix 23 are highlighted with a grey background and here the match of values is accurate when the same illuminant is used for both cameras and within 1% when illuminants 'C' and D65 are the standard illuminants. The values given for the two cameras are referenced to equal reflectances white being equal to 1.00, as selecting ER White as the sample colour will illustrate.

J.5.6   Worksheet 18, The Television Lighting Consistency Index

Worksheet 18 is split into five sections by grey vertical bars; on the left are the summarised results of the sheet calculations, i.e. the Correlated Colour Temperature (CCT), the ΔE and the Q_a values. In the second section are the calculations associated with establishing the CCT of the test illuminant and for deriving the SPD of the reference illuminant, these calculations follow the identical pattern for the same task in Worksheets 7(a) and 7(b); in the third section is the derivation of the camera spectral sensitivities associated with the TLCI standard model; in the fourth section are the calculations associated with the remainder of the TLCI standard model, the CIEDE2000 and TLCI calculations are located beneath this section and in the fifth section are the charts illustrating the results of the calculations.

To use the worksheet the SPD of the test illuminant together with its column heading illuminant name should be either entered or copied into the worksheet at location E7 using 'paste values'. This column is given a green marker to ease the copying process when rapid changes are required. Sample test illuminant SPDs are available for copying into the 'CEI CRI' worksheet' from the 'Illuminants' worksheet.

This worksheet uses several formula some of which appear in the text of the book and others which have been established by workers in the field and also yet others specified by the CIE. These formulae are outlined in text boxes located adjacent to the area of the worksheet where they are used.

In the second section headed 'Illuminants' it is necessary to establish a reference illuminant with a matching CCT to the CCT of the test illuminant. Thus it is necessary to first establish the CCT of the test illuminant since this figure is used to calculate the SPD of the reference illuminant. The u,v chromaticities of the test illuminant are first derived from its SPD using the CIE XYZ curves, then the McCamy formula is used to ascertain the CCT. (The EBU TLCI uses a different method for calculating the CCT which leads to minor differences in result on some illuminants.)

In order to provide a continuous relatively smooth spectrum, the reference illuminant SPD is a Planckian radiator when the test illuminant CCT is below 5000K, as described in Section 6.1 of the book, and a CIE defined daylight radiator, as described in Section 7.3, for other temperatures. Thus it is necessary to have calculations which support the derivation of the SPD for either of these two situations and then select the required SPD for calculating the camera RGB values for the reference illuminant. The selected SPD figures for both the test and reference illuminants are normalised to enable a chart to be drawn which compares their spectral distributions.

In the third section the TLCI standard camera spectral sensitivities are generated from the data in Appendix 5 of EBU Tech3355 and the data is matrixed using the matrix defined in the same appendix. Also in this section the functions required for calculating the RGB and XYZ values derived from the camera are laid out and the matrix for deriving the XYZ values from the RGB values is calculated.

In the top of section 4 the RGB values from the TLCI model camera are calculated by convolving the spectral values of the camera, the illuminant and the reflectivities of the ColorChecker samples for both the reference and the test illuminant. These RGB values are then subjected to the mathematical processes defined in the standard model to obtain the XYZ values of the displayed samples and from them the u',v' and $L*a*b*$ values are calculated. These values for the two illuminants are used to create chromaticity charts which illustrate in section 5 the difference vectors for each of the ColorChecker chart sample colours.

In the lower part of Section 4 both sets of $L*a*b*$ values are used in the CIEDE2000 colour metric calculator to establish the $\Delta E*_{00}$ values for each colour sample.

Finally at the bottom of section 4 the mean ΔE value is calculated and from this value the TLCI is derived.

J.5.7   *Worksheet 19, Complete Television Signal Chain Colorimetric Performance*

The Layout of the Worksheet

Worksheet 19 brings together the functionality of several of the previous worksheets in a manner which mathematically emulates the performance of a complete television signal chain, from the scene, through the camera to the reproduced image appearing on the display. Each element of the workflow is provided with its own section in the worksheet, separated for ease of identification from each other, by a vertical grey bar. The scene is comprised of the first 18 colour samples of the ColorChecker chart illuminated by the selected scene illuminant. The final section of the worksheet contains the calculations which provide the levels of colour differences between the XYZ values of the chart samples illuminated by the system white and the same samples generated by the display. The colour differences in chromaticity terms are displayed on the $u'v'$ chromaticity chart and in terms of the three dimensions of colour are expressed as ΔE^*_{00} values.

The elements of the worksheet, in the order in which they appear horizontally across the worksheet are as follows:

- Scene Illuminant Selection
- Derivation of Ideal Camera Spectral Sensitivities based upon selection of system primaries
- Selection of Ideal or positive only lobes, camera spectral sensitivities
- Selection of Camera Correcting Matrix
- Selection of OETF or Gamma Correction Characteristic
- Selection of Display Gamma
- Selection of Display Primaries Chromaticities
- Colour difference calculations

Each section contains a number of control buttons which enable one of a number of sets of parameter values to be selected, which sets the characteristics of the section element.

The resulting calculation from each element is made to be the input to the following element in the signal chain such that the final element in the chain, that is the display, takes the display input RGB values and calculates the XYZ values of the light generated by the display. However sometimes the sequential calculations take place within the section and sometimes the section merely produces the selected characteristics which are then used in the calculation block in the end section of the worksheet. The Illuminant, Gamma Corrector and Display Characteristics fall into this latter category.

The calculations in the final section of the worksheet take the XYZ values from the display and those it calculates directly from the selected system white illuminating the ColorChecker chart and derives u',v' and $L^*a^*b^*$ values for each colour sample. These values are used both in the chromaticity chart to illustrate the change in chromaticity and in generating the colour difference ΔE^*_{00} values respectively.

Copies of the chromaticity chart and the ΔE^*_{00} value are pasted into each section in order that the effect of the selection on the performance of the system can be immediately seen without having to locate to a different area of the worksheet. These charts obscure the calculations associated with that section, which may be revealed by using the mouse to temporarily relocate the chart.

The Mathematical Functioning of the Worksheet Sections

As distributed, the parameter values for each section are set to neutral, i.e. the button selections provide all elements of the signal chain with the ideal and matched characteristics for a Rec 709 specification signal chain, with the gamma set to linear in both the camera and the display device, which will result in the chromaticity charts illustrating dots rather than vectors and all

ΔE^*_{00} values will be close to zero, with the exception of the cyan sample which is outside of the Rec 709 chromaticity gamut.

Reference Illuminant Selection

Before changing the individual element parameters in the signal chain it is important to first select the system white chromaticity. This is the illuminant used to calculate the reference values of the ColorChecker chart against which the reproduced values from the system are compared. The current system white is noted in the extension to the worksheet title at the top left of the worksheet. Should a different system white be required it is necessary to locate to cell DR where there are three control buttons which enable the system white illuminant to be selected from the three CIE illuminants, D50, D65 and C.

Scene Illuminant Selection

The reference spectral power distribution (SPD) of the system white selection is illustrated as a brown line on the 'Illuminants' chart. The scene illumination may be selected either by clicking on one of the illuminant buttons or copying an illuminant column of values from the 'Illuminants' worksheet and pasting it, together with its heading cell into cell B6, whereupon its SPD will be illustrated by the blue line on the chart. These SPD values will be used in a convolved calculation with the spectral reflectances of the ColorChecker chart samples and the camera spectral sensitivities to obtain the RGB values from the camera in the calculation section.

Deriving the Ideal Camera Spectral sensitivities

This section enables the selection of the notional 'system primaries' from a number of system primaries sets previously defined by clicking on the appropriate primaries button, alternatively the chromaticity of other primary sets maybe copied from the 'Primaries' worksheet and pasted into cell range J14:K19, or the x,y, chromaticity values of a new set of primaries and white point may be entered directly into the cells located at J14:K19. The appropriate matrix for deriving the camera spectral sensitivities is calculated from the primaries chromaticities and the XYZ colour matching functions are used with the matrix to calculate the spectral sensitivities, which are located adjacent and also copied into the spectral sensitivities location in the next section.

The resulting colour gamut and camera analyses characteristics will be portrayed on the adjacent charts. This section also calculates the positive lobes only characteristics of the selected spectral sensitivities.

Selecting the Camera Spectral sensitivities

Three different cameraspectral sensitivities are available for selection: the ideal characteristics derived in the previous section, the positive lobes only of the ideal characteristics or the Television Lighting Consistency Index (TLCI) characteristics derived in EBU 3353, which are contained in the Camera worksheet. Clicking on the desired button will paste the appropriate characteristics into the adjacent cell range which forms the input to the matrix located in the next section.

Select Camera Correction Matrix

The 'active' matrix located at the top of this section operates on the selected camera spectral sensitivities of the previous selection and produces a corrected set of camera spectral sensitivity values in the adjacent range of cells.

There are four different matrices laid out beneath the charts in this section and clicking on one of the matrix selection buttons will copy the appropriate matrix to the 'active' matrix position. The 'No Matrix' is actually a matrix with unity values in the appropriate matrix positions which ensures that when this option is selected as the active matrix no adjustment of the analyses characteristics takes place. Other options include a correction matrix proposed by the BBC in early work to match the characteristics of the plumbicon image sensors in use at that time, the matrix proposed for use with the TLCI characteristics by the EBU and also an adjustable matrix.

The adjustable matrix is located in the green cells just below the chromaticity chart in the cell range AY56:BA58, and immediately below this range is a further matrix of values which may be copied into the adjustable cell range as a good starting point for the adjustments. Also located here is a further chart which illustrates the ideal and the adjusted camera spectral sensitivities to enable a best match to be compared.

If the 'Adjustable' button is selected then one may make adjustments to the green cell values and see dynamically the effect on the spectral sensitivities, the difference values on the chromaticity chart and the ΔE^*_{00} values.

Select Gamma Correction

To select the required gamma law click on the name of the law and the law constants will be generated in the adjacent cells and illustrated on the adjacent chart. If a law not listed is required, insert the appropriate values for the inverse of the display gamma and the gain of the linear section of the combined curve in the green cells and the worksheet will calculate the remaining constants and display them below the green cells. DO NOT INSERT VALUES BELOW THE GREEN CELLS; SUCH ACTION WILL DELETE THE COMPLEX CALCULATIONS CONTAINED THERE.

The constants in cells BW13:BW17, defined by the selected gamma law, are used in the calculation block in the last section for both the camera gamma correction at cells CT44:DQ46 and the display regenerated gamma correction, following the linear correction matrix, at cells CT58:DQ60.

Display Characteristics

The display may or may not have the same characteristics as the 'system' characteristics defined in earlier elements of the signal flow. The display gamma may be set directly by entering the required exponent in cell CE2; entering '1' into this cell will produce a linear characteristic. This characteristic law is used in the calculation block of the next section at cells CT62:DQ64.

In addition this section calculates the primaries correction matrix, in the event that different primaries are used in the display from the system primaries selected earlier, and also the matrix for calculating the XYZ values of the light produced from the RGB values fed to the display, which in turn are dependent upon the display primaries chromaticity values.

The display primaries may be selected by clicking on the required primaries button which copies and pastes the appropriate primaries from the 'Primaries' worksheet to the highlighted cells where they are used together with the primary chromaticities of the system primaries to calculate the correction matrix. These calculations take place lower down in the worksheet at cells BQ126:CM165.

Note that if the display primaries match the system primaries the appropriate values in the correcting matrix go to unity.

Colour Calculations

At the top of the Colour Calculations section are the calculations for deriving the reference XYZ values for each of the colour samples of the ColorChecker chart by convolving the SPDs of the sample with the system white SPDs contained at cell range DU7:DU87 with the XYZ colour matching functions contained in the CIE worksheet. From the XYZ values the other values required are calculated.

In the lower part of this section, starting at cell CT27, are the calculations associated with the elements comprising the complete signal path, commencing with the RGB values following the camera spectral sensitivities correction matrix, obtained by convolving the selected and corrected camera spectral sensitivities with the scene lighting SPD, and the ColorChecker chart samples spectral reflectance distributions.

The characteristics selected for the remainder of the signal path elements are then applied in turn to the RGB signal values, culminating in the values of RGB which drives the display. The matrix derived in the previous section is then used to calculate the XYZ values of the light of the display and from these values the other colour values are calculated for each sample.

The colour differences between the direct ColorChecker chart values and the values derived from the display are then tabled for use in the chromaticity chart and for the ΔE^*_{00} calculator, which is a self-contained calculator downloaded from the website of Sharma et al., already cited in Chapter 18.

References

Amidror, I. & Hersch, R. D. 2000. Neugebauer and Demichel: dependence and independence in n-screen superpositions for colour printing. *Color Res. Appl.*, 25(4), 267–277.

Bala, R. 2003. Device characterisation. In: G. Sharma, ed. *Digital Color Imaging Handbook*. CRC Press, pp. 269–384.

Baron, S. & Wood, D. 2005. *Rec. 601 – The Origins of the 4:2:2 DTV Standard*. EBU Technical Review, October 2005.

Barten, P. G. J. 1999. *Contrast Sensitivity of the Human Eye and Its Effects on Image Quality*. SPIE Optical Engineering Press.

Bartleson, C. J. & Breneman, E. J. 1967. Brightness perception in complex fields. *J. Opt. Soc. Am.*, 57(7), 953–956.

Berlin, B. & Kay, P. 1969. *Basic Color Terms: Their Universality and Evolution*. Berkeley and Los Angeles: University of California Press.

Böhler, P., Emmett, J. & Roberts, A. 2013. Toward a "standard" television camera color model. *SMPTE Motion Imaging J.*, 122(3), 30–36.

Carnt, P. S. & Townsend, G. B. 1961. *Colour Television – The NTSC System, Principles, and Practice*. Illife Books Ltd.

Carnt, P. S. & Townsend, G. B. 1969. *Colour Television Volume 2 – PAL, SECAM and Other Systems*. Illife Books Ltd.

Craford, M. G. 1997. High brightness light emitting diodes. *Semiconduct. Semimet.*, 48.

Demichel, M. E. 1924. *Le procede*, 26, 17–21.

Estévez, O. 1979. *On the Fundamental Data-base of Normal and Dichromatic Colour Vision*. University of Amsterdam.

Fairman, H. S., Brill, M. H. & Hemmendinger, H. 1997. How the CIE 1931 color-matching functions were derived from Wright-Guild data. *Color Res. Appl.*, 22(1), 11–23.

Fielding, G. & Maier, T. 2009. Rendering of scene data in digital cinema workflows. *SMPTE Motion Imaging J.*, 118(3), 32–36.

Fink, D. G. 1955. *Color Television Standards – Selected Papers and Records of the National Television Systems Committee*. McGraw-Hill Book Company Inc.

Frohlich, J., Schilling, A. & Eberhardt, B. 2014. Gamut mapping for digital cinema. *Motion Imaging J.*, 123(8), 41–48.

Gilbert, P. U. & Haeberli, W. 2007. Experiments on subtractive color mixing with a spectrophotometer. *Am. J. Phys.*, 75, 313.

Grassman, H. 1853. Zur Theorie der Farbenmischung. *Poggendorff's Annalen der Physik und Chemie*, 89, 69–84.

Guild, J. 1932. The colorimetric properties of the spectrum. *Philos. T. Roy. Soc. A*, 230, 149–187.

Henderson, S. T. 1970. *Daylight and Its Spectrum*. American Elsevier Pub. Co.

Hunt, R. W. G. 2004. *The Reproduction of Colour*, 6th ed. John Wiley & Sons, Ltd.

Hunt, R. W. & Pointer, M. R. 2011. *Measuring Colour*, 4th ed. John Wiley & Sons, Ltd.

Hyder, D. 2009. *The Determinate World: Kant and Helmholtz on the Physical Meaning of Geometry*. Walter de Gruyter & Co.

Colour Reproduction in Electronic Imaging Systems: Photography, Television, Cinematography, First Edition. Michael S Tooms.
© 2016 John Wiley & Sons, Ltd. Published 2016 by John Wiley & Sons, Ltd.
Companion Website: www.wiley.com/go/toomscolour

Iwaki, Y. & Uchida, M. 2013. Accurate ACES rendering in systems using small 3D LUTs. *SMPTE Motion Imaging J.*, 122(3), 41–46.

Judd, D. B. 1925. Chromatic visibility coefficients by the method of least squares. *J. Opt. Soc. Am.*, 10, 635–647.

Sayanagi, K, 1989. *Neugebauer Memorial Seminar on Color Reproduction*. Tokyo: SPIE, pp. 194–202.

Kitsinelis, S. 2010. *Light Sources: Technologies and Applications*. Taylor & Francis.

Knight, R. 2014. *Considering Colour Mixing & Picture Matching*. Manuscript awaiting publication.

Li, C., Luo, M. R., Pointer, M. & Green, P. 2013. Comparison of real colour gamuts using a new reflectance database. *Color Res. Appl.*, 39(5), 442–451.

Liu, C. & Fairchild, M. D. 2007. *Re-measuring and Modeling Perceived Image Contrast under Different Levels of Surround Illumination*. New Mexico: Albuquerque, pp. 66–70.

Luo, M. R., Lo, M.-C. & Kuo, W.-G. 1998. The LLAB(l:c) colour model. *Color Res. Appl.*, 21(6), 412–429.

MacAdam, D. 1937. Projective transform of the ICI color specifications. *J. Opt. Soc. Am.*, 27, 294–299.

MacAdam, D. L. 1935a. The theory of maximum visual efficiency of colored materials. *J. Opt. Soc. Am.*, 25(8), 361–367.

MacAdam, D. L. 1935b. Maximum visual efficiency of colored materials. *J. Opt. Soc. Am.*, 25(11), 361–367.

Mahy, M. & Delabastita, P. 1998. Inversion of the Neugebauer equations. *Color Res. Appl.*, 21(6), 404–411.

Maier, T. O. 2007a. Color processing for digital cinema 1: background, encoding and decoding equations, encoding primaries and the exponent (1/2.6). *SMPTE Motion Imaging J.*, 116, 439–446.

Maier, T. O. 2007b. Color processing for digital cinema 2: explanation of the constants 4095 and 52.37, the white points, and the black. *SMPTE Motion Imaging J.*, 116, 447–455.

Maier, T. O. 2007c. Color processing for digital cinema 3: minimum linear bit depth, encoding of near blacks, gamut mapping, and colorimetry for tolerances in the standards. *SMPTE Motion Imaging J.*, 116, 500–509.

Maier, T. O. 2007d. Color processing for digital cinema 4: measurement and tolerances. *SMPTE Motion Imaging J.*, 116, 510–517.

Maier, T. O. 2008a. Color processing for digital cinema 5: calculation of the matrices needed for processing digital cinema images. *SMPTE Motion Imaging J.*, 117, 42–48.

Maier, T. O. 2008b. Color processing for digital cinema 6: an example of color processing through the entire system. *SMPTE Motion Imaging J.*, 117, 46–53.

Marshall, W. H. & Talbot, S. A. 1942. Recent evidence for neural mechanisms in vision leading to a general theory of sensory acuity. *Biological Symposia*, Vol. 7: Visual Mechanisms. The Jaques Cattell Press.

Martinez-Verdu, F., Perales, E., Chorro, E., de Fez, D., Viqueira, V., & Gilabert, E. 2007. Computation and visualisation of the MacAdam limits for any lightness, hue angle, and light source. *J. Opt. Soc. Am.*, 24(6), 1501–1515.

McElavain, J. S., Ruhoff, D., Woodruff, G., Gish, W. & Schnuelle, D. 2012. Direct display of integer ACES content for post-production environments. *SMPTE Motion Imaging J.*, 121(3), 46–51.

Munsell, A. H. 1912. A pigment color system notation. *Am. J. Psychol.*, 23, 236–244.

Murray, A. 1936. Monochrome reproduction in photoengraving. *J. Frankl. Inst. Eng. Appl. Math.*, 221(6), 721–744.

Neugebauer, H. E. J. 1937. Die theoretischen Grundlagen des Mehrfarbendrucks. *Zeitschrift fuer Wissenschaftliche Photographie*, 36, 36–73.

Pointer, M. 1980. The gamut of real surface colours. *Color Res. Appl.*, 5(3), 145–155.

Poynton, C. 2012. *Digital Video and HD Algorithms and Interfaces*, 2nd ed. Burlington, MA: Elsevier/Morgan Kaufmann.

Reinhard, E., Ward, G., Pattanaik, S. & Debevec, P. 2006. *High Dynamic Range Imaging*. Morgan Kaufmann Publishers Inc.

Roberts, A., Böhler, P. & Emmett, J. 2011. *A Television Lighting Consistency Index*. IBC.

Schewe, J. 2013. *The Digital Negative*. Peachpit Press.

Schubert, E. F. 2006. *Light Emitting Diodes*, 2nd ed. Cambridge University Press.

Sharma, G., Wu, W. & Dalal, E. N. 2005. The CIEDE2000 colour-difference formula: implementation notes, supplementary test data, and mathematical observations. *Color Res. Appl.*, 30, 21–30.

Soffer, B. H. & Lych, D. K. 1999. Some paradoxes, errors, and resolutions concerning the spectral optimization of human vision. *Am. J. Phys.*, 67(11), 946–953.

Sproson, W. N. 1983. *Colour Science in Television and Display Systems*. Adam Hilger Ltd.

Stauder, J., Blonde, L., Morvan, P., Schubert, A., Doser, I., Endress, W., Hille, A., Correa, C. & Bancroft, D. 2007. *Gamut ID*. London: European Conference on Visual Media Production.

Stiles, W. S. 1978. *Mechanisms of Colour Vision*. Academic Press.

Thomson, L. C. & Wright, W. D. 1947. The colour sensitivity of the retina within the central fovea of man. *J. Physiol.*, 105(4), 316–331.

Tooms, M. S. 1981. *Systems Engineering Considerations in the All Digital Television Production and Transmission Centre*. In: 15th Annual SMPTE Television Conference: Production & Post Production in the Eighties, San Francisco, pp. 177–199.

Watkinson, J. R. 1999. *MPEG-2*. Focal Press.

Wentworth, J. W. 1955. *Colour Television Engineering*. McGraw-Hill Book Company.

Wright, W. D. 1928. A trichromatic colorimeter with spectral primaries. *Trans. Opt. Soc.*, 29, 225.

Wright, W. D. 1929. A re-determination of the trichromatic coefficients of the spectral colours. *Trans. Opt. Soc.*, 30, 141–164.

Wright, W. D. 1939. A colorimetric equipment for research on vision. *J. Sci. Instrum.*, 16(1), 10.

Wright, W. D. 1969. *The Measurement of Colour*, 4th ed. London: Adam Hilger.

Yule, J. A. C. & Neilsen, W. J. 1951. The penetration of light into paper and its effect on halftone reproductions. *TAGA Proc.*, 3, 65–76.

Index

Colour Reproduction in Electronic Imaging Systems: Photography, Television, Cinematography, First Edition. Michael S Tooms.
© 2016 John Wiley & Sons, Ltd. Published 2016 by John Wiley & Sons, Ltd.
Companion Website: www.wiley.com/go/toomscolour

Printed and bound by CPI Group (UK) Ltd, Croydon, CR0 4YY

27/10/2024

14580355-0002